# COST-EFFECTIVE SPACE
# MISSION OPERATIONS

# SPACE TECHNOLOGY SERIES

*This book is published as part of the Space Technology Series,
a cooperative activity of the United States Department of Defense
and the National Aeronautics and Space Administration.*

## Wiley J. Larson
*Managing Editor*

From Kluwer and Microcosm Publishers:

**Space Mission Analysis and Design** Second Edition by Larson and Wertz.

**Spacecraft Structures and Mechanisms: From Concept to Launch** by Sarafin.

From McGraw-Hill:

**Understanding Space: An Introduction to Astronautics** by Sellers.

**Space Propulsion Analysis and Design** by Humble, Henry, and Larson.

**Cost-Effective Space Mission Operations** by Boden and Larson.

Future Books in the Series:

**Reducing Costs of Space Missions** by Wertz and Larson.

**Fundamentals of Astrodynamics and Applications** by Vallado.

**Modeling and Simulation: An Integrated Approach
to Development and Operation** by Cloud and Rainey.

**Human Space Mission Analysis and Design** by Connally, Giffen, and Larson.

# COST-EFFECTIVE SPACE
# MISSION OPERATIONS

Edited by
## Daryl G. Boden
*United States Air Force Academy*
*University of Colorado at Colorado Springs*

## Wiley J. Larson
*United States Air Force Academy*
*University of Colorado at Colorado Springs*
*International Space University*

*This book is published as part of the*
*Space Technology Series, a cooperative activity*
*of the United States Department of Defense and the*
*National Aeronautics and Space Administration.*

## McGraw-Hill, Inc.
## College Custom Series

*New York   St. Louis   San Francisco   Auckland   Bogotá*
*Caracas   Lisbon   London   Madrid   Mexico   Milan   Montreal*
*New Delhi   Paris   San Juan   Singapore   Sydney   Tokyo   Toronto*

# COST-EFFECTIVE SPACE MISSION OPERATIONS

1 2 3 4 5 6 7 8 9 0   BKM BKM   9 0 9 8 7 6

ISBN 0-07-006382-6, Softcover
ISBN 0-07-006379-6, Hardcover
Library of Congress Catalog Card Number: 96-075833


Editor: M. A. Hollander
Cover Design: Dale Gay
Text Design: Anita Shute
Technical Editor: Perry Luckett
Printer/Binder: Bookmart Press

# Table of Contents

# List of Authors and Editors

**Karen R. Altunin.** Astronomer, AR Space Enterprises, Incline Village, Nevada. M.S. (Astronomy) University of Virginia; B.A. (Behavioral Sciences) University of California. Chapter 18—*International Space Mission Operations.*

**Valery R. Altunin.** Coordinator, DSN VLBI Radio Astronomy, Jet Propulsion Laboratory, California Institute of Technology, Pasadena, California. Ph.D. (Radio Astronomy) Space Research Institute; M.S. and B.S. (Physics) Gorky University. Chapter 18—*International Space Mission Operations.*

**Mikhail Artukhov.** Deputy Manager for Flight Operations, Lavochkin Association, Moscow, Russia. M.S. (Engineering) Moscow Aviation Technology Institute. Section 18.3—*Russian Space Mission Operations.*

**Viktor D. Blagov.** Flight Operations Director, NPO Energia, Moscow, Russia. M.S. (Engineering) Moscow Aviation Technology Institute. Section 18.3—*Russian Space Mission Operations.*

**Carolyn Blacknall.** Operations Integration Officer, NASA Johnson Space Center, Houston, Texas. M.S. (Aerospace Engineering) University of Texas at Austin; B.S. (Astronomy) University of Texas at Austin. Chapter 20—*Human Space Flight Operations.*

**Daryl G. Boden.** Visiting Professor, United States Air Force Academy, Colorado Springs, Colorado. Ph.D. (Aeronautical and Astronautical Engineering) University of Illinois; M.S. (Astronautical Engineering) Air Force Institute of Technology; B.S. (Aerospace Engineering) University of Colorado. *Co-Editor*, Chapter 1—*Space Mission Operations*; Chapter 2—*Designing Space Mission Operations*; Chapter 10—*Space Navigation and Maneuvering.*

**John Carraway.** Senior Member of the Technical Staff, Jet Propulsion Laboratory, California Institute of Technology, Pasadena, California. B.S. (Electrical Engineering) Massachusetts Institute of Technology. Chapter 5—*Assessing Operations Complexity.*

**Gary M. Comparetto.** Senior Principal Engineer, The MITRE Corporation, Colorado Springs, Colorado. M.S. (Nuclear Engineering) Pennsylvania State University; M.S. (Electrical Engineering) Georgia Institute of Technology; B.S. (Electrical Engineering) Bucknell University. Chapter 11—*Communications Architecture.*

**Richard S. Davies.** Technical Director, Stanford Telecommunications, Inc., Sunnyvale, California. Engineer, Stanford University. M.S. and B.S. (Electrical Engineering) University of Pennsylvania. Chapter 11—*Communications Architecture.*

**Eileen Dukes.** Senior Staff Engineer, Mars Global Surveyor MGS Program, Lockheed Martin Corporation, Denver Astronautics, Denver, Colorado. B.S. (Aeronautics and Astronautics) Massachusetts Institute of Technology. Chapter 8—*Conducting Space Mission Operations.*

**William J. Emery.** Professor, University of Colorado, Boulder, Colorado. Ph.D. (Physical Oceanography) University of Hawaii; B.S. (Mechanical Engineering) Brigham Young University. Section 13.2—*Data Systems*; Section 13.3—*Data-Processing Requirements.*

**Michael Fatig.** Manager, Flight Operations, Allied Signal Technical Services Corporation, Lanham, Maryland. B.S. (Technical Management) University of Maryland. Chapter 9—*Launch and Early-Orbit Operations.*

**Viktor I. Glebov.** Senior Engineer, All Union Research Institute of Electromechanics, Moscow, Russia. M.S. (Engineering) Dzerginsky Military Academy. Section 18.3—*Russian Space Mission Operations.*

**Felix Godwin.** Senior Systems Analyst, Teledyne Brown Engineering, Huntsville, Alabama. B.S. (Physics) University of London. Chapter 20—*Human Space Flight Operations.*

**Paul Graziani.** President and Chief Executive Officer, Analytical Graphics, Inc. King of Prussia, Pennsylvania. B.S. (Biology and Computer Science) LaSalle University. Section 10.7—*Mission Geometry.*

**Jim Green.** Chief, Space Science Data Operations Office, NASA/Goddard Space Flight Center. Ph.D. and M.S. (Physics) University of Iowa; B.A. (Astronomy) University of Iowa. Section 13.1—*End-to-End Data Flow.*

**David E. Kaslow.** Chief Engineer, Lockheed Martin Corporation, King of Prussia, Pennsylvania. Ph.D. (Physics) University of Michigan; M.S. (Physics) Indiana University; B.A. (Mathematics) Indiana University. Chapter 6—*Defining and Developing the Mission Operations System*; Chapter 7—*Activity Planning.*

**Wiley J. Larson.** Visiting Professor, U.S. Air Force Academy, Colorado Springs, Colorado. Professor of Engineering, International Space University, Illkirch, France. D.E. (Spacecraft Design) Texas A&M University; M.S. and B.S. (Electrical Engineering), University of Michigan. Co-Editor, Chapter 1—*Space Mission Operations*; Chapter 2—*Designing Space Mission Operations.*

**Matthew J. Lord.** Senior R&D Engineer, Loral Space and Range Systems, Sunnyvale, California. B.S. (Electrical Engineering) University of California at Santa Barbara. Chapter 12—*Ground Systems.*

**Perry D. Luckett.** Director and Chief Consultant, Executive Writing Associates, Colorado Springs, Colorado. Ph.D. (American Studies) University of North Carolina at Chapel Hill; M.A. and B.A. (English) Florida State University. Technical Editor.

**Mac Morrison.** Deputy Manager, Ground Development and Operations, TRW, Redondo Beach, California. B.S. (Physics) Ohio University. Chapter 15—*Spacecraft Performance and Analysis.*

**Dave W. Murrow.** Member of Technical Staff, Systems Division, Jet Propulsion Laboratory, California Institute of Technology, Pasadena, California. M.S. (Aerospace Engineering) University of Texas; B.S. (Aerospace Engineering) University of Colorado. Chapter 17—*Interplanetary Space Mission Operations.*

**Paul Ondrus.** Head, Mission Operations Systems Office, Goddard Space Flight Center, Greenbelt, Maryland. M.B.A. University College; B.S. (Electrical Engineering) University of Akron. Chapter 14—*Assessing Payload Operations.*

**Emery Reeves.** Aerospace Consultant, Palos Verdes Estate, California. M.S. (Electrical Engineering) Massachusetts Institute of Technology; B.E. (Electrical Engineering) Yale University. Chapter 16—*Spacecraft Anomalies.*

**Jeffrey K. Shupp.** Associate Chief Engineer, Rearchitecture Programs, Lockheed Martin, King of Prussia, Pennsylvania. M.S. (Nuclear Engineering) University of California at Berkeley; B.S. (Nuclear Engineering) Pennsylvania State University. Chapter 6—*Defining and Developing the Mission Operations System;* Chapter 7—*Activity Planning.*

**Gael F. Squibb.** Manager, Mission Operations Development, Jet Propulsion Laboratory, California Institute of Technology, Pasadena, California. M.S. (Systems Management) University of Southern California; B.S. (Physics) Harvey Mudd College. Chapter 3—*Mission Operations Functions;* Chapter 4—*Developing a Mission Operations Concept.*

**Konstantin G. Sukhanov.** Deputy Director, Lavochkin Association, Moscow, Russia. Ph.D. (Aeronautical Engineering) Moscow Aviation Technology Institute; M.S. (Engineering) Moscow Aviation Technology Institute. Section 18.3—*Russian Space Mission Operations.*

**Dennis Taylor.** Managing Director, Space Systems Engineering, Ltd., West Sussex, England. M.S. (Engineering) University of Manchester, Institute of Science and Technology; B.S. (Engineering) Imperial College. Section 18.4—*European Space Mission Operations.*

**Craig I. Underwood.** Lecturer in Spacecraft Engineering, C.S.E.R. University of Surrey, Guildford, Surrey, England. B.S. (P.G.C.E.) University of York. Chapter 19—*Microsatellite Mission Operations.*

**Jeffrey W. Ward.** Technical Director, Surrey Satellite Technology, Ltd., Guildford, Surrey, England. Ph.D. University of Surrey; B.S. (Computer Engineering) University of Michigan. Chapter 19—*Microsatellite Mission Operations.*

# Preface

Cost-Effective Space Mission Operations is part of a collection of books in the Air Force Academy's Space Technology Series. The series intends to provide practical approaches to designing and operating space systems; capture experiences, wisdom, and lessons learned by key people in our space programs; and help educate space professionals and students in the "big picture" of space missions. Carefully selected experts from around the world write each book in the series.

The goal of this book is to provide processes, tools, and data that can help us do space mission operations better and more cost effectively. It imagines you're a mission operations manger who is responsible for the early planning of mission operations, designing the mission operations system, and conducting daily operations. This book defines a process that helps you translate mission objectives and requirements into a viable mission operations concept. As mission operations manager, you must develop this operations concept early enough, during concept exploration, so the project manager can trade future cost of operations with current development costs.

The first two chapters describe how the mission operations element blends in with the other elements of a space mission. Chapters three through seven describe the thirteen mission operations functions, provide a process for developing a mission operations concept, present a model for evaluating the cost and complexity of conducting operations, and explain how we define and develop a mission operations system. Chapters eight through sixteen describe how we conduct routine, launch and early orbit, and special operations; plan and analyze the mission; and transport and process data. Finally, the last four chapters describe how we conduct interplanetary, international, microsatellite, and crewed mission operations, as well as the main differences between these missions and Earth-orbiting, uncrewed missions.

Leadership, funding, and support essential to developing the book came from the Air Force Space Missile Systems Center, Air Force Space Command, Naval Research Laboratory, Office of Naval Research, National Aeronautics and Space Administration, Goddard Space Flight Center, Johnson Space Flight Center, Lewis Research Center, the Jet Propulsion Laboratory, the Advanced Projects Research Agency, and the Department of Energy. The European Space Agency also supported as through their European Space Operations Center. Getting money to develop much-needed reference material is exceptionally difficult in the aerospace community. We're deeply indebted to the sponsoring organizations—particularly the Air Force's Phillips Laboratory—for their support and their recognizing the importance of projects such as this one.

This book is intended for professionals and students involved in space systems engineering and space operations, including program managers, mission operations managers, spacecraft engineers and designers, project scientists, and operators. It's suitable as a textbook for a course in space mission operations, a supplementary text in other courses, or as a professional reference.

Our thanks go to the many people who made this book possible—in particular,

- Robert Giffen and Michael DeLorenzo, Astronautics Department Heads at the Air Force Academy, who furnished the leadership and resources to complete this book.
- Perry Luckett, who made the text much more concise, readable, and grammatically correct.
- Anita Shute, who designed the book, edited and formatted the manuscript, interpreted and incorporated our numerous changes, and prepared drafts and camera-ready copy.
- Joan Aug and Connie Bryant, who cheerfully provided administrative support at the Air Force Academy.
- Several people who spent many hours reviewing the book: Curt Heftman, Dave Linick (Jet Propulsion Laboratory), Felix Godwin (Teledyne Brown Engineering), Tim Gillespie (Onizuka Air Force Base), Mike Violet (Falcon Air Force Base), Ron Humble (United States Air Force Academy), and the students who used a draft version of this book in the space mission operations course at the University of Colorado at Colorado Springs.
- Our families for their support during this project.

Through many iterations of reviewing, editing, and revising, we've tried to eliminate all ambiguities and errors. If you find any, please contact us.

*Daryl G. Boden*　　　　　*HQ USAFA/DFAS*
*Wiley J. Larson*　　　　　*2354 Fairchild Dr., Suite 6J71*
　　　　　　　　　　　　*USAF Academy, CO   80840-6224*
　　　　　　　　　　　　*FAX:  (719) 472-3723*

# Chapter 1

# Space Mission Operations

Daryl G. Boden, *United States Air Force Academy*
Wiley J. Larson, *United States Air Force Academy*

So you ask, "Why would anyone write a book about space mission opera-tions?" We've asked ourselves the same question—many times. Much of mission operations is subjective, with no absolute laws, like laws of physics, to guide the way. In many cases there are no concrete answers, only opinions. We often depend on a "seat of the pants" approach and rely solely on the "seat" of the person chosen to do the work. A key goal is to be objective and to identify processes, tools, and data that can help do space mission operations better and more cost-effectively. We want to baseline this material so people doing the work—perhaps, someone like you—can start from higher ground and benefit from previous experience. Remember, we stand on the shoulders of giants!

This book is written for the hypothetical *mission operations manager* (MOM) who is responsible for early planning of mission operations, designing the mission operations system, and conducting day-to-day operations. Our MOM can repre-sent the operations or project organization because many of the users are the same for both organizations. We take an aggressive approach to reducing the cost of mission operations, tempered with a perspective on life-cycle cost.

We advocate a concurrent approach to designing mission operations versus the traditional approach. In the traditional approach, operations develops after the mission and spacecraft are designed. In many cases, this approach restricts cost-reduction strategies to mission operations, so our ability to reduce life-cycle cost is overly constrained. Table 1.1 shows a timeline for traditional mission operations planning. Note that operational planning begins well after the spacecraft is designed—too late to constructively change the operations concept and spacecraft design. In contrast, the concurrent approach to designing mission operations forces planners to address key issues in these areas with other mission elements. The proposed concurrent approach considers many of the traditional issues but gives us *more time* to make decisions that reduce operational complexity and, ultimately, life-cycle cost. Issues about operations and spacecraft design arise early, when changes are easier and less costly to implement. Figure 1.1 shows a timeline for the concurrent approach.

**Table 1.1. Functional Timeline for Spacecraft Operations.** Status of the spacecraft and ground system determines the level of detail of these activities. For new spacecraft or a new generation of spacecraft, test and integration become quite detailed. Integrating operations activities, a new ground system, or a major ground-system upgrade requires much lead time, typically driven by software development.

| Prelaunch | Launch | Early Orbit Checkout | Normal Operations |
|---|---|---|---|
| | | 2 Days – 6 Months | Several Weeks – 30 Years |
| 1 – 2 Years<br>• Develop flight plan<br>  - Spacecraft<br>  - Payload<br>  - Ground system<br>• Develop training plan<br>• Identify simulator requirements<br>• Integrate and test support systems<br>6 Months<br>  • Assemble operations team<br>  • Validate ground-system database<br>  • Validate ground-system hardware<br>  • Validate flight software<br>  3 Months<br>    • Start prelaunch training<br>    • Rehearse launch<br>    • Demonstrate communications protocol<br>    1 Month<br>      • Simulate launch operations<br>      • Review readiness | • Support launch team<br>• Transfer spacecraft to initial orbit | • Validate components<br>• Validate subsystems<br>• Validate subsystems interfaces<br>• Validate systems<br>• Detect and analyze anomalies<br>• Calibrate instruments<br>• Validate instrument processing<br>• Validate protocol for external interfaces<br>• Maneuver spacecraft to mission orbit | • Perform real-time spacecraft operations<br>• Process and distribute payload data<br>• Translate requirements into operational activities<br>• Resolve anomalies<br>• Maintain ground-system database<br>• Maintain flight software<br>• Maintain ground software<br>• Continue operator training<br>• Recover and repair spacecraft<br>• Dispose of non-operational spacecraft |

Over nearly 40 years in the United States, Europe, and the former Soviet Union, we've learned a lot of lessons—some the hard way—while doing commercial, military, and civil space missions. We hope to merge the lessons learned and wisdom gained over these years, analyze what we've done, identify what has

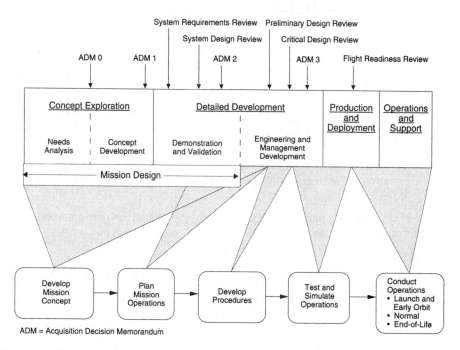

**Fig. 1.1.** **Concurrent Approach to Space Mission Operations.** We must design mission operations in parallel with all other elements of the space mission. Mission phases are defined in the next section.

worked well, and try to develop better ways of planning and conducting space mission operations.

Because we're addressing such a diverse group of people, missions, and cultures, we begin by providing key definitions used in this book. The terms may not be exactly what you use, but we define them well enough so you can understand and apply them to your situation. We focus on Earth-orbiting, uncrewed missions in the US and discuss the differences between them and interplanetary missions. We also compare and contrast approaches in the US, Europe, and Russia. Finally, we describe additional operational requirements when considering a human mission. We hope to give you a broad perspective on mission operations and how different cultures do business in space.

# 1.1 The Space Mission Life Cycle[*]

Table 1.2 illustrates the life cycle of a space mission, which typically progresses through four phases:

*Concept exploration*, the initial study phase, which results in a broad definition of the space mission and its components.

*Detailed development*, the formal design phase, during which we define the system components and, in larger programs, develop test hardware or software.

*Production and deployment*, which includes constructing the ground and flight hardware, writing software procedures, and launching the first full constellation of spacecraft.

*Operations and support*, the day-to-day operation of the space mission, its maintenance and support, and finally its deorbit or recovery at the end of the mission.

**Table 1.2. Development Phases for Space Systems.** Every space program progresses through the top-level phases. Subphases may or may not be part of a given program. The time required to complete the process varies with the program's scope. Major programs take as many as 15 years from concept exploration to initial launch, whereas some programs may require only 12–18 months.

| Phase | Concept Exploration | | Detailed Development | | | |
|---|---|---|---|---|---|---|
| Subphase | Needs Analysis | Concept Development | Demonstration and Validation | Engineering and Management Development | Production and Deployment | Operations and Support |
| Typical DOD Milestones | ADM 0 | ADM 1 | ◣ SRR ◣ SDR | ◣ PDR ◣ CDR | | |
| | | | ADM 2 ◢ | ADM 3 ◢ | Launch ◢ | Deorbit ◢ |
| Typical Products | Statement of needs; studies | Breadboards and studies | Advanced prototypes | Engineering prototypes; detailed design | Production hardware | Operational system |
| Time required in major program | Continuous | 1 – 2 years | 2 – 3 years | 3 – 5 years | 4 – 6 years | 5 – 15 years |

◢ Program Milestones
   ADM - Acquisition Decision Memorandum

◣ Program Reviews
   SRR - System Requirements Review
   SDR - System Design Review
   PDR - Preliminary Design Review
   CDR - Critical Design Review

---

[*] Adapted with permission from *Space Mission Analysis and Design*, Larson and Wertz [1992].

This book discusses concept exploration and detailed development with emphasis on preparing for operations and support. Mission operations costs make up 12%–50% of a space mission's life-cycle cost—the percentage varies with duration and complexity of the mission. In Chap. 5 we'll discuss how to assess mission operations complexity and cost.

These phases may be divided and named differently depending on whether the *sponsor*—the group which provides and controls the program budget—is the Department of Defense (DOD), National Aeronautics and Space Administration (NASA), foreign agencies like the European Space Agency (ESA), or a commercial enterprise. The time required to progress from initial concept to deorbiting or end of the mission appears to be independent of the sponsor. Large, complex space missions typically require 10–15 years to develop and operate from 5–15 years, whereas small, relatively simple missions require as few as 12–18 months to develop and operate for 1–6 months.

Procurement, operating procedures, and risk policies vary with sponsoring organizations, but the key players are the same: the space mission operator, end user, and developer. *Operators* control and maintain the space and ground assets, and are typically applied-engineering organizations. *End users* receive and use the mission's products. They include astronomers and physicists for science missions, meteorologists for weather missions, the general public for communication and navigation missions, geologists and agronomists for Earth resources missions, and the war fighter for offensive and defensive military space missions. The *developer* is the procuring agent, be it DOD, NASA, ESA, or a commercial enterprise, and includes the contractors, subcontractors, and government organizations who handle development and testing. The operators and users must generate technically and fiscally responsible requirements; the developer must provide the necessary product or capability on time and within the changing constraints of politics and funding.

Three basic activities occur during the Concept Exploration Phase (see Fig. 1.2). Users and operators develop and coordinate a set of broad needs and performance objectives based on an overall concept of operations. At the same time, developers generate alternative concepts to meet the perceived needs of the user and operating communities. In addition, the sponsor does long-range planning, develops program structure, budgets, and estimates available funding to meet the needs of the users, operators, and developers. To successfully produce and deploy a cost-effective space capability, the four key players must closely integrate their activities.

The goal during concept exploration is to assess the need for a space mission and to develop affordable alternatives that meet the operators' and end users' requirements. The *Needs Analysis* continues until it culminates in a new program start. Operators and end users develop potential mission requirements based on the considerations shown in the left-hand column of Table 1.3. The process is different for each organization, but at some point a new program begins with a set of mission objectives, concept of operations, and desired schedule. In DOD, the *Mission Needs Statement* documents this information and becomes part of the planning,

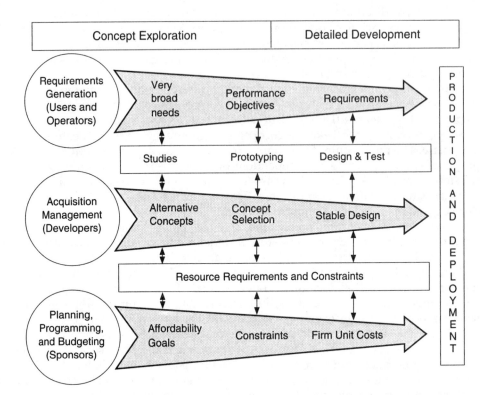

**Fig. 1.2.    Interaction Between Users and Operators, Developers, and Sponsors.** All players
should be involved in the early and continuing development of the mission.

programming, and budgeting system [Defense Systems Management College,
1990]. If approved, the program receives funding and proceeds to concept
development.

At the *Program Initiation* milestone, the funding organization commits to concept development. The program will receive different levels of scrutiny depending
on its scope, political interest, and funding requirements. In DOD, *major programs*
receive the utmost attention at the highest levels. Military components use distinct
criteria to identify major programs [Defense Systems Management College, 1990].
A DOD program is "major" if it requires more than $200 million for research,
development, test, and evaluation or more than $1 billion for production. Pro-
grams that require participation by more than one component of the armed forces
or have congressional interest may also be classified as major programs.

During *Concept Development* the developer must generate alternative ways to
meet the operator's and end user's needs. This procedure includes developing and
assessing different concepts and components for mission operations, as well as
estimating the factors shown in the right-hand column of Table 1.3. The

**Table 1.3.    Further Breakdown of the Concept-Exploration Phase.** During concept exploration
the operator and user define their requirements and pass them to the developing
organization for concept development. We assert that the operator, user, and developer
must work together to develop realistic, affordable mission objectives and requirements
that meet the user's needs.

| Concept Exploration and Definition | |
| --- | --- |
| **Needs Analysis** | **Concept Development** |
| Generate potential requirements based on   Mission objectives   Concept of operations   Schedule   Life-cycle cost and affordability   Changing marketplace   Research needs   National space policy   Long-range plan for space   Changing threats to national defense   Military doctrine   New technology developments | Reassess potential requirements generated during   needs analysis<br><br>Develop and assess alternative concepts for mission   operations<br><br>Develop and assess alternative architectures for the   space mission<br><br>Estimate<br>  performance                       supportability<br>  operational complexity        produceability<br>  schedule                           funding profiles<br>  risk                                 life-cycle cost |

information becomes part of the system concept. High-level managers in the user, operator, and development communities evaluate whether the concepts, initial mission objectives, and potential requirements meet the mission's intentions. If the program satisfies the need at a reasonable cost, it passes the *Requirements Validation* milestone and proceeds into the Detailed Development Phase.

This book provides the technical processes and information necessary to explore concepts for many space missions. Table 1.3 identifies a major concern that can undermine the entire process: in many cases, users and operators analyze the needs and formulate mission requirements apart from the development community. Then they pass these requirements "over the wall" without negotiating. The developer often generates alternatives without the operators and users. These unilateral actions produce minimum performance at maximum cost.

> *To explore a concept successfully, we must remove the walls between the*
> *sponsor, space operators, users, and developers and become a team.*

A good team considers the mission's operations, objectives, and requirements as well as the available technology to develop the best possible mission concept at the lowest possible life-cycle cost.

## 1.2   Elements of a Space Mission[*]

All space missions consist of a set of *elements* or *components* as shown in Fig. 1.3. Arranging these elements forms a *space mission architecture*. Organizations and programs define their mission elements differently, although all elements shown in Fig. 1.3 are normally present in any space mission.

The *subject* of the mission is the phenomenon that interacts with or is sensed by the space payload: moisture content, atmospheric temperature or pressure for weather missions; types of vegetation, water, or geological formations for Earth-sensing missions; or a rocket or intercontinental ballistic missile for defense missions. We must decide what part of the electromagnetic spectrum to use in order to sense the subject, thus determining the type of sensor as well as payload weight, size, and power. In many missions, we may trade off the subject. For example, if we are trying to track a missile during powered flight, the subject could be the rocket body or exhaust plume, or both.

For communications and navigation missions, the subject is a set of equipment on the Earth or on another spacecraft, including communication terminals, televisions, receiving equipment for navigation using the global positioning system, or other user-furnished equipment. The key parameters of this equipment characterize the subject for these types of missions.

The *payload* consists of the hardware and software that sense or interact with the subject. Typically, we trade off and combine several sensors and experiments to form the payload, which largely determines the mission's cost, complexity, and effectiveness. The subsystems of the *spacecraft bus* support the payload by providing orbit and attitude maintenance, power, command, telemetry and data handling, structure and rigidity, and temperature control. The payload and spacecraft bus together are called the *spacecraft, space element,* or *launch-vehicle payload*.

The *launch element* includes the launch facility, launch vehicle, and any upper stage required to place the spacecraft in orbit, as well as interfaces, payload fairing, and ground-support equipment. The selected launch system constrains the spacecraft's size, shape, and mass.

The *orbit* is the spacecraft's trajectory or path and typically is different for initial parking, transfer, and the final mission. There may also be an end-of-life or disposal orbit. The mission orbit significantly influences every mission element and provides many options for trades in the mission architecture.

The *command, control, and communications architecture* is the arrangement of components which satisfy the mission's communication, command, and control requirements. It depends strongly on the amount and timing requirements of data to be transferred, as well as the number, location, availability, and communicating ability of the space and ground assets.

---

[*] Adapted with permission from *Space Mission Analysis and Design*, Larson and Wertz [1992].

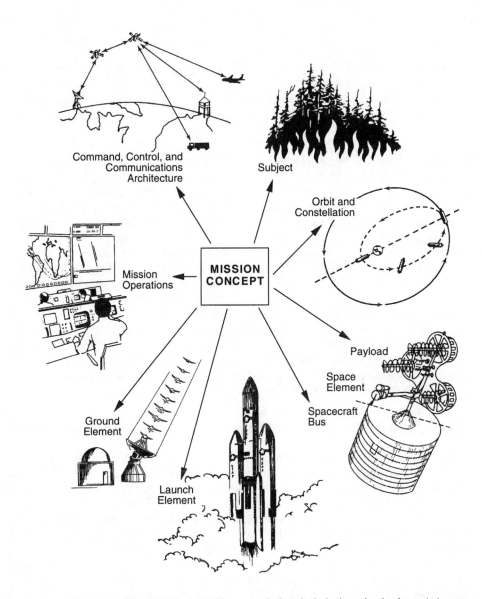

**Fig. 1.3.**    **Space Mission Architecture.** All space missions include these basic elements to some degree. See the text for definitions. Requirements for the system flow from the operator, end user, and developer and are allocated to the mission elements based on the mission concepts.

The *ground element* consists of control centers and fixed and mobile ground stations around the globe, as well as their data links. They allow us to command and track the spacecraft, receive and process telemetry and mission data, and distribute the information to the operators and users.

Finally, *mission operations* consists of the people occupying the ground and space assets, as well as the hardware, software, facilities, policies, and procedures that support the mission operations concept. A key consideration is the *command, control, and communications* ($C^3$) architecture, which connects the spacecraft, ground elements, and mission operations elements.

In this book we'll focus on three pieces of the space mission architecture that form the basis of the mission operations concept: mission operations; ground systems; and command, control, and communications. We will propose alternative approaches for operating a mission and trade among all the mission elements.

We may combine the elements shown in Fig. 1.3 in different ways to form alternative concepts for a space mission. A mission concept describes how all elements work together to meet the mission's objectives. To develop a reasonable and complete mission concept, all players—users, operators, developers, and sponsors—must actively develop alternatives and select the baseline mission concept.

The *mission operations system (MOS)* is the collection of people, procedures, hardware, and software associated with elements shown in Fig. 1.3 for command, control, and communications; ground systems; and mission operations. The MOS also includes parts of the space element needed to do mission operations. Because the spacecraft is part of mission operations, we must consider the MOS early in the spacecraft and mission design. The *mission operations concept*, referred to in Chap. 4, describes how the mission operations system will carry out the mission according to the mission concept and supporting architecture.

## 1.3  Cost-Effective Strategies for Mission Operations

Much of the motivation for this book is based on today's trends of reduced funding for space systems and fewer projects. We believe these trends will continue well into the next decade—into the 21st Century. In addition, almost half of the world's new space systems and upgrades to existing systems will be designed and developed within the next three to five years. This means we have a wonderful chance to design more cost-effective operations for space missions. We will truly live with our successes and failures for the next several decades.

Speaking of decades, this country has emphasized the life-cycle cost of space systems for over a decade, yet very little has been done about accounting for mission operations costs in DOD or NASA. We tend to give the mission operations infrastructure a pot of money to support space missions without allocating operations costs to particular missions. The infrastructure has grown and serviced our missions well under the stewardship of its managers. Now, as the infrastructure is forced to downsize, we must look for ways to accurately assess its cost and the cost of operating space missions. The idea is to charge projects directly for their use of the operations infrastructure. This single activity will profoundly affect the way we trade between spacecraft design and mission operations, ultimately helping us better predict and control the life-cycle cost of individual missions.

Additionally, funding for mission operations is decreasing while operations become more complex. As budgets shrink, NASA and DOD are looking for money to support new and existing programs. One way to get this money is to save it from operations.

We recommend strategies we can use to develop cost-effective operations of space missions with emphasis on reducing the life-cycle cost of space systems. Table 1.4 lists these strategies and provides references.

**Get the mission operations manager actively engaged in developing the mission and mission operations concepts from the very beginning of the project.** Many of the cost-effective approaches we're discussing stem from efficient concepts for the mission and mission operations. Usually, we can reuse existing or evolving resources and affect spacecraft designs only if we try early in conceptual design. Later, when concepts, hardware, and software exist, changing systems to make their operation easier and more cost-effective is expensive.

We must train more people on all aspects of mission operations, so early interaction among users, operators, developers, and sponsors is effective. Organizations must strive to develop more people with the perspective of a mission operations manager.

**Standardize functions for mission operations.** Many spacecraft and mission designers don't understand what happens during operations. This phenomenon isn't unique to space. For decades, designers, builders, and users have had difficulty communicating what they need and how to achieve the lowest cost and best performance. If our industry can focus on the key (most costly) operations tasks, we can save money. Reviewing what operators do during operations can have significant benefits. Once functions are understood, we can combine them, reduce their scope, or even eliminate them all together—ultimately reducing operations costs.

**Standardize communication techniques and protocols.** Standard interfaces provide fertile ground for reducing cost. The communication link is the main interface between spacecraft and ground, users and operators, and data providers and data reducers. If we can standardize these interfaces, we'll be able to streamline our missions. We may be able to use more cost-effective commercial capabilities to carry out daily communications.

**Directly account for operations costs.** We've been doing space operations for three and a half decades; yet, our industry doesn't know what operations cost. We do know, in total, how much funding goes to operations organizations, but we don't have a way to track how much a particular space mission spends for mission support. Thus, inefficient missions can hide among all the other operations.

Table 1.4.    **Strategies for Cost-Effective Operation of Space Missions.** Select a strategy or combination of strategies you can use on your mission to reduce operations and life-cycle costs.

| Cost-Effective Strategy | Key Features | Where Discussed |
|---|---|---|
| Get the mission operations manager actively engaged in developing the mission and mission operations concepts | • Focus development and operations efforts on key cost-effective approaches early in the process, while negotiation and change are practical<br>• Do space/ground trade-offs early<br>• Consider centralized vs. distributed data processing and control early | Chaps. 1, 2, 3, and 4 |
| Standardize functions for mission operations | • Standardized functions allow direct comparison—this may facilitate, combine, or eliminate functions | Chap. 3 |
| Standardize communication techniques and protocols | • One way to standardize interfaces | Chaps. 11, 12, and 13 |
| Directly account for operations costs | • Assess actual cost of operations<br>• Provides better information for trade-offs of life-cycle cost | Chaps. 1 and 5 |
| Promote reuse of procedures, software, hardware, and people | • Reuse works well if planned from the beginning<br>• Keep abreast of changes in the infrastructure because it evolves with time | Chaps. 5 and 8 |
| Make spacecraft operability a primary concern early in designing the mission and spacecraft | • Assess operability using key factors<br>• Maintain margin in all spacecraft systems | Chaps. 5, 14, and 15 |
| Combine technical and operational demonstrations to improve technical and operational performance | • Use operational demonstrations to validate and improve mission operations concepts and reduce operations cost | Chaps. 1 and 2 |
| Carefully consider using automation and autonomy | • Remember that automation and autonomy have mixed results<br>• Use autonomy and automation to improve technical and operational performance, reduce risk, and reduce life-cycle cost<br>• Don't automate just to automate—automate to reduce life-cycle cost | Chap. 5 and 8 |

If we're sincerely trying to reduce operating costs, we must directly account for mission support. Doing so will allow us to trade costs intelligently between space and ground assets. Managers of development programs can also use this accounting to justify spending more or less money to develop the spacecraft.

One particularly effective approach is to use *zero-base operations*—starting a project with no operations people and having to justify adding people and capability on the ground. We believe the more acceptable approach is to begin accounting for operations costs and use the information to make sensible decisions about how to operate.

**Promote reuse of procedures, software, hardware, and people.** The mission operations infrastructures within NASA and DOD have developed incredible capabilities over the years. In fact, they're struggling under the weight of these capabilities because we can no longer support the size of these infrastructures. The stewards of our infrastructures must trim down by identifying and maintaining the best procedures, software, hardware, and people they have to offer.

Then, representatives like the MOM must interact with new projects early in the design stage of the mission and spacecraft. During these early discussions, developers can determine which procedures, software, and hardware might be appropriate, so we can re-use many capabilities and save the developers money. If this interaction doesn't take place early, developers will design systems using their own procedures, hardware, and software. Then, the operators will offer capabilities that are unacceptable because they've become changes that will drive up cost.

**Make spacecraft operability a primary concern early in designing the mission and spacecraft.** For centuries designers have created designs that looked great on paper but were very difficult and expensive to construct and operate. Our recent experiences in space suggest that things haven't changed. We still design spacecraft that are one of a kind and are therefore difficult to build and operate. The Global Positioning System in DOD and Earth Observing System in NASA devoted about 11% and 51% of their 1994 budgets to operate their respective space and ground systems. That's an incredible amount of money!

We must assess spacecraft and mission operability before we release the spacecraft design to production. We can no longer afford to spend money in operations that isn't absolutely justified.

**Combine technical and operational demonstrations to improve technical and operational performance.** Our industry tends to focus on technical demonstrations to show that technology is ready to fly in space and enhance technical performance. Today we need to focus on technological **and** operational approaches that reduce the cost of operations and, ultimately, life-cycle cost. The focus here should be on technologies that can reduce the cost of operations. We can do so by reducing the number of ground contacts; minimizing (or eliminating) tracking requirements; and associating attitude, position, and time information with the mission data on the spacecraft. We must try new operational approaches, such as allowing the spacecraft computer to schedule ground contacts only when necessary. We can save operations money by using new technologies and innovative operations concepts.

**Carefully consider using automation and autonomy.** Ground-station automation and spacecraft autonomy are phrases we hear whenever we consider reducing the cost of mission operations. But, we must carefully consider the overall life-cycle costs before we blindly move forward with automation and autonomy. Automation and autonomy increase development costs. If we don't apply them correctly, they can also increase operational costs, especially if management adds procedures to either verify or prevent automatic or autonomous operations. Still, for long missions that require frequent, repetitive actions, automation and autonomy can save a lot of money.

We begin developing cost-effective mission operations by describing how to design a mission operations system in Chap. 2. This process takes us from stating a mission objective to describing the operations hardware, software, people, and procedures needed to meet the mission's objectives while ensuring its safety. In Chap. 3 and 4, respectively, we describe integral parts of the mission operations system: how we do each of the mission operations functions and the mission operations concept.

## Reference

Defense Systems Management College. 1990. *System Engineering Management Guide*. Ft. Belvoir, VA:U.S. Government Printing Office.

Larson, Wiley J. and James R. Wertz. 1992. *Space Mission Analysis and Design*. Second Edition. Netherlands: Kluwer Publishing.

Rechtin, Eberhardt. 1991. *Systems Architecting*. Englewood Cliffs, NJ: Prentice Hall.

# Chapter 2

# Designing Space Mission Operations

Daryl G. Boden, *United States Air Force Academy*
Wiley J. Larson, *United States Air Force Academy*

Space designers typically consider the mission operations concept after designing the mission and the spacecraft, so mission operations only slightly affect the design of other mission elements. In many cases the mission operations element develops costly and complex approaches to work around inadequate mission and spacecraft designs. By including mission operations early in the mission design, we can evaluate how design decisions affect mission operations and then trade between operations and other elements of the space mission.

The concept of concurrent design for mission operations sounds like a good idea*, but how do you do it? This chapter defines the process that takes us from a statement of the mission objective to a mission operations concept. The mission objectives broadly state the goals the system must achieve to be productive. Typically, space missions have several primary and secondary objectives. The mission operations concept describes how operations will work to meet these objectives.

Table 2.1 lists the steps for designing space mission operations and where to find more information in our book explaining each step. The key word in this table is *process*. In this chapter we walk through the process step by step, describing what needs to be done, how to do each step, the inputs and outputs of each step, and where we provide more detail. To develop a mission operations concept, we must continually update and refine the concept as the mission design matures and then review and revise decisions from earlier steps.

---

* Some people think this confusion on how to implement concurrent design drives cost up, not down.

16

**Table 2.1.** **Developing a Space Mission Concept.** This process takes us from a statement of the mission objective to a mission operations concept. This process is iterative and we must continually update the concept as the design matures.

| Step | Issues | Where Discussed |
|---|---|---|
| 1. Determine the mission's objective, key requirements and constraints, and type. | State the mission objectives. Identify requirements and constraints for the mission, program, hardware, and software. | Chap. 1, 2, 10, 11, and 12; Larson and Wertz [1992] Chap. 1–4 |
| 2. Develop alternative mission concepts that support step 1. | Describe the mission and any underlying philosophies, strategies, and tactics. Key is to identify significantly different ways of doing the mission using a mix of space, ground, and air assets. Identify key organizations and interfaces. | Chap. 1 and 2; Larson and Wertz [1992] Chap. 2 and 22 |
| 3. Identify and do key trades:<br>• Among mission elements for each of the selected concepts<br>• Among organizations. | Inter-element trades are the biggest cost savers. Do trades early in the design to be most effective. Minimize the number of organizations and interfaces if possible. | Chap. 1, 2, 3, and 9–16; Larson and Wertz [1992] Chap. 2 and 3 |
| 4. Characterize acceptable mission concepts and associated space mission architectures.<br>• Information-system characteristics<br>• Payload characteristics<br>• Spacecraft characteristics<br>• Data products<br>• Ground-system characteristics | Summarize needed payload, spacecraft bus, ground-system, and communications capabilities and characteristics. Describe products needed by users and operators for each mission concept. | Chap. 4, and 9–16 |
| 5. Assess items in step 4 and select a baseline mission concept with supporting space mission architecture for future development. | Use the baseline mission concept to develop the mission operations element, ground element, and architecture for command, control, communication, and computers. | Chap. 1 and 2; Larson and Wertz [1992] Chap. 3 |
| 6. Develop alternative mission operations concepts to support the mission concept. | Step 4 characterizes inputs to developing a mission operations concept. | Chap. 4 |
| 7. Do key trades within mission operations. | Typically involves trading people tasks with hardware and software | Chap. 3, 4, 5, 6, 8, 12, and 15 |
| 8. Allocate resources to functions. | Identify operationally difficult requirements and requirements derived from the mission operations concept. | Chap. 3, 4, and 6–9 |
| 9. Assess mission utility, life-cycle costs, and relative complexity and cost of operations. | -- | Chap. 5; Larson and Wertz [1992] Chap. 3 and 20 |
| 10. Iterate and document reasons for choices. | -- | -- |

**Step 1.    Identify Mission Objectives, Functional Requirements, Constraints, and Type of Mission**

When developing a mission concept we must consider the end-user of the data, or mission objective, and the user's requirements. The user may be the military command and control system for early warning data, project scientists for remote sensing data, or the general public for communications and navigation data. A particular mission may have several different users with different requirements for a single payload or multiple payloads with multiple users. The user's requirements define the data volume, timeliness, quality, and level of processing. For example, if the user is the project scientist on an exploratory or scientific mission, the scientist may require level-zero (unprocessed) mission data plus engineering data from the spacecraft. But the project scientist may not need the data immediately. Conversely, a missile warning crew needs the data as soon as an event occurs, but they need only the processed data displaying the event. The mission operations manager (MOM) must develop an operations plan that will meet the user's needs within the mission constraints. The MOM must also be prepared to question a user's requirements, estimate their cost, and show how to spend less on mission operations. A requirement that specifies unprocessed and uncompressed data for scientific analysis increases data volume and may strain the communications architecture selected for the mission.

In addition to mission objectives and user's requirements, we must also determine how mission constraints such as cost, schedule, politics, or the environment will affect our operations. Political constraints may require us to use an existing ground network or place our ground station at a particular location. We balance constraints on the mission with mission objectives and the user's requirements to develop a workable operations plan.

We define space missions based on three categories: trajectory, type of payload, and payload complexity. Most trajectories fall into one of the following categories: low-Earth orbits (LEO), semi-synchronous orbits, geostationary orbits (GEO), and interplanetary trajectories. Most spacecraft are in the LEO category because LEO requires the least amount of launch-vehicle energy to achieve the orbit. The altitudes for LEO typically range from 150 km to 1000 km, and inclinations vary from 0° (equatorial) to 180°. Some typical inclinations for LEO are 28°–55° for Space Shuttle missions, 90° (polar) for Earth mapping, and 95°–105° for sun-synchronous (see Chap. 10 for a definition of sun-synchronous orbits). LEO has several advantages over GEO and other higher orbits. LEO is easier to achieve because it requires less energy for the launch vehicle per kilogram of payload. Thus, we can launch a heavier spacecraft to LEO using a given launch vehicle or we can use a smaller launch vehicle to launch our spacecraft into LEO. Also, the distance from the spacecraft to a point on the Earth's surface is much shorter for LEO than GEO (a few hundred km compared to 36,000 km). Shorter path lengths allow for better sensor resolution and lower space losses in the communications system. Finally, the natural radiation hazards are lower in LEO than in higher orbits.

However, LEO increases operational complexity, mostly because of reduced coverage and short pass times. A spacecraft in LEO may be visible to any one ground tracking station only during a small percentage of revolutions and it may be invisible to all ground stations for several orbits. We overcome this problem with relay satellites or multiple ground stations, both of which increase operations complexity. Also, even when a ground station can see the spacecraft, coverage time is extremely short and the angular rate of the spacecraft relative to the tracking station is high. For example, a spacecraft in a circular orbit at an altitude of 200 km has a maximum time in view of a ground station of slightly less than seven minutes and a maximum angular rate of two degrees per second [Larson and Wertz, 1992]. The operators must establish a communications link, determine the spacecraft's state of health, upload commands, and download the payload data—all in less than seven minutes. Again, we overcome this problem by using relay satellites or multiple ground stations. Finally, in LEO, atmospheric drag perturbs the orbit, making orbit prediction difficult and limiting the spacecraft's lifetime. Without periodic orbit maneuvers, spacecraft in LEO will eventually re-enter the atmosphere and burn up. Or, at the very least, atmospheric drag will change the orbit parameters (semi-major axis and period), and we may need periodic maneuvers to maintain our mission orbit.

*Semi-synchronous* spacecraft orbit the Earth two times per day, yielding an orbit period of about twelve hours. Two classes of semi-synchronous orbits are *Global Positioning System (GPS) orbits*—circular orbits inclined at about 55°, and *Molniya orbits*—highly elliptical orbits inclined at 63.4°. We define both orbits in more detail in Chap. 10. Semi-synchronous orbits exhibit long viewing times, low relative angular velocities, and immunity to atmospheric drag. These characteristics make operations easier, but it's more difficult (requires more launch energy) to achieve these orbits and the radiation environment is more severe than for LEO.

*Geostationary (GEO) spacecraft* orbit the Earth once per day, exactly matching the Earth's rotational velocity. In addition, the orbit plane is the Earth's equatorial plane. In this orbit, the spacecraft remains stationary relative to a point on the Earth's surface. This position provides for continuous coverage of the Earth and continuous contact with the ground site, making scheduling less complicated. Also, tracking is easy because the relative velocity between the ground site and the spacecraft is zero. But GEO does have some disadvantages. First, the spacecraft requires more launch energy than LEO or semi-synchronous to reach this orbit. Also, the path lengths between the spacecraft and the target or the ground station are long (36,000 km). Finally, perturbations caused by a non-spherical Earth, the Moon, and the Sun cause the spacecraft to drift from its assigned position. We correct for this drift by doing periodic station-keeping maneuvers which increase operational complexity.

Some people argue that spacecraft in GEO orbit cost more to operate than spacecraft in LEO. The main reason is that because GEO spacecraft are continuously visible, operators interact with the spacecraft continuously—driving up operations costs. We must carefully consider how much time is needed to interact with the spacecraft.

The final category of trajectories is *interplanetary trajectories*, in which spacecraft escape Earth's orbit, traverse the solar system, and either fly by or rendezvous with other bodies in the solar system. Interplanetary trajectories have large launch energies, precise navigation requirements, extremely long path lengths for communications, and long flight times. All of these characteristics increase the complexity of mission operations.

The second category we use to define a mission is the type of payload carried on the spacecraft. Most common payloads are for

- Communications
- Navigation
- Remote Sensing
- Research, Development, Testing, and Evaluation (RDT&E)

- Scientific
- Interplanetary Exploration
- Technology Demonstrations

*Communications spacecraft* transfer information from one point to another. Most communication spacecraft are in GEO, taking advantage of the hemi-spherical coverage and continuous availability. A new class of communications spacecraft is being developed for LEO to provide continuous, global coverage for personal communications with hand-held phones. Although LEO reduces power and launch requirements, this system is much more operationally complex than a small number of spacecraft in GEO. Space communications systems give us high volumes of data and continuous coverage.

*Navigation spacecraft* transmit a signal that contains highly accurate spacecraft position, velocity, and time data, which allows us to determine the position and velocity of users anywhere on, or near, the Earth. Navigation systems such as GPS, a constellation of 24 satellites in semi-synchronous orbits, offer highly accurate orbit determination and prediction, as well as continuous, worldwide coverage.

*Remote-sensing spacecraft* normally carry a sensor for observing the Earth or some other object from space. Some typical remote-sensing spacecraft are weather observation, photo reconnaissance, missile detection, and Earth-resources mapping. Remote-sensing spacecraft in LEO take advantage of short path lengths for high resolution, whereas remote-sensing spacecraft in GEO take advantage of global coverage. Remote-sensing spacecraft are characterized by high data volume, precise orbit navigation, plus frequent station-keeping maneuvers and payload calibrations.

*Scientific spacecraft* are similar to remote-sensing spacecraft and exhibit many of the same characteristics. Additionally, scientific spacecraft like the Hubble Telescope may observe objects in space rather than on Earth. Besides the requirements for remote sensing, scientific missions may also require frequent slewing maneuvers, extremely accurate attitude determination, and precise pointing control.

Exploratory *interplanetary spacecraft* are actually remote-sensing (Magellan) or scientific (Voyager), but because of their trajectories we treat them separately. Chapter 17 describes the operational differences between interplanetary and Earth-orbiting spacecraft.

*Technology demonstration* and *RDT&E spacecraft* are experimental and typically high-risk; they may place more stress and demands on operations than other missions do.

The final category used to define the type of mission is payload complexity. One factor we use to evaluate payload complexity is how many separate payloads the spacecraft carries. Other factors are the frequency and volume of payload commands, type and number of payload constraints, timeliness and volume of data, and criticality of payload operations. Chapter 14 discusses payload operability and relates the payload's design and operations concept to the complexity of mission operations.

Our plan for mission operations is incomplete until we've identified the hardware, software, and interfaces required to support the mission. Before we develop the operations concept to support a particular mission architecture, we must decide whether we can use existing hardware, software, and ground facilities. Mission architectures that require daily eight-hour tracks using the Deep Space Network may not work, and developing new software and dedicated ground-support facilities will drive up costs.

## Step 2.  Develop Alternative Mission Concepts that Support Step 1

A *mission concept* is a broad statement of how the mission will work in practice [Larson and Wertz, 1992]. While not assuming a particular approach, we should identify significantly different ways of carrying out the mission with a mix of ground, air, and space assets. For example, suppose our mission objective were to travel from Los Angeles to New York. Our alternative mission concepts may include traveling by car, bus, train, or airplane, or some combination of the four. We may also choose to hitch hike or ride a bicycle across country.

We can eliminate several options for meeting the overall mission objective by applying the constraints—typically schedule, cost, politics, and technology. For example, on our planned trip from Los Angeles to New York, schedule constraints may keep us from hitch hiking or riding a bicycle.

Most space missions involve getting and transferring information from one place to another, so our mission concepts should focus on how we obtain and transfer data. Therefore, we must consider issues like data acquisition and processing, communications architecture, ground-systems tasking and scheduling, and mission timeline.

### Step 3. Identify and Do Key Trades among Elements of the Space-Mission Architecture for Each Concept

In this step the mission operations manager identifies and offers ways of controlling system drivers that increase the cost and complexity of mission operations. Doing trades with other mission elements can reduce the cost of mission operations without increasing life-cycle costs or mission risk. We must do these trades as early as possible to avoid costly re-designs and to keep from recommending changes to parts of the mission design that are essentially frozen. In this section we isolate trades between mission operations and other elements. In actual programs these trades may have more complex effects on other mission elements. For example, a trade between orbit altitude and operational complexity affects several elements. Orbit altitude is a key input to communications (propagation path length), space (distance to the target and space environment), and launch (launch-vehicle energy needed to attain orbit). Table 2.2 summarizes some typical trades between mission operations and other mission elements.

**Table 2.2.    Typical Trades between Mission Operations and Other Elements of a Space Mission.** We must do these trades early in the mission design before element designs are frozen.

| Element | Trade | Issues |
|---------|-------|--------|
| Orbit | LEO versus GEO | • LEO reduces launch costs and communications path length but increases operational complexity and reduces coverage compared to GEO |
| Spacecraft (payload and bus) | Resource margins versus development costs | • Positive resource margins reduce operational complexity but increase development cost and spacecraft mass |
|  | Spacecraft autonomy versus development costs | • Autonomy can reduce operations cost but increases development costs |
|  | Software re-use versus development cost | • Re-using software from mission to mission lowers development costs but reduces flexibility<br>• Only works when operators and designers do trades early |
|  | Real-time versus stored commands | • Real-time commands increase flexibility but also increase operational complexity |

**Table 2.2.** **Typical Trades between Mission Operations and Other Elements of a Space Mission. (Continued)** We must do these trades early in the mission design before element designs are frozen.

| Element | Trade | Issues |
|---------|-------|--------|
| Ground Systems | Centralized versus distributed data processing | • Centralized data processing decreases interfaces but increases staffing costs |
| | Dedicated versus existing facilities | • Using existing facilities reduces development costs but leads to more complex scheduling and less flexibility |
| | Automation versus development cost | • Automated ground stations reduce operational cost (staffing) but increase development costs |
| | Software re-use versus development cost | • Re-using software from mission to mission lowers development costs but reduces flexibility |
| Communications | Stored versus real-time data transfer | • Real-time data transfer increases timeliness of data and reduces onboard memory but also increases operational complexity |
| | Data rate versus pass duration | • For a given volume of data, short passes require less ground-network support but also require higher data rates over the communication link |

This trade between operational complexity and development costs has been ignored. Usually, designers have reduced development costs (current dollars) and driven up operational costs (future dollars). Of course, it's easier to spend fewer dollars now and more later than to spend more dollars now with the promise of reducing future costs. But if we are to reduce life-cycle costs, we must do informed trades between current development costs and future operational costs.

## Step 4. Characterize Acceptable Mission Concepts and Associated Space-Mission Architectures

Describe operations for each element of the mission (see Sec. 1.2):

- Subject
- Launch and replenishment
- Orbit and coverage
- Spacecraft (payload and bus) operation and requirements
- Communications architecture and data flow
- Ground element and data flow
- Mission operations

If you are the MOM, work closely with the program manager during this phase, developing plans of operations for each element of the mission as the mission design matures. Be alert for potential operational conflicts among the mission elements and be prepared to identify operational system drivers and propose alternatives to make operations simpler.

### Step 5. Select a Baseline Mission Concept and Supporting Space-Mission Architecture for Future Development

Justify and document the "best" approach. Base it on life-cycle costs and ability to meet mission objectives and top-level requirements. Evaluate operational costs and complexity for each mission concept and make sure cheaper operational alternatives don't drive up development costs. Evaluate the mission concepts and ensure mission operations can meet objectives.

### Step 6. Develop Alternative Concepts for Mission Operations to Support the Mission Concept

The *mission operations concept* explains how the mission will be flown through all phases. It combines the hardware, software, people, and procedures needed to meet mission objectives. At this step in the design process, we concern ourselves with operations concepts that support each mission architecture. If this is truly a concurrent design, we must develop the operations concept with other mission elements so the mission is operable. We describe how to develop a mission operations concept in Chap. 4.

The inputs to the operations concept, shown in Fig. 2.1, vary with each mission and change as the mission design matures. We begin by stating the mission objective and mission concept. Normally, the concept includes the mission type, user, and plan for fulfilling the mission objectives.

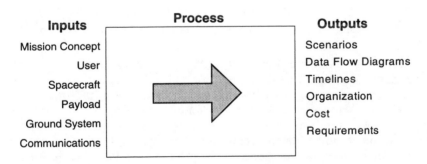

| Inputs | Process | Outputs |
|---|---|---|
| Mission Concept | | Scenarios |
| User | | Data Flow Diagrams |
| Spacecraft | | Timelines |
| Payload | | Organization |
| Ground System | | Cost |
| Communications | | Requirements |

**Fig. 2.1.** **Developing an Operations Concept.** Concept development begins with a mission concept and continues throughout the mission. (See Chap. 4)

Payload operability, discussed in Chap. 14, is key to developing an operations concept. Factors include scheduling, commanding, data processing, data flow, spacecraft support, and physical constraints needed to complete the mission. The payload schedule can be either time driven, event driven, or adaptive. A time-driven schedule requires the payload to execute a given set of commands at specified times. An event-driven schedule requires the payload to execute a given set of commands following an event, such as acquisition of a target. An adaptive schedule is the most complex form of payload scheduling because it requires the payload (or operator) to determine which commands to execute based on observed data. Commanding the payload includes several issues. Simple payloads use repeated commands (several chances to execute) and don't require spacecraft support to carry them out. Complex payloads use commands that are time critical (only one chance to execute), require spacecraft support (maneuvers), are frequent, and don't repeat.

Data processing and data flow affect the communications architecture of the mission. Payloads with high data volume, multiple data formats, and short latency demand strong operational support. Payloads with low data volume, single data formats, and delayed transmission are relatively easier to operate. Finally, payloads with physical constraints such as temperature limits or pointing restrictions make operations more complex. Table 2.3 summarizes factors that influence payload operability.

**Table 2.3. Payload Operability.** Payload operability is key to developing an operations concept for space mission. (See Chap. 14)

| Factor | Issues |
|---|---|
| Scheduling | Time-driven, event-driven, or adaptive |
| Commanding | Repetitive commands, time critical, spacecraft support required |
| Data Processing | Multiple formats, data volume, data latency, data storage |
| Payload constraints | Pointing accuracy, temperature limits, pointing restrictions |

Several factors affect the spacecraft's operability, but perhaps the two most important are adequate resource margins and autonomy. Typical resources monitored on the spacecraft are communications link, power, thermal, pointing, and memory. If we have adequate margins for all resources, we need to model spacecraft operations only at the systems level. If any margins are negative, or nearly zero, we must model the spacecraft at the sub-system level to ensure safe operations. For example, if the spacecraft's peak power load exceeds the available power, some systems must draw power at different times. Thus, we have to model all spacecraft operations at the subsystem level to determine if a particular operation will exceed available power. Other issues affecting spacecraft operability are frequency and timing of maneuvers, autonomy of spacecraft operations, number of telemetry channels monitored, requirements for spacecraft commanding and validation, monitoring and analyzing the spacecraft's state of health, engineering calibrations, and software maintenance. We discuss spacecraft operability further in Chap. 15.

The final inputs to an operations concept are the communications architecture and ground element needed to support the mission. The communications architecture represents the data flow (uplink and downlink) between the spacecraft, the operator, and the user. The communications architecture depends on data volume, ground-station coverage, latency requirements, and data reliability. A spacecraft in LEO that must deliver near-real-time, processed data to the user will require extensive operational support. But, a spacecraft in GEO that can delay data transmission will need less support. Early in the mission we must decide whether to use an existing ground-support network, such as the Deep Space Network or the Air Force Satellite Control Network, or to develop a new, dedicated network. Using an existing network may reduce development costs but may also make operations more complex because of potential scheduling conflicts with other missions and the need to match the network's interfaces. We describe existing ground-support networks on Chap. 12.

Table 2.4 lists the steps in developing an operations concept for a space mission. The results of developing an operations concept and the contents of a concept document are shown in Fig. 2.1 and described in Chap. 4.

**Table 2.4.    Developing an Operations Concept for a Space Mission.** Chapter 4 describes this process in detail; it's an important part of the space mission concept.

| Step | Key Items | Where Discussed |
|---|---|---|
| 1. Identify the mission concept and supporting space mission architecture and gather information | • Information system characteristics <br> • Payload characteristics <br> • Spacecraft bus characteristics <br> • Definition of data product <br> • Ground system | Chap. 2; Larson and Wertz [1992] Chap. 2 |
| 2. Determine what mission operations must do | Key functions usually vary for each mission concept and architecture. We can combine or eliminate some functions. | Chap. 3 |
| 3. Identify ways to carry out operations and determine whether capability exists or must be developed | • Where accomplished (space or ground) <br> • Degree of automation on ground <br> • Degree of autonomy on spacecraft <br> • Software reuse (space and ground) | Chap. 1, 2, 3, and 8 |
| 4. Do trades for items identified in Step 3 | Options are often selected before developing operational scenarios. These trades are done within the operations element, which includes the flight software. | Chap. 1, 2, 3, 4, and 5 Table 3.1 |

**Table 2.4.    Developing an Operations Concept for a Space Mission. (Continued)** Chapter 4 describes this process in detail; it's an important part of the space mission concept.

| Step | Key Items | Where Discussed |
|---|---|---|
| 5. Develop scenarios for operations determined in Step 2 and the options selected in Step 4 | Operational scenarios are step-by-step descriptions of how to do mission operations. Key issues and drivers on the operations system are identified during this step. | Chap. 4, 6–8 |
| 6. Develop timelines for each scenario | These timelines establish performance parameters for each function within mission operations. Identifies what, how fast, and when events occur. | Chap. 4, 6–8 |
| 7. Determine the type of resources (hardware, software, or people) needed to do each step of each scenario | Allocation to hardware, software, or people depends on what is done, how quickly it must be done, and how long the mission lasts. | Chap. 4 |
| 8. Develop data-flow diagrams | These diagrams underpin the ground and flight data systems and the command, control, and communications architecture. | Chap. 4 and 13 |
| 9. Characterize the organization and team responsibilities | Identify type of organization, product responsibility, interfaces, and number of people. For cost-effective operations, minimize the number of organizations and interfaces | Chap. 4 and 5 |
| 10. Assess mission utility and complexity and cost of mission operations | The cost estimates include costs for development and operations. We refine them each time we update the mission operations concept. | Chap. 5; Larson and Wertz [1992] Chap. 20 |
| 11. Identify derived requirements and cost and complexity drivers and negotiate changes to mission concept | Document derived requirements and ensure consistency with top-level requirements. | Chap. 1, 2, and 4 |
| 12. Generate technology development plan | Technology may or may not exist to support the mission concept | Chap. 4 and 6 |
| 13. Iterate | -- | -- |

## Step 7.    Do Key Trades within Mission Operations

Typically, trades within mission operations involve trading people tasks with hardware and software. Pertinent factors are development costs, operational complexity and cost, level of acceptable risk to the mission, and flexibility of operations. Table 2.5 lists typical trades and issues associated with mission operations and where to find more information in this book.

Table 2.5. **Trades with Mission Operations.** These trades represent areas that add most to the complexity of space mission operations.

| Trade | Issues | Where Discussed |
|---|---|---|
| Spacecraft autonomy | Development cost, mission duration, modeling required, "trust" given to autonomous operations | Chap. 3 and 15 |
| Ground-station automation | Development cost, mission duration, software support required | Chap. 3 and 12 |
| Ground-station staffing | Daytime operations or around the clock, acceptable risk, data latency acceptable | Chap. 4, 5, and 8 |
| Anomaly response | Acceptable risk, down time acceptable, level of spacecraft safing required | Chap. 4, 5, 15, and 16 |
| State-of-health monitoring | Alarming and trending, acceptable risk | Chap. 8 and 15 |
| New or existing hardware | Development cost, flexibility, maintenance | Chap. 4, 5, 6, and 8 |
| New or existing software | Development cost, software maintenance, flexibility | Chap. 4, 5, 6, and 8 |

## Step 8. Allocate Resources to Functions for Each Mission Phase

Our mission operations concept should translate user requirements to a process that works—a system that safely and reliably provides the required data to the user.

Table 2.6 lists what we must do for all space missions. We describe each of the thirteen functions in Chap. 3 and explain further how to do them in the chapters listed in the table. We're not prescribing how to organize mission operations. For some missions, especially smaller ones, we can combine several functions into one activity. In other missions, we may divide the operations organization into uplink and downlink sections and operations functions into uplink and downlink components.

Operations tasking and demand for resources vary as the mission matures. We divide the mission into four phases:

- Launch and early-orbit (L&EO) operations
- Routine operations
- Anomaly operations
- Extended operations

We describe launch and early-orbit operations in detail in Chap. 9. This phase is high risk because the time to make decisions is short, the pace is fast, and deci-

**Table 2.6. What Operations People Do Throughout a Space Mission.** These functions are typical of most space missions and are not intended to imply any organizational structure.

| Function | Where Discussed |
|---|---|
| 1. Mission planning | Chap. 3, 6, and 8 |
| 2. Activity planning and development | Chap. 3, 4, and 7 |
| 3. Mission control | Chap. 3, 4, and 8 |
| 4. Data transport and delivery | Chap. 3, 12, and 13 |
| 5. Navigation planning and analysis | Chap. 3, 7, 8, and 10 |
| 6. Spacecraft planning and analysis | Chap. 3, 4, 7, and 15 |
| 7. Payload planning and analysis | Chap. 3, 4, 7, and 14 |
| 8. Payload data processing | Chap. 3, 13, and 14 |
| 9. Archiving and maintaining the mission database | Chap. 3, 13, and 14 |
| 10. Systems engineering, integration, and support | Chap. 3, 4, and 6 |
| 11. Computers and communications support | Chap. 3, 4, 8, 11, 12, and 13 |
| 12. Developing and maintaining software | Chap. 3, 4, and 8 |
| 13. Managing mission operations | Chap. 3, 5, 6, 7, and 8 |

sions may determine whether the mission succeeds or fails. In addition, we're operating in a harsh and dynamic environment, where many activities take place for the first time in the mission.

As we transition to routine operations, our attention moves from the health and safety of the spacecraft to supporting the mission. (See Chap. 8.) Routine operations typically involve well established activities, static procedures, and fully trained operators.

If something unexpected occurs, we begin anomaly operations. (See Chap. 16.) The approach we follow to resolve the anomaly and return to normal operations depends on the nature of the anomaly, the type of mission, the user's requirements, and the operators' qualifications.

Finally, missions that continue beyond the design life of the mission enter into extended operations. Extended operations usually involve reduced staffing, limited onboard resources, and well understood operations. Even though operations are well understood, operator complacency and reduced resources can increase risk and potential for operator errors during extended operations.

## Step 9. Assess Relative Complexity and Cost of Operations

We use the cost model described in Chap. 5 and summarized in Table 2.7 to assess the complexity and cost of our mission. The model is divided into four factors: operability of the mission design, operability of the spacecraft and payload, risk policies, and inheritance and complexity of the ground system. Each of the four factors divides into approximately twenty sub-factors, and the sub-factors divide into low, medium, and high tasking levels. We determine the tasking level (high, medium, or low) using the criteria described in the cost model. To determine the values required for the sub-factors, we turn to the appropriate chapter in the book. For example, turn to Chap. 10 to determine the frequency and number of maneuvers our mission requires. The cost model we use in this book estimates only the cost of mission operations, not development costs. For example, this model predicts lower operational costs if we use a dedicated ground station because we don't have to adapt our mission to an existing system. It doesn't predict the cost of developing the dedicated ground station.

**Table 2.7.    Complexity and Cost Factors for Assessing the Mission Operations Concept.** This model estimates the cost of operating a space mission based on top-level requirements. (See Chap. 5)

| Mission Operations Cost Model Factors | |
|---|---|
| **Complexity Metrics for Mission Design and Planning**<br><br>Science Events<br>Engineering Events<br>Maneuvers<br>Tracking Events | **Complexity Metrics for the Flight System**<br><br>Commanding<br>Monitoring<br>Pointing<br>Resources and Margins<br>Automation |
| **Complexity Metrics for Risk Avoidance**<br><br>Command and Control<br>Data Return<br>Performance Analysis<br>Fault Recovery | **Complexity Metrics for the Ground Systems**<br><br>Interfaces<br>Complexity of Data<br>Automation<br>Organization and Staffing |

After we evaluate all of the sub-factors in the cost model, we convert the tasking levels to numerical values, called scale factors, and average them for each of the four factors. Finally, we combine these four scaling factors and compare them to a similar mission in a look-up table. We multiply the number of full-time equivalents (FTEs) needed to operate the nominal mission by the scale factor to estimate the FTEs needed to operate the new mission.

The four factors in the cost model cover all aspects of mission operations and include the thirteen mission operations functions described in Chap. 3, but functions don't map directly into cost factors. For example, each of the four cost factors includes different aspects of navigation. After we've determined the number of FTEs needed to support our mission, we can estimate the number of FTEs necessary to support each function by comparing our mission to a similar, existing mission. Chapter 3 contains data for allocating a percentage of resources to operations functions.

## Step 10. Iterate and Document Reasons for Choices

We present two extremely important points in this step:

- The operations concept continuously matures over the mission's lifetime
- We must document and justify our decisions

The operations concept is a growing and changing document. As the mission design matures and changes, we must evaluate the effect of these changes on mission operations. For example, someone may decide to save weight by placing fewer batteries on the spacecraft. But this decision may create a negative power margin for the spacecraft during eclipse periods. Whenever we reduce resource margins on the spacecraft, we increase operational complexity and cost. Similarly, someone may decide to remove or replace one of the payloads on the spacecraft. Removing a payload may increase resource margins or reduce maneuver requirements, thus reducing operational complexity. Each time the mission design changes, we have to determine whether the operations concept must change.

It's hard to evaluate the need for these changes if we haven't documented and justified our previous choices and decisions. This step is especially important if a change to the mission design significantly increases operational complexity and cost. If we can't justify our estimate of operational costs, it will be difficult to trade against other mission elements in order to reduce life-cycle costs.

### References

American National Standard. 1993. *Guide for the Preparation of Operational Concept Documents*. Washington, DC: American Institute of Aeronautics and Astronautics.

Larson, Wiley J. and James R. Wertz. 1992. *Space Mission Analysis and Design*. Second Edition. Netherlands: Kluwer Academic Publishers.

# Chapter 3

# Mission Operations Functions

Gael F. Squibb, *Jet Propulsion Laboratory,*
*California Institute of Technology*

We've discussed the big picture and how mission operations fits into the overall scheme. But you may still not know what space mission operators do. You're not alone—many people don't. Different countries, and indeed different organizations within a country, may organize differently to operate space missions. For example, in the United States, NASA and DOD organize their operations differently, and people within each organization differ on philosophy and implementation. In this book, *mission operations* includes actions needed to prepare for launch, activities that take place after launch, and activities required to maintain the infrastructure that supports space missions.

Recall that we must distinguish between the *mission concept*—how the elements of the mission fit together (Chap. 1) and the *mission operations concept*—how we'll do mission operations to carry out the mission concept (Chap. 4). In this chapter we'll define the 13 functions key to mission operations everywhere and discuss how we can combine them to meet the mission operations concept. Hardware, software, and people work together to complete these 13 functions. The mission operations manager (MOM) must carefully trade automation against human operations because automating some of these functions can lead to lower operations costs and, in most cases, lower life-cycle costs. Organizations may group or name those functions differently, but we believe they capture the tasks essential to mission operations.

The MOM must decide *which functions* to do, as well as their *scope* and *how* they'll be done. Depending on the size of the mission, a manager may even have to add functions to our list. In addition, the MOM must address organizational, hardware, and software interfaces between the functions. For example, he or she has to decide whether to use an existing mission operations infrastructure or to create one for this mission. In any event the discussion will revolve around what needs to be done and what is the most cost-effective way to do it.

Note that the 13 functions don't normally correspond to operational teams. The number and size of the operations teams depend on the mission, its complexity, and the operations organization's philosophy. For example, if we examine a communications mission using a geostationary satellite, a

- Commercial operation requires about 12 people per spacecraft to do everything
- DOD operation requires about 27 people per spacecraft to do the same tasks
- European Space Agency operation requires about 22 people per spacecraft

These examples share a common mission with minor differences in complexity and tremendous differences in philosophy. We'll examine these philosophical differences throughout this book and, in Chap. 5, we'll discuss how to assess the complexity and cost of space mission operations.

Figure 3.1 overviews a mission operations system, which implements the mission concept and mission operations concept. It processes information from, and controls, the ground and space assets, so users and operators get needed information and services. Most missions today—science, communications, navigation, and remote sensing—focus on providing information to users or customers.

Note that we divide mission operations into two key pieces:

- *Data* —the hardware and software on the ground and in the spacecraft that help us operate the mission
- *Operations Organization*—the people and procedures that carry out the mission operations concept

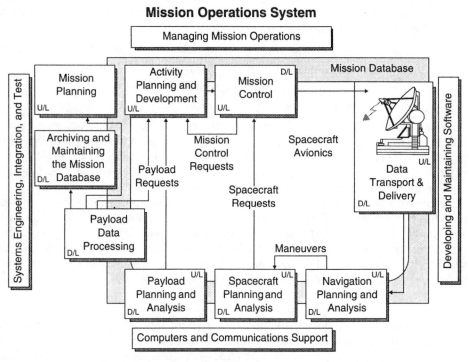

**Fig. 3.1.** **The 13 Functions of Mission Operations.** The figure shows how the functions interact and whether they're involved in downlink (D/L) or uplink (U/L). Functions in the shaded area share data within the mission database.

Mission operators should help develop the mission concept, but they often get involved too late. We intend to aggressively attack this problem in the future to reduce the life-cycle costs of space missions. Mission operators *do* generate the concept that leads to successful mission operations.

Figure 3.1 and Table 3.1 describe the 13 functions mission operators do for space missions. We define the functions so we can discuss them clearly. We can combine or eliminate functions to support the desired mission operations concept. The first nine functions, shown in the mission database box in Fig. 3.1, are basic to all space missions. We can tie them together in order to support the two essential processes for mission operations—uplink and downlink (U/L and D/L) in the boxes. Chapter 8 discusses uplink and downlink, which operate interactively for many Earth-orbiting spacecraft. However, for a planetary mission with long, one-way light times (>10 minutes), they operate separately. The nine functions share data through the mission database and the spacecraft's avionics and require extensive data processing. The order from upper left (mission planning) clockwise to lower left reflects the usual emphasis in processing: from uplink to downlink analysis, and then to planning new uplink activities. The other four functions, which give us an infrastructure for mission operations, are

- Managing mission operations
- Developing and maintaining software
- Computers and communications support
- Systems engineering, integration, and test

The arrows in Fig. 3.1 highlight key interactions. For example, payload data processing provides data for archiving and is a key input to mission planning. The payload data can profoundly affect mission planning, especially if we're doing an *adaptive mission*, in which operators learn from the data they see and change the mission to get different information or do other activities.

Note from Fig. 3.1 that the spacecraft and ground systems must work together on at least seven of these functions. Thus, mission operations people must take part early in decisions that divide responsibility for designing, developing, testing, and operating them. Indeed, we believe mission operations should share control of spacecraft avionics to meet mission needs and constrain life-cycle costs. Key space-to-ground trade-offs in these areas are integral to the mission and mission operations concepts.

Table 3.1 lists the 13 functions and provides key issues and trades we must consider for each. We'll describe them in detail in the following sections. We'll also discuss the information required to design them, recommend a way to carry them out, and describe what they should provide—inputs, process, and outputs. Note we must do some functions, or parts of them, before launch, after launch in real time, and after launch in non-real time. Timing often depends on the mission type. For example, planning and developing activities for Earth orbiters often occurs before launch because we usually carry out these activities right after launch. But for interplanetary missions, we may develop many activities after launch—while the spacecraft is traveling to its target planet.

All 13 functions are important, and achieving mission objectives depends on parts of each. Some are also more expensive, so we must emphasize them during design to lower life-cycle costs. Table 3.2 highlights five functions, each of which consumes more than 12% of total human resources. Four of these five vary more than 6% from low to high percentages of cost. They therefore demand more of our attention to keep life-cycle costs down.

The following sections describe the information and steps needed to carry out all 13 functions, as well as the products they generate. We use background shading for post-launch activities and no shading for pre-launch activities in the tables throughout this chapter.

**Table 3.1.    Mission Operations Functions.** This table shows key issues, trades, and where we discuss functions in more detail.

| Function | Key Issues and Trades | Where Discussed |
|---|---|---|
| 1. Mission Planning | • Provide positive spacecraft resource margins<br>• Automate certain space and ground activities<br>• Restrict the number of mission and flight rules<br>• Focus on operability | Chap. 7, 8, 14, and 15<br>Sec. 3.1 |
| 2. Activity Planning and Development | • Reduce the required number of command loads<br>• Automate validation of command loads and keep approval at lowest possible level<br>• Avoid late changes to plan | Chap. 7 and 8<br>Sec. 3.2 |
| 3. Mission Control | • Consider sharing operations between missions or multi-tasking operators within a mission<br>• Design ground and flight systems that can be easily upgraded<br>• Automate analysis of ground system and spacecraft performance | Chap. 8, 12, and 15<br>Sec. 3.3 |
| 4. Data Transport and Delivery | • Keep tracking requirements to a minimum<br>• Design data structures and formats to use standard services<br>• Use variable-length data packets | Chap. 11, 12, and 13<br>Sec. 3.4 |
| 5. Navigation Planning and Analysis | • Minimize number of maneuvers<br>• Match mission requirements to accuracy of orbit determination method<br>• Match mission requirements with standard services provided by tracking network | Chap. 8 and 10<br>Sec. 3.5 |
| 6. Spacecraft Planning and Analysis | • Maintain positive resource margins<br>• Minimize subsystem interactions<br>• Automate spacecraft safe modes and analysis | Chap. 15 and 16<br>Sec. 3.6 |
| 7. Payload Planning and Analysis | • Use existing planning tools<br>• Automate payload data gathering and calibrations<br>• Insist on payload operability<br>• Minimize payload operations that require knowledge of previously acquired data | Chap. 14<br>Sec. 3.7 |
| 8. Payload Data Processing | • Use existing tools for data processing<br>• Address data processing requirements early in mission design<br>• Understand need for and availability of ancillary data | Chap. 13<br>Sec. 3.8 |
| 9. Archiving and Maintaining the Mission Database | • Consider data push vs. data pull approach<br>• Develop plan early and ensure capabilities exist<br>• Understand need for and availability of ancillary data | Chap. 13<br>Sec. 3.9 |
| 10. Systems Engineering, Integration, and Test | • Involve users, developers, and operators early in mission design<br>• Do key trades on mission operations concept early<br>• Maintain the big picture perspective<br>• Use rapid development and prototyping processes | Chap. 4 and 6<br>Sec. 3.10 |

**Table 3.1. Mission Operations Functions. (Continued)** This table shows key issues, trades, and where we discuss functions in more detail.

| Function | Key Issues and Trades | Where Discussed |
|---|---|---|
| 11. Computers and Communications Support | • Understand existing capabilities and define additional requirements <br> • Consider maintenance and upgrades when defining system | Chap. 11 and 12 <br> Sec. 3.11 |
| 12. Developing and Maintaining Software | • Make software re-use a priority <br> • Involve operators and users early in the design | Chap. 6 <br> Sec. 3.12 |
| 13. Managing Mission Operations | • Implement automation and autonomy carefully <br> • Consider multi-tasking staff <br> • Keep organization flat and reduce interfaces | Sec. 3.13 |

**Table 3.2. Relative Cost of Mission Operations Functions.** We've compiled data for four missions and expressed the cost for each function as a percentage of the annual mission operations cost. The missions included are Magellan, Mars Observer, Galileo, and TOPEX. The functions highlighted with gray background use the most resources. [CSP Associated, Inc., 1993]

| Function | % Cost* | | % FTE† | |
|---|---|---|---|---|
| | Low % | High % | Low % | High % |
| 1. Mission planning | 1 | 3 | 2 | 4 |
| 2. Activity planning and development | 4 | 9 | 7 | 13 |
| 3. Mission control | 6 | 9 | 10 | 11 |
| 4. Data transport and delivery | 4 | 9 | 5 | 11 |
| 5. Navigation planning and analysis | 3 | 6 | 4 | 7 |
| 6. Spacecraft planning and analysis | 13 | 20 | 18 | 25 |
| 7. Payload planning and analysis | 3 | 19 | 3 | 14 |
| 8. Payload data processing | 7 | 25 | 6 | 13 |
| 9. Archiving and maintaining the mission database | 0 | 3 | 0 | 3 |
| 10. Systems engineering, integration, and test | 3 | 6 | 3 | 7 |
| 11. Computers and communications support | 3 | 9 | 2 | 5 |
| 12. Developing and maintaining software | 5 | 8 | 5 | 8 |
| 13. Managing mission operations | 9 | 11 | 11 | 14 |

\* Percent of average annual mission operations costs
† Percent of annual, average, full-time equivalents each function uses

# 3.1   Mission Planning

The *mission plan* is a top-level description which spans the mission's life-cycle. It describes the way the mission will be flown, expresses objectives in operational terms, and sets in place major activities. Mission planning concentrates on uplink but takes into account the downlink abilities of the spacecraft and ground system. The mission plan is consistent with and generated after the mission and mission operations concepts. In this activity we develop mission objectives and plan how to meet them. Mission planners create a plan before launch and then change the plan as required. Before launch, mission planners play a key role in trades between mission functions. During operations, they respond to unforeseen events with updated mission plans and make sure operations meet objectives.

### 3.1.1   Information Required for Mission Planning

Table 3.3 lists information needed for mission planning—most during mission development. Early in development we may have only mission objectives, a mission concept, and a mission operations concept. We have to get the other inputs or assume them so we can begin mission planning. As a design matures, we learn enough about the mission requirements, spacecraft, and mission operations to update our plan.

Table 3.3.   **Information Required for Mission Planning.** Before launch, we know the mission objectives, concept, requirements, and abilities of the spacecraft (bus and payload) and mission operations system. We then compare current status and mission plan throughout the mission. Background shading indicates a post-launch mission phase.

| Information Required | Comments | Where Discussed |
|---|---|---|
| Mission objectives | • State in terms of science or payload return<br>• Usually, state in qualitative terms | Chap. 1 |
| Mission concept | • State how the mission elements will work together | Chap. 1 |
| Mission operations concept | • Emphasize mission operations; ground systems; and command, control, and communications system<br>• Focus on goals and vision of customer, program management, and users of payload data | Chap. 4 |
| Mission requirements | • Make sure they are quantitative, measurable, and consistent with mission objectives | Chap. 6 |
| Spacecraft capabilities (Includes spacecraft bus and payload) | • Include capabilities of spacecraft bus<br>• Focus on areas that are changeable to simplify operations | Chap. 14 and 15 |

**Table 3.3.** **Information Required for Mission Planning. (Continued)** Before launch, we know the mission objectives, concept, requirements, and abilities of the spacecraft (bus and payload) and mission operations system. We then compare current status and mission plan throughout the mission. Background shading indicates a post-launch mission phase.

| Information Required | Comments | Where Discussed |
|---|---|---|
| Mission operations system capabilities | • Identify current and planned abilities of the existing operations system, including tracking stations, ground stations, hardware, and software | Chap. 12 |
| Current status versus plan (ongoing activity) | • Obtain reports from other 12 functions<br>• Identify deviations caused by changes in spacecraft performance, customer requirements, or data requirements | Chap. 3 |

### 3.1.2 How We Plan a Mission

**Table 3.4.** **How We Plan a Mission.** We develop a plan during mission development and change it after launch only when spacecraft capabilities or mission requirements change.

| Steps | Where Discussed |
|---|---|
| 1. Quantify mission objectives and goals so they are meaningful to operators | Chap. 1 |
| 2. Define orbit or trajectory and calculate launch windows, number and frequency of maneuvers, and viewing periods | Chap. 10; Chap. 7 [Larson and Wertz, 1992] |
| 3. Describe the payload and define operational characteristics | Chap. 14; Chap. 9 [Larson and Wertz, 1992] |
| 4. Describe spacecraft bus and define operational characteristics | Chap. 15; Chap. 10 and 11 [Larson and Wertz, 1992] |
| 5. Define mission phases, allocate activities to phases, and establish a mission timeline—basis for first high-level activity plan | Chap. 7 and 8 |
| 6. Evaluate requirements for operations and identify those that are difficult to meet | Chap. 6, 11, and 12 |
| 7. Decide whether to use existing mission operations system or develop a new one | Chap. 12 |
| 8. Identify mission rules not related to health and safety imposed by program office, users, or operators and express rules in quantitative terms that software can check | Chap. 4 and 6 Sec. 3.1.3 |
| 9. Document and iterate as necessary | |

In mission planning we prepare the baseline set of documents described in Sec. 3.1.3. The mission plan describes how the mission will be flown until in-flight events require it to change. The plan establishes operational criteria to meet objectives, as well as major activities and events. Although planning produces a document or set of documents, it intends to understand the mission and to begin trading between the mission functions to get the best return for the customer's money. Much of this work begins when we're developing the mission concept.

After launch, mission planners periodically compare achievements to the mission plan. We learn about these achievements from payload or spacecraft planning and analysis. We then change the mission plan, mission rules, and orbital plan based on the flight system's changing abilities or processed data from the payload.

### 3.1.3    Products of Mission Planning

Table 3.5.    **Products of Mission Planning.** A document describing the mission plan, mission rules, and orbital phases. We complete the document during mission development and update it as needed after launch.

| Products | Contains |
|---|---|
| Mission Plan | 1. Mission objectives and goals<br>2. Orbit or trajectory description and orbital plan<br>3. Payload description and operation<br>4. Spacecraft bus description and operation<br>5. Mission phases<br>6. Description and techniques of mission operations<br>7. Mission rules and method of verification |

The mission plan matures as launch approaches and continually reflects trades across the project's elements. The mission plan is usually placed under configuration control before the first review of project requirements. Configuration control allows changes only after everyone they affect has had a chance to review them and their costs. A proper authority decides whether or not to change the plan. Each plan contains:

**Objectives and Goals.** Quantify the objectives and goals of the mission so operators can understand them. For example, "discovering and understanding the relationship between newborn stars and cores of molecular clouds" is meaningful to a scientist, but "observing 1000 stars over two years with a repeat cycle of once every five months using each of the four payload instruments" is much more meaningful to an operator. Describing the objectives and goals in these terms requires experienced operators and payload users to interact. This interaction becomes very detailed while generating the mission operations concept, as discussed in Chap. 4, but the plan must include top-level objectives and goals.

**Trajectory or Orbit Requirements and Description.** This section establishes a context for the mission by describing the number and frequency of maneuvers, as well as constraints on the view periods of tracking facilities and target, launch time, and launch windows. (See Chap. 10)

**Payload Requirements and Description.** Make sure this section includes the number of instruments and how they'll gather the data needed for the mission. Also include the payload's operational characteristics, which derive from the mission operations concept. (See Sec. 4.2.2 and Chap. 14)

**Spacecraft Bus Requirements and Description.** Define performance characteristics; allocate to the spacecraft mass, power, and flight information—such as memory reserve as a function of time from launch. During mission design, make sure you show the spacecraft is operable. (See Sec. 4.2.2 and Chap. 15)

**Definition of Mission Phases.** Define the mission phases associated with the orbit or trajectory while developing the mission concept; then, refine and modify them while generating the mission operations concept. Normally, specify the duration and function of these activities:

- Orbit insertion
- Spacecraft checkout
- Payload checkout

- Cruise (interplanetary)
- Maneuvers
- Orbit operations

The orbital plan or mission-phase plan defines in more detail what takes place during each of the mission phases described in the mission plan. In describing the payload checkout phase, you may define the priority of checking out the payload instruments relative to calibrating the spacecraft's attitude-control sensors, as well as the order of checking out these instruments. The mission-phase plan reflects agreements between spacecraft and payload designers and users. It therefore defines, at a very high level, the operational activities that the spacecraft and ground will carry out during the mission timeline.

**Mission Operations System Requirements and Description.** Describe and include requirements for key functions. Include allocations of performance parameters by end-to-end information system (EEIS) engineering. The EEIS distributes and defines the nature of bit-error rates between the spacecraft and the ground, as well as between the ground antenna and the mission database. Most NASA missions follow standards of the Consultative Committee for Space Data Systems (CCSDS) for protocols regarding data transmission. These protocols define the way packets of information are assembled for transport. Designers of the information system must understand how packets will react to bit-error distributions, especially if the information in the packets is compressed.

**Mission Rules (Guidelines and Constraints).** Developing mission rules is key to mission planning. *Mission rules* describe how to conduct the mission and are usually not related to the spacecraft's health and safety. The health and safety rules are called *flight rules*; we discuss them under the section on spacecraft performance and analysis. Mission rules set policy for what should be done, not for how to do it. Examples of mission rules include:

- Use a specific payload sensor only over the ocean and turn it off ten miles before crossing any land

- Schedule spacecraft events that require real-time monitoring during prime shifts
- Record telemetry 15 minutes before and after critical activities
- Don't overwrite critical data stored on the flight recorder until you verify playback data

The program office may impose mission rules, but in general they represent agreements between the users and operators of the mission operations system. Thus, the rules tend to depend on the mission operation team's background and experiences.

Many mission rules are checked through processes while developing and generating *activities*—a time-ordered set of contiguous events. Thus, you must quantify mission rules so operators can understand and translate them into software that can check spacecraft activities automatically for compliance with the mission rules.

### 3.1.4    Key Considerations

Depending on the complexity of the mission, the spacecraft, and the payload, we find that integration of engineering and payload requests can be time consuming and difficult. Complexity is a function of the

- Number of payload instruments
- Operability of the spacecraft
- Adaptability required because of the data received from the payload instruments

Mission planners, working together with designers, may be able to lower the cost of operations by providing positive margins, automating certain space or ground functions, restricting the number of mission or flight rules, and focusing on operability.

**Positive Margins**. Planners participating in trades across the mission elements during the design phase make sure the spacecraft and instruments have positive margins. A healthy positive margin may mean we don't have to model, evaluate, and constantly monitor a spacecraft. A spacecraft designed with positive margins is easier and less expensive to operate.

An example is the power system. When the power available to the spacecraft is less than the power the spacecraft uses with all instruments and subsystems turned on, we must be very careful what we do to the spacecraft. Imagine you're at home, and you could have only half your lights, your clothes dryer **or** your stove, and your garbage disposal **or** your refrigerator on at any given time. You don't control your refrigerator, so you would have to model its performance to predict when it would be turning on and off. Then, you would have to plan to use your disposal only when the refrigerator is off, or turn off your refrigerator every time you turn on your disposal. You would also have to plan your life so you wouldn't need to dry your clothes for a dinner party at the same time you'd be cooking dinner for that party. When positive margins exist, we don't need to pay for modeling and the tools used in modeling the spacecraft.

Earth-orbiting spacecraft typically have positive margins. Providing positive margins is more difficult for planetary missions beyond Mars, but we can do it. Of course these margins often reduce the scientific return of the mission, and we must trade-off operational complexity and cost against the additional scientific return.

**Spacecraft Autonomy Versus Ground Generation of Appropriate Functions.** Planners seldom trade well between what the spacecraft and ground units do. Often decisions are based on the cost of designing and building the spacecraft, and afterward ground operations must absorb the cost of accommodating the decision. The trade-offs include cost of making the spacecraft autonomous versus doing the same thing on the ground for the life of the mission. The mission duration is an important consideration in making this trade.

**The Number of Mission and Flight Rules.** Make sure you keep the mission and flight rules to a minimum during the design of the mission and the spacecraft. Each rule that must be checked involves either software or people, or both, and adds to the expense of operations.

**Operability and Interaction Between Subsystems.** During the design phase and during the element trades, make the spacecraft as operable as possible and keep interactions between spacecraft subsystems to a minimum.

## 3.2 Activity Planning and Development

In this function we convert each orbital-phase plan generated in mission planning into detailed commands that are ready for uplinking to the spacecraft bus as command mnemonics. Normally, we script the spacecraft's actions so each command takes place at a particular time. Many tools within the ground data system have been developed to generate these activities. Each of the NASA centers involved in flight operations has these tools. For example, JPL uses a software program called SEQGEN. Commercial packages are also available and are adequate for many missions. The next generation of spacecraft are beginning to look at process control as a way to store rules on board and to have actions occur when events satisfy these rules. These rule-based actions have usually been limited to fault-protection rules related to the health and safety of the spacecraft and payload. As the technologies mature, putting process-control rules onboard spacecraft should save operational costs and increase the payload return. Certain missions planned by NASA, such as returning asteroid samples, will require this type of onboard technology; it won't be possible to plan the spacecraft events or control them from the ground because of long delays in transmission over great distances. (See Chap. 17)

### 3.2.1 Information Required to Plan and Develop Activities

Table 3.6 lists the information we need. Mission planning provides most of this information for the initial activities. After launch, the mission operators' activity requests become increasingly important to achieving the mission objectives.

We also plan activities before launch for the spacecraft—during its integration and test phase—so we can make sure it will work the way we expect it to work.

Table 3.6.    **Information Required to Plan and Develop Activities.** The mission plan (Sec. 3.1) gives us most of this information. Shaded activities occur after launch.

| Information Required | Comments |
| --- | --- |
| Mission Plan, including the Mission Rules and the Orbital Phase Plan | • Generated by mission planning<br>• Changes occur during the mission as the mission deviates from the nominal plan |
| Requests for Test and Integration Activities | • Used to support spacecraft integration and test<br>• Some of these activities may occur while verifying the spacecraft after launch<br>• Comparing integration and test data to the in-flight data enables rapid understanding of the spacecraft's health |
| Requests for Post-Launch Activities | • Requests come mainly from spacecraft and mission control<br>• The activity requests may have been generated before launch, but most are generated in detail after launch using building blocks, or macros of tested and validated groups of commands<br>• The activity requests are the details which implement the higher-level plan for the orbital phase |

## 3.2.2    How to Plan and Develop Activities

The steps for planning and developing activities, as listed in Table 3.7, provide a command load for uplink that is ready to transmit to the spacecraft. In the following pages, we'll amplify these steps. Chapter 7 provides even more detail.

Table 3.7.    **How to Plan and Develop Activities.** This process converts the mission plan into activities, timelines, and commands.

| Post-Launch Steps | Comments | Where Discussed |
| --- | --- | --- |
| 1. Define Activities | • Next level of detail from the orbital-phase plan<br>• Planning usually considers a series of activities, each of which relates directly to spacecraft commands<br>• Timelines of these activities are generated<br>• Function works closely with spacecraft and payload engineers and scientists | Chap. 7, 14, and 15<br>Sec. 3.2 |
| 2. Generate and Integrate Activities | • Next level of detail after activity planning<br>• Integrates new requests with the activity-planning timeline<br>• Ensures shared resources don't conflict<br>• Automated tools are required to do this process efficiently | Chap. 7<br>Sec. 3.2 |
| 3. Check Mission and Flight Rules | • Final check of mission and flight rules<br>• Software helps generate activities more efficiently | Chap. 7<br>Sec. 3.1 and 3.2 |

Table 3.7.   **How to Plan and Develop Activities. (Continued)** This process converts the mission plan into activities, timelines, and commands.

| Post-Launch Steps | Comments | Where Discussed |
|---|---|---|
| 4. Generate Time-lines | • The timeline displays the activities of the spacecraft and the ground | Chap. 7 and 8 Sec. 3.2 |
| 5. Validate Activities | • Verifies the safe interaction of planned activities<br>• Should be automated<br>• Checks constraints for health and safety<br>• Checks to ensure the activities do what is necessary to support the payload<br>• Review and approval should be at the lowest possible level and add value to the process | Chap. 7 and 8 Sec. 3.2 |
| 6. Translate Activi-ties | • Converts the command mnemonics and associated parameters to a binary stream packaged as required by the network being used | Sec. 3.2 |

**Define Activities.** Activity planning supports payload and spacecraft-bus engineers in their early planning of the flight system. The result is one or more activities on the spacecraft and the ground needed to return certain sets of data. Each activity relates directly to many spacecraft commands. Payload-planning tools are provided that allow the payload user to superimpose the field of view of the instrument on the target body of interest. These tools correct for spacecraft trajectory, target body rotation, instrument location, boresight offsets, allowable scan platform, and mirror motion. They also provide a realistic planning footprint. As a result, an investigator can look at different ways of gathering data and determine how to set up observation patterns so they fully cover the target. Later, mission operators will carry out these activities at the scheduled times.

**Generate and Integrate Activities.** Before launch, define how long it takes to generate and complete an activity. Use increments of an orbit for Earth-orbiting spacecraft, or weeks to months for a planetary spacecraft. In any event the mission phase plan lays out pre-launch priorities for the given period as the starting point for generating activities.

The Infrared Space Observatory (an ESA mission) generates a command load for seven orbits (seven days) at a time. The process is started 21 days and frozen three days before uplink. Other missions have plans tailored to their mission attributes.

New requests typically come from four functions: mission control, navigation, and planning and analysis for the spacecraft bus or payload. New and unplanned requests are a part of any mission for several reasons:

- We need an unplanned calibration (for example the star tracker, the attitude-control gyros, or the movement of a platform) to understand some unexpected data

- Unexpected changes in the availability of tracking facilities
- The spacecraft's abilities have changed, so all the standard activities for a certain function need changing before the next command period
- We need to send a new set of tables up to the spacecraft based on the last calibration

These new activities may be merely executing an activity at a different time. If they're truly new, you'll have to design and test them.

Using the mission-phase plan as the baseline, take new requests from the other mission operations functions and integrate them into the command load. Remove conflicts based on rules and priorities or negotiation with the various requestors. First, integrate activity periods at a planning level and then expand these activity periods into groups of commands and individual commands. Keep integrating activities and resolving conflicts at increasing levels of detail so individual commands never conflict.

Make sure shared resources don't conflict. The desire to maximize the use of the spacecraft typically results in many activities placed close together. Late changes to activities or adding new activities can cause oversubscription of spacecraft resources such as power, time, data storage, command and telemetry links, and memory. The resource most often exceeded and hardest to resolve late in the process is time—an attempt to do too much, either on the ground or the spacecraft.

Usually, we fix allocations; that is, during a specific period, each payload instrument can use a pre-defined maximum amount of a given resource. As events begin to drive activities, we have to consider dynamic allocation of resources. Of course, we need to allocate resources only when a particular resource doesn't have enough margin. Thus, to avoid some of these conflicts, make sure resources have enough margins.

Then, ensure the margins are suitably defined, applied, and consistently enforced. If we set margins too small or use them too quickly, the uplink will be overwhelmed with conflicts, requiring last-minute deletions or rework of activities. Spacecraft commanding is one of the few areas where the final deadline is firm. If the commands aren't ready when the spacecraft needs them, there is a BIG problem—probably a complete rework of the command set while the spacecraft is idle.

Some integration occurs during mission design. We usually need to do more after the design is complete because certain information, such as final tracking schedules, may not be available earlier, or we must incorporate new activities. Many activities must happen at specific times, whereas others, such as various engineering activities, can fit in wherever possible. Except for very simple activities, you can improve integration by using automated tools that identify conflicts and missing dependencies. At this point, you'll have an integrated command load with no conflicts at the activity level. The commands are mnemonics with their required parameters.

**Check Mission and Flight Rules.** You can check many of the rules with the same tools used to generate the activities. But you can check other rules only after you've generated the final command load. Use software as much as possible to make activity development faster and therefore less expensive.

**Generate Timelines.** These timelines display at a high level—and usually in ground-received time—the activities of the spacecraft and the ground. This display allows the controllers and analysts to monitor the spacecraft's activities as the ground station receives data. This distinction isn't necessary for Earth-orbiting spacecraft, but those on interplanetary missions require it to make up for the effects of one-way delays in light time.

**Validate Activities.** Verify the interactions of the planned activities and make sure the command load meets the intended goals of the activities while posing no risk to the spacecraft. Automate this step as much as possible. Software tools can check to see that the command load doesn't harm the spacecraft and often visualize the sequence to make sure it gathers the information requested. Early understanding of the need and method for validation is essential to minimize the operations costs.

Because we can achieve the same end in many ways, the risk for an activity or group of activities can be difficult to quantify. Each approach has good and bad, as well as unknown, aspects. Fear of the unknown pushes perceived risks higher and drives constant searching to find a perceived "safer" way. People who know the interactions on the ground and spacecraft can accurately judge the risk without complicated analysis, but the results still must be quantified. For these reasons, projects often use hardware or software to evaluate the activity interactions. Electronic or manual reviews, hardware or software simulations, or a combination of these techniques can carry out these evolutions.

Operations are constrained for all flight and ground systems. Good spacecraft design and automation can make this task extremely easy and reduce staff. Hardware constraints (known as *flight rules*) are those which, if violated, may damage or stress a piece of the flight system, such as exceeding thermal, electrical, or radiation limits. There are obvious constraints, such as Sun impingement on a sensor array designed for viewing deep space, and less obvious constraints on the abilities of power subsystems for spacecraft in certain radiation environments. Other constraints (known as *mission rules*) provide guidelines on how to operate the flight hardware and ground systems within acceptable bounds. Violating mission rules typically won't cause permanent damage but may exceed management guidelines on effort or money expended, or compromise project goals. These rules may also set restrictions on instrument operating margins or limit operation to modes that have been tested and validated.

If you properly document constraints, you can verify compliance simply and readily automate constraint checking. Having to interpret the rules can delay the process, especially if different interpretations are supportable. For manual reviews, make statements clear, so reviewers don't need special knowledge to understand them. For electronic reviews, make the constraints statement precise and detailed enough to convert it into code.

Because we must understand constraints early and try hard to automatically validate them, we often run command loads through simulators to make sure they're coherent and correct. We can make spacecraft simulators by setting up flight computers and flight hardware in a testbed and running the actual command load through the system. Hybrid simulators use software to simulate some of the components, such as the attitude-control subsystem. Of course, we can also build simulators entirely through software, such as modeling the flight computer in software and then running it on a workstation or other computer. But simulation costs money because we must build, operate, and check the output of the simulator. These steps require time and people.

The amount of simulation required varies tremendously within the NASA missions. We use less for Earth-orbiting spacecraft because they usually have power and telecommunication margins and round-trip transmission times of a fraction of a second. Planetary missions simulate many of their activities because they have negative margins and their data transmissions take tens of minutes to tens of hours.

Complexity of the command load and acceptable risk determine whether review and approval is long or short. Today, mission operators often risk the loss of some data and therefore shorten reviews to save money. Design each review to require the fewest people, meet technical specifications, and add to the command load's integrity. If the review is merely to have a manager sign off on the command load, you should probably eliminate the review.

**Translate Activities.** At this point the *command load* is in the form of words and parameters—typically called a command mnemonic for the spacecraft and some key words for the tracking net. A *command dictionary* has translation tables to convert each command mnemonic to a binary bit pattern that is sent to the spacecraft. Translating a command load means converting activities into a format the ground transport and uplink system (such as the Deep Space Network or TDRSS) and the spacecraft can handle. For a simple flight system, this translation can convert the commands into a binary stream for ground transport and subsequent uplink. For a more sophisticated system, it can use a predicted state of the onboard memory and compile the commands into a direct memory load with required memory management.

### 3.2.3    Products of Activity Planning and Development

Table 3.8 describes these products. We have converted the mission plan into a series of commands or instructions that the spacecraft will execute for the next activity period. We have checked the command load to ensure that it's safe and will return the desired payload data.

**Table 3.8.    Products of Activity Planning and Development.** These three products enable the spacecraft and ground elements to operate for the next activity period.

| Post-Launch Products | Comments |
|---|---|
| Detailed command loads in mnemonic form | • Spacecraft and payload analysts review the command loads in detail<br>• The mnemonic form is close to an English language, in that the commands are names and parameters which relate to the command |
| Timeline | • The timeline graphically represents the command load. The x-axis is time, with different spacecraft, payload, and ground actions plotted in parallel against this time.<br>• Different timelines emphasize different aspects of the mission and display at varying degrees of resolution: at minutes, hours, days, or weeks<br>• Chapter 8 shows an example of a timeline |
| Command load | • The command load, ready for transmission to the spacecraft, is a series of bits generated as described in the translation process |

### 3.2.4    Key Considerations

The most difficult step in planning and developing activities is avoiding conflicts while integrating the mission plan with current requests. More complex spacecraft and planning tools, as well as many automatic checks of ground data, increase this difficulty.

Larger mission margins make activity planning easier because they mean we don't have to manage and simulate resources as accurately. Spacecraft autonomy is also key; in fact, a completely autonomous spacecraft would require no activity planning and development. Finally, a mission plan that depends heavily on receiving and analyzing data to determine what to do makes activity planning and development more difficult.

To make your activity design more cost effective, follow these concepts:

- Reduce the required number of command loads. Having many command loads and dense activities within each of them will add to your staff and therefore to your budget. Whenever possible, generate command loads one after the other instead of in parallel. You need parallel generation only when the time to generate a command load is longer than the time to the execute it.

- Validate command loads electronically and automate validation as much as possible.

- Make sure approval of command loads occurs at the lowest possible level while maintaining acceptable levels of risk. Always ask, how does this review and approval add value?

- Minimize changes to the command load once development starts. If mission objectives require late changes, define when they can occur and limit the amount of change at each time window. Document these constraints and make them part of the mission rules.

The speed, memory size (random access memory), and storage size (for example, hard disk space) of today's computers allows us to integrate tools for planning and developing activities. Integrating these tools will usually automate file interfaces, make the process faster, and reduce staff.

## 3.3   Mission Control

*Mission control* runs the mission. By following a script, it directs a spacecraft's operation in real time, mainly to ensure that ground crews safely receive the required data. We use mission control whenever mission needs demand it and spacecraft tracking coverage allows it. Mission control includes carrying out the detailed activity plan and ensuring the spacecraft's health and safety during the station contact. It also means setting up and verifying the ground configuration, and then briefing everyone involved before we contact the spacecraft. Mission control then sends commands to the spacecraft, monitors its performance and that of the ground data system, and directs its recovery from any nonstandard conditions.

### 3.3.1    Information Required for Mission Control

We get needed information mainly from mission planning and activity planning and development. Other information comes from navigation and spacecraft planning and analysis. Table 3.9 summarizes this information.

Table 3.9.   **Information Required for Mission Control.** Mission control receives information from mission planning, activity planning, spacecraft planning and analysis, and navigation planning and analysis. Shading indicates information required after launch.

| Information Required | Where Discussed |
|---|---|
| Mission Plan and Mission Rules | Sec. 3.1 |
| Command Loads and Command Files | Sec. 3.2 |
| Real-time Telemetry | Chap. 14 and 15; Sec. 3.6 and 3.7 |
| Tracking Schedules and Data | Chap. 10; Sec. 3.5 |
| Pass Plan | Chap. 8 |

### 3.3.2    How to Do Mission Control

Mission control uses the information in Table 3.9, along with procedures, to operate the mission in real time. We develop procedures for spacecraft contacts, make sure data transfers take place for all uplinks and downlinks during the contact, monitor all systems during the contact, and document all activities that occur during the contact. As listed in Table 3.10, if you were on a mission-control team, you would

**Table 3.10. How to Do Mission Control.** Mission control develops procedures and supports integration testing before launch and directs the real-time operation of the mission after launch. Mission control supports several planning and analysis functions throughout the mission. Shading indicates steps we do after launch.

| Steps | Comments | Where Discussed |
|---|---|---|
| 1. Develop Procedures for Controllers to Configure the Spacecraft and the Ground System | • Step-by-step instructions for pass support | Chap. 7 and 8 |
| 2. Support Spacecraft Integration and Test | • Verifies uplink and downlink systems<br>• Tests end-to-end compatibility | Chap. 8, 9, and 12 |
| 3. Configure Ground System to Support Passes | • Real-time with some pre-launch planning | Chap. 8 and 12 |
| 4. Transmit Commands to the Spacecraft | • Done in real time during pass<br>• Real-time commands and command loads | Chap. 8<br>Sec. 3.4 |
| 5. Verify the Spacecraft's Receipt of Commands | • Done in real time during pass<br>• Usually automated | Chap. 7 and 8 |
| 6. Monitor the Spacecraft's Health and Safety | • Monitor telemetry in real time during the pass<br>• Compare predicted and actual spacecraft states | Chap. 7, 8, and 15 |
| 7. Monitor the Ground System Operations | • Monitor performance in real time | Chap. 7, 8, and 12 |
| 8. Coordinate Mission-Control Functions | • Planning and scheduling before and after passes | Chap. 7 and 8 |
| 9. Generate an Integrated Plan for "As-Flown" Activities | • Post-pass report<br>• Document deviations from pass plan | Chap. 8 |
| 10. Support Activity Planning and Development | • Post-launch support for new activities | Chap. 7 |
| 11. Generate Operations Schedules and Plans for Future Passes | • Post-launch planning and scheduling for passes | Chap. 7 and 8 |
| 12. Negotiate and Schedule Tracking Support | • Coordinate with activity planning and development to generate pass plans | Chap. 8<br>Sec. 3.5 |
| 13. Support Planning and Analysis Teams | • Help investigate anomalies | Chap. 16 |

1. **Develop Procedures.** Controllers fly the mission using your baseline procedures. These procedures direct controllers on what to do before, during, and after a contact with the spacecraft. They cover how to configure and control the ground system and the spacecraft.

2. **Support Spacecraft Integration and Test.** By helping integrate and test the spacecraft bus and payload before launch, you'll see how the spacecraft and ground system work and get hands-on training.

   You'll also do end-to-end tests, during which the ground element is integrated with the spacecraft to verify uplink and downlink. These tests validate the operations system's ability to command, deliver telemetry, and monitor alarms in real time.

3. **Configure the Ground System to Support Passes.** Mission controllers configure and control ground-system tools and procedures. Like the conductor of an orchestra, you use this step to make sure everything happens at the right time.

4. **Transmit Commands to the Spacecraft.** During the pass, mission controllers transmit scheduled real-time commands and the spacecraft's command loads. These commands are nearly always pre-planned, but procedures or oral orders may authorize mission controllers to issue real-time commands under certain predefined conditions. These predefined conditions typically cover situations in which the spacecraft has violated limits or is in a non-standard state.

5. **Verify the Spacecraft's Receipt of Commands.** As a mission controller, you must use spacecraft telemetry and information on the ground system's status to verify that the spacecraft has correctly received its commands. Verification is usually automatic, and detected errors produce messages and alarms.

6. **Monitor the Spacecraft's Health and Safety.** Monitor telemetry measurements versus alarms, the onboard memory readout and verification, and the spacecraft's predicted and actual states. Use graphical displays and text which the ground system presents to you based on processed information from the spacecraft.

7. **Monitor the Ground System's Operations.** During a pass, you must monitor the ground system's performance just as you do that of the spacecraft, so the mission can meet its objectives.

8. **Coordinate Mission-Control Functions.** Before the pass you must coordinate some activities needed to support real-time contact between the ground and the spacecraft. For example, you may need to

   - Schedule short-term tracking support
   - Generate an integrated plan for flight and ground activities
   - Schedule institutional support and make sure it's available
   - Generate logs and reports
   - Update databases and files

Efficient mission operations require coordination. It ensures that the ground system is able to transmit planned commands to the spacecraft and receive, process, and store the spacecraft's telemetry data.

9.  **Generate the Integrated Plan of "As-Flown" Activities.** This activity plan is identical to the one prepared before the pass if everything went according to plan. But if you deviate from the plan, you must do an "as-flown" plan to record what was done and when. Typically, mission controllers use it to analyze anomalies.

10. **Support Activity Planning and Development.** Mission controllers help generate and approve the activities described earlier. They typically bring to the process the interface issues between the spacecraft and the ground.

11. **Generate Operations Schedules and Plans for Future Passes.** To do plans for future passes, you must get ready and check out all information and data needed for the next spacecraft contact.

12. **Negotiate and Schedule Facility and Tracking Support.** These schedules for tracking stations become part of the activity plans described earlier in this chapter.

13. **Support Planning and Analysis Teams.** Help teams analyze anomalies in the spacecraft bus and payload. Experience and information from real-time controllers is invaluable to the analysts who understand the spacecraft and its subsystems but don't operate it.

### 3.3.3    Products of Mission Control

Pass plans, pass reports, database files, and alarm notices are the post-launch products of real-time operations. We list these products in Table 3.11 and describe them below.

Pass plans describe what will be done during each pass, whereas procedures describe how controllers will do it. A typical pass plan will describe the planned times for acquisition, as well as key events anticipated or required on the spacecraft and the ground. It is a level higher than a command load and is used to brief everyone in the ground system before the pass.

A pass report describes the activities which took place during a contact between the ground and the spacecraft. It typically includes the commands sent, the telemetry channels that were in alarm, and any unplanned action that procedures or on-duty analysts authorized.

Database files generated and validated by the mission control team are the command validation files and the files or updates to telemetry and tracking data received during the pass.

When alarms occur, mission control notifies the people that their procedures identify. Although controllers have issued notices manually in the past, systems are now in place that beep someone who is on duty but not at the control center.

Table 3.11. **Products of Mission Control.** All products are completed after launch and are related to real-time mission operations.

| Products | Comments |
|---|---|
| Procedures | • Describe in detail what the operators do<br>• Describe what commands are authorized to be sent and under what anomaly conditions |
| Pass Plans | • Describe what is scheduled to happen during pass |
| Pass Report | • Describes activities which took place during a pass<br>• Notes deviations from the pass plan |
| Database Files | • Uplink-command and command-verification files<br>• Downlink telemetry files<br>• Ancillary data |
| Alarm Notices | • Alarm notices and responses occur in real time |

The person can then use a computer to connect with the ground system. After proper identification and authorization, this person can receive and analyze the spacecraft telemetry on his or her computer. This technology keeps teams at the control center small during spacecraft contacts.

### 3.3.4 Key Considerations for Mission Control

Mission control is easier when we can easily modify ground and flight systems to react to unplanned changes or conditions. For example, workstations and networks are more robust than a central system. Graphical user interfaces that allow operators to change configurations by point and click are easier to use than those that take lines and lines of commands.

We can lower mission control costs in several ways. Possibilities include sharing operators between missions, multi-tasking operators within a given mission, designing flexible ground and flight systems, and increasing automation in the ground system.

When several missions have common features in flight at the same time, we can train people so they may work in more than one mission at a time, thus lowering the total staff required. For individual missions, multi-tasking lowers the total staff required and makes jobs more interesting. Many of the specialists typical of early space flight now do related, but new, tasks to reduce costs.

If we design the spacecraft to remain flexible to change, we'll need fewer resources to make these changes. By automating analysis of the performance of the ground system and the spacecraft, we can reduce staff and probably improve performance and reliability.

## 3.4 Data Transport and Delivery

This function transmits data to the spacecraft, receives data from the spacecraft, and processes tracking data. It accepts commands or command files from mission control and prepares the data for transmission, ultimately modulating the carrier signal and radiating the telecommands to the spacecraft. This function also captures the signal from the spacecraft and processes the spacecraft's engineering and mission data. Finally, it gathers and processes radiometric data used for planning and analyzing navigation.

### 3.4.1 Information Needed to Transport and Deliver Data

Mission control and navigation planning and analysis provide information before launch to validate the system's capabilities and after launch to support real-time transmission of data to and from the spacecraft. Table 3.12 summarizes this information.

**Table 3.12. Information Needed to Transport and Deliver Data.** Mission control and navigation planning and analysis provide this information before and after launch.

| Information Required | Where Discussed |
|---|---|
| Antenna pointing predictions | Chap. 10; Sec. 3.5 |
| Binary commands or command files | Chap. 8; Sec. 3.4 |
| Ground activities | Chap. 7, 8, and 12; Sec. 3.4 |
| Radio frequency predictions | Sec. 3.5 |
| Pre-track calibration data | Chap. 8 and 12; Sec. 3.4 |

People in navigation planning and analysis issue antenna pointing predictions based on either pre-flight nominals or orbit determination after launch. These predictions help us point the ground-tracking or in-orbit-tracking antennas to the spacecraft during a scheduled pass. Also, if a spacecraft uses other spacecraft for orbital tracking and relay, it will have the ephemeris of these spacecraft on board and employ it to point a high-gain antenna at the tracking spacecraft during the pass.

Navigation planning and analysis generates predictions of radio frequencies (rf) and frequency shifts expected during each pass. Activity planners generate commands for rf transmission. Mission control then transmits commands to staff who handle data delivery and transport. Mission control also configures the ground system by using activities that set up the ground system to support passes.

### 3.4.2    How to Transport and Deliver Data

We must test and validate our approach to handling data before launch so we can be sure the spacecraft and ground system are compatible. After launch, this function uplinks commands to the spacecraft and receives and delivers the data downlinked from the spacecraft. Table 3.13 lists the steps in this process.

**Table 3.13.  How to Transport and Deliver Data.** This process establishes our communications link to the spacecraft.

| Steps | Comments | Where Discussed |
|---|---|---|
| 1. Validate Each Function's Abilities | • Done before launch<br>• Validates uplink and downlink tasks<br>• Part of end-to-end test of the system | Chap. 8<br>Sec. 3.8 |
| 2. Send Commands to Spacecraft | • Tested before launch and carried out in real time after launch<br>• Commands uplinked to spacecraft | Chap. 8 and 11<br>Sec. 3.3 |
| 3. Manage Data Flow | • Tested before launch and carried out in real time after launch<br>• Data downlinked to ground station<br>• Standard formats recommended | Chap. 8, 11, 12, and 13 |
| 4. Determine Data Quality, Continuity, and Completeness | • Evaluates uplink and downlink data after activities are complete | Chap. 11 and 13 |

1.  **Validate Each Function's Abilities.** Do pre-launch steps similar to those of mission control. Test and validate facilities for transporting and delivering data. Participate in end-to-end tests of the spacecraft to validate the uplink and downlink functions. Generate procedures used for station's setup, calibration, and pass support.

2.  **Send Commands to Spacecraft.** Telecommanding is the ground-to-spacecraft step used to instruct a spacecraft, its subsystems, or scientific instruments. To do telecommanding, you modulate telecommand data on an rf carrier. When received by the spacecraft's rf subsystem and distributed to the appropriate device, this data starts, changes, or ends an action.

    Mission control supplies telecommands. Once the tracking network receives a telecommand message, determine the telecommand's destination and route it to the appropriate tracking station (or spacecraft). Check the telecommand's format to verify that it is acceptable. Often, destination codes are reversed and the telecommand message is returned to the control center for a bit-by-bit comparison with the original. Once you've found the telecommand to be acceptable, handle it by one of these methods:

*Thruput telecommanding.* In this mode, route the telecommand immediately to the transmitting station's modulator and transmit it to the spacecraft without delay.

*Store-and-forward telecommanding.* Route the telecommand to a specific tracking complex together with a release time. At the appointed time, the telecommand will be modulated on the rf carrier and transmitted to the spacecraft.

*Telecommand.* Once some or all of the above steps have taken place, modulate the telecommand on a sine wave subcarrier; then, phase-modulate it on an rf carrier, amplify it, and transmit (radiate) it to the receiving spacecraft.

3.  **Manage Data Flow.** This step returns information from a spacecraft to Earth-based users. Data can include the results of scientific measurements or information about the spacecraft or its subsystems. Package telemetry data to conform to one of the acceptable data formats recommended by the Consultative Committee for Space Data Systems (CCSDS). Then, encode it (optional), phase-modulate it on an rf carrier, and transmit it to a receiving tracking station or relay satellite. At the tracking station, systems or people:

*Capture data.* Once a tracking station's receiver is locked to a spacecraft's rf carrier, you'll demodulate telemetry data and synchronize symbols.

*Extract data.* Depending on whether or not coding is used and its type, there may be several times as many symbol bits as there are original data bits. To extract data, you'll need to take several typical steps. For data with convolutional encoding, use a maximum-likelihood algorithm to do convolutional decoding while detecting and correcting errors. Supply partially decoded data to a frame synchronizer, which continually tests for a unique, synchronization marker. When the synchronizer finds this sync word at successive periodic intervals, frames are synchronized. Finally, decode the Reed-Solomon block code at the end of the transfer frame correct errors before archiving transfer frames. Discard any data the system couldn't correct or, if the project requests, pass it through to the project's facility for capturing data.

*Time tag transfer frames.* Once decoding is complete and you consider the transfer frame valid, time tag the frame. For NASA, this time stamp is typically accurate to within 1.5 microseconds with respect to the time that a specific bit enters the antenna's feed horn. This time stamp appears in the secondary header of the standard formatted data unit (SFDU).

*Add valued services to transfer frames.* Plans predetermine how to read headers of decoded transfer frames and handle different virtual channels. For example, you can return a virtual channel containing spacecraft engineering data in real time. But you may want to record a channel containing imaging data at the stations and trickle the data back as line capacity permits or mail it back. Finally, you can route a third virtual channel containing a specific instrument's data directly to the payload-processing center.

*Deliver data.* Finally, you must deliver the data to the project as agreed. This usually involves sending the telemetry data from the tracking facility to the project control center and populating a project or mission database. As described above, you may also send the data directly to a payload center.

*Process spacecraft data.* This processing is necessary if the spacecraft sends data to the ground faster than the tracking facility can send it to the control center.

4.  **Determine Data Quality, Continuity, and Completeness.** If possible, complete this set after the pass and collect missing data from archives in the tracking station. This task is more automated now, but in most operations centers before 1992, people had to do it.

### 3.4.3    Products of Data Transport and Delivery

Our main products are the telemetry and tracking data collected during a pass. We also produce reports on tracking passes and on the quantity, quality, and continuity (QQC) of stations.

**Table 3.14.  Products of Data Transport and Delivery .** We generate these products in real time during a station pass and when time permits following a station pass.

| Product | Comments |
|---|---|
| Telemetry data | • Generate in real time during station pass |
| Tracking data | • Generate in real time during station pass |
| Tracking pass report | • Post-pass, when time permits<br>• Generate with mission control |
| Station QQC reports | • Post-pass, when time permits |

We process the telemetry data as described in Chap. 13 and Step 3 above and then deliver it into the project or mission database.

Tracking data, also defined as radiometric data, is used to determine a spacecraft's position as well as for scientific investigations. These data quantify the relative position or motion of the spacecraft and the Earth. Several types of data are available:

- Doppler
- Ranging
- Angle
- Very long baseline interferometry
- Radio science measurements

Station pass reports typically capture the configuration of the tracking station or spacecraft during the pass and concentrate on deviations from normal conditions. We list these reports as a project deliverable and include them in the mission database.

Station QQC reports describe the quantity, quality, and continuity of the telemetry gathered for the project during the contact. They include the statistics on both the real-time data and the high-rate data received but not necessarily processed during the pass. Most networks include this information electronically, but we also deliver it to the project database.

### 3.4.4    Key Considerations

Network loading and out-of-date systems cause most of our problems in handling data. For example, sending data from the tracking stations to the control centers with protocols of the 1970s causes errors in the mission database. NASA developed these protocols before commercial standards were in place. Now, we need to follow CCSDS standards or those of the network tracking our spacecraft, or we'll have problems getting standard services.

We can reduce the cost of the handling data by first keeping tracking coverage to a minimum. Tracking consumes resources from every part of mission operations. Overloaded tracking networks already have trouble allocating these resources to space projects.

Next, we can design structures and formats for flight data to match the abilities of our transport and delivery system. Using standard services is reliable and cost effective. Having an engineer design a new spacecraft characteristic to enhance a mission may result in spending a lot of money on developing and operating ground systems that must support this characteristic.

Finally, we should use variable-length packets instead of many unique, predetermined formats. Today's standards and technology allow us to identify a data packet's length in the header, so we don't need to use inefficient fixed-length packets.

# 3.5   Navigation Planning and Analysis

We must determine the spacecraft's position and predicted flight path, as well as how to correct that flight path to achieve mission objectives. To do so, we must

- Acquire radiometric, tracking, or optical measurements

- Determine the statistically best estimate of the trajectory based on these measurements
- Compute trajectory-correction maneuvers (TCMs) to achieve the desired targeting objectives or orbit changes
- Plan and analyze performance to ensure it meets mission and payload objectives

### 3.5.1    Information Needed to Plan and Analyze Navigation

Because determining and predicting spacecraft orbits is uncertain, we must continually update our estimate of the spacecraft orbits by analyzing tracking data. We also have to know the locations of other celestial bodies, such as the Sun, Moon, or Jupiter, to predict their effects on the spacecraft's trajectory. Finally, when orbital maneuvers are necessary, we must have an accurate model of the spacecraft propulsion system to plan the maneuver. Table 3.15 summarizes these requirements.

**Table 3.15. Information Needed to Plan and Analyze Navigation.** We need this information to determine and predict orbits, as well as to plan spacecraft maneuvers. Shading indicates information required after launch.

| Information Required | Where Discussed |
|---|---|
| Ephemeris data of celestial bodies | Chap. 10 |
| Planetary atmospheric and dynamic models | Chap. 10 |
| Tracking data | Chap. 10; Sec. 3.4 |
| Information on spacecraft propulsion | Chap. 15; Sec. 3.6 |

### 3.5.2    How to Plan and Analyze Navigation

We must convert raw tracking data to current and future estimates of the spacecraft's orbit or trajectory. We also calculate the maneuvers necessary to maintain our spacecraft orbit or change the orbit to meet mission objectives. We've listed steps in Table 3.16 and described them below.

**Table 3.16. How to Plan and Analyze Navigation.** We statistically combine tracking data to estimate the current orbit and then use this estimate to generate trajectories and plan maneuvers.

| Steps | Comments | Where Discussed |
|---|---|---|
| 1. Support Pre-Launch Mission Planning | • Demonstrate the orbit propagator's abilities before launch | Chap. 10 Sec. 3.1 |
| 2. Determine Orbit(s) | • Statistically determine best estimate of orbit using tracking data, onboard data, or other sources | Chap. 8 and 10 |

**Table 3.16. How to Plan and Analyze Navigation. (Continued)** We statistically combine tracking data to estimate the current orbit and then use this estimate to generate trajectories and plan maneuvers.

| Steps | Comments | Where Discussed |
|---|---|---|
| 3. Design and Analyze Maneuvers | • Use to maintain and modify orbit | Chap. 10 |
| 4. Determine and Plan Spacecraft Attitude | • Done with spacecraft planning and analysis<br>• Combine with position information to accurately determine pointing information | Chap. 10 and 15 |
| 5. Generate and Regenerate Trajectories | • Use to generate future spacecraft location predictions | Chap. 10 |

1. **Support Pre-Launch Mission Planning.** Use predictive tools and other tests to show the navigation equipment can do what the mission requires.

2. **Determine Orbit.** Extract radio-tracking data (Doppler, range, range-rate, and VLBI) at the tracking stations, transmit it over high-speed data lines to the control center, and buffer it on computer-disk storage. Newly acquired data of different types from individual tracking stations for various spacecraft are automatically sorted and merged into a single time-ordered array for the spacecraft of interest. After removing data of poor quality, you have a data file ready for orbit determination.

   Chapter 10 explains this process in detail. Missions are experimenting now with Earth-orbiting spacecraft that can use onboard GPS receivers and reduce or eliminate the need for radiometric tracking data. This is an example of moving the function completely inside the avionics part of the mission operations diagram (Fig. 3.1).

3. **Design and Analyze Maneuvers.** Propulsive maneuvers are required to maintain a spacecraft's orbit and to navigate a deep-space mission to its target. Before launch, determine how to place these maneuvers to satisfy mission objectives and operational constraints while using as little fuel as possible. During the mission, factor in the spacecraft's actual performance. To make future maneuvers more accurate, analyze each maneuver to improve orbit determination and the spacecraft's performance. Chap. 10 discusses spacecraft maneuvers in more detail.

4. **Determine and Plan Spacecraft Attitude.** For missions which require the spacecraft to point at an object on Earth, in space, or on the surface of another planet or asteroid, you need a predictive tool to help generate commands. People handling the spacecraft or payload

usually operate these tools, but you give them information on the spacecraft's attitude and the payload's pointing direction. By calculating these values, you give payload users inputs that are more accurate than the predictions used to generate the commands.

5. **Generate and Regenerate Trajectories.** Each time the orbit or trajectory is updated, you must do new predictions, which are used by nearly everyone in mission operations. Onboard the spacecraft, they help us point the payload instruments, antennas, and communications.

### 3.5.3    Products of Navigation Planning and Analysis

Accurate, post-launch predictions of the spacecraft's trajectory, required maneuvers, and attitude history make the products listed in Table 3.17 important to other mission operations functions, especially in mission planning and data processing.

**Table 3.17. Products of Navigation Planning and Analysis.** Products are generated post-launch and used as inputs to other mission operations functions.

| Product | Comments |
|---|---|
| Trajectories (past, present, and future) | • Input to data processing, mission planning, and mission control |
| Maneuver designs | • Input to spacecraft planning and analysis and mission planning |
| Attitude history | • Input to data processing |

### 3.5.4    Key Considerations

Accuracy drives the cost of orbit determination. We have to make sure the mission needs certain levels of accuracy before spending money on navigation systems that must support them.

Having fewer propulsive maneuvers simplifies navigation because we must plan them and then determine and analyze new orbits or trajectories after each maneuver is complete.

After selecting a tracking network, we should try to match our mission requirements to their standard services for determining orbits and predicting trajectories. Sometimes, we can change our requirements so they're compatible with the network's services.

Because onboard computational abilities and supporting technologies are improving, we must understand the cost and accuracy trade between determining orbits and propagating trajectories onboard versus doing the same things on the ground. GPS's capabilities to support onboard navigation may well be sufficient for many LEO missions.

# 3.6　Spacecraft Planning and Analysis

Here, planning and analysis make sure we maintain the spacecraft's health and safety and get back its mission and scientific data. Spacecraft engineers

- Prepare telemetry predictions
- Analyze the real-time and processed data
- Identify and resolve anomalies
- Design maneuvers
- Maintain attitude-control and flight software
- Develop, analyze, and test the engineering commands for uploads and real-time commanding

Spacecraft engineers also use the analysis software and hardware (testbed) on the ground to analyze data, prepare predictions, and simulate commands. Finally, they maintain and update various documents or databases (dictionaries, maps, procedures, flight rules and constraints, plans) and reports (consumables, trends, and in-flight performance) needed to complete a space mission.

### 3.6.1　Information Needed to Plan and Analyze a Spacecraft

To do this function, we get inputs from the five previously discussed mission operations functions: mission planning, activity planning and development, mission control, data transport and delivery, and navigation planning and analysis. The information describes how we expect the spacecraft bus to operate, the spacecraft bus's actual performance, and the spacecraft's upcoming maneuvers and planned activities. We need telemetry data, plans, and activities before launch to validate the spacecraft bus's capabilities and all five inputs after launch to plan the spacecraft bus's activities and analyze its performance. Table 3.18 summarizes this information.

Table 3.18. **Information Needed to Plan and Analyze a Spacecraft.** We get information needed to plan and analyze spacecraft-bus operations from previously described functions.

| Information Required | Where Discussed |
|---|---|
| Plans | Sec. 3.1 |
| Command files | Sec. 3.2 |
| Alarms and pass reports | Sec. 3.3 |
| Channelized telemetry data | Sec. 3.4 |
| Maneuver designs | Sec. 3.5 |

### 3.6.2    How to Plan and Analyze Performance of a Spacecraft

In this function we compare the spacecraft bus's actual performance with the expected performance to make sure it meets mission objectives. Many of the steps listed in Table 3.19 begin before launch, and all but the first step, validation, continue until the end of the mission. We further describe these steps below.

**Table 3.19. How to Plan and Analyze a Spacecraft.** Steps are not necessarily sequential. We do them as needed before and after launch to maintain the spacecraft's health and safety and ensure we get back all mission data.

| Steps | Comments | Where Discussed |
|---|---|---|
| 1. Validate Processing Abilities of the Spacecraft and Ground System | • Done before launch during end-to-end tests of the system<br>• Helps train operators | Chap. 12 and 15 |
| 2. Generate Reports and Maintain Database | • Reports describe the spacecraft bus and its operation<br>• After launch databases document actual performance | Chap. 15<br>Sec. 3.1 and 3.4 |
| 3. Plan Calibrations | • Pre-launch planning of calibrations<br>• Post-launch updates based on spacecraft performance | Chap. 7 and 15 |
| 4. Generate Reports on Consumables, Trends, and Performance | • Compare actual and expected performance of spacecraft hardware and resources | Chap. 15 |
| 5. Plan Spacecraft Bus Commands | • Generate all commands uplinked to spacecraft bus<br>• Coordinate with mission planning and activity planning and development | Chap. 15<br>Sec. 3.1 and 3.2 |
| 6. Operate and Maintain the Flight Simulator | • Used to verify command loads before uplinking commands to spacecraft | Chap. 8<br>Sec. 3.3 |
| 7. Maintain the Spacecraft Bus's Flight Software | • Manage flight software<br>• Generate software memory loads<br>• Control software configuration | Chap. 15<br>Sec. 3.12 |
| 8. Analyze Engineering Data | • Determine health, safety, and performance of spacecraft<br>• Investigate spacecraft anomalies | Chap. 15 and 16 |

1. **Validate Processing Abilities of the Spacecraft and Ground System.** Before launch, test and validate the ground processing system and its compatibility with the spacecraft. Show compatibility mainly through end-to-end testing, with the spacecraft and ground station flowing data into the control center. Often, you'll do these tests while the spacecraft is in the thermal-vacuum chamber.

You can also participate in spacecraft system tests to ensure that the part of the ground system used for spacecraft planning and analysis is operating properly. This step also best trains people in mission operations on the spacecraft's characteristics before launch.

2. **Generate Reports and Databases.** Include:

   *Flight rules and constraints.* List all in-flight operational limitations imposed by the spacecraft hardware and software.

   *Decommutation maps.* List the telemetry channels contained in each of the different maps, which define the variable part of the telemetry commutator. Nearly all missions are now using CCSDS standards which define packets of information. Each packet is self identifying and, when used properly, replaces decommutation maps.

   *Telemetry dictionary.* Describe in detail the engineering and science telemetry measurements, as well as the operational limits and parameters needed to understand each measurement.

   *Command dictionary.* Describe and show bit patterns of the messages and commands that may be sent to the spacecraft; cross-reference the flight rules and telemetry dictionary.

   *Spacecraft idiosyncrasies.* Include unusual or anomalous performance characteristics.

   *Operating procedures.* Define how to do spacecraft planning and analysis.

   *Spacecraft contingency plan.* Identify potential anomalies that would require ground response and plan corrective actions.

3. **Plan Calibrations.** Show the strategies and activities for calibrating subsystems of the spacecraft that affect its performance. Subsystems you must usually calibrate after launch (and sometimes during the mission) include attitude control, star trackers, and moveable platforms.

4. **Generate Reports on Consumables, Trends, and Performance.** Periodically identify uses, changes with time and operation, and performance of the spacecraft. Although you'll do most of this after launch, start reporting during the system test and then maintain and update reports during the mission.

5. **Plan Spacecraft Bus Commands.** Generate all command inputs and review all system commands for uplink to the spacecraft. Command requests (planned real-time and command loads) are to calibrate spacecraft subsystems and maintain the spacecraft's health. Review planned command uploads for correct engineering and completeness.

   Develop maneuver designs using inputs from navigation planning and analysis and determine appropriate commands to carry out the maneuver. In the future, the ground system may transmit a new state vector to the spacecraft, which would then compute and execute the maneuver. If so, detailed ground commands won't be necessary.

Spacecraft planning and analysis supports mission planning and activity planning and development as required during the mission.

6. **Operate and Maintain the Flight Simulator.** A flight simulator (such as a testbed, simulator, or testlab) is often required to verify commands before radiation to the spacecraft.

7. **Maintain the Spacecraft Bus's Flight Software.** Maintain flight software, generate memory loads for this software, and control all changes to flight programs, including databases.

8. **Analyze Engineering Data.** Analyze the spacecraft's engineering data (real-time and non-real-time) and determine its health, safety, and performance. Develop performance models before launch. These models may be software programs or hardware models that predict the spacecraft's performance and analyze its data. Assess performance by analyzing the spacecraft's system data. Investigate spacecraft anomalies and correct them. Help investigate payload anomalies as requested. Maintain trends and supply information to mission planning about the spacecraft's deviations from pre-launch assumptions of its abilities.

### 3.6.3     Products of Spacecraft Planning and Analysis

We use the products from this function to develop future spacecraft requests. We coordinate with mission control, mission planning, and activity planning and development to schedule necessary commands during future spacecraft passes. We also use the products listed in Table 3.20 to support payload planning and analysis and data processing.

**Table 3.20. Products of Spacecraft Planning and Analysis.** We complete these products after launch and input them to other mission operations functions.

| Product | Comments |
|---|---|
| Spacecraft status versus plan | • Generate post-launch<br>• Use to track performance and modify mission plan |
| Spacecraft limits | • Generate pre-launch and update post-launch based on actual performance |
| Spacecraft activity requests | • Generate pre-launch and update post-launch<br>• Input to mission control |
| Processed spacecraft data | • Generate post-launch<br>• Input to payload planning and analysis and data processing |
| Reports | • See Step 2 for a list of the most important reports |

### 3.6.4    Key Considerations

The key driver of operational complexity is the level of resource margins available on the spacecraft. A large positive power margin means we don't need detailed analysis of planned activities. If there is a negative power margin with all spacecraft and payload sources on, analysis becomes more difficult and detailed. If subsystems interact under negative margins, spacecraft planning and analysis becomes even more difficult.

Other factors are level of spacecraft analysis required, number of interactions, spacecraft safing, the need for real-time engineering analysis, and the level of automation available. Properly designing the spacecraft and the associated telemetry measurements will make it possible to analyze the spacecraft as a system. There is still a tendency to design the "best" subsystems, integrate them into a spacecraft, and then attempt to figure out how to analyze the spacecraft in flight. This approach will usually mean analyzing each of the subsystems for proper operation and then integrating this information at the system level. Subsystem analysis of a spacecraft is labor intensive.

The spacecraft that minimizes interactions between spacecraft components, between spacecraft and payload components, and between payload components will be easier to operate and require fewer resources from planning and analysis.

A spacecraft that will go to a safe state when an error occurs is easier to operate than one that is fragile and needs a lot of monitoring.

Many of today's spacecraft can go unattended for a week or more at a time, but operators tend to want to observe and monitor the spacecraft in real time. Frequent real-time operations require more resources than real-time passes once a day or once a week.

Finally, using automated analysis tools to decrease staffing lowers operations costs. Artificial-intelligence techniques are now becoming useful for mission operations.

## 3.7    Payload Planning and Analysis

We need to identify and prioritize payload opportunities in order to design observations and activities and to correctly carry out the command load transmitted to the spacecraft. We analyze payload data to assess the payload's performance and to change planned observations when necessary to get better data.

### 3.7.1    Information Needed to Plan and Analyze Payload

We get information from mission planning, activity planning and development, mission control, and data transport and delivery. We use it before launch to make sure the payload is operable and the data system is compatible. After launch, the information helps us assess the payload's performance and plan observations. Table 3.21 lists the inputs for this function.

**Table 3.21. Information Needed to Plan and Analyze a Payload.** We need information before launch to validate payload operability and after launch (shaded rows) to plan observations and assess performance.

| Information Required | Where Discussed |
|---|---|
| Plans | Sec. 3.1 |
| Activities | Sec. 3.2 |
| Alarms and pass reports | Sec. 3.3 |
| Payload data | Sec. 3.4 |

### 3.7.2    How to Plan and Analyze a Payload

Table 3.22 lists the steps for payload planning and analysis. Before launch, monitor the design and development phases to ensure you have a payload operations concept that will work and meet mission objectives. After launch, monitor the payload performance, help plan future operations, and plan necessary payload calibrations.

**Table 3.22. How to Plan and Analyze a Payload.** These steps are similar to those done in spacecraft planning and analysis. They focus on the payload's proper and accurate operation.

| Steps | Comments | Where Discussed |
|---|---|---|
| 1. Validate Processing Abilities of the Payload and Ground System | • Conduct pre-launch during end-to-end tests of the spacecraft<br>• Helps train operators | Chap. 12 and 14 |
| 2. Generate Reports and Databases | • Write reports pre-launch with mission planning<br>• Maintain databases post-launch | Chap. 14<br>Sec. 3.1 and 3.4 |
| 3. Plan Payload Observations | • Identify opportunities<br>• Assign priorities and resolve conflicts<br>• Generate and validate command loads | Chap. 7 and 14 |
| 4. Plan Payload Calibrations | • Use to correctly interpret data<br>• Use for calibrations based on analysis of payload data | Chap. 7 and 14 |
| 5. Analyze Payload Performance | • Quick-look analysis<br>• Calibration analysis<br>• Trend analysis | Chap. 14 |
| 6. Assess and Maintain Payload Flight Software | • Manage payload software<br>• Generate software memory loads<br>• Control configuration of payload software | Chap. 14<br>Sec. 3.12 |
| 7. Investigate Anomalies | • Do as required<br>• Support spacecraft planning and analysis | Chap. 16 |

1. **Validate Processing Abilities of the Payload and Ground System.** Before launch, these activities are much like those for spacecraft planning and analysis. You will usually participate in the same tests as those described for the spacecraft. A key difference is that some of the instruments may not be turned on, except during the thermal-vacuum testing with its simulations of space environments.

2. **Generate Reports and Databases.** Similar to those for spacecraft planning and analysis.

3. **Plan Payload Observations.** Identify payload opportunities by evaluating trajectory information and the ephemerides of the body being investigated, be it Earth or a planet. These opportunities represent times in the mission during which the payload's observations will achieve mission objectives. Once opportunities have been identified, design, implement, and integrate these observations to form the command load and transmit it to the spacecraft. Depending on the payload, you may divide the observations into discipline groups, so that each discipline group identifies observations to meet the mission's objectives. If the payload has only one instrument, this process is considerably simpler than if it has a dozen instruments. Each discipline group then prioritizes its observations and activities.

   Once priorities are set, combine into a single file the inputs from each group and the observations requested from the payload and spacecraft engineering disciplines. Priorities help resolve conflicts between the discipline groups and produce a conflict-free timeline of activities that will generate commands for the spacecraft and instruments. The degree to which this process can be automated varies with mission and spacecraft design. At NASA, mapping missions are highly automated because mapping strategies are automated and observations don't require decisions. The astrophysics community uses a very automated process to determine observations for each orbit. On the other hand, planetary missions tend to require a lot of interaction and many human decisions.

   Payload planning often begins months or even years before final command load goes to the spacecraft. As such, observations may need changing to account for discoveries or new information about the observation. Thus, the ground system allocates resources (people, hardware, and spacecraft) to these changes. For essential observations, you can also slightly modify the command load onboard the spacecraft.

   Payload planning ends when you validate internal commands for the spacecraft and instruments. Base these commands on scientific needs, the spacecraft's abilities, and mission guidelines and flight rules.

Submit commands to activity planning and development, where people will expand, constraint-check, and compile them. Then, validate them again to ensure the final commands don't harm the spacecraft and accurately reflect the initial observation requests.

4. **Plan Payload Calibrations.** You need calibrations to correctly interpret the payload data. First, analyze the payload telemetry during downlink and then design and do changes to the calibration plan, or do more calibrations, to ensure accurate processing and interpretation of the received data. Early definition of this process during payload design reduces the time you must spend on calibration.

5. **Analyze Payload Performance.** Operate and calibrate the payload hardware and analyze trends. This step ensures the instrument is within specifications and no trend is developing that would keep it from meeting future demands.

To analyze the payload, you can do

*Quick-look analysis.* Inquire into the health and quality of the instrument data.

*Calibration analysis.* Analyze calibration observations and selected observations by the instrument. If necessary, ask for new calibration observations or even a change to the basic calibration plan written before launch.

*Trend analysis.* Analyze mainly the engineering measurements on the instrument that show the health and performance trends during the mission. Note carefully anything that suggests you must change the instrument's operating plan.

6. **Assess and Maintain Payload Flight Software.** Many of the instruments flown today have processors as large as the spacecraft's processors. Maintain this software just as you would the spacecraft's software.

7. **Investigate Anomalies.** Do these as required and help the spacecraft group analyze spacecraft anomalies as appropriate.

### 3.7.3 Products of Payload Planning and Analysis

As with spacecraft planning and analysis, products from this function help mission planning and mission control plan and conduct future operations. Table 3.23 lists these products.

Table 3.23. **Products of Payload Planning and Analysis.** Products are generated after launch and are inputs to mission planning, mission control, and data processing.

| Product | Comments |
|---------|----------|
| Payload status versus plan | • Input to mission planning<br>• Use to monitor payload performance |
| Payload limits | • Input to mission planning and mission control |
| Payload activity requests | • Input to activity planning and development |
| Processed payload data | • Input to payload data processing |

### 3.7.4    Key Considerations

The spacecraft and data-handling enable us to get payload data to the user for planning and analysis. When these systems introduce errors or don't meet specifications, we must reevaluate payload planning and often change it to solve these problems. We must also do more payload planning when the mission plan asks us to generate payload activities based on analysis of received data. When the ground system has to be more adaptable, mission operations costs go up.

To reduce these costs, use existing planning tools, automate payload data-gathering, emphasize payload operability during design, and minimize required adaptivity of the payload in the operations concept.

Many planning tools exist, and others are "new" designs of present capabilities. We tend to re-invent capabilities instead of finding tools we can use "as is" or with slight changes.

Next, consider automating the payload data-gathering and calibration. Onboard automation of data-gathering reduces the resources required to operate the payload but usually demands resources up front for planning the automation.

Keep insisting on payload operability during the design phase. Ask questions of the instrument builders: "How are the instrument commands going to be determined? What information is required, and how is this information converted into instrument commands?" Make sure the team addresses the number and complexity of calibrations before completing the payload design.

Finally, minimize the amount of adaptivity required to achieve mission success. A mission that requires information from the payload **before** the next command load can be generated is more expensive to operate than one that collects data in a standard manner. Adaptivity is sometimes necessary but designers often use it to put off understanding how to establish observations or command loads.

## 3.8    Payload Data Processing

First, we must bring together instrument data packets, engineering data, and ancillary data (e.g., orbital/navigation data) into instrument data records. We may also need to do higher-order processing to support payload analysis and generate digital and photography products of archival quality. In fact, we may need to do any of the processing steps described below—at either the control center, a dedicated payload-processing facility, or on the spacecraft.

The software for processing payload data is often unique (not commercially available) because specialized routines are required for some scientific processing that has few commercial customers. Examples include

- Cartographic projections for bodies other than the Earth
- Radiometric reconstruction of color imagery from multiple images acquired through spectral filters
- Image registration to less than one pixel accuracy

After this—and perhaps more specialized—processing, the data is ready to be archived for use by other scientists in the case of NASA missions. Chapter 13 further discusses processing of payload data.

As mentioned earlier, processed payload data may change the mission plan, but at this processing level, we usually pick up very subtle but important instrument characteristics.

### 3.8.1   Information Needed to Process Payload Data

**Table 3.24. Information Needed to Process Payload Data.** To process payload data, we need it plus other data regarding the spacecraft and ground system. We collect all information after launch.

| Information Required | Comments | Where Discussed |
|---|---|---|
| Payload data | • Payload data in the form sent to the spacecraft avionics system (usually packets) is received by payload data processing <br> • Instrument-level testing, payload integration testing, spacecraft integration testing, and flight data all make up the payload data | Sec. 3.4 |
| Ancillary data | • Ancillary data is information about the spacecraft and ground system required to properly process and analyze the payload data <br> • The specific data and the formats are specified before launch. These data include spacecraft and payload engineering information. | Sec. 3.4 |
| Orbital/navigation/ attitude-predict data | • This data is sometimes included as ancillary data but is important enough to be called out separately <br> • Most observations require information relative to where the payload is pointing to be able to interpret the data <br> • Sometimes, as information about the orbit or trajectory improves, it may appear in several versions, which we must be able to identify separately | Sec. 3.5 |

## 3.8.2 How to Process Payload Data

Take the data received from transport and delivery, sometimes through the mission database, and construct the sensor information, which may be an image, a spectra, or other meaningful product. This process transforms the data into a product usable by a discipline expert, as opposed to a sensor expert. You'll start processing payload data before launch and continue throughout the mission and often well past the end of the mission.

Payload data processing often occurs in a facility separate from the mission control center and under the control of scientists and specialists. Depending on the mission, organizations separate from mission operations may do some of this processing. Table 3.25 shows the steps for processing payload data. We discuss it in more detail in Chap. 13.

**Table 3.25. Steps for Processing Payload Data.** The steps will vary depending on the payload, but these are typical for scientific missions.

| Steps | Comments | Where Discussed |
|---|---|---|
| 1. Validate Payload-Processing System | • Process data before launch from one or all of these sources:<br>  – Bench-level testing<br>  – System-level testing<br>  – Simulated data or real ground based data in a format similar to the flight instrument<br>Note: Participation in system-level tests and end-to-end tests is helpful, but usually there are so many constraints from the non-space environment that the tests don't fully check out the processing capabilities. The "real" test often doesn't occur until after launch, when the instrument is in space. | Chap. 13 and 14 |
| 2. Generate Payload Data Records | • Data transport and delivery provides the payload data—either to the mission database or directly to the payload-processing facility—in the form of packets separated by instrument type<br>• This data is then aggregated into payload data records—typically called level 0 processing<br>• No value is added to the data at this point. It is the best set of data from the instrument that data transport and delivery can produce. | Chap. 13 and 14 Sec. 3.4 |
| 3. Process Sensor-Specific Data | • Process instrument engineering data<br>• Decompress data<br>• Remove sensor signature<br>• Apply calibration information to the raw data | Chap. 13 |
| 4. Correlate Ancillary Data with Sensor Data | • Add to the sensor-specific data the ancillary data needed to process it into meaningful payload products | Chap. 13 |
| 5. Generate Products | • Generate a data record that contains the instrument data and its ancillary data with proper identification of the processing completed | Chap. 13 |

**Table 3.25.  Steps for Processing Payload Data. (Continued)** The steps will vary depending on the payload, but these are typical for scientific missions.

| Steps | Comments | Where Discussed |
|---|---|---|
| 6. Analyze Products | • Product analysis is usually separate from, but associated with, product generation<br>• Before the data goes to the user, the operations staff makes sure it's meaningful and doesn't contain incomplete information or processing errors. This is especially true for scientific missions, in which the spacecraft is an observatory used by hundreds of astronomers. | Chap. 13 |
| 7. Manage Data | • Deliver products to the payload user or store them in a database for further processing and product generation<br>• We need to manage data so we can locate for analysis products of the same region but taken at different times<br>• The archiving function also uses data management | Chap. 13<br>Sec. 3.9 |

### 3.8.3    Products of Payload Data Processing

Although these products vary from mission to mission, Table 3.26 lists the most common ones. The objective of any mission is to obtain the sensor output. As we process data, we may detect deviations from the mission plan even when all other mission indications are normal. The Viking project reached Mars and had to wait several weeks for a dust storm to diminish before mapping the landing sites and then landing on the surface. All engineering aspects of the mission were normal. Only the payload processing showed that the mission plan had to change.

**Table 3.26.  Products of Payload Data Processing.** These are typical post-launch products for a scientific mission. Commercial and defense missions will have similar products, except for the press releases.

| Product | Comments |
|---|---|
| Deviations from plan | • The mission plan describes the expected conditions. When these conditions aren't met, and sensor data isn't meaningful, users request changes to the plan. |
| Files of instrument calibration | • A separate file usually records how data is recorded, so future users of the data may use different calibration techniques based on later information and understanding<br>• Including this file and the data to which it was applied in the archival data records makes sure we don't lose information |
| Records of archival data | • The main product of this function ensures others can use the information for generations to come. The archival data record includes the basic sensor data, the ancillary data, and the calibration files. |

**Table 3.26. Products of Payload Data Processing. (Continued)** These are typical post-launch
products for a scientific mission. Commercial and defense missions will have similar
products, except for the press releases.

| Product | Comments |
|---|---|
| Hard-copy products | • Hard-copy products are still used in many missions. Electronic versions are becoming more common, but missions still use high quality, specialized photo processing. |
| Press releases (for science missions) | • NASA missions use the output from payload processing to inform the public. Although military missions don't usually have press releases for their payloads, they do have similar requirements for briefings.<br>• These products are for a specialized audience so they require a different approach compared to those we use to understand information from payload sensors. We must understand these requirements early rather than after launch. |

### 3.8.4    Key Considerations

You can lower costs for processing payload data by addressing the topic early
in the mission-concept study and then emphasizing them while developing the
mission operations concept. During the MOS conceptual design (see Chap. 4),
include these concepts for processing payload data:

- Required final products
- Diagrams of end-to-end data flow
- The ancillary data required to process the payload data and its
  sources
- Robustness of the payload processing to data loss or drop out.
  How will the proposed data compression and data formatting
  react to the noisy environment?

The mission operations system can make the analysis more difficult if it loses,
or doesn't collect, data. This means we need to pay special attention to the size,
quality, and completeness of our ground system and how these characteristics will
affect processing of the payload data. For many payloads, it's good enough for the
ground system to complete 95% of the processing, but that depends on where the
5% loss occurs. Data lost as a block may be acceptable, but if the 5% loss is from
compressed data across all the instruments, it may be unacceptable.

When designing and doing payload data processing,

- **Use Existing Tools.** As discussed under payload planning and
  analysis, using existing tools is an obvious way to save resources.
  Within the scientific community, many payload users now take
  advantage of standard processing packages.
- **Understand Requirements for Ancillary Data.** Make sure you
  understand and have available the ancillary data needed to

interpret payload data. If you discover after launch that this data isn't readily available, solving this problem will be expensive.

* **Ensure Data Processing Robustness.** Understand early how well the system can process data without losses. Make sure it can transport data smoothly, taking into account such factors as data compression, data formatting, and the characteristics of the link performances. It is now common to simulate these characteristics early in the design phase to ensure the data-processing system is robust.

# 3.9    Archiving and Maintaining the Mission Database

The payload archive for processed data is usually separate from the mission database that supports controlling and acquiring mission data. But because these entities are closely related, we treat them as one element of the MOS. The *mission database* receives, stores, and delivers data between the nine MOS functions. It can be centralized or distributed, usually receives data in real time, and can be accessed in either real time or in at some later time. The *archive* receives selected data for permanent storage and either immediate or historical review.

NASA scientific missions require the payload data to be archived so scientists other than the ones who designed the observation or built the instrument can use it. This data must be in a form others can use without specialized knowledge of the instrument. NASA archives cover several scientific disciplines.

The National Academy of Science did a study on archiving in 1986 and issued their recommendations in the CODMAC report [National Research Council, 1986]. NASA has followed several important recommendations in their archiving program. The report stressed that an active scientific archive should be located where scientists are using and improving the data. An example is the center for infrared processing and analysis on the California Institute of Technology campus. This archive holds the data from the Infrared Astronomical Satellite (IRAS), whose mission ended in 1984. The archive is still actively used. Another example is the Space Telescope Science Institute on the Johns Hopkins University campus. The Science Institute holds the archive data from the Hubble Space Telescope. The mission is still in progress and will be past the end of the century. Scientists around the world use this archive.

The mission database is the repository for all data collected and delivered to the project by data transport and delivery, as well as for data prepared for sending to that transport function. This database contains the controlled files of data required for processing the spacecraft data, plus the raw and processed spacecraft data. The database receives data in near-real time, but people in planning and analysis can use it at any time. Mission databases in the past have been centralized, but with today's technology, they're often distributed—around the world in some cases involving NASA scientific missions.

*Active archives* contain the instrument data generated during instrument testing and calibration before launch. They provide a valuable source of data for comparison with flight data. The degree of pre-flight testing varies from mission to mission but is often very extensive. After launch, they contain the data products sent to payload analysis from mission operations and the higher-quality data payload analysis produces.

*Dormant archives* store data from past missions that is either not used or very rarely used. Active scientists aren't at the storage location.

### 3.9.1     Information Required to Archive and Maintain the Mission Database

Table 3.27.  **Information Required to Archive and Maintain the Mission Database.** Archives enable use of the data long after the mission is over.

| Information Required | Comments | Where Discussed |
|---|---|---|
| Telemetry and tracking data | • The data received by the mission database is generally referred to as level 0. The mission database then processes and stores this data as level 1, so we can use it to analyze the mission.<br>• The level 1 data may also be part of the data archived. | Chap. 13<br>Sec. 3.8 |
| Data from other functions in the mission operations system (MOS) | • Other MOS elements also generate data and store it in the mission database. Examples are command files and pass reports from mission control.<br>• Certain of these products are also packaged into data to be stored in the archive. Analysis often requires engineering data to effectively understand the payload data. We may correlate this ancillary data to each payload product. | Sec. 3.1–3.8 |
| Final payload products | • Products the MOS generates for payload analysis—called level 3 data—are stored in the mission database and transferred to the project archive | Chap. 13 |
| Higher-level products from the payload | • Data processed by the payload centers—called level 4 data—are also transmitted to the project archive for use by others | Chap. 13 |

### 3.9.2    Process for Archiving and Maintaining the Mission Database

**Table 3.28. Process for Archiving and Maintaining the Mission Database.** The archive allows users to access data from previous missions and enables comparative analysis as well as more detailed analysis of the data.

| Steps | Comments | Where Discussed |
|---|---|---|
| 1. Manage and Retrieve Data | • Acquire or receive the data from the mission database or payload users<br>• Provide cataloging and retrieval, which references the data sets and allows retrieval by many different types of queries<br>• Ensure the data received adheres to defined, agreed-to, standard formats | Chap. 13<br>Sec. 3.8 |
| 2. Secure Data | • Data security is required as with the overall mission operations system<br>• For scientific missions, we must give certain people access to some data for a time before it becomes available to other researchers. We call this data proprietary because it belongs to the scientist for a specified period. | |
| 3. Notify Users of Data's Arrival | • After placing data products into the archive, identify the new products and notify users of their availability<br>• Usually automated for missions that generate large amounts of data | |

### 3.9.3    Products of Archiving and Maintaining the Mission Database

**Table 3.29. Products of Archiving and Maintaining the Mission Database.** The products are data—and more data.

| Products | Comments |
|---|---|
| Archived data | • At the top level, two types of data are in the archive: payload data and the ancillary data that describes its attributes<br>• Back up and store data in physically separate facilities |
| Operational data to and from functions in the mission operations system | • The mission database is a two-way database that allows the mission operations functions to share data and pass data from one function to another<br>• The database is populated in real time, when we contact a spacecraft, as well as after the contact, when we add information received during the contact |

### 3.9.4    Key Considerations

The technology for archives and mission databases is changing rapidly, so our challenge is to keep up with the technology while efficiently and cost-effectively providing access to data. The medium of choice for many scientific missions is now CD-ROMs, whereas several years ago it was still tape. We also need to reconsider the cost/performance ratio of delivering a data set by electronic means instead of by Federal Express.

Besides considering the technology issues in the mission operations concept (see Chap. 4), we also must ask ourselves how the data gets to the user? Does the archive send or does the user request? We must develop the archive plan, including the required ancillary data, as early as possible, so the ground team and spacecraft designers can more easily build a cost-effective archiving system.

# 3.10 Systems Engineering, Integration, and Test

These next four areas are most often on-going functions that apply equally well before and after launch. These disciplines appear in many fields, but we'll discuss only those which apply to mission operations.

The engineering function for the ground system receives support from the nine MOS functions previously discussed. Thus, systems engineers understand the overall system and the interfaces, and each of the nine MOS functions provide details. Remember that what may be a sub-system at one level (e.g., the nine MOS functions) will appear as a system at the next level (e.g., activity planning and development). Thus, systems engineering occurs at all levels. But here we're talking about engineering of the mission operations system.

### 3.10.1    Information Required for Systems Engineering, Integration, and Test

Table 3.30. **Information Required for Systems Engineering, Integration, and Test.** The information listed enables this function to work across the nine main MOS functions from which it draws support.

| Information Required | Comments | Where Discussed |
|---|---|---|
| Project requirements | • Systems engineering keeps the project "vision"— making sure the data system and requirements satisfy this vision. | Chap. 1 and 4 |
| Outputs from the operations concept | • Systems engineering uses these key outputs:<br>  – Operations scenarios<br>  – Derived requirements<br>  – Timelines<br>  – Data-flow diagrams | Chap. 4 |

## 3.10.2    Process for Systems Engineering, Integration, and Test

**Table 3.31.   Process for Systems Engineering, Integration, and Test.** These steps are typical for most missions.

| Steps | Comments | Where Discussed |
|---|---|---|
| 1. Consider Typical Systems Engineering Functions | • Consider these typical functions:<br> – Develop system architecture<br> – Generate, review, and control requirements<br> – Define, document, and control interfaces<br> – Monitor the application of standards, such as software walkthrough, which the project requires | Chap. 1 and 4 |
| 2. Integrate | • Usually (today—always), a different group of people does integration<br>• Integrate the software programs and program sets into a system<br>• Integrating software can be a large task depending on the complexity of the system being developed, or the multi-mission system being adapted for a new project<br>• Validate the interfaces<br>• Certify the system ready for operational testing, training, or use | Chap. 4 and 6 |
| 3. Simulate, Test, and Train People for the Mission | • Preparing simulations and testing can be as much work as carrying out the mission itself<br>• Develop training plans<br>• Develop simulation capabilities<br>• Conduct training<br>• Certify operational readiness | Chap. 4 and 6 |
| 4. Evaluate System | • Based on the mission operations concept, discussed in Chap. 4, develop and validate performance requirements<br>• Test the MOS to verify that system performance meets requirements | Chap. 4 |
| 5. Ensure Network Is Secure | • Network security is becoming more important as the mission operations systems are becoming distributed and, in many cases, are using public networks<br>• Write and implement the security plan; monitor the network | Chap. 12 |

### 3.10.3    Products from Systems Engineering, Integration, and Test

Table 3.32. **Products from Systems Engineering, Integration, and Test.** Make sure all MOS people review and understand these products.

| Products | Comments |
|---|---|
| System architecture | • In the system architecture, describe the vision of the sponsor and project management. Developing this vision early and keeping it updated are important because it provides a framework for daily decisions. |
| Specification of software interfaces | • This specification is one of the most important products of systems engineering—especially if the project requires distributed processing or integration of the payload users' software into mission operations. |
| Integration and test plan | • The integration and test plan describes what to do when integrating new deliveries into the system. Having recursive tests and automated testing is important. Tools today allow even graphical user interfaces to be tested automatically, without a person sitting in front of a console. |
| Training plan | • The training plan developed for the mission operations manager allows mission operators to show they're ready for launch |
| Procedures | • Cover integration and test plus training |
| Security plans and tests | • Increasingly important, system security needs good plans and regular testing |

### 3.10.4    Key Considerations

While the systems engineering processes of mission operations are not especially different from that of any large software system, the requirements are often fuzzy. They also change during the development cycle as the payload users learn more about what they really need. Especially on scientific missions, these payload users have a lot of power with upper management in the project and NASA. Thus, we must involve the payload users early and use rapid development and prototyping, rather than following the classic process of requirements, design, development, and delivery. Prototyping allows users to see capabilities.

The mission operations manager (MOM) should also

- Ensure the key concept trades for mission operations occur early
- Maintain the big picture
- Use rapid development and prototyping processes whenever possible

# 3.11 Computers and Communications Support

Designing and implementing mission operations hardware may involve staff within the project and from other organizations. For example, if the project uses an existing system, hardware development becomes a joint effort between the project and the organization that owns the basic system. Of course, the project manager must oversee hardware development by an outside organization to make sure it operates properly. Other textbooks treat designing and building hardware for information systems, so we don't discuss it in detail in this book.

### 3.11.1 Information Needed to Support Computers and Communications

Table 3.33. Information Needed to Support Computers and Communications. These requirements enable engineers to develop the hardware systems a mission needs.

| Information Required | Comments | Where Discussed |
|---|---|---|
| Data-flow diagrams | • Develop data-flow diagrams first while developing the operations concept, as discussed in Chap. 4. Allocate processes to hardware and software at the same time. This is the starting point for designing the hardware. | Chap. 4 and 13 |
| Requirements for computers or workstations | • Most mission operations now use distributed workstations or are moving toward them.<br>• The number and location of the workstations depend on data flow, the mission operations architecture, and organization and staffing levels. Systems engineering typically specifies these requirements.<br>• Many of the scientific missions have staff around the world who need workstations and, therefore, computer networking. | Chap. 7 and 8 |
| Requirements for networking and data communication | • The specific and derived requirements for flowing data around the control center, country, or world demand much more networking and communications equipment.<br>• Commercial open networks such as internet require attention to ensure that critical command and control have enough security to protect them from unauthorized users. | Chap. 13 |
| Requirements for voice communications | • Voice communications and video-conference capabilities are important to understand early in the design phase. Designers often overlook these requirements, which are becoming much more important as we move from centralized to distributed control. | Chap. 9 and 12 |

### 3.11.2    How to Support Computers and Communications

Because these processes are typical for any computer hardware, we don't discuss them in detail.

**Table 3.34. How to Support Computers and Communications.** These steps are similar to those for any information system.

| Steps | Comments | Where Discussed |
|---|---|---|
| Design and Build Computers and Communications Systems | • Understand existing capabilities and define what you must add to meet requirements<br>• Get the data-system architecture from the EEIS engineer who helped develop the mission concept and from MOS engineers who helped generate the mission operations concept<br>• Based on the hardware architecture and what you must add to the system(s), design and build the added capabilities<br>• Test the mission operations system | Chap. 4 |
| Maintain Computers and Communications Systems | • Maintain all computers and network equipment throughout the mission | -- |

### 3.11.3    Products from Computer and Communications Support

The products listed in Table 3.35 are again typical of hardware in any information system.

**Table 3.35. Products from Computer and Communications Support.** The hardware and communications systems allow us to develop software and, later, carry out the mission.

| Products | Comments |
|---|---|
| Computer and communications systems | • The ground hardware for mission operations |
| Networks and network access to functions of the mission operations system | • The networks and network access for mission operations<br>• Identify open networks and those restricted for command and control functions |
| Voice-communication system | • The voice and video conference systems for mission operations |

### 3.11.4    Key Considerations

Understand and include current technology in your design to get the best performance from hardware. Look at maintenance and its alternative sources to keep costs down.

# 3.12  Developing and Maintaining Software

We must develop and maintain software throughout a project's life cycle:

- The ground system before launch
- Software to correct errors after launch
- Software to capture changes in mission requirements after launch

On many of today's missions—especially interplanetary missions which take 2–10 years to reach the outer planets—we plan to develop more software capabilities after launch in order to make the mission successful. In some cases, mission operations people develop flight software before and after launch.

We must understand in advance how much development and maintenance will be necessary after launch, so we can organize and build procedures that will meet operational requirements.

### 3.12.1    Information Needed to Develop and Maintain Software

**Table 3.36.    Information Needed to Develop and Maintain Software.** This information is typical for any software-development project.

| Information Required | Comments |
|---|---|
| New requirements | • Most mission operations systems build on existing ones, but requirements will often emphasize new aspects of the system and be written as if nothing existed. Usually, we can refer to existing capabilities and documentation, but if you're asking for new capabilities, clearly state your minimum requirements. |
| Error reports | • Error reports start maintenance |
| Control authority for change | • To control costs, make sure your change authorizations are efficient. Deciding to approve a change after detailed design is done wastes resources. |
| As-built documentation | • An output of software development required for maintenance |

### 3.12.2    How to Develop and Maintain Software

The steps listed in Table 3.37 are typical of any software-development effort. Many other texts discuss this development in more detail. For scientific space missions, requirements often aren't well defined, and the end user (often the scientist developing one of the instruments on the spacecraft bus) doesn't pay attention to mission operations until late in the development cycle. To interact with the scientist early and clarify requirements at the same time, try rapid prototyping or try delivering the ground data system in increments. The Mars Pathfinder project demonstrated data flow for downlink during concept development. They built on this capability in concept development, so a basic system existed when the project was approved.

**Table 3.37. How to Develop and Maintain Software.** The mission operations manager must understand that developing a mission operations system mainly means developing a big information system with a lot of software.

| Steps | Comments |
|---|---|
| Manage Software Development | • Understand the requirements and ensure they are testable and agree with the mission operations concept<br>• Develop the software; use rapid prototyping to ensure the system's users understand how fuzzy requirements will translate into hardware<br>• Deliver the ground data system in increments<br>• Plan for and understand that you'll be developing and maintaining software while the mission is operating<br>• Ensure that schedules are compatible with schedules for developing and testing the spacecraft bus<br>• Test the software—often it is developed in several increments<br>• Deliver software to operations<br>• Train on the new capabilities as required<br>• Maintain the software |
| Maintain Software | • Changes to correct errors<br>• Changes needed because mission requirements change |

### 3.12.3 Products of Software Development and Maintenance

The outputs listed in Table 3.38 are typical of any software project.

**Table 3.38. Products of Software Development and Maintenance.**

| Products | Comments |
|---|---|
| Plan for software development and maintenance | • The sponsor usually requires this plan, which document and communicate to the mission operations staff how the project will proceed |
| Schedules | • Schedules are a part of any software development. Make sure you understand the project schedules and how the software schedule supports the project's milestones:<br>  – Launch date<br>  – Date the mission operations system (MOS) must be ready<br>  – Start of MOS training<br>  – End-to-end tests of the system<br>  – System tests of the spacecraft bus |
| Test plans and reports | • These are required for any software development, but make sure you test interfaces and defined functions of a spacecraft or ground-system element that other agencies will supply |
| Working software and as-built documentation | • The result of the plans, schedules, and tests<br>• Maintenance requires documentation<br>• The level of documentation varies depending on whether the developers or a different organization does the maintenance |
| Reports of performance and errors | • Tracking of errors and authorizing their fixes must be efficient |

### 3.12.4 Key Considerations

This area is also rapidly changing. What was developed as unique on the last mission may very well be available off the shelf as commercial software today. Understanding what is available commercially will save money and time. Looking forward—by trying to design software for reuse on future missions—is important. Finally, developers must communicate existing and planned capabilities to the system's users and operators. Find a way to involve them in discussions, reviews, demonstrations, and tests. Ask for their inputs. The earlier and more involved operators become with the users, the more operable the developed system will be.

## 3.13 Managing Mission Operations

Management functions for mission operations differ before and after launch. Before launch, we must develop and train so the entire mission operations system is ready for launch. Sometimes, one manager oversees both; sometimes, responsibility transfers formally from one organization to another. Names also vary depending on the organization. Before launch, the mission operations manager may be called the ground system or ground-data-system manager. At this point, the MOM works with other project functions to make sure operations are compatible with mission and spacecraft design. This process begins during the study phase and during development of the mission concept and the mission operations concept, but it continues throughout the development phase. After launch, manager names include mission director, mission operations manager, and chief of mission operations.

This section focuses on management functions unique to mission operations and discusses unconventional management practices.

### 3.13.1 Information Needed to Manage Mission Operations

Table 3.39 lists the information management needs to be successful. We use the first three inputs to guide development of the mission operations system. Before launch they guide changes to the mission activities that arise when status deviates from plans.

### 3.13.2 How to Manage Mission Operations

Table 3.40 lists management steps, all of which take place before and after launch. Before launch, we focus on developing the ability to support mission operations. After launch, we maintain the system, develop changes as required by changing mission conditions, and—on projects such as long planetary missions—finish developing capabilities needed to operate over longer periods.

**Table 3.39. Information Needed to Manage Mission Operations.** This information enables the mission operations manager to make decisions during the mission.

| Information Required | Comments | Where Discussed |
|---|---|---|
| Requirements | Includes those placed on the mission operations system and those derived and generated by the mission operations system. The requirements not only are important for developing the MOS but also aid operational decisions during the mission. | Chap. 1 |
| Operations concept | The operations concept is developed early in the life cycle (study phase) and kept updated. This concept represents a common understanding between all project people and the sponsor(s). Although the mission operations manager is key in generating the operations concept, it's also important to follow the concept every day. | Chap. 4 |
| Sponsor's goals and vision | You must understand what your sponsor is expecting. How does he or she view schedule versus cost versus product quality? Can this project stand alone, or is it an enabling project for a new program? How does the sponsor view operational risk? The MOM's decisions and actions need to complement and enforce the sponsor's goals and vision. | Chap. 1 and 4 |
| Status versus plans | After launch, managers must make operational decisions when deviations from plans or expectations occur. Understanding the current status versus the current plan is essential to making timely decisions. | Sec. 3.1 |

**Table 3.40. How to Manage Mission Operations.** These steps are typical of those mission operations managers do. Each mission will emphasize certain steps. Shading indicates post-launch steps.

| Steps | Comments | Where Discussed |
|---|---|---|
| 1. Define and Develop Operations Organization | • Generate and update the operations concept<br>• Define an operations organization that minimizes interactions between groups and is process oriented<br>• Define clearly the responsibilities for each position and make sure all members understand the operations organization<br>• Monitor the organization's development as it matures<br>• Make changes to improve the efficiency of operations | Chap. 4 |
| 2. Manage Interfaces | • Define each of the interfaces for the development organization<br>• Define due dates for the receivables and deliverables that flow across the interface for each area<br>• Manage the performance of the development areas by updating the receivables and deliverables schedule weekly or monthly, as appropriate | Chap. 4 |

**Table 3.40. How to Manage Mission Operations. (Continued)** These steps are typical of those mission operations managers do. Each mission will emphasize certain steps. Shading indicates post-launch steps.

| Steps | Comments | Where Discussed |
|---|---|---|
| 3. Manage Change Control and Program Control | • Standard management texts describe these processes<br>• Make sure the ground system evaluates the effects of changes to the spacecraft bus before you consider them for approval | -- |
| 4. Administer Contract Procurement | • Standard management texts describe these processes | -- |
| 5. Manage Contingency Reserve | • Before launch, include performance reserve for tools and processes, in addition to typical reserves for resource development<br>• After launch, manage contingencies to ensure that added requirements fit within timelines and capabilities | Sec. 3.1 |
| 6. Manage Resources of the Mission Operations System (MOS) | • Manage these resources to meet mission objectives within the budget<br>• Manage the MOS to provide a satisfactory return for a fixed or agreed-on cost. Five years ago, we managed the MOS to return the very best payload data. | Chap. 4 |
| 7. Manage Operations Schedule | • Manage daily activities to use minimum staff while maintaining a proper and acceptable set of checks and balances<br>• Ensure review and approval of planned command activities are at the lowest appropriate level and contribute value to the planned activity | Chap. 7 |
| 8. Manage Miscellaneous Activities | • Approve ad-hoc analysis and allocate resources to these activities<br>• Manage development and maintenance activities needed during operations<br>• Develop processes for placing decisions into a risk matrix to make sure the risk isn't too high compared to the effect on cost or performance | -- |

### 3.13.3    Products of Mission Operations Management

The products of management aren't documents like those from many of the first nine functions. The management products listed in Table 3.41 keep mission operations focused on conducting the mission. Often, we can't analyze a problem for weeks or months before reaching a decision. We may have to decide in hours or days how to keep mission operations on track, despite problems or surprises.

### 3.13.4    Key Considerations

The mission operations manager must understand all aspects of the mission operations well enough to make effective decisions. For example automation and

Table 3.41. **Products of Mission Operations Management.** The operations manager's most important products are timely decisions and approvals.

| Products | Comments |
|---|---|
| Project policies and guidelines | • The management team, or the MOM, issues policies and guidelines during the mission<br>• These statements encompass all aspects of the mission, but the focus is on guidelines that respond to changes in planned capabilities or performance |
| Decisions and approvals | • Decisions and approvals cover standard and non-standard aspects of the mission<br>• Making timely decisions about problems or situations that arise during the mission is extremely important<br>• Make sure an approval adds value to the product or recommended decision. The lower the level of approval, always considering risks to the project, the more efficient the organization. |
| Reports and status to customers and sponsors | • The MOM reports upward and outward<br>• Sponsors and customers shouldn't be surprised by hearing about problems or successes from third parties<br>• Make sure reports are meaningful and useful |
| Goals and visions of the project | • The mission operations manager's goals and vision are just as important to mission operations as the sponsor's goals and vision are to the mission operations concept |

autonomy can save enormous costs during operation, so managers must understand how to use them.

The operations staff at nearly all facilities and programs are being multi-tasked. Specialized operational positions remain on only the most complex missions. A flat organization allows multi-tasking to take place more naturally while also saving money.

Once you've established the first conceptual organization, identify the organizational interfaces. What are the inputs and outputs of each organizational element and where do they begin or go to? If the diagram looks like a spider web, you don't have an efficient organization.

## References

CSP Associates, Inc. August 19, 1993. Mission Operations Cost Study Results of Phase I. Prepared for the Flight Projects Office, Jet Propulsion Laboratory, Pasadena, CA.

National Research Council Committee on Data Management and Computation, Space Sciences Board, Commission on Physical Sciences, Mathematics, and Resources. 1986. *Issues and Recommendations Associated with Distributed Computation and Data Management Systems for Space Sciences.* National Academy Press, Washington, D.C.

Chapter 4

# Developing a Mission Operations Concept

Gael F. Squibb, *Jet Propulsion Laboratory,*
*California Institute of Technology*

4.1   Process for Developing a Mission
        Operations Concept
4.2   Steps for Developing the Mission
        Operations Concept

A *mission operations concept* describes—in the operators' and users' terms—the operational attributes of the mission's flight and ground elements. It results from the cooperative work of several disciplines. Its development is similar to that of a space mission concept, as discussed in Chap. 1, but the mission operations concept is more detailed and emphasizes the way we will **operate** the mission and **use** the space element (operational characteristics). It is generated in phases and becomes more detailed as our design of the operations system progresses.

The mission operations concept follows from, and is consistent with, the mission concept. It's the most important deliverable from the mission operations manager (MOM) before launch and is key to keeping system costs down. It's also the main way a MOM influences the design and operability of the mission and spacecraft. It often results in changes to the mission concept to reduce life-cycle costs. The MOM must understand what a mission operations concept is, how to develop one, and what it contains.

We generate a mission operations concept in ever more detailed layers. It doesn't detail a mission's development or operational costs, but it's primary to cost estimation. By combining the operations concept with assessments of operational complexity (Chap. 5), we can determine the probable costs of operations early in mission design.

## 4.1   Process for Developing a Mission Operations Concept

Figures 4.1 and 4.2 show the process for developing a mission operations concept, along with the information it requires and its products. Table 4.2 shows the process steps and states where we discuss these steps further.

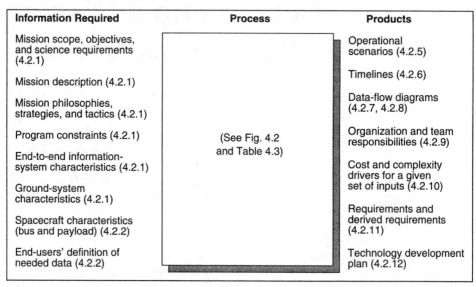

| Information Required | Process | Products |
| --- | --- | --- |
| Mission scope, objectives, and science requirements (4.2.1) | | Operational scenarios (4.2.5) |
| Mission description (4.2.1) | | Timelines (4.2.6) |
| Mission philosophies, strategies, and tactics (4.2.1) | | Data-flow diagrams (4.2.7, 4.2.8) |
| Program constraints (4.2.1) | (See Fig. 4.2 and Table 4.3) | Organization and team responsibilities (4.2.9) |
| End-to-end information-system characteristics (4.2.1) | | Cost and complexity drivers for a given set of inputs (4.2.10) |
| Ground-system characteristics (4.2.1) | | |
| Spacecraft characteristics (bus and payload) (4.2.2) | | Requirements and derived requirements (4.2.11) |
| End-users' definition of needed data (4.2.2) | | Technology development plan (4.2.12) |

**Fig. 4.1.   Developing a Mission Operations Concept.** We describe the mission operations concept using a standard set of products.

Developing a mission operations concept requires different disciplines to communicate with each other. These disciplines include

- Designers of the mission, spacecraft, payload, and ground system
- Operators of the ground system
- Users (those who receive data from mission operations)

A space vehicle often operates differently from what its designers had in mind. This early communication between designers, operators, and users shortens the development time because fewer changes are required during development or testing.

The process emphasizes areas in these disciplines for which we should study trades to minimize life-cycle costs and get better information from the mission. The operations concept, when documented, provides derived requirements for developing functions of the mission operations system. Because the mission operations

PRODUCTS

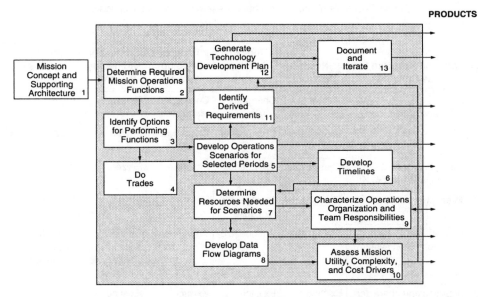

**Fig. 4.2.    Process for Developing a Mission Operations Concept.** The process is iterative and the products contain increasing levels of detail as we move from the study phase to the development phase. This process represents steps 6–8 in the process for developing a space mission concept, as described in Chap. 2.

concept responds to top-level mission requirements, we can easily trace these derived requirements to top levels while designing the mission. Each mission concept should have a corresponding mission operations concept. If we develop the two concepts in parallel, we can save time and shorten a project's conceptual and definition phases. That's because the mission operations concept can quickly feed back operational cost drivers into the mission-concept development.

To influence development, we need to complete the mission operations concept early, when the interfaces and attributes of the inputs are less defined and trade-offs are possible.

Completing the mission operations concept early often forces us to resolve design incompatibilities which operations developments or procedures would otherwise have to solve or minimize. Table 4.1 shows how required information solidifies during the project phases. Usually, by the time production and deployment begin, only the ground system, people and procedures, or flight software can change.

Because the mission is continually changing and maturing, updating the mission operations concept during the project is also important.

Chapter 3 describes the 13 mission operations functions. Eventually, people representing these functions need to help develop a mission operations concept.

**Table 4.1. Changes to Information Required for the Operations Concept Versus Mission Phase.** This table shows when inputs freeze within the project phases.

| Information Required for Operations Concept | Project Phase | | | | |
|---|---|---|---|---|---|
| | Needs Analysis | Concept Development | Detailed Development | Production and Deployment | Operations and Support |
| Mission scope, objectives, and science requirements | Changeable | Changeable | Changeable for cause | Frozen | Frozen |
| Mission description | Changeable | Changeable | Changeable for cause | Frozen | Frozen |
| Mission philosophies, strategies, and tactics | Changeable | Changeable | Frozen | Frozen | Frozen |
| Program constraints | Changeable | Changeable for cause | Frozen | Frozen | Frozen |
| Characteristics of the end-to-end information system | Changeable | Changeable | Changeable | Existing frozen; New changeable for cause | Existing frozen; New changeable for cause |
| Capabilities and characteristics of the ground system | Existing frozen; New changeable | Existing frozen; New changeable | Existing frozen; New changeable | Existing frozen; New changeable for cause | Existing frozen; New changeable for cause |
| Spacecraft capabilities and characteristics (bus and payload) | Changeable | Changeable | Changeable for cause | Frozen | Frozen |
| Flight software | Changeable | Changeable | Changeable | Changeable for cause | Existing frozen; New changeable for cause |
| End-users' definition of needed data | Changeable | Changeable | Changeable | Frozen | Existing frozen; New changeable for cause |

Often, in the early phases (conceptual), one person will represent several functions. One representative will usually lead discussions and keep the concept. This person often comes from the system engineering or mission planning functions and typically has experience working with information systems.

Because the mission and mission operations concepts are closely linked (see Sec. 4.2.5), some people participate in both activities. Table 4.2 relates the mission operations functions described in Chap. 3 to the main engineering discipline involved in developing them and to the inputs to and outputs from the mission operations concept. This table follows the order of putting products into the mission operations concept, as shown in Fig. 4.1.

**Table 4.2.** **Providers of Information Required to Develop a Mission Operations Concept.** Functions of the Mission Operations System (MOS) are listed in the order they are needed. Products include use and description of new or existing capabilities, timelines, and requirements and derived requirements for each function.

| MOS Function | Discipline | Information Provided by MOS Function | Where Discussed |
|---|---|---|---|
| Mission planning | Mission-planning engineer | • Mission scope, objectives, and payload requirements<br>• Mission description<br>• Mission philosophies, strategies, and tactics | Sec. 3.1 |
| Managing mission operations | Project manager | • Program constraints and requirements<br>• Sponsor's goals and vision | Sec. 3.13 |
| Systems engineering, integration, and test | End-to-end information systems (EEIS) engineer<br>Ground systems (GS) engineer | • EEIS characteristics<br>• Characteristics of GS | Chap. 6, 11, and 12<br>Sec. 3.10 |
| Payload planning and analysis | Payload designer<br>Payload-operations engineer | • Payload characteristics<br>• Payload operations | Chap. 8 and 14<br>Sec. 3.7 |
| Payload data processing | End-users (recipients of the data products) | • Data-product definition<br>• Payload operations | Chap. 8 and 13<br>Sec. 3.8 |
| Spacecraft planning and analysis | Spacecraft designer<br>Spacecraft operations engineer | • Spacecraft characteristics<br>• Spacecraft operations | Chap. 8 and 15<br>Sec. 3.6 |
| Activity planning and development | Sequence designer and operations engineer | • Mission plan<br>• Mission rules<br>• Test and integration requirements | Chap. 7 and 8<br>Sec. 3.2 |
| Mission control | Mission-control designer and operations engineer | • Mission plan<br>• Mission rules | Chap. 8<br>Sec. 3.3 |
| Data transport and delivery | Data-transport designer and operations engineer | • GS capabilities<br>• User requirements<br>• Data format and volume | Chap. 11 and 12<br>Sec. 3.4 |
| Navigation planning and analysis | Navigation designer and navigation operations engineer | • GS capabilities<br>• Requirements for trajectory and attitude data | Chap. 8 and 10<br>Sec. 3.5 |

**Table 4.2.** **Providers of Information Required to Develop a Mission Operations Concept. (Continued)** Functions of the Mission Operations System (MOS) are listed in the order they are needed. Products include use and description of new or existing capabilities, timelines, and requirements and derived requirements for each function.

| MOS Function | Discipline | Information Provided by MOS Function | Where Discussed |
|---|---|---|---|
| Archiving and maintaining the mission database | Archive designer and operations engineer | • GS capabilities<br>• User requirements<br>• Data format and volume | Chap. 8 and 13<br>Sec. 3.9 |
| Computers and communications support | Computers and communications engineer | • Data-flow diagrams<br>• Computer, network, and voice requirements<br>• GS capabilities | Chap. 11<br>Sec. 3.11 |
| Developing and maintaining software | Software-development engineer | • Software requirements<br>• Existing software capabilities<br>• Change control authority | Chap. 6<br>Sec. 3.12 |

## 4.2  Steps for Developing the Mission Operations Concept

Figure 4.2 diagrammed how to generate a mission operations concept; Table 4.3 lists the major steps. We describe the process in detail below.

**Table 4.3.** **Developing a Mission Operations Concept.** This table expands Fig. 4.2 by identifying key items and pointing to chapters with additional information.

| Step | Key Items | Where Discussed |
|---|---|---|
| 1. Identify the mission concept and supporting architecture; gather information | • Characteristics of information system<br>• Characteristics of payload<br>• Characteristics of spacecraft bus<br>• Definition of data product<br>• Ground system | Chap. 1, and 2 |
| 2. Determine functions needed for mission operations | Key functions usually vary for each mission concept and related architecture. You can combine or eliminate some functions | Chap. 3 |
| 3. Identify ways to accomplish functions and whether capability exists or must be developed | • Where accomplished (space or ground)<br>• Degree of automation on ground<br>• Degree of autonomy on spacecraft<br>• Software reuse (space and ground) | Chap. 1, 2, and 3 |
| 4. Do trades for items identified in Step 3 | Options are often selected before developing operational scenarios. These trades are done within the operations element which includes the flight software. | Chaps. 1, 2, 3, 4, and 5<br>Table 3.1 |

**Table 4.3.     Developing a Mission Operations Concept. (Continued)** This table expands Fig. 4.2 by identifying key items and pointing to chapters with additional information.

| Step | Key Items | Where Discussed |
|---|---|---|
| 5. Develop operational scenarios for the functions determined in Step 2 and the options selected in Step 4. | *Operations scenarios* are step-by-step descriptions of how to do integrated activities. Identify key issues and drivers on the operations system. | Chap. 4, 6–8 |
| 6. Develop timelines for each scenario | Timelines identify events and how fast and when they occur. They drive the performance parameters for each mission operations function. | Chap. 4, 6–8 |
| 7. Determine the type of resources needed to perform each step of each scenario | The allocation to hardware, software, or people is based on what, how quickly, and for how long steps must be done | Chap. 4 |
| 8. Develop data-flow diagrams | Data-flow diagrams form the basis for the ground- and flight-data systems and the command, control, and communications architecture | Chap. 4 and 13 |
| 9. Characterize responsibilities of the organization and team | Identify organizations involved and their structure, product responsibility, interfaces and number of people. For cost-effective operations, minimize the number of organizations and interfaces. | Chap. 4 and 5 |
| 10. Assess mission utility and complexity, as well as the cost of mission operations | The cost estimates include both development and operations costs. Refine them each time the mission operations concept is updated. | Chap. 5; Chap. 3 and 20 [Larson and Wertz, 1992] |
| 11. Identify derived requirements and cost and complexity drivers; negotiate changes to mission concept if necessary | Document derived requirements and ensure consistency with top-level requirements | Chap. 1, 2, and 4 |
| 12. Generate technology development plan | The technology to support the mission concept may or may not exist | Chap. 6 Sec. 4.12 |
| 13. Iterate and document | Iteration may occur at each step | -- |

## 4.2.1     Identify the Mission Concept and Supporting Architecture

We begin developing the mission operations concept by examining the mission concept and supporting architecture. By obtaining the information listed below, or by making assumptions, we can describe the mission in users' and operators' terms. When information is not available—the design is not yet started or not specified in the mission architecture—the operations-concept team assumes information and gives this input to the person responsible for it. The operations concept is then valid until the assumed information changes. We can determine the cost of this change by modifying the operations concept as required and then re-evaluating the complexity and cost of the mission operations using the model described in Chap. 5.

**Mission Scope, Objectives, and Payload Requirements.** We must state mission objectives in terms of data the payload can get through effective operations. These objectives capture what the spacecraft must do to achieve the mission's scientific, commercial, or defense goals. This input defines and describes the users of the payload data and their level of sophistication. Our system will be different for a scientist who can use state of the art computer tools for analysis versus a politician who wants to access information from the Earth Observation System. We need to know how people will use the payload's processed data. Potential uses are research, commercial, defense, education, or public information. We also need to know the timeliness requirements for the payload's processed data and the success criteria (percentage of total data actually returned).

**Mission Description.** This input contains information about the trajectory, such as launch date(s) and window, trajectory profile, maneuver profile needed to meet mission objectives, mission phases, and a description of activities required during each phase. Observation strategies describe how we'll gather the mission data. We either define and finalize the observation strategies and mission description before launch or adapt them to data gathered during the mission.

**Mission Philosophies, Strategies, and Tactics.** These items are rules not associated with the health and safety of the mission. They may relate to the mission objectives or to the background of the mission designers and those developing the mission concept. It's important to try to determine if the item is associated with the mission or a person's preferences.

Examples are

- Maximize real-time contact and commanding versus maximize onboard autonomy and data-storage capability
- Maximize the involvement of the educational institutions and teach science students key aspects of issues like operations or space physics
- Make sure a central authority approves all commands
- Limit the image budget to 50,000 images

**Program Constraints.** The sponsor of the mission and the project manager impose these non-technical constraints. Operators must follow them until the project manager is convinced they are increasing the cost of the mission. Examples of program constraints are

- Limit mission cost and cost profiles
- Use a specific tracking network
- Use existing flight hardware

- Use existing ground-system capabilities and design the spacecraft to be compatible with them
- Centralize or distribute operational teams
- Use multi-mission versus project-dedicated teams
- Involve students versus dedicated professionals
- Involve educators and the educational community

**Capabilities and Characteristics of the End-to-End Information System (EEIS).** The EEIS engineer will have helped develop the mission concept, which will describe the spacecraft and ground elements in terms of information flow and processes. System-level requirements include

- Using information standards (layered protocols or CCSDS standards)
- Locating capabilities and processes (includes both space and ground)
- Characterizing the information system's inputs and outputs

The EEIS engineer specifies these requirements in terms operators and users can understand, not in terms used by computer scientists and programmers.

**Capabilities and Characteristics of the Ground Element.** Most missions are designed around a specific agency's ground system, such as the AFSCN, NASA, or European Space Agency. Each ground system has standard services which will support the mission at low (or even no) cost if the mission meets specified interfaces and standards. We follow these requirements on the flight system until they keep us from meeting mission objectives. We must ask

- What standards does the ground system require the spacecraft to meet?
- How do we carry out the standard and what requirements does this place on the spacecraft?

Some standards, such as CCSDS, are broad and allow for interpretation. Thus, we can meet CCSDS standards for telemetry and commanding but choose a way of meeting them that our ground system won't support.

### 4.2.2 Determine Functions Needed for Mission Operations

The mission concept drives top-level functions, but the spacecraft's capabilities determine the detailed ones we must carry out. A completely autonomous spacecraft requires few operations, whereas a spacecraft payload that can't compute or store data onboard requires continuous ground control or ground automation. Thus, to determine what we must do, we have to understand the spacecraft's characteristics. The mission concept may state these characteristics if the mission concept team includes operators. Otherwise, we'll develop them as part of the mission operations concept. Through discussions, the operators and developers define the characteristics described below.

**Capabilities and Characteristics of the Payload.** Including payload designers and mission planners in the mission-concept process leads to timely definition of the payload characteristics. People working on the concept need to develop information that answers the types of questions we've asked below. These questions come from past missions and often start payload designers thinking early about how the payload will operate and the differences between testing it on the ground and operating it in space. (See Chap. 14 for more information.)

To understand how a payload will operate, we must describe what the payload will do during an observation period.

- What are the payload's attributes?
- What is the commanding philosophy? Is the payload commanded by using
  - Buffers?
  - Macro commands?
  - Tables?
- Does the payload use default values?
- Will some commands degrade or damage the payload?
- Do some mechanisms depend on previous commands?
- Does the payload use position commands or incremental commands to control rotating or stepping mechanisms?
- What classes of commands does the payload use?
  - Real-time?
  - Stored program?
- What processing occurs within the instrument?
- How can we describe the payload in terms of
  - CPU/Memory?
  - Closed-loop functions?
  - Predictive commanding versus event-driven commanding?
  - Requirements placed on the spacecraft bus for instrument control?
  - Mechanical power and thermal attributes?
  - Avoidance areas (Sun, Earth, South Atlantic Anomaly, Venus, or Moon)?
  - Requirements for controlling the spacecraft bus?
- Are the payload apertures larger than the spacecraft bus's pointing control?
- What are the user-specified parameters for observation?
  - How are these converted into instrument commands?
- What is the instrument heritage?

- What ground processing and analysis is required to support the instrument's operation?

The above questions may seem obvious, but they come from missions in which the spacecraft had been designed and, in some cases, built before anyone asked them.

The example below shows how considering the payload's operation will help us define the mission concept.

---

An instrument is designed with an aperture that has a field of view of ten arc-seconds. At the same time, the spacecraft designers have designed a spacecraft bus that has a control authority (the ability to point with a certain accuracy) of 20 arc-seconds. The operators can never be sure the celestial object they command the spacecraft bus to observe will capture the object in the instrument's field of view. The solution was to

- Command the spacecraft to the desired position
- Observe with a different instrument that has a wider field of view and see how far the spacecraft is off from the desired position
- Generate attitude commands to move the spacecraft just a bit (tweak commands) until the actual attitude corresponds to the desired attitude
- Verify the attitude errors are gone
- Select the instrument with the narrow aperture and observe as specified

---

This one design issue caused real-time operations, decision making, and commanding to be routinely involved for this mission. This involvement required more ground software, controllers trained in routinely commanding the attitude-control system, and more people whenever the instrument was used. The goal of generating a mission operations concept is to identify incompatibilities and cost drivers early, before designing and building any hardware.

It's a good idea to ask the designer of a payload instrument how you'll go from an observer's requirement to a set of commands for the instrument. Sometimes, the answer is simple, but in other cases instruments have been designed for a laboratory rather than for space and the process is complex. Let's look at another example

---

An astronomer knows the celestial object we need to observe and its estimated brightness within the wavelength we'll look at. A spectrometer is the specified instrument. What must the operations staff do to generate the commands for this observation?

- Make sure the previous observation has an intensity lower than the one we'll observe
- Add a calibration before the observation

- If the observation is longer than five minutes, add short, internal, calibrations every five minutes until the observation is complete
- Add a calibration after the observation

This instrument design requires us to sort the observations and do them in order of increasing intensity, even though it may not be the most efficient way to move the telescope from one object to another.

For this design, the astronomer's requested observation time drives how long we must look at the object, which in turn determines the total number of commands our spacecraft requires. Also, until we've established the instrument's characteristics in space, we can't set the calibrations and durations. So we'll probably need to recalculate after launch.

**Capabilities and Characteristics of the Spacecraft Bus.** As is true for the payload's capabilities and characteristics, timely definition of the spacecraft's characteristics depends on including spacecraft designers and mission planners in the mission concept process. (See Chap. 15 for more information.) Again, people working on the concept have to answer the following types of questions for the overall spacecraft and its subsystems:

- What are the spacecraft's operational attributes?
  - What commands and parameters are sent from the ground?
  - How are the values of these commands determined?
  - How many commandable states are there?
  - Are engineering calibrations required? What are the purpose, frequency, and schedule constraints of the calibrations?
- How many engineering channels need monitoring?
- Do these channels provide subsystem-level information to the operators, or must operators derive information about subsystems?
- For the attitude-control system
  - Are guide stars used? If so, how are they selected?
  - How does the pointing-control accuracy compare to instrument requirements?
- What types of margins exist and what must be monitored and controlled?
- What expendables need monitoring during flight?
- Does the spacecraft subsystem use any onboard, closed-loop functions?
- What are the attributes of the spacecraft's data system?
- What processing must we do on the ground to support spacecraft operations?

- What is the heritage for each of the spacecraft subsystems?

Consider an example of how a design decision affects operations.

---

The Galileo spacecraft was designed to take heat from the Radioisotopic Thermoelectric Generators (RTGs) and use it to warm the propulsion system. This design saved weight and power and cost less to develop. But the spacecraft's operational characteristics tied together subsystems for propulsion health and safety, thermal transfer, and power. Operators had to check each command load to see how it changed power states and affected the propulsion subsystem. Engineers from power, thermal, and propulsion had to check each sequence, even if only the payload instrument's states changed. For example, turning an instrument off (or on) caused the heat output of the RTG to change.

---

**End-Users' Definition of Needed Data.** End-users inform mission operations by defining how, and how often, they will use data from the payload. By understanding these data products, ground-system engineers can start designing how they'll get them.

We must understand how confident the end-users are about the products. Often, they don't know what they want until they see how the instrument works in flight and what it observes. In these cases, engineers can prototype processes before launch and finish them after launch, once definitions are complete.

We also need to define the products' relationship to the payload data by answering the following questions:

- Is the product based on the payload's raw data or must we remove the payload instrument's signatures?
- Must the data be calibrated? How? Does it involve processing special calibration observations? At what rate do we expect the calibration files to change?
- Does the data need to be converted into geophysical units? How? Where do the algorithms for this conversion come from? Must the project generate them and update the mission database as they become more refined?
- What are the formats and media of the payload's data products? Is there a community standard, such as the Astrophysical community's use of an Flexible Image Transport System (FITS) format on all of NASA's astrophysics missions?
- What ancillary data must we provide so the end-user can interpret the payload data?
  - Spacecraft position?
  - Spacecraft attitude?

- Ground truth data?
- Who processes the payload data?
  - Project?
  - End-user?
- How does the processed data get into the archives?
  - Through the project?
  - Through the end-user?
- What, if anything, must the project archive after the flight phase is over?

**Generate a List of Top-Level Functions Operators Must Do.** Referring to Chap. 3, we can address each process within the 13 operational functions to generate this list using the information we've developed in Sec. 4.2.1 and 4.2.2. When time constraints or requirements are available, we may add them to our early design of a mission operations system for a given mission concept.

### 4.2.3  Identify Ways to Accomplish Functions and Whether Capability Exists or Must Be Developed

At this level, many steps—and the ways to do them—will be straightforward. For example, to track an interplanetary spacecraft, we use NASA's Deep Space Network. For other steps, we'll have to identify several options and describe them. For example, to determine a spacecraft's orbit, we might use global-positioning satellites and automated procedures onboard, or we might track the spacecraft and calculate its orbit on the ground.

To understand options, try building a table which contains the operations functions that apply to the mission's database and avionics. Then, identify whether the avionics (automated) or the ground system will do each function. If on the ground, further determine whether the hardware, software, or operators will complete it. If a check goes in more than one place, describe which functions are done in each place and whether options exist. Table 4.4 shows how such a table would look.

If the ground hardware and software do something, ask, "Could the avionics partially or completely do this function and lower the mission's life-cycle costs?" For example, if you were considering orbit determination, you'd ask, "Could we plan and analyze position location within the avionics?" Then, you'd look at the accuracies of the GPS system, check the cost of GPS receivers that are flight qualified, and do a first-order estimate of the change in life-cycle costs. Don't forget that the costs of tracking facilities are important in this type of trade.

### 4.2.4  Do Trades for Items Identified in Step 3

For the options identified in step 3 that will drive either performance or cost, a small group of engineers needs to do trades and decide how to carry them out.

**Table 4.4.    Identifying Where to Carry Out Functions.** Using a table similar to this one will help the MOM identify options for carrying out mission operations. We assume functions not included in table are done on the ground. As you evaluate each function, place a check mark in the table to indicate where you accomplish the function. The functions listed correspond to the functions in Fig. 3.1 that may be part of the spacecraft avionics.

| MOS Function | Where to Do the Function | | |
| --- | --- | --- | --- |
| | Spacecraft Avionics | Ground Hardware/ Software | Operators |
| Activity planning and development | | | |
| Mission control | | | |
| Data transport and delivery | | | |
| Navigation planning and analysis | | | |
| Spacecraft planning and analysis | | | |
| Payload planning and analysis | | | |
| Payload data processing | | | |

In some cases, engineers may develop an operations scenario for each option to describe it in detail.

### 4.2.5    Develop Operational Scenarios

Operational scenarios are key to an operations concept. A *scenario* is a list of steps and we can describe an operations concept with about a dozen top-level scenarios. Typically, planners generate three types of scenarios during study and design, with each increasing in detail:

- User scenario
- System scenario
- Element scenario

Each of these scenarios has a corresponding timeline and data-flow diagram, as discussed in Sec. 4.2.6 and 4.2.8.

During the early study phases of a mission, users develop a scenario to show how they want to acquire data and receive products from the payload. For a NASA mission, the user would be the principal investigator or science group or, for a facility spacecraft such as the Hubble Space Telescope, an individual observer.

We create a system scenario after we've developed the operations architecture—during the operations-system design. Here, we emphasize the steps within a process needed to conduct the mission. Finally, during element design, we expand these system scenarios to include more detailed elements and subsystems.

The mission concept, mission operations concept, and design of the space and ground elements are closely related. As design proceeds, costs are always in planners' minds. The cost of development and operations are both important, so we must estimate them during study and design. As designs for the space and ground elements mature, we have to do trade studies to get the lowest cost and best design within some cost cap. Whenever we expect to exceed this cap or believe we can reduce costs by changing the mission concept, we study possible mission trades. Figure 4.3 depicts this process.

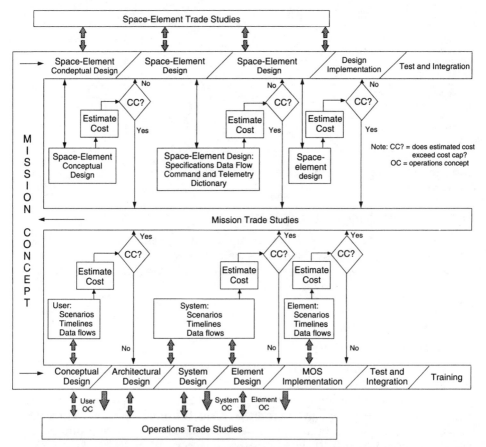

**Fig. 4.3.    Trade Studies for Mission Operations and Space Element.** Designing to cost requires frequent cost estimates and trade studies within mission operations and the space element.

Typical operational scenarios, along with questions to ask or items to include, appear below.

**Develop a scenario for total mission based on the trajectory and the mission goals.** Define the trajectory in terms of activity periods and mission phases. (See Sec. 3.1)

- Launch
- Spacecraft bus checkout
- Payload checkout
- Key payload periods, data gathering, and inactive or quiet periods
- Maneuvers
- End of mission

For each phase specified, describe in words the steps (inputs, processes, and outputs) required to go from the user(s) request for an observation to the payload's acquisition of requested data. Also address how you'll handle review and approval.

The scenarios discussed below are part of the uplink process:

**How Does User(s) Plan the Request?** A typical request from an astronomer could be: Use a particular instrument to observe a celestial object with an estimated flux and located at a given position; obtain a 4×4 mosaic image.

- What parameters are used to describe the request?
- What tools are required to help the user make the request?
- Where is the user?
  - At the control center?
  - At a remote site?
- What are the form and content of the request?
- Is the request part of a mission plan that has constrained the user?
- How does the mission operations system support the user?
- What type of user(s) are involved in the mission?
  - Single user?
  - Multiple users?
- Where and how is the request is delivered?
  - Spacecraft?
  - Activity planning group?
  - Mission control?

For many missions today, users stay at home but submit requests to the control center. Commands are interactive or non-interactive. The non-interactive commands don't need monitoring or checking. The control center can create and send them directly to the spacecraft during the current or next track.

**How Are the Activity Requests Integrated?** Integration resolves conflicts among users, tracking facilities, and spacecraft resources. Fewer requirements for integration mean lower costs plus shorter and more adaptable sequences.

- Who is requesting the activity?
- What are the relative priorities for these requesters?
- Where do inputs come from?
  - User(s)?
  - Mission control?
  - Spacecraft analysts?
  - Payload analysts?
- What are the flight rules?

The spacecraft and payload designers should list flight rules they expect activity planners and developers to validate before sending commands to the spacecraft.

**How Are the Integrated Requests Converted to Command Mnemonics and Binary Files?** We may need to create one command for each activity, or we may need to expand blocks into many commands. We also must describe how to generate and maintain the command dictionary that will control all conversions.

**How Are the Binary Files Wrapped with Ground and Spacecraft Protocols and Transmitted to the Spacecraft?** Getting messages from the ground to the spacecraft will be more difficult if the ground system uses a standard that the project hasn't imposed on the spacecraft.

- Who is responsible and what steps are required to get the commands into the spacecraft?
- What steps will get the commands from the control center to the tracking station?
- What ground resources are used for this transmission?
- What are the uplink characteristics?
  - Frequencies?
  - Command rates?
- What are the allowable bit and expected-error rates?

We have now transitioned into activities that occur onboard the spacecraft:

**How Does the Spacecraft Receive and Verify the Command Files and Store or Execute the Commands?**

- What are the interfaces and processes between the spacecraft's avionics system and the payload's data system for various commands?
  - Immediate execution?

- – Stored commands (within the spacecraft)?
- – Commands stored by the payload instruments?
- How are the commands verified?
  - – Command receipt?
  - – Command execution?
- Is the verification real time or delayed?
- Who supplies the information which will enable the commands to be verified?

The following scenarios are for downlink—relaying data from the payload to the recipient:

### What Are the Processing Functions and Interfaces From Payload to User?

- What happens to the data as it goes from the payload instrument to spacecraft storage?
  - – Is the data compressed?
  - – What are the payload output rates?
  - – How does the spacecraft collect the payload data? Synchronous or asynchronous?
  - – What type of storage is used? (Solid-state memory is replacing tape recorders on most missions.)
  - – What are the storage and retrieval characteristics? Is the data retrieved by location or by a request for a file?

Today most data is retrieved by location or first-in, first-out (FIFO) from an area of storage. The next generation of missions will store and retrieve data by files, just as we do today on our personal computers.

- What happens to the data as it goes from the storage area to the spacecraft transmitter?
  - – Is the data encoded to improve signal-to-noise characteristics of the link? If so, what type of encoding?
  - – Is the data encrypted for security?
  - – What transmit rates are available on the spacecraft?
  - – Can stored data be transmitted at the same time as real-time data?
- How do we transmit data to storage in the ground database (level 0 processing)?
  - – Must the spacecraft or spacecraft antenna be pointed prior to transmission? How?
  - – What transmitter and receiver characteristics determine the link performance?

- With these characteristics plus the encoding, what are the error rates and allowable data-transmission rates?
- What are the transmission rates for the various mission phases?
- What are the processing and storage steps from receiving data at the ground station to storing this data in a project database?

We have now transitioned from the spacecraft to the ground system.

- What steps are necessary from receiving the data in the project database to having the content ready for delivery to the end user? We must specify processing volume and speed to properly size the system—usually an extensive section that's different for each mission.

### How Is the Data Formatted, Analyzed, and Delivered to Meet the User's Requirements?

- What data must we analyze in real time for
  - Health and safety?
  - Payload quality?
- What data must we analyze later to
  - Control the mission?
  - Determine quality and content of payload data?

### How Is the Data Archived? What processing must we do to meet the requirements for

- Format?
- Quantity, quality, and continuity?
- Frequency of updating the archive?

### How Does the Processed Data Change the Mission Plan or Activity Plan? (Closing the Loop)

- How is the processed data compared to the mission plan?
- What are the criteria for changing the mission plan?
- How do we change the plan changed?
- How do we deliver the modified plan to the operations-system element for implementation?

### How Does the Payload-Calibration Plan get Generated, Verified, and Modified?

- What is the payload-calibration plan?
- How are descriptions of calibration activities generated? Is specialized software necessary? Does the payload instrument have special modes only for calibration?

- How are the calibration observations analyzed? By whom? How fast?
- How are new observations generated?

**What Are the Scenarios for Top-Level Contingencies?** These scenarios address what happens when spacecraft-bus or payload anomalies occur. We must assign categories of anomalies, primary responsibility, (ground or space element), and required response times.

Other scenarios may be helpful. Use your imagination and knowledge of your mission to develop them.

### 4.2.6    Develop Timelines for Each Scenario

Now you can add times needed to do each set of steps and determine which steps can be run in parallel or must be serial. This information becomes a source of derived requirements for the mission operations system's performance.

Various agencies, universities, and companies own timeline tools. There are no standard tools, but many are modified from commercial, off-the-shelf software. Most missions use the same timeline tools for operational scenarios and activity planning.

### 4.2.7    Determine the Types of Resources (Hardware, Software, and People) Needed to Perform Each Step of Each Scenario

Once you've developed scenarios you may assign machines or people to do each step. This choice will be obvious for many steps, but others may be done by people or machines, depending on performance requirements and available technology.

Having allocated resources, turn steps assigned to hardware and software into data-flow diagrams. For steps assigned to people, develop an operational organization and assign steps and functions to teams.

At this point, you should examine each step to which you've assigned an operator and ask, "Can this process be automated to eliminate the operator? How?" Allow a person to do something only when life-cycle costs mandate human involvement. Don't accept the idea that we need a person because we've always done it that way. Technology is advancing so rapidly that a machine may very well do now what a person had to do on the last mission.

Through this process, you'll also discover areas on which to focus R&D funding for possible automation in future missions. This concept of "justified operation" was developed at a NASA workshop held in January, 1995. [NASA, 1995]

### 4.2.8    Diagram Data Flow

System-engineering tools can convert your machine steps into data-flow diagrams showing processes, points for data storage, and interrelationships.These tools also generate a data dictionary which ensures a unique name for each process or storage point in the data flow. These computer-aided systems engineering (CASE) tools then generate information you can use for development.

### 4.2.9    Characterize Responsibilities of the Organization and Team

Once you've defined processes, gather the people steps and form an organization around them. Assign teams to the steps and analyze the organization to establish operational interfaces. Generally, the more inputs required from different teams, the more complicated, costly, and slow the operations organizations will be.

### 4.2.10    Assess Mission Utility, Complexity, and Cost Drivers of Mission Operations

Chapter 5 describes methods and concepts for identifying mission complexity and cost drivers. Development and operations (post-launch) costs for a given set of inputs are key aspects of an operations concept. Chapter 3 describes the top-level cost drivers for the 13 operations-system functions.

### 4.2.11    Identify Derived Requirements

Relate the scenarios we've described to the top-level requirements. You can then use the steps within a scenario as the source of derived requirements. The operations concept is a good place to keep and document the relationship between the top-level requirements and the derived requirements related to these scenarios.

### 4.2.12    Generate a Technology Development Plan if Appropriate

The technology to support a mission concept may not exist or may not be focused and prototyped to a level that is appropriate for the mission approval. Identifying the technology needed to support a mission operations concept, along with the schedule and needed funding, is an important output of generating the mission operations concept.

### 4.2.13    Iterate and Document

Document results of your effort to develop an operational concept so others can benefit from your information. Following is a suggested structure for this document. While developing the concept, keep this information in electronic form and available to all members of the project, so people can review and critique it.

- Inputs
- Scenarios
  - Describe operational scenarios—those listed in this chapter or more detailed ones based on them
- Timelines
- People and procedure functions
  - Organization and team responsibilities
- Hardware and software functions
  - Data-flow diagrams
- Requirements and derived requirements

The mission operations concept is the operations manager's most important product before launch. It enables the manager to discuss changes to the mission concept based on quantitative data. The earlier the first mission operations concept appears, the greater the leverage for minimizing life-cycle costs. It's important to keep the mission operations concept current because it's the best top-level description of how the mission will be flown and the tools we need to fly it.

### References

Future NASA Miniature Spacecraft Technology Workshop. Feb. 8–10, 1995. Pasadena, CA.

# Chapter 5

# Assessing Operations Complexity

John Carraway, *Jet Propulsion Laboratory,*
*California Institute of Technology*

Trying to understand operations costs based on measurements of operations complexity is of interest to a mission operations manager (MOM) who wants to see how design requirements from designs for the spacecraft, payload, mission, and ground system drive operations costs. If you also want to lower costs, we list 95 complexity parameters you can negotiate with operations users (sources of operations requirements) to do so. Finally, we present a model for converting operations complexity to predicted operations costs.

If we understand operations complexity we can see why operations costs are so much higher for some missions than for others. During the project-requirements phase, understanding operations complexity helps us define and negotiate requirements that are compatible with low costs. Concurrent design can be effective if project engineers can show how their requirements affect operations complexity and costs. During design, assessing operations complexity gives us rules for trade studies to reduce operations costs with less expensive designs for the spacecraft, instruments, or mission. Even after launch, operations complexity metrics suggest changes in designs for the mission, flight software, and ground system.

Quantifying metrics of operations complexity establishes numbers for comparing costs between missions, helps identify areas for trades to reduce operations costs, and gives us a numerical model that predicts the operations costs of a mission.

This chapter presents 95 metrics for operational complexity, organized into four categories:

1. Mission design and planning
2. Flight system
3. Risk avoidance
4. Ground system

These four categories correspond to the four main sources of requirements on operations. We want to relate operations costs to requirements placed on operations by the mission plan, the spacecraft and instrument design, the project's policies on operational risk, and the ground system. We've placed the metrics into high, medium, and low ranges so we can do early estimates while the project design is still flexible enough to lower operational costs.

We present a cost model at the end of this chapter. It uses these complexity metrics to compute four complexity factors which correspond to the four categories above. These complexity factors allow us to predict the workforces for 14 Jet Propulsion Laboratory missions. For each mission, we show the metrics, computed values for the complexity factors, and predicts for the workforces. Model predicts are accurate within 25% of actuals for 13 of the 14 cases studied. Figure 5.1 is a block diagram of the model.

Sections 5.1 through 5.4 present tables of metrics for operations complexity, interpret them, and suggest design rules for low-cost operations based on them. Section 5.5 presents metric data from characterized missions and prediction results from the cost model. We can use the model to estimate the operational cost of a new mission if the new mission corresponds to one of the four mission types listed in Table 5.5. If it doesn't, we can still use the model to compute relative costs and to determine how changing one or more requirements changes our estimated cost.

## 5.1 Complexity Metrics for Mission Design and Planning

How we design and plan a mission can strongly influence operations costs. Mission design imposes requirements on operations in areas such as levels of mission activity, timeline margins, and planning horizons. It can lower operations costs by choosing orbits or mission events that match an operations work day, or

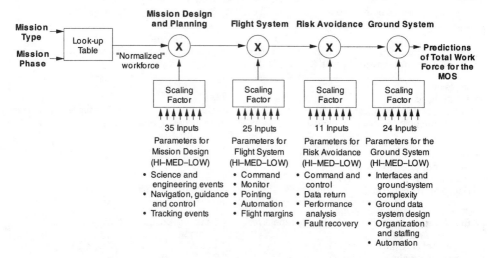

**Fig. 5.1.　Block Diagram of a Cost Model for Mission Operations.** Metrics predict operations complexity for four categories. By comparing predicted complexity with similar missions, we estimate the workforce for our mission.

better yet, a single shift. The metrics in this section suggest ideas like designing timelines with margins, mission events with re-try opportunities, and planning strategies for efficient, flexible operations.

Mission activities that are important to operations costs include engineering and science events for the flight system, guidance and control events, and tracking events. The complexity metrics in this section characterize activity levels, criticality, and planning requirements for these events. Table 5.1 presents 35 questions that provide complexity metrics for mission design and planning.

**Table 5.1.　Complexity of Mission Design and Planning.** Generous margins in the timeline, multiple re-try opportunities, and planning flexibility reduce complexity.

| | Low | Medium | High |
|---|---|---|---|
| **Science and Engineering Events** | | | |
| Frequency<br>• Number per day? | Fewer than 10 | 10 to 20 | More than 20 |
| Criticality<br>• Science re-try opportunities?<br>• Engineering re-try opportunities? | More than 2<br>More than 2 | 1 or 2<br>1 or 2 | None<br>None |

**Table 5.1.    Complexity of Mission Design and Planning. (Continued)** Generous margins in the timeline, multiple re-try opportunities, and planning flexibility reduce complexity.

| | Low | Medium | High |
|---|---|---|---|
| **Science and Engineering Events (Continued)** | | | |
| Complexity<br>• Timing accuracy?<br>• Instrument pointing events?<br>• Spacecraft pointing events? | Hours or more<br>None<br>Fixed | Minutes<br>Platform<br>Articulated antenna, panels | Seconds or less<br>Spacecraft maneuver<br>Spacecraft maneuver |
| Data return<br>• Number of routine downlink data modes?<br>• Real time vs. playback? | 1 or 2<br><br>Playback: Fifo* | 3 to 6<br><br>Playback: Non-fifo | 7 or more<br><br>Real time |
| Science & engineering planning<br>• Event repetitiveness?<br>• Number of consumables and margin constraints routinely planned?<br>• Timeline duty cycle margin?<br>• Plan execution time?<br>• Plan development time?<br>• Response time to late change request? | Highly repetitive<br><br>None<br><br><br><br>More than 50%<br>Less than 1 day<br>Less than execution time<br>More than 50% of development time | Some repeat, some unique<br>1 or 2<br><br><br><br>5 – 50%<br>1 day to 1 week<br>1 to 2 times execution time<br>10 to 50% of development time | Many unique<br><br>3 or more<br><br><br><br>Less than 5%<br>More than 1 week<br>3 times or more than execution time<br>Less than10% of development time |
| **Navigation, Guidance, and Control** | | | |
| Maneuver frequency<br>• Slews per day?<br>• Propulsive maneuvers per quarter? | None<br>1 or less | 1 to 5<br>2 | More than 5<br>3 or more |
| Maneuver criticality<br>• Re-try opportunities? | More than 3 | 1 or 2 | None |
| Maneuver complexity<br>• Timing accuracy?<br>• Maneuver accuracy? | Days<br>1 sigma or less | Hours<br>2 sigma | Minutes or less<br>3 sigma or more |
| Navigation data return<br>• Number of navigation data types?<br>• Navigation data per day? | 1 or 2<br><br>Less than 1 hour | 3 or 4<br><br>1 to 4 hours | 5 or more<br><br>More than 4 hours |
| Maneuver planning<br>• Maneuver design repetitiveness?<br>• Number of consumables and margin constraints routinely planned?<br>• Mission delta V margin? | Highly repetitive<br><br>None<br><br><br><br>More than 50% | Some repeat, some unique<br>1 or 2<br><br><br><br>5 – 50% | Many unique<br><br>3 or more<br><br><br><br>Less than 5% |

**Table 5.1.    Complexity of Mission Design and Planning. (Continued)** Generous margins in the timeline, multiple re-try opportunities, and planning flexibility reduce complexity.

|  | Low | Medium | High |
|---|---|---|---|
| **Navigation, Guidance, and Control (Continued)** | | | |
| Ephemerides<br>• Number of objects needing ephemerides? | Fewer than 3 | 3 to 5 | More than 5 |
| **Tracking Events** | | | |
| Frequency<br>• Number of passes per day?<br>• Hours of link coverage per week? | Fewer than 2<br>Fewer than 8 hours | 2 or 3<br>8 to 56 hours | More than 3<br>More than 56 hours |
| Criticality<br>• Re-try opportunities? | Playback data with replay option | Routine real-time data | Critical real-time data |
| Complexity<br>• Station configuration changes per pass?<br>• Peak data return rate?<br>• Simultaneous, multi-station coordination? | None<br><br>Less than 1 kbps<br>None | 1 or 2<br><br>1 kbps–100 kbps<br>Less than 5% of tracks | More than 2<br><br>More than 100 kbps<br>More than 5% of tracks |
| Tracking facility planning and scheduling<br>• Coverage repetitiveness?<br><br>• Track duration time margin?<br>• Schedule iterations?<br>• Toleration to late losses in scheduled coverage? | Repeating weekly pattern<br>More than 50%<br>1 or 2<br>More than 20% of tracks | Some repeat, some unique<br>5 – 50%<br>3 or 4<br>5 to 20% of tracks | Every day unique<br><br>Less than 5%<br>More than 4<br>Less than 5% of tracks |

\* Fifo = first in, first out. Data stored first is played back first.

Let's define some terms and look at examples that will help us understand and use Table 5.1.

- **Events.** We define a *science or engineering event* to be a complete set of self-contained actions involving the spacecraft, instruments, and (sometimes) ground system. It involves configuring the spacecraft or instruments and collecting the data we expect to collect. A science observation or an engineering calibration are typical examples. Maneuvers and slews are examples of navigation, guidance, and control events; specified, scheduled DSN or TDRSS passes are examples of tracking events.
- **Frequency.** Activity level or number of events per unit time.
- **Criticality.** Fewer opportunities to retry mean higher criticality.

- **Complexity.** Specification accuracy and requirements for pointing, maneuvering, or tracking.
- **Data Return.** How complex is the data-return system?
- **Planning.** Influences on event-planning complexity include the event's repetitiveness, constraints on planning, timeline margin, plan duration, plan-development time, and response time for replanning.

The design rules in Table 5.1 suggest that, to operate at low cost, our mission design must

1. Consider the effect of activity levels on operations staffing
2. Allow multiple chances to retry events
3. Allow margins in the accuracy of event timing
4. Allow simple pointing for the spacecraft and engineering instruments
5. Allow simple data capture and return
6. Use repetitive events as much as possible
7. Minimize the number of design rules, constraints, and parameters that we must plan for and manage as resources
8. Provide margins in the planning timeline
9. Plan over short durations
10. Use comparable amounts of time to develop and carry out the plan
11. Allow late requests for replanning, not by changing the current plan at the last minute, but by incorporating the requests into the next plan

## 5.2 Complexity Metrics for the Flight Systems

Designs for spacecraft and instruments strongly affect operations cost. Designs that result in tight coupling and high interaction between spacecraft subsystems, or between instruments, are more complex and more expensive to operate. Designs also cost more to operate if they have in-flight consumables with low margins that operators must carefully plan for, protect, and manage. Designs with numerous flight rules and constraints complicate operations; reduce operational options and robustness; and require expensive, close to zero-defect, techniques for simulating and validating commands.

We don't want operational margins to go forever unused. Rather, we want operators to allocate them during flight to respond to in-flight surprises and faults, to complete bonus science or mission goals, or to allow more efficient or lower-cost operations. Flight margins are to operations teams as design margins are to design teams. Lack of margins results in limited options, inflexibility, and high costs.

Understanding operational complexity for a spacecraft leads to design rules and recommendations for better operations. These rules make spacecraft and instrument designers more sensitive to operability issues. The problem is that most of the spacecraft attributes recommended to achieve operability cost money. Spacecraft designers can estimate with fair precision how operability adds to mass, development dollars, development schedule, and test time. Before we can decide to spend these resources, we must predict operations cost savings with the same precision. Until we get credible models of how spacecraft attributes affect operations cost, most missions will tend to reject spending more money now for the promise of poorly understood savings in the long term.

Table 5.2 presents 25 questions that provide metrics for the flight system.

**Table 5.2.  Complexity of the Flight-System Design.** The complexity metrics in this table characterize our ability to command, monitor, point, automate, and develop flight margins for the spacecraft and instruments.

|  | Low | Medium | High |
|---|---|---|---|
| **Command** | | | |
| Commandable states? | Fewer than 50 | 50 – 200 | More than 200 |
| Flight rules and constraints? | Fewer than 10 | 10 – 20 | More than 20 |
| History-dependent commandable states? | None | 1 – 2 | More than 2 |
| Onboard tables routinely updated? | Fewer than 10 | 10 – 20 | More than 20 |
| Number of instruments? | Fewer than 3 | 3 – 5 | More than 5 |
| **Monitor** | | | |
| Telemetry channels? | Fewer than 100 | 100 – 1000 | More than 1000 |
| Ambiguous states? | None | 1 – 2 | More than 2 |
| **Pointing** | | | |
| Attitude control? | Gravity gradient | Spinner | 3-axis |
| Pointing accuracy? | More than 0.1° | 0.1° – 0.01° | Less than 0.01° |
| Independent fields-of-view? | Fewer than 3 | 3 or 4 | More than 4 |
| Articulating devices? | None | 1 – 3 | More than 3 |
| Constraints on hazard pointing of instruments? | None | 1 – 2 | More than 2 |
| **Automation** | | | |
| Unattended safing of flight system? | More than 1 week | 2 – 7 days | Less than 2 days |
| Flight-system command states requiring routine ground command for safing? | None | 1 – 10 | More than 10 |

**Table 5.2.    Complexity of the Flight-System Design. (Continued)** The complexity metrics in this table characterize our ability to command, monitor, point, automate, and develop flight margins for the spacecraft and instruments.

|  | Low | Medium | High |
|---|---|---|---|
| **Flight Margins** | | | |
| Onboard consumables managed by operations? | 1 or 2 | 3 – 5 | More than 5 |
| Onboard data storage? | More than 3 times the DL* period | 2 – 3 times the DL period | Less than 2 times the DL period |
| Onboard command-file memory? | More than 3 times the plan duration | 2 – 3 times the plan duration | Less than 2 times the plan duration |
| Onboard flight-software memory? | More than 30% margin at launch | 20%–30% margin at launch | Less than 20% margin at launch |
| Speed of the flight computer? | More than 30% margin at launch | 20%–30% margin at launch | Less than 20% margin at launch |
| Time to uplink full plan? | A fraction of a pass | Approx 1 full pass | More than 1 pass |
| Time to downlink planned data storage? | A fraction of a pass | Approx 1 full pass | More than 1 pass |
| Real-time downlink bandwidth (data rate)? | More than twice the data-capture rate | 1 – 2 times the data-capture rate | Less than the data capture rate |
| Telecom link margin? | More than 3 dB | 1.5 – 3 dB | Less than 1.5 dB |
| Power margin? | Power available more than peak load | Power available approx equal to peak load | Power available less than peak load |
| Thermal margin? | No thermal constraints on pointing or power | 1 or 2 thermal constraints on pointing or power | More than 2 thermal constraints on pointing or power |

\* downlink

Let's define some terms and show examples that will help us understand and use Table 5.2.

- **Commandable states.** The number of flight-system states that are commandable. The number of commandable state specifications we'd need to define a unique operational configuration of the spacecraft and instruments. For example, one commandable filter wheel with four positions would count as a single commandable state.

- **Flight rules and constraints.** The number of operational flight rules and constraints we must obey in commanding the spacecraft and instruments. (See Sec. 3.1)

- **History-dependent commandable states.** The number of commandable states whose command response depends on initial conditions. A two-position switch controlled by a toggle command is one example. A filter wheel commanded by "step forward 3 positions" rather than "go to filter # 1" is another example.

- **Telemetry channels.** The number of onboard measurements routinely downlinked. This number includes status telemetry for the flight system's commandable states, as well as temperature, voltage, current, and other sensor-measurement values throughout the flight system.

- **Ambiguous states.** The number of commandable states for the spacecraft and instruments that telemetry channel values don't uniquely specify.
- **Independent fields-of-view.** The number of flight-system elements having independent pointing requirements or constraints. Examples are a solar panel, a high-gain antenna, instrument sensors, and radiator fields-of-view. Several instruments sharing a common boresight would count for just one field-of-view.
- **Articulating devices.** Things we must move routinely during operations. Solar panels, articulated antennas, scan platforms, filter wheels, and instrument covers are examples.
- **Constraints on hazard pointing of instruments.** Instruments that have flight rules disallowing us from pointing them in certain directions (e.g., at the Sun).
- **Unattended safing.** The amount of time the flight system, without ground intervention, can safe itself and not suffer irreversible damage.
- **Routine safing commands.** The number of states the ground staff must routinely command to prevent irreversible damage.
- **Onboard consumables.** *Flight consumables* include propellent, number of thruster firings, start-stop cycles of the tape recorder, tape across the head, and battery charge-discharge cycles. *Replenishable consumables* include battery charge, wheel momentum, and onboard memory.
- **Onboard data storage.** Low-cost missions have enough margin in onboard data storage to make up for occasional missed downlinks.
- **Onboard command-file memory.** The margin for command-file storage is based on the capacity required to store a typical full uplink plan.
- **Onboard software memory.** We define software margin in terms of unallocated memory at launch.
- **Speed of the flight computer.** The onboard processing margin for the flight computer(s) calculated in terms of millions of instructions per second (MIPS) margin at launch.
- **Time to uplink full plan.** The uplink bandwidth margin calculated in terms of how much time it takes to uplink a full plan.
- **Time to downlink planned data storage.** The downlink bandwidth margin calculated in terms of how much of a pass it takes to downlink a full quantity of planned data storage.

- **Realtime downlink bandwidth.** The downlink bandwidth (downlink bit rate) margin calculated in terms of the flight system's nominal data-capture rate.
- **Telecom margin.** The downlink margin calculated in terms of the downlink's signal-to-noise ratio. (See Sec. 11.3)
- **Power margin.** Power margin calculated as a function of peak load and total power available—both generated and stored.
- **Thermal margin.** Thermal margin established in terms of the number of power and pointing constraints.

The design rules in Table 5.2 suggest that, to operate at low cost, our spacecraft and instrument design must:

1. Minimize the flight system's controllables and observables (command states and telemetry points) that operators must routinely manage
2. Minimize the flight system's operational rules and constraints
3. Minimize history-dependent command states
4. Minimize ambiguous states
5. Simplify devices for pointing, controlling attitude, and articulation
6. Provide onboard automation that safes the spacecraft and instruments for long periods without ground interaction
7. Provide operational flight margins for consumables, onboard data storage, onboard sequencing memory, flight-software memory, flight-computer speed, time to uplink full command load, time to downlink full memory, telecom link, power, and thermal

# 5.3   Complexity Metrics for Operational Risk Avoidance

Policies for avoiding operational risks can strongly drive up operations cost. If we decide not to tolerate command errors, we may need a zero-defect uplink with elaborate procedures for simulation, constraint checking, validation, and approval. If we can't tolerate data losses, we'll need a zero-defect downlink with elaborate procedures for scheduling and for capturing, detecting, recalling, validating, and archiving data.

Policies on using onboard automation and fault protection can drive operations costs in several ways. Sometimes, planners so mistrust certain event-triggered algorithms on the spacecraft that they instruct operators to fly the spacecraft so these algorithms won't get triggered. Now modeling, analysis, simulation, and human oversight intensifies because mission policy considers triggering of onboard automation an operational failure.

Risk policy requirements on operations to predict and prevent onboard failures, rather than to simply detect and respond to failures once they have occurred, results in more modeling, trending, in-flight testing, and performance analysis. These tasks drive up operations costs significantly.

Table 5.3 presents 11 questions that capture complexity metrics for risk avoidance.

**Table 5.3. Project Operational Risk Avoidance.** Metrics are organized into command and control, data return, performance analysis, and fault recovery.

| | Low | Medium | High |
|---|---|---|---|
| **Command and Control** | | | |
| Command errors tolerated per week? | More than 2 | 1 or 2 | None |
| Onboard adaptive algorithms—tolerated entries per week? | More than 2 | 1 or 2 | None |
| Activity simulation? | None | Simulation by software model | Simulation by flight-system testbed |
| Command reviews and approvals? | Fewer than 2 per command file | 2 or 3 per command file | More than 3 per command file |
| **Data Return** | | | |
| Amount of lost science data tolerated? | More than 5% | 0.5 – 5% | Less than 0.5% |
| Amount of lost engineering data tolerated? | More than 5% | 0.5 – 5% | Less than 0.5% |
| **Performance Analysis** | | | |
| Routine performance validation? | Fewer than 50 parameters | 50 – 200 parameters | More than 200 parameters |
| Routine trend analysis and prediction? | None | For up to 5 performance parameters | For more than 5 performance parameters |
| Routine performance modeling? | Simple algorithms | Configuration-controlled software models | Spacecraft-system testbed |
| Model calibration? | Fewer than 2 times per year | 2 – 4 times per year | More than 4 times per year |
| **Fault Recovery** | | | |
| Tolerated timeliness in ground response to an anomaly? | More than 1 day | 8 – 24 hours | Fewer than 8 hours |

Let's present some terms and examples that may help us understand and use Table 5.3.

- **Command errors.** Includes commands that inadvertently violate a flight rule or constraint, cause unintended consequences, and wouldn't have been approved under more rigorous constraint checking and review.

- **Number of onboard adaptive algorithms tolerated per week.** This metric attempts to quantify operational use of onboard automation. It distinguishes between missions that use automation and missions that have onboard automation but avoid risk by not using it, thus creating an uplink process that tries to avoid triggering these algorithms.

- **Activity simulation.** A software-simulation model emulates the flight system's behavior and allows activities to be run on the ground (often in faster than real time) to validate expected results before being uplinked. An alternative approach is to validate activities using a testbed of flight-like hardware.

- **Command reviews and approvals.** The number of authorities or teams that must review and approve a command file before it may be uplinked. For this metric, review by several members on the same team counts as just one review.

- **Amount of lost data tolerated.** The percentage of data we plan to capture and downlink but lose because we don't have zero-defect commanding and downlinking. The loss may be permanent or temporary (until a recovery activity and downlink is scheduled and carried out).

- **Performance validation.** The number of onboard parameters operators routinely monitor and report on including onboard sensor measurements, as well as configuration and status measurements.

- **Trend analysis and prediction.** The number of onboard parameters for which we routinely predict and analyze future states.

- **Performance models.** The number of onboard parameters whose performance we routinely analyze by running formal ground models or hardware-simulation models.

- **Model calibration.** The number of special in-flight tests we do to provide a performance model. A maneuver activity to calibrate an antenna is an example.

- **Anomaly response time.** The time in which operations must respond to an onboard anomaly measured from the time of the anomaly to the time of an uplinked response. This time includes two-light time delays as well as detection delays due to scheduled no-track durations.

The design rules in Table 5.3 suggest that, to operate at low cost, we must

1. Tolerate faults in the uplink
2. Trust onboard automation and allow its use
3. Permit simple simulating and validating of activities and commands
4. Minimize review and approval of commands

5. Tolerate faults in the downlink

6. Reduce the number of onboard parameters operations must routinely analyze for performance

7. Permit operators to respond to observed failures, rather than requiring them to predict and prevent failures

8. Require minimum modeling and engineering calibration to ensure performance

9. Allow reasonable ground response times to observed anomalies in flight

## 5.4 Complexity Metrics for the Ground System

A complex ground system certainly influences operations costs. Operations tools and displays are part of this complexity, but so are the geographical, program, and institutional environments. Having to schedule and compete for limited ground resources may save development or institutional money, but it costs more for operations. In the same way, inherited ground systems can save development dollars, but they drive up costs for operations, which must adapt to new users' requirements. Adapting to new flight software or supporting new flight-system can also cost more for operations.

Table 5.4 presents 24 questions that establish complexity metrics for ground systems.

**Table 5.4. Complexity of Ground Systems (GS).** Metrics are grouped by interfaces and shared resources, data-system design, organization, and how much key operational tasks are automated.

| | Low | Medium | High |
|---|---|---|---|
| **Interfaces and GS Complexity** | | | |
| Geographical distribution of operations and ground system? | 1 site | 2 or 3 sites | More than 3 sites |
| Institutions needed to operate the GS? | 1 – 3 | 4 – 6 | More than 6 |
| Science teams? | 1 or 2 | 3 – 6 | More than 6 |
| Shared GS components? | None | 1 or 2 | More than 2 |
| Users sharing instruments? | 1 or 2 | 3 – 5 | More than 5 |
| Shared operations teams? | None | 1 or 2 | More than 2 |
| Shared resource-scheduling epoch? | Hours | Days | Weeks or more |
| Command and control data? | High-order command language | Combination of blocks and individual commands | Individual command level |

**Table 5.4.    Complexity of Ground Systems (GS). (Continued)** Metrics are grouped by interfaces and shared resources, data-system design, organization, and how much key operational tasks are automated.

| | Low | Medium | High |
|---|---|---|---|
| **Design of the Ground System** | | | |
| GS designed to what requirements? | Designed for single user project | Designed for many users, tailored for single user | Designed for many user projects |
| GS built, maintained, upgraded by? | Project - user | Combination of project and other | Another institution |
| Development of flight-system software after launch? | None planned | Up to 10% | More than 10% |
| **Organization and Staffing** | | | |
| Staffing schedule for most operations positions? | Prime shift only | 2 shifts per day | 3 shifts per day |
| Number of management levels? | 1 level | 2 levels | More than 2 levels |
| Tasking strategy? | Project specialists, multi-tasked | Combination of multi-task and multiproject | Task specialists, multi-project |
| **Automation** | | | |
| **Number of separate steps requiring project operator action(s) in:** | | | |
| Alarm monitoring from decommutation through notification? | 1 or 2 | 3 – 6 | More than 6 |
| Level 0 data capture from receipt from tracking facility to project data base? | 1 or 2 | 3 – 6 | More than 6 |
| Real-time command from entry to radiation? | 1 or 2 | 3 – 6 | More than 6 |
| Activity generation from entered requests through command file generation? | 1 or 2 | 3 – 6 | More than 6 |
| Scheduling of tracking coverage? | 1 or 2 | 3 – 6 | More than 6 |
| **Number of transfers of non-electronic data in processes defined above** | | | |
| Alarm monitoring? | 1 or 2 | 3 – 6 | More than 6 |
| Data capture? | 1 or 2 | 3 – 6 | More than 6 |
| Real-time command? | 1 or 2 | 3 – 6 | More than 6 |
| Activity generation? | 1 or 2 | 3 – 6 | More than 6 |
| Scheduling of tracking coverage? | 1 or 2 | 3 – 6 | More than 6 |

The metrics in Table 5.4 are intended to lower operations costs for a single project; they may or may not do so across a large set of projects. For instance, shared operations may compensate for an individual project's higher complexity costs because the project has to schedule and compete for shared resources, accept unneeded capabilities, and pay its share of overhead for institutional management.

Here are some terms and examples that may help you understand and use Table 5.4:

- **Ground system.** The hardware, software, people, and procedures needed on the ground to operate a mission.
- **Institutions needed to operate the ground system.** The number of geographically separate operations sites (5 miles or more apart) that must cooperate to carry out routine operations. This metric doesn't account for travel and per diem costs that may be required to centralize operations.
- **Science teams.** The number of science teams that participate in routine operations.
- **Shared components in the ground system (GS).** The number of major elements in the ground system shared among multiple users and controlled by organizations other than the project.
- **Users sharing instruments.** The number of science teams that don't control dedicated instruments, but rather share instruments controlled by someone else. In our model, this metric is the number of science teams, not scientists.
- **Shared operations teams.** The number of operations teams shared among multiple users and controlled by organizations other than the project.
- **Shared resource scheduling.** The lead time required for scheduling shared resources.
- **Command and control language.** The level of command and control language for the flight system. High-level language contains reusable block or macro command expansions that carry out complex functions. Low-order language requires us to specify each executable command.
- **GS design heritage.** This metric categorizes ground systems by purpose: built for the user project, built for various user projects but tailored to each user, or built for various user projects with no tailoring permitted.
- **GS maintenance.** Who maintains the ground system? Besides monitoring and repairing the ground system, maintenance includes deciding what and when to upgrade, scheduling rebuilds, and delivering new versions or configurations.
- **Postponed development on the flight system.** Has all the planned flight-system software been developed, integrated, and tested before launch. Or must the ground system develop or change a lot of flight software during operations?
- **Staffing schedule.** The number of 8-hour shifts staffed per day.

- **Management levels.** The number of hierarchical levels in the operations organization chart. An example of a three-level hierarchy would be operators who report to team chiefs, who report to office managers, who report to the mission operations manager.
- **Tasking strategy.** Do operators work several tasks for this project only. Or are they task specialists who support multiple projects?
- **Automation—operator actions.** How many actions must an operator routinely take for an operation task? For example, an alarm-monitoring design could require only one action from the operator if it had incoming data automatically compared to standard alarm limits and, when a limit was violated, had an auto-dialer phone call people until it gets an acknowledgment (the single human action). An alternative, less automated, design might require an operator to (1) activate alarm monitoring; (2) load that day's special alarm limits; (3) see when an alarm triggers; (4) look up whom to notify for the particular alarm; (5) look up the phone number; (6) dial the call; (7) report the alarm; and (8) obtain acknowledgment.
- **Automation—non-electronic data transfers.** How many transfers of data aren't electronic? As automation decreases, data will require reading, keyboard entry, or physical handling and logging (like hard-copy schedules, timelines, or tapes)—all examples of non-electronic interfaces.

The design rules in Table 5.4 suggest that, to operate at low cost, our design for a ground system must

1. Minimize geographical distribution. This rule doesn't consider travel and perdiem costs required to co-locate operations.
2. Minimize institutional interfaces
3. Minimize constrained resources such as shared data-system elements and shared operations teams
4. Permit quick scheduling to turn around shared resources
5. Use a command language that exploits high-order, reusable, commands (block or macro)
6. Develop as little flight software as possible after launch
7. Schedule operations to keep staff low on off-prime shifts
8. Minimize management levels in the operations organization
9. Use multi-task experts within a project rather than task experts working several projects
10. Automate by decreasing the number of human steps needed in routine operations
11. Automate by lowering the number of non-electronic transfers of data for routine operations

## 5.5   Using Complexity Metrics to Predict Operations Costs

Figure 5.1, at the beginning of this chapter, is a model for predicting operations costs based on complexity metrics. The model predicts operations costs in terms of workforce size. This approach allows each project to compute dollar costs by multiplying the predicted workforce by their unique salary rates, burden rates, inflation, and duration.

We obtain the workforce prediction from the product of four numerical factors for operations complexity, scaled by an empirically determined constant that depends on mission and phase type. We compute each complexity factor by counting the project's high, medium, and low complexity metrics in each category; multiplying by high, medium, and low weighting factors (empirically determined); and then computing the average value. We compute the average value by adding the products of the complexity metrics and weighting factors and dividing by the total number of metrics. The weighting factors are the same for all mission types and are determined empirically by adjusting them to achieve a best fit of the data used to calibrate the model. Table 5.5 lists the mission-type constants and weighting factors.

The model user specifies the mission type and phase and then answers the 95 questions on operations complexity arranged under high, medium, and low parameters. Based on these user inputs, the model predicts an operations workforce that matches the project's operational complexity.

The model predicts costs for operating large or small uncrewed missions and either Earth-orbiter or planetary missions. Model calibration data is based on NASA's JPL missions. Table 5.5 shows each mission's complexity metrics and computed values for the complexity factors. The last three columns of Table 5.5 show model predicts to be within 25% of the actual mission workforce except for one mission (Galileo Cruise). In the future, we'll expand the calibration database to include Goddard and DOD missions.

---

For example, we predict the FTEs for the Voyager mission by identifying the mission type. Voyager is a cruise mission, so the Mission Type Constant is nine. We next determine the 95 complexity metrics (high, medium or low) for Voyager. We find the numerical complexity factor for the four categories:

Mission Design = $(7 \times 1.4 + 18 \times 1.6 + 7 \times 4) / (7 + 18 + 7) = 2.08$
Flight System = $(6 \times 1.2 + 11 \times 2.5 + 8 \times 3.6) / (6 + 11 + 8) = 2.54$
Risk Avoidance = $(1 \times 0.7 + 5 \times 1.4 + 5 \times 2) / (1 + 5 + 5) = 1.61$
Ground System = $(8 \times 0.7 + 12 \times 1.7 + 3 \times 2) / (8 + 12 + 3) = 1.39$

Next, we calculate the overall complexity factor by multiplying the four individual factors

Factor = $2.08 \times 2.54 \times 1.61 \times 1.39 = 11.83$

Finally, we calculate the FTEs predicted by multiplying the Mission Type Constant with the complexity factor:

FTEs = 9 × 11.83 = 107

---

The model and the complexity metrics will work throughout the project-development cycle. While analyzing preliminary requirements, we could use them to predict how science and mission requirements would affect operations cost, as well as to support descoping studies for cost-capped missions. They could support establishing an operations concept early in the project design, when we have the best chance to influence designs for the payload, spacecraft, and mission. While designing a project, we could use the model to support trade studies such as evaluating design options that save money on spacecraft development but reduce operability. Finally, during the systems-design phase, when more accurate cost estimates based on detailed design are available, we could use it to support "what-if" trade studies, employing a fraction of the time and effort required by detailed cost estimates based on point design.

The model doesn't predict costs for:

- Pre-launch development or for postponed development after launch. The model applies to operations, not development costs.
- Processing science data beyond level 0 because instruments tend to be unique and require unique data processing. The model **does** include costs for processing science data by producing on the ground a complete, error-corrected, time-ordered replica of the science data produced on the spacecraft.
- Space-link operations, such as DSN and TDRSS, which traditionally have been free to the project. It **does** include project costs of scheduling and interfacing with these services.
- Major in-flight anomalies and failures
- Managing the project, program, or science. It **does** include costs for managing the project's operations organization.
- The project's Public Information Office

**The model is causal.** Although cost-modeling techniques vary, two fundamental approaches are associative and causal.

- *Associative models* predict costs based on parameters we correlate or associate with costs. One traditional associative operations model predicts costs based on spacecraft and payload mass. Because associative parameters may have little or no direct causal influence on operations costs, such models aren't very useful when we're trying to decide how to reduce costs. For the mass-model example, we can't use it to claim that operations costs will go down if the spacecraft structure

**Table 5.5. Cost Model for Mission Operations.** Predicted versus actual operations staffing in full-time equivalents (FTE). Note: The numbers of project parameters (low, medium, and high) for each metric aren't always the same for different projects. Some project counts are less than 95 because they were unable to answer all the questions. Other project counts are more than 95 because they used an earlier version of the model with more questions.

| | Mission Design (MD) | | | Flight Systems (FS) | | | Risk Avoid (RA) | | | Ground System (GS) | | | Mission Type Constant | Complexity | | | | Factor | FTE Predict | FTE Actual | Error of Predict |
|---|---|---|---|---|---|---|---|---|---|---|---|---|---|---|---|---|---|---|---|---|---|
| | Low | Med | High | Low | Med | High | Low | Med | High | Low | Med | High | | MD | FS | RA | GS | | | | |
| Weighting | 1.4 | 1.6 | 4 | 1.2 | 2.5 | 3.6 | 0.7 | 1.4 | 2 | 0.7 | 1.7 | 2 | | | | | | | | | |
| **Cruise** | | | | | | | | | | | | | | | | | | | | | |
| Voyager pre-enc | 7 | 18 | 7 | 6 | 11 | 8 | 1 | 5 | 5 | 8 | 12 | 3 | 9 | 2.08 | 2.54 | 1.61 | 1.39 | 11.83 | 107 | 107 | 0% |
| Ulysses | 17 | 12 | 6 | 6 | 17 | 2 | 5 | 6 | 0 | 11 | 12 | 1 | 9 | 1.91 | 2.28 | 1.08 | 1.25 | 5.91 | 53 | 65 | -22% |
| Ulysses-polar pass | 9 | 14 | 12 | 5 | 16 | 4 | 5 | 5 | 1 | 9 | 13 | 9 | 9 | 2.37 | 2.42 | 1.14 | 1.5 | 9.74 | 88 | 70 | 20% |
| Voyager intersteller | 13 | 16 | 6 | 6 | 11 | 8 | 4 | 3 | 4 | 15 | 8 | 1 | 9 | 1.94 | 2.54 | 1.36 | 1.09 | 7.3 | 66 | 52 | 21% |
| Galileo | 12 | 12 | 11 | 2 | 7 | 16 | 3 | 4 | 4 | 7 | 10 | 7 | 9 | 2.29 | 3.1 | 1.43 | 1.5 | 15.13 | 136 | 330 | -142% |
| **Austere Cruise** | | | | | | | | | | | | | | | | | | | | | |
| Voyager-Austere | 13 | 16 | 6 | 5 | 12 | 8 | 7 | 2 | 2 | 16 | 7 | 1 | 5 | 1.94 | 2.59 | 1.06 | 1.05 | 5.59 | 28 | 33 | -18% |
| Pluto | 18 | 9 | 9 | 15 | 9 | 6 | 16 | 4 | 0 | 15 | 13 | 2 | 5 | 2.1 | 2.07 | 0.84 | 1.22 | 4.45 | 22 | 18 | 19% |
| **Orbital*** | | | | | | | | | | | | | | | | | | | | | |
| MGLN-radar map | 7 | 9 | 19 | 7 | 6 | 12 | 1 | 5 | 5 | 7 | 15 | 2 | 11 | 2.86 | 2.66 | 1.61 | 1.43 | 17.59 | 193 | 230 | -19% |
| MGLN-gravity | 9 | 14 | 12 | 8 | 6 | 9 | 4 | 3 | 4 | 9 | 15 | 0 | 11 | 2.37 | 2.48 | 1.36 | 1.33 | 10.62 | 117 | 125 | -7% |
| TOPEX | 9 | 12 | 17 | 6 | 8 | 13 | 7 | 4 | 0 | 6 | 9 | 6 | 11 | 2.63 | 2.74 | 0.95 | 1.5 | 10.31 | 113 | 85 | 25% |
| IRAS | 17 | 13 | 5 | 11 | 11 | 3 | 2 | 8 | 1 | 11 | 9 | 3 | 11 | 1.85 | 2.06 | 1.33 | 1.26 | 6.36 | 70 | 72 | -3% |
| SIRTF | 17 | 7 | 7 | 11 | 9 | 4 | 6 | 3 | 2 | 15 | 5 | 3 | 11 | 2.03 | 2.09 | 1.13 | 1.09 | 5.2 | 57 | 45 | 21% |
| **Flyby** | | | | | | | | | | | | | | | | | | | | | |
| Galileo-prime | 5 | 8 | 22 | 3 | 5 | 17 | 2 | 3 | 6 | 5 | 9 | 6 | 18 | 3.08 | 3.09 | 1.6 | 1.62 | 24.63 | 443 | 420 | 5% |
| Galileo-playback | 6 | 11 | 18 | 3 | 6 | 16 | 2 | 3 | 6 | 5 | 9 | 6 | 18 | 2.8 | 3.05 | 1.6 | 1.62 | 22.08 | 397 | 420 | -6% |
| Your Mission | | | | | | | | | | | | | | | | | | | | | |

* Note: MGLN = Magellan, TOPEX = Ocean Topography Experiment, IRAS = Infrared Astronomical Satellite, SIRTF = Space Infrared Telescope Facility.

is redesigned to a lower mass. Another more trivial example would be that we could probably predict operations costs as a function of the operations manager's age because more expensive missions tend to use older, more senior personnel. But we couldn't claim for a given project that the way to reduce operations costs would be to replace the operations manager with a three-year-old.

- *Causal models* predict outcomes based on inputs that directly influence the outcome. An example of a causal model from the world of health and medicine is a quiz that attempts to predict your life expectancy based on how much you exercise, how many packs of cigarettes you smoke per day, how many pounds overweight you are, etc. Each of these questions identifies parameters that are believed to be causally associated with life expectancy. Besides using these "design rules" for predictions, we can also apply them to change our lifestyles and live longer. The operations-complexity inputs in this model are all believed to be causally related to operations costs. Thus, reducing the complexity value of any parameter should reduce operations costs. Because of this attribute, we can use the operations-complexity parameters not only to predict costs but also as an organized set of design rules to achieve low-cost operations.

The model is based on requirements. We intend it to work for various operations designs, team structures, staffing strategies, and ground data systems. It does so by defining the operations-complexity metrics to **model operations costs in terms of the requirements on, rather than the design of, the operations system.** It's designed this way for several reasons. One reason is that it must be able to predict costs in very early design phases, before the operations design has been baselined. A second, more important reason is that operations costs are driven the hardest by requirements imposed on operations, and only to a lesser extent by the efficiency of the operations-system design. The difference between a 200-person operations team and a 20-person team traces mainly to a difference in performance requirements, and only secondarily to a difference in the efficiency of an operations design. We can reduce operations costs most by negotiating cost-effective operations requirements early in a project. This model supports such negotiating by covering the four main sources of requirements on the operations system: the mission design, the flight-systems design, the project's policies on operational risk, and inheritance and constraints on the ground. It's no accident that the model's four factors correspond to these four most significant sources of operations requirements.

## References

Adams, M. and W. B. Gray. "Design and Implementation of the Mission Operations System Cost Model." Jet Propulsion Laboratory D-3119 March 15, 1986.

Carraway, John. "FY'94 End-of-Year Report for Technical Infrastructure Task #962-93201-0-3170-MOS Cost Model." Jet Propulsion Laboratory IOM #JBC-317-02-16-95, February 16, 1995.

Carraway, John. "Lowering Operations Costs Through Complexity Metrics." November 15, 1994. Space Ops '94. Greenbelt, Maryland.

Carraway, John. "MO&DA Cost Estimates for Three 2012 Mission Set Scenarios." Jet Propulsion Laboratory IOM # JBC-317-12-11-92, December 11, 1992.

Kohlhase, C. E. "Criteria for Identifying and Funding Cassini Operability Improvements." Jet Propulsion Laboratory IOM # CAS-CEK-04-93, January 21, 1993.

Squibb, Gael. "MO&DA Cost Model Technical Inputs." Jet Propulsion Laboratory IOM # TMOD.GFS.94.003:jmm, July 15, 1994.

# Chapter 6

# Defining and Developing the Mission Operations System

David E. Kaslow, *Lockheed Martin*
Jeffrey K. Shupp, *Lockheed Martin*

6.1   Definition and Development Process
6.2   Monitoring Definition and Development
6.3   Scenarios and Subsystems to Monitor

The mission operations system consists of the space and ground assets needed to accomplish mission objectives. The system divides into components we call elements—in categories such as space, ground, and communication. One element is mission operations, which includes tasking to meet mission objectives, carrying out the mission, and processing and distributing mission data [Larson and Wertz, 1992].

In this chapter we describe how to develop concepts, requirements, and designs for the system and its elements. Mission concepts, plus specified requirements for the system and interfaces, describe how the system and its elements interact. In the same way, each element has mission operations concepts and element specifications. The focus in our chapter is on defining and developing the ground element for mission operations, with examples of mission execution. Mission execution covers activities to collect mission data and to maintain the spacecraft and ground equipment.

Getting requirements and design for mission operations and the entire system under control early is critical if we want to develop and operate at the lowest effective cost. Delays in understanding and stating requirements for mission operations become much harder, and more costly, to correct as the project moves into development. Yet, such delays are typical. In this chapter, we present a process the mission operations manager (MOM) can use to successfully define and develop a space system.

The MOM leads the ground element through definition and development into operations to produce:

- Definition and development phases that are within cost and schedule constraints
- Efficient and cost-effective operations
- Operations that meet mission objectives

This chapter presents the information MOMs need to manage and judge the adequacy of definition and development, based on their review and approval of two main aspects:

- The definition and development plans, written at the beginning of the definition and development phases. These plans present in detail the management and technical work that must be done, along with the corresponding schedules.
- The major design reviews, which include lower-level reviews and formal presentations to establish approved baselines for concepts, requirements, and design specifications.

We have organized this chapter into the following three major sections covering the four topics illustrated in Fig. 6.1:

- Section 6.1: Definition and Development Process

  In this section we look at how the system and mission operations element evolve through definition and development. We outline the process, showing the mission statement evolving into system requirements and concepts evolving into element design and development. To clarify, we exemplify the definition, decomposition, and design trades needed to achieve a successful preliminary design review for mission operations.

- Section 6.2: Monitoring Definition and Development

  In this section we discuss how the MOM monitors definition and development of mission operations and the system. This monitoring includes ensuring the system requirements are properly incorporated into the design of the mission operations element as well as discussing which attributes should be monitored.

- Section 6.3: Scenarios and Subsystems to Monitor

  In this section we show and discuss example scenarios and subsystems that need monitoring.

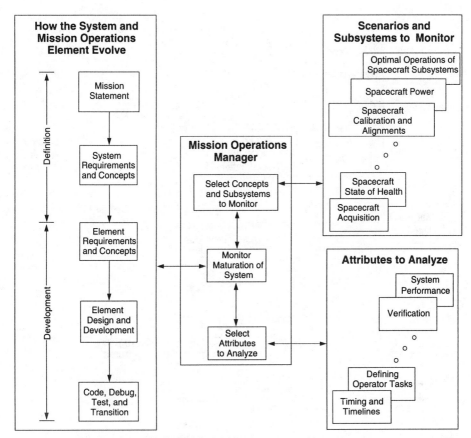

**Fig. 6.1.     Monitoring Definition and Development of a System.** Analyzing attributes of key scenarios and subsystems underpins the mission operations element's evolution throughout definition and development.

## 6.1     Definition and Development Process

Definition and development for both the system and ground operations element begin with a mission statement and end with the design of the software, hardware, and operations. We'll review the process as outlined in Fig. 6.1 and then present a lower-level overview and more detailed discussion of the components illustrated in Fig. 6.2.

Taking a system from a mission statement to designing software executables and hardware components spans two very different periods in a system's or element's development. The first period, as illustrated in Fig. 6.1, covers the definition phase for the system and mission element. We start this phase by issuing the mis-

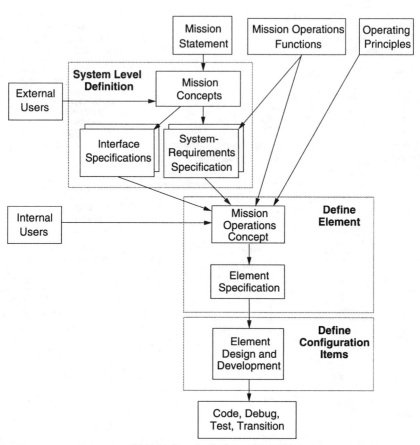

**Fig. 6.2.    Evolution of System and Mission Operations Elements.** The mission concept and mission operations concept drive how we define the system and mission operations element, respectively. The high-level functions establish the basic processing entities of the element. The operating principles provide the basic approach to mission operations on the ground.

sion statement. We end with the system defined well enough so requirements for the system and interfaces are contractually binding and development of the mission operations element can be accurately costed. In other words, definition proceeds until the system-requirements specification articulates all system objectives defined in the mission concept. A *requirements specification*, as used throughout this chapter, is a document that organizes requirements and contains information on verification of the requirements, verification method, and any performance results to be met. The requirements in the system specification have to be clear enough so designers of the mission operations element can unambiguously interpret and break them out in a way that maintains the system's intent. This step is necessary to put together a not-to-exceed cost for the element's required capabilities. It ends when we review the element's functions and validate the designer's interpretation and response to system requirements.

Although we don't discuss costing in this chapter, the mission operations manager must have accurate and complete costs. High estimates can discourage projects, whereas low estimates will result in overruns which demand changes to requirements and designs in order to recover the overrun costs. Estimating the cost to develop a ground element for mission operations is very difficult. First, we develop an element architecture, based on element requirements. The architecture includes hardware, software, and operations components; operations and support-computer architecture; support infrastructure such as maintaining software and managing configurations; and operations and support facilities. Then, we estimate the cost to develop the element architecture. The cost includes managers, engineers, and development personnel and facilities.

The second period covers development of the mission operations element. We start with the element specification and end with software, procedures, and hardware installed and transitioned to operations at the ground station. This process evolves based on results of identified design trades, as concepts and designs mature. Many distinct sub-phases are part of this phase. At a high level, the sub-phases are preliminary design, which ends in the preliminary design review (PDR); critical design, which ends in the critical design review (CDR); code, debug, and test; and installation, checkout, and transition to operations. This chapter focuses on the PDR and CDR because they encompass the design trades and analysis needed for cost-effective design of the mission operations element.

In Fig. 6.2 we illustrate a classical waterfall approach to definition and development. In this paradigm, understanding the mission capabilities moves from the system level, to the element level, and to the configuration item level. The advantage of this approach is that concepts and requirements mature at one level before flowing down to a lower level. But compared to the concurrent approach presented in Sec. 6.1.4, it requires more time for definition and development and it is difficult to feed back changes from lower levels to higher levels. At the top of Fig. 6.2 is the mission statement, which defines the system's purpose. In the system-level definition, the mission concepts describe how the system's elements react to various stimuli to meet the mission objectives defined in the mission statement. In doing so, they drive out the relationships between elements, as well as the roles of the elements themselves in meeting the mission objective. Recall that the system is made up of several elements, with our focus being on the ground element for mission operations. For each element, mission concepts treat the other elements as external users, with the roles of the element defined in high-level terms. We capture the role of each element in the system's requirements specification for that element, whereas we capture its relationship to the other elements in the interface specifications. We can construct requirement specifications for each element in a topical or a functional organization, each with its own advantages and disadvantages. We provide examples in this section to illustrate the issues involved in capturing the system's abilities in the system-level specifications.

In defining the elements, system designers use the mission operations concept to turn the system's requirements specification into the element design. This concept guides how designers define the context within which the element meets the system-level concepts—stated another way, we use the mission operations concepts for element interpretations of upper-tier specifications. As we show in Fig. 6.2, the mission operations concepts derive not only from the upper-tier specifications but also from the needs of the element's internal users (operators and maintainers), facility constraints, and basic processing paradigms captured as operating principles. Through analysis, the element designers determine how the element design will satisfy the higher-level definition of the system's requirements. This is true regardless of whether a single task within the element or several tasks working together capture the upper-tier requirement. The mission concepts also define the role of each of the element's tasks (software, hardware, and operators) in meeting the upper-tier requirements. Equally important, these concepts define the interfaces between those tasks. The element specification then captures all roles and relationships.

A *configuration item* (CI) is a logical grouping of tasks that we develop and test as a unit. When element designers and developers define configuration items, they produce a more mature element design, designating tasks as capabilities and finally developing them into software executables and hardware components. They complete development through a long period of coding, debugging, and testing before handing the CIs off to operations.

Definition and development proceeds by phases but is also characterized by a see-saw effect of "what" versus "how" steps. In general, steps that involve requirement specifications describe what the system, element, or executable must do; steps that involve concepts and design describe how the system, element, or executable responds to meet those requirements. To clarify how these steps operate, we offer a concept for operating spacecraft's state of health from the spacecraft and ground perspectives, for the system and its elements.

## 6.1.1   Define System Concepts and Requirements

As we highlighted in Fig. 6.2, the system-level definition covers the mission concepts and the requirements specifications for systems and element interfaces. In Table 6.1 we illustrate the attributes and characteristics of this definition, which breaks out into five main parts:

- Establish Mission Concept
- Specify System Requirements
- Organize System Requirements—Topically

- Organize System Requirements—Functionally
- Specify Interfaces

In this subsection, we'll discuss each of these parts and then run through an example showing how to develop concepts and specifications.

Table 6.1.    **System-Level Definition.** The mission concepts, system-requirements specifications, and interface specifications define all the elements at the system level.

| Document | What the Document Does |
|---|---|
| Mission Concept<br>• Definition of elements<br>• Inputs and outputs of system | Allocates tasks and timelines to elements |
| System Requirements Specification | Allocates requirements to each element |
| Interface Specification<br>• Inputs and outputs among elements<br>• Data format, frequency, and content<br>• Constraint information | Describes data flow, control flow, and constraints among elements |

**Establish Mission Concepts.** The mission concepts establish the system design at the highest level. Here, system designers break the system into elements and determine a high-level concept of the role each element is to play within the space-mission architecture. Designers logically organize elements within data and control boundaries, as well as along space and ground boundaries. The mission concepts describe the inputs and outputs of the system and specify at a high level how the system responds to satisfy the mission statement.

**Specify System Requirements.** System designers must specify system requirements, based on the mission concepts, for each of the elements in the system. To each element, each other element is an external user—either providing input to, or receiving output from, another element. The mission operations element on which this chapter focuses is a central clearing house of data and control for the mission architecture. In this role, it communicates with space and ground assets to conduct the spacecraft's mission and state of health operations. It receives tasking data which drives control of the payload and it prepares data for correlating the spacecraft-mission data to the tasking data. Virtually all mission operations elements share high-level responsibilities, or functions, as discussed in Chap. 3. The system-requirements specification for the mission operations element captures all of these functions, but we may organize the requirements levied against them in several different ways.

**Organize System Requirements Topically.** The topical approach states how the element must respond to each system stimulus. It complements a systems-analysis technique developed by McMenamin and Palmer and known as *Essential Requirements Analysis.* [McMenamin, 1984] This technique involves analyzing the system's inputs and outputs to determine the essential nature (no design constraints imposed) of the processing required to respond to each of them. This approach gives us the advantage of leaving the internal-element design more open and not implying control and data relationships between the element's functions. It also tends to define the element requirements at a higher level, which allows the element designers more freedom in developing the element without constraining the technology of the solution. Its main disadvantage is that the requirements against an element's function are distributed throughout the document. Also, a requirement assigned to a function within a particular element may be repeated in the system-requirements specification under different topics. For example, two completely different input events to the mission operations element may require us to extract data from the downlinked telemetry. A topical organization would place a telemetry-processing requirement under two topics. But this requirement redundancy may add choices to the element design. Although the need for telemetry-based data may be the same under the two topical areas, one of the events may require a much faster response than the other. This difference in performance requirements provides a design option to the element that might not be apparent if we organize the system requirements another way.

Designers could take the topical approach to the extreme by specifying system requirements for an element based only on inputs and outputs of the functions relative to external interfaces and operator actions. This kind of topical specification views the element as a black box, discussing only the element's responsibility in meeting external interfaces, without explicitly placing requirements on the element functions to generate the data. Rather than specifying the element's internal operations, it leaves all of these issues to element design. For example, the mission operations element might be required to provide its mission-plan information to another ground element at a particular time each day. The system-requirements specification will discuss the timing of that message, the recipient element, and the responsibility to deliver the plan during defined contingencies. But the specification wouldn't get into the expected timelines for producing the plan or the inputs that influence it. These details are left for the element designers to resolve.

**Organize System Requirements Functionally.** In this type of organization, system designers capture in one place all system-level requirements against a particular function. Organizing the specification this way allows everyone to see easily how much processing each function requires. But the functional approach has two main disadvantages. First, it tends to specify the relationship between functions, a task system designers normally leave to element designers. By specifying the relationship, we ensure all the inputs and outputs to a function are captured. Second, the specification tends to lose the reasoning for a particular requirement because the requirement becomes divorced from the topic which drove it. Going back to the telemetry-processing example above, a functional specification may stipulate only the most stringent processing timeline, instead of the multiple timelines for specific driving events, thereby eliminating the possibility of a design option for the element.

On some occasions the system-requirements specification will have topical and functional organization. But designers should avoid this hybrid form if possible, for it carries the disadvantages of both schemes without any added value. Further, because the upper-tier specification looks and feels like a functional approach, element designers may lose requirements against functions embedded in the topical sections.

**Specify Interfaces.** System designers capture in the interface specifications all requirements for data and control passing into or out of the element, as shown in Table 6.1. Stated another way, the mission operations element shouldn't need access to, or knowledge of, any other element's requirements to operate properly within the system. These specifications tell exactly what data is shared between the element and each of its external users. They show how often, and at what specific times, data is passed, and they specify how the data is to look, so the interfacing elements can read and write the information in a format everyone agrees to.

The interface specification also contains constraint information so one element's processing doesn't cause a processing problem for its interfacing element. This is especially true for the interface specification between mission operations and the spacecraft. The specification includes constraints dealing with command timing, rate and acceleration, and antenna motion. If sensors onboard the spacecraft detect that these constraints have been violated, the spacecraft could be permanently damaged if we don't place it into a safe operating condition. Operating constraints in the interface specification will help the spacecraft meet its on-orbit life expectancy by specifying limits on certain items, such as duty cycle or switching frequencies for equipment that will keep it operating safely. Designers achieve this kind of detail by placing high-level concepts for each element in the mission concept and then specifying constraints as designs for each pair of elements mature.

**Example Showing How to Develop Concepts and Specifications.** Looking at the spacecraft's state of health is a good way to see how the mission concept helps mold relationships among elements throughout the system. Less sophisticated spacecraft, like TDRSS, are based more on real-time command and control, with very little onboard processing of detected errors. In these cases, the mission operations' role is to recognize the error condition for the spacecraft by interpreting the telemetry as it arrives and to take the appropriate actions, such as quickly sending the proper commands to place the spacecraft into a safe condition. Recovery to normal operations is simple as well. More sophisticated spacecraft, such as the Hubble Space Telescope or the Magellan spacecraft, can detect more error conditions onboard and place the spacecraft autonomously into a safe condition. Mission operations' role with this kind of spacecraft is to ascertain, after the fact, what the spacecraft detected that caused the error, to operate the spacecraft in the safe condition for an extended time, and to recover the spacecraft to normal operations. Also, because the spacecraft may put itself into a particular kind of safe condition for each type of error, the recovery is more complex. System designers must analyze the abilities and limits of each spacecraft configuration to determine how much sophistication the spacecraft needs to meet mission and maintenance objectives for the system.

If systems analysis of the spacecraft's state of health shows that the more sophisticated spacecraft better meets the system's objectives, designers must expand the system-requirements specifications and interface requirements to support it. The element specification requires processing onboard the spacecraft to detect error conditions, such as overlapping commands and violations of the spacecraft's acceleration limits or temperature ranges. The onboard processing must autonomously put the spacecraft into one of several potential safe operating conditions and transmit the needed information to the ground in the telemetry. The requirements specification for mission operations includes requirements to detect when the spacecraft has entered a safe operating condition and to operate the spacecraft in that condition indefinitely. Mission operations must recover the spacecraft to its normal operating condition within a specified time of the go-ahead signal. The interface specification contains details of all of the spacecraft's safe operating conditions as well as the kinds of error conditions the spacecraft must detect.

### 6.1.2    Define Element Concepts and Specifications

As we show in Fig. 6.2, the system-requirements specification, along with interface specifications for external users, drives what the element does. Element designers capture the element's response to these drivers by defining the element while developing the mission operations concept and the element specification. Table 6.2 lists the attributes and characteristics of the element-level definition. This subsection addresses

- Mission Operations Concept
- Internal Users and Mission Operations Concept
- Definition of High-Level Functions and Mission Operations Concept
- Operating Principles and Mission Operations Concept
- Design Trades and Mission Operations Concept
- Element Specification

**Mission Operations Concept.** Element designers use the mission operations concept to drive out the highest-level design at the element level. Here, they show generally how each processing and operator task will work within the high-level element functions. The tasks, which we'll discuss later, are grouped logically into the Computer Software Configuration Item (CSCI) and Hardware Configuration Item (HWCI) definitions, with operator tasks and procedures (Ops) added wherever necessary. In general, designers group tasks into CIs to minimize data and control flow between tasks of different CIs while keeping tasks of the same CI relevant to each other. Mission operations concepts describe the element's inputs to,

**Table 6.2.     Defining the Mission Operations Element.** The mission operations concept and element-requirements specifications define the mission operations element.

| Document | What the Document Does |
|---|---|
| Mission Operations Concepts | Allocates tasks and timelines to hardware and software configuration items and operations<br>• Establishes processing and operations tasks for the element<br>• Lists information required and products<br>• Responds to internal users<br>  – Mission operators<br>  – Data-systems operators<br>• Captures high-level functions through processing tasks<br>• Defines key operating principles<br>• Acts as source of design trades |
| Element-Requirements Specification | Allocates requirements to hardware and software configuration items and operations<br>• Captures requirements from concepts<br>• Assigns tasks to configuration items<br>• Reduces system requirements to individual configuration items<br>• Captures interface specifications and constraints |

and outputs from, each CI. They also describe, at a high level, how the element's tasks work together to meet its system-requirements specification. For example, one concept will define how plan update, a task assigned to activity planning, tells command generation, a task assigned to activity scheduling, that the plan has changed. The mission operations concepts also overview the operators' roles and the kind of process interaction the operators will see in the operational element. To provide enough insight into the processing and operator tasks, these concepts delve into characteristics within and between CIs, as necessary, to drive out the element's response to a system requirement.

**Internal Users and Mission Operations Concept.** As we show in Fig. 6.2, another influence on the element's requirements, and ultimately its design, is how the element responds to its internal users. For a ground element, these users include operators, people who maintain hardware and software, engineers who tune the element's performance in operations, and support services, such as platform hosting and layout, network management, and data-center operations.

The most visible of a ground element's internal users is its mission operators. Again, as in the case of the interface specifications between two elements, designers lay out internal-user interfaces very early and then iterate them as the element evolves. Even though these interfaces are extremely important to cost-effective element design and requirements, designers often don't take them as seriously as the external-user interfaces. Although design drives the ultimate look and feel of an element's operations, they also depend on concepts for internal-user interfaces seeded at this level. Additionally, items such as limits on station staffing for operations and maintenance, training for station staff, and the amount of operator involvement versus software automation, will strongly influence the ground element's design.

Technologies for computing and database management are changing rapidly, and systems take up to five or more years to define and develop. As a result, the computers and data-systems architectures established in the definition phase drive a ground element's design. Thus, element designers must decide whether to use technologies that meet general industry standards or to develop new ones, keeping a careful eye on life-cycle costs.

**Definition of High-Level Functions and Mission Operations Concepts.** By defining functions at a high level, designers provide the seeds for a good mission operations concept, which must capture all of the element's processing tasks. Then, using these definitions of high-level functions to define tasks, element designers can determine the scope, control, and flow of data for each function. That means they can analyze and trade design attributes of the element long before we create software executables and hardware components. Further, by putting together element tasks, the mission operations concept helps to complete the element definition. This definition is complete when it covers the system requirements plus the requirements levied by or derived from the needs of internal and external users.

The processing tasks for mission operations depend directly on the language that describes the element's requirements in the system-requirements specification. As an example, mission operations calculates the latest spacecraft ephemeris and may also have to send the ephemeris to other elements. From this very straightforward requirement, we can quickly identify or derive several tasks across the element functions—for example, orbit determination, ephemeris propagation, and external-message handling. Assuming the updated ephemeris is based on information passed through telemetry, we also need tasks for processing telemetry, handling internal messages, and managing databases. The mission operations concept then ties all of these tasks together to describe how the element generates the ephemeris data, constructs the outgoing message, and sends it out.

**Operating Principles and Mission Operations Concept.** To develop an effective mission operations concept, element designers must also identify key operating principles that the element will follow as it responds to the system requirements. An example is the relationship of activity planning to scheduling. They look at planning and scheduling from three different perspectives. They can distinguish the two functions by timeframe—for example planning takes care of all activities up to four hours from the present time, and scheduling operates only the first four hours. Or planning handles all general aspects of operations, and scheduling does all detailing. Or planning handles activities that require a lot of operator negotiation before approval, whereas scheduling deals with activities that don't require this negotiation. Designers must work out these high-level operating principles before the mission operations concept begins to mature, because they structure how we identify and analyze tasks that will become part of this concept. If they don't define and thoroughly describe a principle early on, we'll need a lot of money to correct this omission later.

**Design Trades and Mission Operations Concept.** Besides showing how an element's tasks are related, the mission operations concept defines situations and then presents responses, including a timeline for each response. Thus, developing the mission operations concept provides the definitive source of design trades for the element, which in turn may greatly affect requirements for the element's software, hardware, and operations.

Trades based on the mission operations concept can also spawn extensive analysis of software algorithms and changes to requirements. Working with the software engineers, designers of the mission operations concept describe generally how the element will react to a given situation; continuing analysis of the software methods must validate that concept. Failure to do the necessary analysis results in one of two equally expensive situations for a program. The element will oversolve the problem, putting in complex functions that will rarely, if ever, be used. Or the element won't put enough into the design, unrealistically relying on operator procedures to handle what was believed to be a small-chance situation but turns out to be a regular occurrence. The second circumstance is more expensive than the first because we spend more to fix a finished product than to include a new capability during development. The results of the trades are incorporated as specific design concepts and scenarios into the mission operations concept.

Trades on the mission operations concept fall into five categories:

1. Operator- versus software-directed reactions to situations
2. Mission objectives versus timeline margin
3. Task-unique versus generic processing
4. Database versus message-based control and data flow
5. Ability to operate during and recover from a contingency versus loss of the mission objectives during that contingency

These categories are typically the most sensitive areas of the mission operations element's development and life-cycle costs because the trades involve increasing or reducing the mission software's complexities, processor speed and sizing, and data-management volume and complexities. We discuss each of these trade categories below.

*Trade 1:   Operator- Versus Software-Directed Reactions to Situations*

Allocating tasks to operators or software to detect errors and recover from software-executable faults is a prime example of a design trade. Designers can choose software or operators to detect and recover from certain faults, with software providing data to the operator. The trade is between allowing software to gather data and display tasks and then relying on operator tasks to assimilate and isolate the error to its source versus a more automatic approach, which allows the operator to stay focused on the mission. Designers must analyze the error condition, discussing each option in an operations concept in enough detail to fully understand the issues of each option. For an operator-driven solution, the concept states the kind of data the software should collect and display to the operator—in this case, having missed receiving an output. It describes how operators determine the source of the error. It also describes any follow-on actions the operator may take to recover from the problem and reconfigure the receiving process to go with an old version of the input data. It establishes online procedures to help the operator isolate the source of an error. Of course, designers have to consider how much they're loading the operator, what's happening with the main task while an operator isolates the error condition, and whether there's enough time to recover mission processing.

*Trade 2:   Mission Objectives Versus Timeline Margin*

Certain tasks may require solutions of very high quality. In these cases we need high-fidelity modeling or complex algorithms for mathematical searches. Defining a detailed, finite-element thermal model to predict a spacecraft subsystem's temperature is an example of such a complexity. Complexity, plus the need for a quick solution, creates an opportunity for a concept trade. One way to resolve the issue is to add data-processor speed or capacity by using parallel processors instead of serial processors. Another way is to use a two-stage solution: a coarser model does most of the processing, and a detailed model does final processing. A third way is to trade processing time from another task done earlier or later, adding this time to the task requiring the detailed model. Element designers weigh the pros and cons of each of these approaches to arrive at the most effective concept.

*Trade 3:   Task-Unique Versus Generic Processing*

Many times, designers don't take advantage of synergistic processing when presented with tasks that have only some unique qualities. For example, the processing tasks that respond to a request for a high-priority activity in activity planning are very similar to those supporting normal-priority requests, although response times and optimizing techniques may differ. Unique tasks offer independence and allow us to tune both processing tasks so we take advantage of these unique attributes. But there's a trade between duplicating tasks to handle variants and creating a common processing task with unique drivers. In general, the more generic the processing tasks, the more cost effective the overall design because duplicating similar tasks often results in duplicating the developed software. Designers must weigh generic processing against the overhead (including timeline and control) for modifying the software to fit the tasks.

*Trade 4:   Database Versus Message-Based Control and Data Flow*

In the software-processing architectures of the 1970s, the data was persistent, and processing tasks existed long enough to read in the data, manipulate it, and put out new results. Data didn't "flow" much because it really didn't move. In the 1990s, the roles of data and processing tasks have reversed. The processing tasks are now persistent and distributed, whereas the data is temporary. The data goes from one processing task to another over a LAN. The volume of this data can become so immense that it can drive a design because of delays in transmitting the data or the complexity required to keep track of all destinations. Still, a lot of data is extracted from a database instead of being routed. The trade involves determining what data is best served in a message, and what is best kept in a database. For example, many processing tasks require spacecraft ephemeris. The frequency of the ephemeris time points varies from task to task: some require points at no more than one-minute intervals, while others require time-point intervals of one second or less. The more volume put into a message the longer the delay time from source to destination. Element designers must also consider any limits on storing archives if the element logs and stores its message traffic for troubleshooting.

*Trade 5:   Ability to Operate During and Recover from a Contingency Versus Loss of the Mission Objectives During that Contingency*

Depending on how critical a mission objective is, mission operations may need to remain highly available by providing rapid recovery to backup data processors. The trade is between satisfying a mission objective and adding development and operations effort to provide recovery. For example, designers may need to develop ways to maintain process state, detect contingency situations, and reconfigure software and process state on backup data processors. For operations, software maintains process state while completing normal tasks and provides backup data processors. An element designer must trade off the severity of losing a mission objective with the chance of its occurring. Another factor in the trade is the response time needed to detect the failed primary processors and recover to the backup processors.

**Element Specification.** Element designers use requirements language in the element specification to encapsulate roles and relationships from the mission operations concept. The specification captures each task, except for operator tasks. First, they assign tasks to software, hardware, and operators. Then, they analyze the software and hardware tasks to group them into CIs and the operator tasks to group them into preliminary positions.

Many CIs may participate in meeting one system-level requirement. The requirements in the element specification are not simply restated from the system-requirements specification. Element designers analyze requirements based on the mission operations concept, show how each CI meets the system requirements, and show how data and control flow between CIs. If done properly, this step will contribute to a program's success by keeping its costs down. Otherwise, the program will require a lot of effort to redefine the tasks later because further design details will lack the proper insight into what is needed to support the element concepts. This situation doesn't occur until testing begins at the element level. The subsequent requirements and design steps, which we discuss in Sec. 6.1.3, can concentrate on requirements for the individual CIs, rather than constantly backtracking to the element and system-level specifications to understand what a requirement meant.

Note that the element specification, as well as the lower-level specifications we'll discuss next, must explicitly reflect the requirements for interface specifications at the system level, as well as the constraints for which the element is responsible. Furthermore, designers must break down these requirements and constraints whenever necessary to fully understand what the element must do. For example, details about the content of messages exchanged between elements may impart requirements on functions that generate or respond to the external message. If they don't reflect these interface specifications in the element specification and lower documents, they'll have to rework the element's design at the worst time in development—the final stages of element testing.

### 6.1.3    Define Configuration Items

As highlighted in Fig. 6.2, we further design and develop mission operations by defining an operations concept, requirements specifications, and interface specifications for all CIs. Table 6.3 shows the attributes and characteristics of this definition. This subsection covers

- Operations concepts and capabilities for CIs
- Software and hardware components
- Element-design maturity

**Operations Concepts and Capabilities for Configuration Items (CI).** Design and requirements for an element see-saw through "how" and "what" steps at two more levels. The first step below the element specification is the CI design, which includes requirements specifications for accompanying software, hardware, and operators, as well as those for the interfaces between CIs. Just as the mission operations concept describes how the element responds to its requirements, so we must describe how a CI responds to its requirements. The CI operations concept helps define the CI design in that it specifies how the hardware, software, and operations components of the CI interact to satisfy the concepts and requirements specification for the mission operations element. This first step builds an understanding of

**Table 6.3.   Definition of Configuration Items (CI).** We develop operational concepts and requirements specifications for each CI.

| Document | What the Document Does |
|---|---|
| Configuration Item Operations Concept | • Allocates requirements to capabilities<br>• Establishes how configuration items meet element requirements<br>• Assigns capabilities to tasks |
| Configuration Item Requirements Specification | • Allocates requirements to executables<br>• Form executables from capabilities<br>• Shows design and methods trades |

what a CI must do to meet its role as defined in the mission operations concept and element specification.

Configuration item designers then break out CI requirements into capabilities. A capability, in this context, refers to a concept for processing that encapsulates several related requirements. For example, we may follow similar criteria to place different calibration activities onto a planning timeline. Each calibration activity will have a time budget reserved in activity planning, with specific calibration tasks generated in activity scheduling. In the mission operations concept, element designers identify the task as placing the activities into a plan and allowing the operator to interact as necessary. The CI designer may derive a generic capability to accomplish many such activities so they meet the task objectives stated at the element level.

A CI-level operations concept is vital. It describes how the capabilities work together to satisfy requirements. If designers don't develop a CI operations concept, they can't turn requirements into capabilities or form executables. Moreover, the operations concept at the CI level provides a common way to discuss the CI requirements, tasks, and ultimately, capabilities, in terms of how they support the mission objectives. In analyzing capabilities, they're not specifying software or hardware design. That will come later. They must start developing the CI operations concept early in the definition phase and complete it before the preliminary design review, so they can develop and allocate corresponding requirements at the PDR.

Usually, the tasks of the CI identified in the mission operations concept reduce to more than one capability. Designers derive more capabilities until the CI can do what is assigned to it. As an illustration, let's go back again to activity placement. While analyzing how this capability works within the CI, designers must determine what triggers it. For example, if the CI must start placing activities based on time, they need a capability to manage this timing, so they should add it to complete the CI operations concept.

**Software and Hardware Components.** The second level of CI design and requirements specification represents the physical layout of the software and hardware. The software executables and hardware components take their requirements from the CI-requirements specification. At this level, we must define very specific data and control flows between executables are defined. Again, as in all preceding steps, we must analyze requirements to fully understand their impact on the design. And we must work through a set of trades to arrive at the most cost-effective design. CI designers must consider the CI's operations concept within the element and not just "bin" requirements from capabilities to CI executables. If the executable software doesn't behave consistently with its higher-level operations concept, the element design won't close.

**Element-Design Maturity.** The element design and requirements specifications take the element to the critical design review, which represents the end of a program's design phase. The final output to support the CDR is the detailed design for software, hardware, and operations, as well as components of the data and computer system. It takes the element design down to individual software modules. Defined at this time are prologues, all variables set or used by the software modules, detailed processing logic, actual displays that operators of the element will use, and all static and dynamic data with initial values. The hardware units and modules are at final design and ready for manufacture. The operator tasks and displays are defined, and operator loading is at acceptable levels. A performance analysis demonstrates that timing and timelines close and that we've designed and properly sized the data and computer system. A risk analysis shows there are no unacceptable design or production risks. The element design must be at this level of maturity by the program's CDR, for during the next 1–2 years, the entire element-design team will be absorbed in converting the detailed design into actual code and then testing from the CI level to the element level. When designers of the mission operations element build several hundred thousand to several million lines of code, they can ill afford to go back to the element- or system-level requirement and renegotiate its interpretation.

### 6.1.4    Concurrent Definition of System and Elements

Figure 6.3 shows a variant, known as the concurrent approach, on the waterfall approach to definition and development we've presented. The concurrent approach has parallel system and element design phases with feedback; the waterfall approach has sequential, non-overlapping system and design phases. One advantage to the concurrent approach is that the time to produce an element design from a mission statement is reduced because the system-level definition matures in parallel with the element-level design. Another advantage is the ability to work system-level design trades as the element design matures, with least-cost solutions defined before the element reaches the critical design phase. But the concurrent approach does risk having an immature element design because functional and performance requirements may not be complete at the system level. We reduce the risk by using conceptual system requirements. These conceptual requirements aren't contractually binding because they're not in system specifications, but they do provide enough direction for the element designers to do analyze and further assign capabilities to requirements. Then, if design difficulties arise, the system and element designers can work together to solve them.

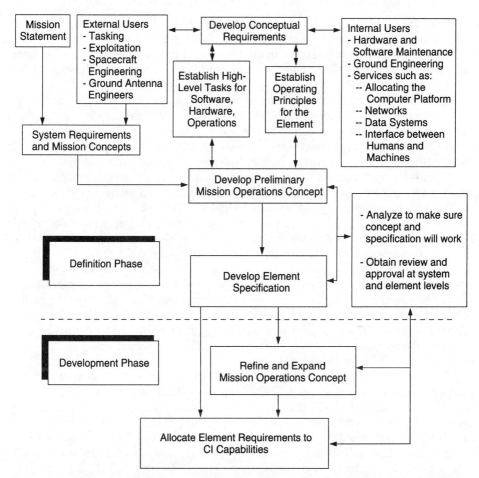

**Fig. 6.3.      Concurrent Development of Requirements for the Mission Operations Element.** The concurrent process is a cost-effective and efficient approach to defining and developing the ground element. Note the parallel design of the system and element, as well as the chances for feedback.

This subsection addresses the following aspects of concurrent definition:

- • Conceptual system requirements and external or internal users
- • Preliminary operations concepts, high-level tasks for elements, and element operating principles

- Element specification
- Completing definition and starting development
- Benefits and risks of concurrent definition and development

**Conceptual System Requirements and External or Internal Users.** The source of the element's requirements in concurrent development are the conceptual system requirements established by the external and internal users of the element. The conceptual system requirements express, in high-level terms, what the system should achieve.

Developing conceptual requirements is iterative, based on the element designers' assessment of requirements that drive the conceptual requirements. Each iteration refines the requirements to the point at which we can generate mission concepts, system-requirements specifications, and interface specifications. We commonly trade on roles and responsibilities between elements.

The conceptual requirements associated with external users, as identified in Fig. 6.3, are easier to develop because they're characterized by the inputs, outputs, and timelines required to meet mission objectives. An example of such a requirement at the system level is the throughput of spacecraft mission data, which affects spacecraft processing and transmission, data processing on the ground, space-to-ground communication paths, and activity scheduling.

For each element involved, system designers might write the conceptual requirement as: "provide a capability to generate, process, or limit processing of up to x bits per second of mission data." They might also describe conditions that inhibit the specified throughput. Rather than specifying the details of how each element responds to this requirement at the system level, the concurrent approach allocates the requirement down into all of the affected elements and allows them to determine how they would meet it.

Conceptual system requirements also reflect the needs of the element's internal user, as identified in Fig. 6.3. This is the more difficult set of requirements to develop because they don't deal with specific mission objectives. Instead, they consider the more esoteric desires of element operations, such as operability, maintainability, and testability, and the element's general capabilities, such as data management. This difficulty applies especially to mission operations because this element interacts so much with operators and engineering support. Therefore, element designers must design carefully to make sure conceptual requirements cover all aspects of mission and support operations.

**Preliminary Operations Concepts, High-Level Tasks for Elements, and Element Operating Principles.** Element designers now define the element's high-level tasks and operating principles, in turn, to support the conceptual requirements. The requirements reflecting the needs of an element's internal and external users significantly influence internal processing. The system's conceptual requirements for speed and complexity determine whether ground processing is highly automated or highly manual. For example, the conceptual requirements from external and internal users will address reacting to a situation that affects the spacecraft's state of health. The speed at which mission operations is to react to the situation and establish contingency communications links to the spacecraft will greatly influence its processing tasks and operating principles for spacecraft command and control.

The preliminary mission operations concept responds to conceptual requirements, high-level tasks for elements, and basic operating principles. Preliminary concepts define the general data flow, processing, and timelines that satisfy the requirements. Element designers then define subsets of these preliminary concepts during the definition phase and develop and finalize the rest in the development phase, as the lower-level design fleshes out the concepts. This approach is risky because, if their subset for early development doesn't drive out the correct architecture and design for the rest of the element, they'll have to do a lot of redesign when they should be finishing the mission operations concept. Thus, they must place most concepts for design and architecture in the preliminary mission operations concept. Later in this chapter, we'll offer ways to identify key concepts that drive the element's design, effectively taking this risk out of development.

The mission operation concept is also preliminary during definition because system designers are developing the system-requirements specification at the same time. Thus, our mission operations concept becomes a tool for analyzing and negotiating the evolving system requirements. To support this effort, the mission operations concept also feeds into, and is fed by, the element designers' analysis of whether or not the requirements are workable. As a result, we can't finalize the mission operations concept until we completely specify the system requirements and make sure they'll work.

**Element Specification.** Next, element designers define the element specification based on the conceptual requirements, the system-level requirements, and the preliminary mission operations concept. Designers analyze requirements at a high level and allocate them to the element's CIs, as discussed previously in Sec. 6.1.2. These requirements define what data and capabilities the element must provide to the external users, to the internal support organizations, and to internal processing. In essence, the element requirements are the contract between the element and the external and internal support organizations.

**Completing Definition and Starting Development.** Once element designers begin development, conceptual system requirements no longer drive element requirements and processing or operator tasks. Instead, they begin focusing on the system-requirements specification and associated interface specifications. The rest of development depends more on analyzing and allocating requirements into functions and tasks, as discussed earlier in Sec. 6.1.3. However, as mentioned previously, the preliminary mission operations concept addresses only the driving subset of the element concepts. Designers must define the rest of the concept, but this step usually overlaps with the early part of steps that establish lower-level requirements and reduce the design to its basic elements. We tolerate this overlap to help us understand all of the tasks driving operations of the element, not just those that influence decisions about architecture and tasks. Therefore, early in the development phase, designers expand and refine the preliminary mission operations concept to support the element requirements.

Finally, the element design begins. It, too, is an iterative process: defining the lower-level design elements (CSCI, HWCI, and operations), analyzing element requirements, deriving additional requirements, and allocating these requirements to the lower-level design elements. Element designers complete it based mainly on analyzing the design at these lower levels. Issues related to speed of processing or algorithmic complexity can drive them back to reexamining the operations concept for other options.

**Benefits and Risks of Concurrent Definition and Development.** The biggest risk with this approach is that system designers may not understand the operations concept of a new system requirement before they allocate it to the elements. System-level trades are also much more difficult because parallel efforts are going on within the system and elements. Ultimately, though, the quality of the analysis put forth in dealing with system and element trades is much higher because the lower-level design steps are involved early to offer insight into a potential problem that a higher-level step might not otherwise uncover. If designers carry out this approach well, they can attain a cost-effective element design in the least amount of time. But if they don't address the right questions, their incomplete knowledge can greatly drive up cost and schedule overruns.

## 6.2   Monitoring Definition and Development

The key to the success of any space-ground system lies in maturing concepts and requirements for the system and elements. If we analyze and design at the right levels, following the process discussed in Sec. 6.1, the system will have fewer issues to resolve as it approaches operation. As illustrated in Fig. 6.1, mission operations managers participate in system definition and development from the mission statement, through the mission concept and requirements specifications, to the operations concept. They do so by focusing on maturing key concepts and configurations or subsystems, with some of those shown in Fig. 6.1.

In this section we'll discuss how, as mission operations managers, we identify early all concepts that drive the element design, analyze them thoroughly, and make them consistent with the conceptual system requirements that drive them. Of special importance are the concepts covering system requirements that are key to satisfying the mission objective and the concepts that span several elements or several CIs within an element. We also discuss criteria which establish a given system or element concept as a key one. Hence, in this section, we're identifying concepts that must be followed more closely and thoroughly during definition and development.

### 6.2.1    Design Attributes that Need Monitoring

Figure 6.1 shows that, as the system and element designs begin to mature, we must select several attributes of the designs to monitor during each phase of definition and development. As we stated earlier, we have to examine each of these attributes in each phase to determine if the element design will fully comply with requirements. Two of these attributes deal with the qualitative characteristics of the element, and three deal with the performance (quantitative) characteristics. One attribute looks at whether the design is operable: another looks at testability and maintenance. In the next subsections, we discuss these attributes:

- Processing Concepts
- Definition of Operator Tasks
- Timing and Timelines
- Definition of System Performance
- Optimal Operations of Spacecraft Subsystems
- Definition of Displays, Alarms, and Procedures
- Verification

We discuss each attribute in three parts. The first overviews the attribute. The second covers the criteria that make the attribute key—one we must monitor. The third discusses the mission operations manager's role in monitoring that attribute.

The introduction to this chapter stated, that, as mission operations managers, we must lead mission operations through definition and development into operations. We do this by first requiring that plans for technical and management tasks be written and approved at the start of definition and development. We must review those plans in detail to see if element designers and engineers have defined all the tasks, schedules, and people needed to carry out missions operations on the ground. Tables 6.4 through 6.10 list some of these tasks. Then, through informal and formal reviews of the elements, we make sure the tasks are done and issues resolved according to the attributes discussed below.

Common to Tables 6.4 through 6.10 are the tasks for developing a preliminary and final design. This design specifies in a set of controlled baseline documents how to build the components for hardware, software, and operations. As mission operations managers, we must approve these documents. We also must require thorough reviews of the design baseline at different phases of the program and approval of the design before proceeding to the next phase.

As part of reviewing the definition and development plans, we should ensure a robust approach to develop the mission operations concepts. These concepts are extremely important—they are the basis for defining, developing, testing, and operating the element. It is very easy to underestimate the amount of work needed to develop a correct and complete set of concepts. We should review the definition and development plans for an approach for developing, reviewing, and approving the concepts; schedules; and manpower estimates. The approach should be detailed enough to cover all aspects of the ground and mission operations element. The schedules should have enough detail so we can see that the concepts and scenarios cover all mission operations on the ground and track incremental progress. The staffing estimates should reflect review and analysis of requirements, discussion with engineers and users, development of alternative concepts, and re-evaluation and refinement of concepts and scenarios as overall knowledge matures.

**Processing Concepts.** Table 6.4 highlights tasks for developing processing concepts throughout the development cycle up to the critical design review. This attribute is often overlooked at the system and element levels. It shows we need to understand how components work together at the system and configuration-item levels.

**Table 6.4.** **Data-Processing Concepts.** We must establish early and then mature the processing concepts, which define how components of the system and ground-mission element work together.

| Phase | Tasks to Cover |
|---|---|
| Definition | • Establish paradigms for system-level processing<br>• Analyze paradigm to clarify element roles and responsibilities |
| Development (Preliminary) | • Develop element-level processing concepts consistent with system paradigms |
| Development (Preliminary Design Review) | • Develop processing concepts for configuration items based on operator task, processing scheme, and concepts for organizing data |
| Development (Critical Design Review) | • Develop final configuration items and element-level concepts |

The element designer must address at least the following areas. They need to establish concepts for starting and stopping processes, as well as for detecting and resolving failures in processes and processors. They must develop criteria for determining if data transfer should be by database, messaging, or memory. Controls based on time or events are necessary for all processes. They need to establish criteria for determining whether we should retain data in files or in relational databases.

*Example of System Paradigm: Data Flow and Access among Elements.* At the system-definition level, as the element requirements are generated, the processing concepts establish how one element deals with the others. The processing concepts show how the element is viewed as a partner of operations with the other elements. They specify how data passes between elements rather than identifying specific data. For example, several elements usually exist in a ground station. Spacecraft engineers may work alongside mission operations engineers but may have developed their own data-processing software to access the same physical data. A system-level processing concept might require mission operations to send any data needed by others (a *data-push paradigm*), or might allow others to come get any data they need (a *data-pull paradigm*). But the elements' roles in providing or obtaining the necessary data are radically different.

*Example of Element-Level Paradigm: Data Flow and Access within an Element.* Element designers have to choose a processing paradigm at the element level. For example, they need to determine how other tasks provide data to the software tasks. One paradigm has data embedded within a message. Another sends notification messages to interested processes and has them query back to the source process. Still another is to have time be the trigger, with either access to a common data element or a query to the generating process. They must decide on a processing paradigm to guide later concepts for processing within configuration items.

*Example of a Paradigm within a Configuration Item (CI): Control and Data-Flow Mechanisms.* At this level, the CI designers determine if they need unique software for processing each activity and constraint or if they can define generic software that is driven by an activity with a unique database and constraints. CI designers also need to decide whether database constructs should define the processing rules, perhaps by using an expert system, or whether these rules should be explicit in the software. Further, element designers need to show how the software or hardware interacts to satisfy a higher-level task, establishing first its capabilities and then its executables. To show interactions within a CI, designers must consider such things as what triggers the processing, how data flows through the CI toward a final product to be passed to another CI, and how various components control the processing. Finally, they must define how the software or hardware reacts to operator control and intervention. For example, operators may require delays in the processing because they know input data isn't yet available. In this case, the software must provide a way for the operator to hold the processing flow.

*When to Monitor Processing Concepts.* This attribute is always key to system and element design. All processing supports some form of work to satisfy element and system requirements. The processing concepts describe how to do that work. Especially for CI design, it's not as important to know that an executable is made up of several components as it is to know how groups of executables work together to meet an objective or performance capability or to react to an operator 's direction.

*How to Monitor Processing Concepts.* As mission operations managers, we monitor this attribute of design from the beginning of system definition to provide system paradigms, where required, that resolve the individual elements' roles in providing data or service. We further monitor the system paradigms during element design to ensure the element meets them as it assigns tasks for the hardware, software, and operators. We also monitor this attribute as the design moves to configuration items. This monitoring ensures that the CI designers are designing all aspects of the executable details, including how database information is constructed and used by the processes, and how the CIs work together to do element-level tasks.

If we don't have enough time or expertise to monitor the processing concepts properly, we should establish an internal program authority and an external expert panel to oversee them.

**Definition of Operator Tasks.** Table 6.5 shows how we develop operator tasking throughout the development phase up to the critical design review.

**Table 6.5.    Definition of Operator Tasks.** We must start determining the operators' roles and staffing levels in the definition phase.

| Process Phase | Tasks to Cover |
|---|---|
| Definition | • Establish the system's conceptual requirements related to operator involvement, training, and number of positions<br>• Establish initial set of operator requirements based on the system's conceptual requirements |
| Development (Preliminary) | • Define operator tasks based on operations requirements<br>• Define operator positions<br>• Assign tasks to operator positions—preliminary |
| Development (Preliminary Design Review) | • Assign tasks to operator positions—final |
| Development (Critical Design Review) | • Evaluate operator loading<br>• Evaluate operations teams |

As we stated in Sec. 6.1.3, the system's conceptual requirements reflect internal and external users of the element. The operators are internal users. At the system level, conceptual requirements may direct a certain maximum number of people for operations, or it may otherwise direct the expected operator involvement in mission operations. Other sources of operator requirements will be from analyzing the element's processing tasks, where all or some of those tasks are assigned to operations. Using this set of initial operation requirements, element designers can define the operator tasks.

*Operator Tasks.* Operators either monitor or control—before, during, or after data processing. For example, if an operator starts processing data when certain conditions occur, these tasks are for pre-processing control. Examining telemetry measurands is monitoring during processing. These categories help identify the operator's relationship to the software executing the mission. The software developed to support these operator tasks has different responsibilities based on the categories. For control tasks, the software waits for the operator's explicit direction at the designated control points before proceeding. For monitor tasks, the software provides information and continues to process; it doesn't wait for an operator input. Therefore, element designers must understand the operator's role in the context of mission operations.

*Operator Positions.* Once element designers define the operator tasks, they combine them into operator positions that cover all aspects of operations, with like tasks assigned to the same operator position. For example, a real-time operator position should have tasks related to spacecraft's state of health, such as monitoring telemetry, uplinking commands, and generating specific trending reports. Designers shouldn't task this position with monitoring or controlling the processing of ephemeris generation.

*Operator Loading.* Element designers also examine the operator positions for operator loading—how much work the operator has to do in a given period. Does the operator constantly monitor, control, and interact, or are breaks scheduled into the routine? Designers may need to change the operator's workload or add people to off-load a single operator during peak periods. For example, they may have required the software to allow an operator to manually generate commands, examine the generated commands, and analyze telemetry to support uplinking a new command load to the spacecraft. Loading analysis may show the operator can't do all of these tasks in the defined timeline. Designers must then choose either to reduce the operator's interactions, relax the timeline constraints, or add a second operator who can do some tasks, such as analyzing the telemetry, while the first operator prepares the new command loads. Some tasks, such as starting and overseeing development of the activity plans, are required only periodically. Designers usually assign these positions only to the day shift as opposed to the continuous shift for real-time operators. Even so, they have to analyze these positions for loading because control and analysis normally dominate activity planning.

*When to Monitor Definition of Operator Tasks.* As mission operations managers, we would monitor this attribute of design when the level of operator staffing is an issue addressed in the conceptual requirements for the system. For systems that continue operating on orbit for a long time, the operations and maintenance costs of a program far outweigh the development costs. In these cases, systems pay very close attention to the staffing required to operate and maintain the ground and space assets. This attribute is also considered key when the element's operations are operator intensive and operability of the system becomes an issue. The relationship between software processing and operators leads to a natural tension. The more tasks an operator must do, the more prone the operations environment is to human error, especially when unusual situations arise. But software maintenance and upgrading become more difficult and require more people whenever we automate a ground element more to off-load the mission operator. As a compromise position, designers can leave the automatic decision processing out of the software but make it easier to control by providing enhanced automated procedures, such as call-up menus that are easy for operators to use.

*Monitor Goals for Staffing.* We must monitor this attribute for several conditions. The first is to make sure staffing meets our goals. Training and maintaining a skilled operations staff for any operational position costs a lot of money for extra people and the infrastructure to keep people trained. As operations managers, we ensure the element designers are sensitive to this issue as they analyze task and position loading.

*Monitor Operability.* We also have to make skilled operators available to thoroughly evaluate the tasks and operability of all operator positions. The operator tasks must match the education and training of the actual operators. For example, operator tasks shouldn't imply that, in order to control the processing, an operator must first know the underlying logic path, or algorithm, that the software took to get to the decision point. Most operators aren't trained to know the details of algorithms; they're trained to know what conditions to look for and what to do next when a particular condition arises.

*Monitor Operator Loading.* Finally, we make sure the operator positions are properly loaded by examining the results of task analysis and determining whether tasks are appropriate and logically grouped into positions. If we find improper loading, we make sure element designers do appropriate design trades to alleviate it. These trades may include converting operator tasks into software requirements, writing software-assisted procedures, adding people, or moving tasks to other operator positions.

**Timing and Timelines.** In Table 6.6 we show tasks for developing timing and timelines throughout the development cycle up to the critical design review.

**Table 6.6.     Timing and Timeline Attributes of Design.** We must allocate to the lowest level each timing and timeline requirement within the ground element of mission operations.

| Process Phase | Tasks to Cover |
|---|---|
| Definition Phase | • Allocate system requirements, events, and timelines to each element<br>• Establish reasonableness of timing and timeline allocation based on high-level functions for the element's hardware, software, and operations |
| Development Phase (Preliminary) | • Convert element's timing and timelines to computer software configuration items, hardware configuration items, and operations<br>• Evaluate the element's ability to meet timing and timelines based on its functions |
| Development Phase (Preliminary Design Review) | • Confirm the ability to meet timing and timelines based on preliminary executables and initial processing traffic |
| Development Phase (Critical Design Review) | • Confirm the ability to meet timing and timelines based on final executables and processing traffic |

*Initial Estimates—Data Volume and Steady-State Processing.* System designers need to determine what elements are involved in meeting required timelines and then allocate part of the system timeline to each of the elements. The timeline may be completely allocated to a single element if we need only one element to carry out the system task. Element designers then have to allocate each element's timeline to its processing and operator tasks and prove this allocation is correct through analysis and design trades. Early in a system and element design, there is a lot of uncertainty as to how much processor power and data flow will be needed to achieve a defined performance objective. Thus, designers must not try to analyze the timeline in detail. Instead, they need "back of the envelope" analysis of the volume of input and output data and steady-state processing. They begin by projecting known data flow and measures of processor use from similar tasks onto our tasks. Even with this immaturity of the data supporting the analysis, obvious choke points in processing and data flow surface quickly. With this information, the element designers can support trades on configuration-item responsibilities, server sizing, and relief for timeline requirements.

*Refined Estimates—Detailed Analysis and Prototype.* As the design matures through preliminary, PDR, and CDR subphases of development, the uncertainties in the timeline analysis decrease, so element designers need to replace static models of processing and data flow with dynamic ones. That is, coarse software and hardware tasks in the static models become executable components, with processing based on more detailed analysis or prototype evaluations. An important transition of static to dynamic models occurs where the element's mission applications interact with the services that support the mission. For example, during the preliminary element design, designers might apply a static overhead to reading a block of data from a database. As the design develops, they replace this static overhead with more realistic data, based on whether the data resides in flat files or a database and on how well software can access the database through a network. With this extra fidelity, the element designers can find hidden choke points in the design or processing architecture not revealed by the simpler static models.

*When to Monitor Timing and Timelines.* Timelines become a key concern (need monitoring) whenever there is a risk of not satisfying timing and timeline requirements, particularly when resources contend in their use of the network or database, or whenever the processing workload forces the design to a top-end machine or to more exotic processor architectures, such as parallel processing machines. If element designers discover these trip points, they must consider assigning another processing paradigm or renegotiating the requirements.

*Monitor Modeling of Timeline Requirements.* As mission operations managers, we also have to monitor how the element design evolves to ensure proper progress in the timeline analysis and trades. We need to identify all timeline issues by the program's PDR and resolve them during the PDR and CDR. This resolution includes transitioning static models to more dynamic ones as the design matures. That is, timeline analysis must track with the element's evolving design. If element designers don't keep the design current with the element's concepts for passing data, control, and processing characteristics, the entire analysis may be useless.

*Monitor Modeling of the Interaction between Mission and Service Layers.* We monitor this aspect to determine if we've modeled mission applications enough to draw out conflicts between, and overhead for, application services. New computer architectures use distributed computing and data servers with an application service layer one notch above the operating system. This service layer insulates mission applications so they don't need to know where the data comes from or which processor holds a particular application. But element designers must model very carefully the time penalty associated with this layering. If an application calls a service many times, and that service requires twice the time originally allocated, the designer of a configuration item may have to reconsider whether the mission can afford to use such a service. Our modeling of this dynamic behavior between the mission and service layers must be detailed enough to discover potential problems in our timelines before we code and test the final product.

*Monitor Modeling of Mission Operations Scenarios.* Finally, timeline analyses of individual requirements and system events must work together in the context of real operations. So element designers have to synchronize their timeline analysis with the element's evolving concepts for processing to yield an accurate "day-in-the-life" scenario for operations. For example, if we have to produce a schedule within a certain time, our timeline model must show data arriving and being processed on the same frequency. If processing is based on ending contact with the spacecraft, the timeline model must script all contacts in a day and set events into its dynamic triggers based on ending the contact. This effort can be quite extensive, but it's our only assurance, before starting operations, that mission operations will work within its specified timelines and within the defined capacities for data flow and processing.

---

*Monitoring Timeline to Update the Spacecraft's Execution of a Mission Activity.* Let's examine a system requirement to update the spacecraft's execution of a mission activity within 30 minutes of a triggering event. Normally, the spacecraft is loaded with mission instructions once per day. We'll assume, for this example, that the triggering event occurs 10 hours after the nominal update.

*30 Min Allocation: 20 Mission Operations + 3 Establish Link + 7 Update Spacecraft.* First, we analyze the requirement at the element level. Because the command-uplink rate is pre-established, it takes 7 minutes to update and enable an existing on-board memory with new instructions. It also takes 3 minutes to establish a space-ground link. Therefore, of the original 30 minutes, mission operations has 20 minutes in which to prepare for the new contact and uplink with the spacecraft.

*20 Min Allocation: 5 Mission Planning + 10 Activity Planning + 5 Mission Control.* Three functions work together to prepare the new sequence: mission planning, activity planning, and mission control. We have to update the mission plan with a new contact time and an activity-level description of what has to be uplinked. We'll begin the analysis by allocating five minutes to mission planning for this effort. Activity planning then is alerted that we've updated the mission plan and now must prepare the actual command-load update. Because we don't know the exact nature of any one update, we allocate ten minutes to activity planning. Finally, the last five minutes of the timeline is allocated to mission control so we can prepare for and establish a space-ground link, and then uplink the new commands.

*Actual: 15 Min for Activity Planning and 10 Min for Mission Control.* After analyzing the individual tasks within the element design, we discover that mission control requires at least ten minutes to support a space-ground link. Further, activity planning requires 15 minutes to prepare the new commands. At first, then, it appears we can't meet the system timeline.

*Solution: Design for Concurrency between Activity Planning and Mission Control.* But after further examination, we discover that, if we send the plan directly to control and activity planning, we can do the real-time processing while generating the new commands. With this small change to the element design, the new requirement is fully satisfied.

---

**Defining System Performance.** Table 6.7 shows how we define tasks related to system performance throughout the development phase up to the critical design review.

**Table 6.7.    Defining System Performance.** We must thoroughly define the work required to design and demonstrate the system's performance requirements.

| Process Phase | Tasks to Cover |
|---|---|
| Definition | • Establish the system's conceptual requirements for performance (quality, throughput, and response)<br>• Investigate potential constraining points in the system that mission operations must deal with<br>• Analyze the system's conceptual requirements to establish the ground-processing capacities necessary to support the desired system performance |
| Development (Preliminary) | • Convert element performance requirements and timelines to computer software configuration items, hardware configuration items, and operations<br>• Identify areas of defined system performance that require complex algorithms, state-of-the-art processing, or data management<br>• Establish performance tests that demonstrate the element's ability to meet objectives for system performance |
| Development (Preliminary Design Review) | • Prototype and analyze processing to ensure we can meet test metrics for performance |
| Development (Critical Design Review) | • Finalize algorithm and processing design based on results of prototyping and analysis |

*Factors that Limit Satisfying the Mission Objective.* Conceptual requirements for the system, which eventually become specifications of system requirements, state the quality and quantity of mission data that the system is to support, as well as responsiveness to input stimuli the system requires to meet the mission objective. The quality of data influences, and is influenced by, characteristics of the spacecraft payload and operational orbit. The quantity and responsiveness of data produced and processed by the system depends on the number of mission spacecraft; the amount of time the spacecraft can communicate with the ground; the data rates (including mission, command, and telemetry) of the space-ground links; and the throughput capacities for ground processing. System and element designers have to fully analyze and understand these factors early in the definition phase.

We may create constraining points while upgrading the system to better meet the mission objective. For example, as a system matures, advances in technology create a capability for increased quality, quantity, or speed in one element that other elements can't handle. Also, at times, we may have more spacecraft within communications line-of-sight with the ground than ground elements to receive or process the data. Managing these constraining points requires more complex mission operations design.

*Identifying Limits.* The system and element designers' initial analyses of the system constellation, spacecraft-bus design, spacecraft payload, and ground processing determine where constraints restrict a system or element from meeting system performance objectives. When they identify a constraint on ground processing, they have to solve it—typically by adding complexity, processing power, or future technology. Proper analysis of the system requirements will identify all potential areas of increased complexity and lead to system approval of an approach the ground element will take when the situation occurs.

Designers have to avoid overdesigning or underdesigning element processing to cope with these situations by balancing the number of occurrences against the severity of any single occurrence. It's sometimes difficult to quantify the severity of a situation for such a trade to take place. But they must try, so they don't choose an exotic solution for situations that don't occur often enough to warrant the cost of developing and maintaining the solution.

*Analyzing Trades.* System designers need to do system-level analysis and simulations as early as possible to determine the system sizing and spacecraft capabilities required to meet the mission objective. These include parametric studies of sizing for the spacecraft constellation and ground receiver, orbit parameters, command rate, telemetry rate, mission-data rate, mission-sensor capabilities, and if appropriate, tasking throughput. These studies compare benefits (better satisfaction of one or more mission objectives) to costs (increased on-orbit or ground hardware, increased processing speed) for these kinds of variables. Usually, the parametric analysis establishes a clear point for the best ratio of benefit to cost— the break-even point on which designers should base requirements for system performance. They need to keep these parameters in mind because technology advances often move this break-even point.

As an example, assume that a key mission objective for a system that examines global deforestation is to minimize the time between looks to a given geographic location. There is a limited altitude at which sensors can resolve an area, given the current technology. The system may decide to study a technology trade on sensor resolution versus number of spacecraft required. Because advanced technology is always a few years ahead of current technology, and systems have a design life of a decade or more, later spacecraft may use the newer technologies.

Figure 6.4 illustrates the results of that parametric study. Curve 1 shows that, with current technology, increasing the number of spacecraft initially achieves large gains in the mission objective. But curve 2 shows that, after a point, the cost of building and concurrently operating multiple spacecraft becomes prohibitive. Advanced resolution sensors, which allow for increased orbit altitudes, are the more cost-effective solution to increased performance, as shown by curves 3 and 4. The advanced sensors allow the same resolution at a higher altitude, and higher altitude increases the spacecraft's coverage. Therefore, fewer spacecraft are required to achieve the same time between looks to a particular area. The system's break-even point could occur at four spacecraft with current technology sensors, but at three spacecraft with advanced sensors. The system designers, who are responsible for the system requirements, must be careful not to overspecify the ground element to operate four spacecraft, all with advanced sensors.

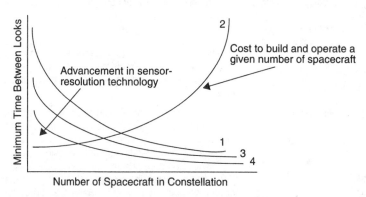

**Fig. 6.4.**      **Illustration of Parametric Analysis for System Performance.** Parametric analysis of system performance results in a design that meets current and future needs. We can increase coverage by increasing the number of spacecraft (curve 1), which increases operating costs (curve 2). But we can also increase coverage by improving sensor resolution and increasing spacecraft altitude (curves 3 and 4).

*When to Monitor Definition of System Performance.* Because of the direct influence this attribute of element design has on the cost of an element, it is always considered key. Further, system and element designers must revisit the need for these system-performance requirements when analysis shows that meeting the specification adds cost (in required processing power or algorithmic complexity).

*Monitor Definition of System Performance.* As mentioned in the introduction to this section, as mission operations managers, we review the definition and development plans to make sure they include the tasks in Table 6.7. We also ensure that tasks are accomplished and issues resolved according to the preceding discussions.

We monitor this attribute of the element design to ensure analysis and simulations are correct and complete. But more importantly, we make sure the mission objective isn't over-satisfied, resulting in an over-designed spacecraft or mission sensor, or leading to a spacecraft constellation and ground-processing capacity that's larger than necessary.

Finally, because monitoring this attribute is so important, we should understand performance requirements, simulation, and analysis. Then, we can develop our own conclusions, rather than relying solely on those of the system and element designers.

**Optimal Operation of Spacecraft Subsystems.** Table 6.8 shows how we achieve optimal operations of spacecraft subsystems throughout the development phase up to the critical design review.

**Table 6.8. Optimal Operation of Spacecraft Elements.** We must thoroughly analyze the space and ground elements and their interactions for limits on optimal operations.

| Process Phase | Tasks to Cover |
|---|---|
| Definition | • Establish conceptual requirements related to system performance (quality, throughput, and response)<br>• Determine which spacecraft subsystems constrain mission operations |
| Development (Preliminary) | • Convert element requirements and timelines to computer software configuration items, hardware configuration items, and operations<br>• Analyze the design and methods of mission operations to fully use spacecraft subsystems<br>• Do trades between optimal operations and complexity of ground processing by way of analysis and prototyping |
| Development (Preliminary Design Review) | • Select "break point" in trade space |
| Development (Critical Design Review) | • Develop final methods based on analysis and prototyping |

*Constraining Subsystems and Operations.* To establish conceptual requirements, element designers need to analyze the spacecraft's components to determine which spacecraft subsystems and operations influence mission objectives. The analysis of system performance (refer to section on System Performance Definition) may show that ground operations must use all of the spacecraft's capabilities for power, loading, and payload to meet the mission objectives for data quality or throughput. In other words, we have to push the spacecraft to its limits without violating constraints. To best use the spacecraft and its subsystems, system and element designers must study trades on gains in system performance versus complexity of ground operations. These trade studies include the affect of increased dependence on processor speed and capacity, as well as on algorithmic complexity.

*Fidelity of System Modeling.* Another area to be traded is the fidelity of the spacecraft subsystem modeling required to get the additional performance from the spacecraft, which must take into account how much uncertainty is in the model's input data. This is probably the parameter least understood in optimizing trades. For example, if the power expenditure has to be planned over a day so that the most important mission objectives are met at all times, the element's planning tasks may have to model power expenditure. However, if the equations necessary for accurate modeling of the subsystem require knowledge of the subsystem's dynamic behavior, we have to statistically generate the data or simplify the equations. Having high-fidelity equations with low-fidelity data leads to false expectations and, potentially, to continual changes in order to isolate more of the statistically outlying data. Through this analysis, designers determine a break point in the trade space—the point at which adding more complexity is not worth pursuing.

*When to Monitor Optimal Operation of Spacecraft Subsystems.* This attribute becomes key to the element design (needs monitoring) whenever the spacecraft's abilities limit meeting the mission's performance objectives. The analysis and methods designed to solve these problems will consume the single largest engineering effort in developing a ground element. Therefore, we must reduce the complexity of the ground element as much as possible by closely monitoring this attribute.

*Monitor Mission-Objective Satisfaction and Design Complexity.* We examine the analysis and trade studies to make sure the system meets mission objectives well enough to satisfy users. As stated before, this may entail revisiting and renegotiating the system requirements. We also ensure that the mission objective isn't oversatisfied because designers have added unnecessary complexity to operations, methods, and design.

To avoid complex design, we must not commit an element's functions to overcome several of the spacecraft's limitations at the same time. The spacecraft's subsystems tend to drive different, almost mutually exclusive, solutions from the ground to compensate for these limitations. Trying to sort out all of these key features at the same time puts tremendous pressure on the organization for element design and often causes development costs and schedule slips.

**Definition of Displays, Alarms, and Procedures.** Table 6.9 shows how we define displays, alarms, and procedures throughout development up to the critical design review. They're the bridge between the operator and the mission software and hardware in the ground element. Operators need to assess, monitor, and control not only the mission but also its underlying software and hardware. How the operator receives data to support mission execution or processing of abnormalities is critical to the overall element design.

*Initial Concepts for Display.* Even in the earliest phases of a system's development, the look and feel of the environment for operational processing begin to take shape. Human Machine Interface (HMI) concepts can help determine if a display can reasonably achieve a system's conceptual requirement. For example, one con-

**Table 6.9.    Definition of Displays, Alarms, and Procedures.** We must start meeting operators' needs early in the definition phase.

| Process Phase | Tasks to Cover |
|---|---|
| Definition | • Establish conceptual requirements related to system performance (quality, throughput, and response)<br>• Investigate driving display characteristics to support conceptual requirements for the system |
| Development (Preliminary) | • Convert element requirements to computer software configuration items, hardware configuration items, and operations<br>• Develop preliminary displays<br>• Prototype and evaluate displays for operability<br>• Develop concepts for alarms and procedures |
| Development (Preliminary Design Review) | • Define operational displays, alarms, and procedures based on defined operator tasks |
| Development (Critical Design Review) | • Refine displays, alarms, and procedures to complement needs of operator tasks and software interfaces for configuration items |

ceptual requirement may be to visually inspect the entire mission plan from a single display. Depending on the size of the constellation and the kind of data to be displayed, this may not be feasible. An early concept of a plan display will show whether it can reasonably present all of the spacecraft activities placed on a timeline in the constellation. This work may change the perspective of how data are presented for review, and hence, influence the interface between the operators and software.

*Display Definition and Operability.* Element designers define operator displays. However, because this attribute is the most visible sign of the element design, the end users—the flight operators—are invited to help review and critique the displays from the beginning of design. They prototype displays as needed to demonstrate features and to get approval from system and operations people. Displays must be designed for operability because the way operators interact with the mission software and hardware truly affects the system's efficiency. Also, end users measure the quality of a ground element's design based on operator interaction. If they see the interaction as inefficient, all software and hardware processing, including complicated algorithms employed in the processing, is suspect. Designers must create displays that allow operators to navigate through them efficiently. They must also establish standards that provide a common look and feel for operator interaction, at least across the tasks associated with a given position, if not across the entire suite of positions.

*Alarm Presentation to Operator.* In general, alarms follow an evolution similar to that of displays. But, because alarms bring urgent information to the operator's attention, element designers give alarm processing special consideration. Alarms must get an operator's attention without obscuring current work. For example, if telemetry indicates a spacecraft's state of health condition is critical, operators need to know which of the spacecraft's subsystems is having the problem. However, they are still monitoring the rest of the spacecraft's operations through other displays. If an operator isn't examining the faulted subsystem's components at the time of the fault, he or she either reduces that display to an icon or places it behind the other displays. When the alarm for that subsystem occurs, it shouldn't immediately obscure the operator's view of other displayed measurands (for example, by automatically bringing the display for the faulted subsystem to the foreground). Changing the color of either the display border or the icon to red, with an adjoining message on a continuously viewable message-center display, gets the operator's attention without jeopardizing other work.

*Alarm Levels and Displays.* Alarm concepts, including the various levels of alarms and the corresponding audio/visual displays, are established early in development. For example, the most visually perceptive way to illustrate a link fault of some kind is to change the link color on a continuously displayed graphic. The graphic may normally show the signal strength numerically next to a line connecting the spacecraft to the ground receiver. Whenever that signal strength goes below some level, the line and value change color. This way the operator's view is immediately focused to the source of the out-of-limit condition. Alarm levels significantly influence the design of software and hardware because element designers must understand the severity of an alarmable event and build an appropriate reaction.

*Alarm Overrides.* Coupled with alarm levels is a feature allowing operators to override or reset alarms. Element designers must carefully examine conditions for overriding an alarm because doing so improperly may jeopardize a spacecraft's state of health operations. Again, the software and hardware underlying these alarms will vary depending on their ability to continue processing when the operator overrules the alarm. Designers can best select these concepts through iterative prototyping and evaluation, in which element designers sit with operators to work out the most instructive ways to show and respond to alarms.

*Procedures.* Procedures are slightly different from displays and alarms. Element designers develop procedures to guide operators through mission activities or problem resolution. The procedure may be as simple as an online handbook that operators refer to when they see a condition they're not familiar with, or it can be as elaborate as a software-based script that can make certain data-based decisions for the operator until it reaches a step that requires human action. For example, an operator may use a dynamically executed procedure to isolate a fault in the telemetry-processing path. All the operator knows at first is that telemetry isn't updating on the measurand displays. The operator carries out the procedure for isolating a telemetry fault. The procedure script may have queries to the last few seconds of telemetry to see if signal strength fell off, indicating a possible spacecraft problem. Or it may have messages sent to the ground software and hardware components, with healthy components responding, thus isolating the faulted components.

*Procedures and Mission Software Design.* All procedures follow a basic flow that is displayed to the operator. The procedure will either find the source of the failure or reach points in the procedure requiring an operator's help. Obviously, the more we tie the procedure to the mission software and hardware executables for information, the more those executables must be aware of what the procedure needs. But even procedures in online handbooks will influence the software design. The handbook may ask the operator to check if a measurand is set one way or another, thus requiring software to retrieve and display this information quickly. Data-retrieval speed, organization, and display characteristics of results may depend on what is in the procedures, especially those that help operators maintain the spacecraft's state of health.

*When to Monitor Displays, Alarms, and Procedures.* Displays, alarms, and procedures are always key to the design of an element because they directly influence the definition of operator tasks. As stated earlier, these attributes bridge what the operator does to control and monitor the element with the underlying processing by software and hardware. If the bridge is missing or improperly thought out, there's little chance the underlying processing will capture the operators' needs. More importantly, displays, alarms, and procedures establish the element's operability in the eyes of the system and the end users. Buyoff for the entire element design lies largely in the operators' ability to believe the element is operable.

*Monitor User Needs.* We monitor this attribute to assure operators, mission engineers, and spacecraft engineers that our prototypes and evaluations satisfy the technical and operability requirements for the human interface. Further, we make sure the mission and service-layer design needed to generate displays is in place at the program's PDR.

**Verification.** Table 6.10 shows how we verify results throughout the development phase up to the critical design review. Even though we don't verify element and system requirements until late in the development cycle, element designers need to engineer the element properly to meet verification needs beginning early in definition and development. Verification and testing of an element are often overlooked in the rush to get the mission operations understood and designed.

*Verification Methods.* We build verification into the design starting with the actual language of requirements in the system definition. For each requirement, designers of the element and its configuration items must determine the effort needed to show the requirement is satisfied. Using verbs such as process, use, and perform disallows verification. For instance, requirements such as "the element shall use ephemeris in its planning functions" is so open-ended that we could think of endless possibilities for verification.

After the requirements are stated in a verifiable language, designers assign verification methods and a time frame. Typical verification methods are test, demonstration, analysis, and inspection. Verification can occur during testing of individual CIs, testing of the integrated element, and system-level demonstrations. Test operates the component under specific conditions and verifies a quantitative

**Table 6.10. Verification as an Attribute of Design.** We must design the element and define testing methods to support verification within the ground element.

| Process Phase | Tasks to Cover |
|---|---|
| Definition | • Establish and assign verification categories for each system and element requirement<br>• Analyze requirements for verification ambiguity and appropriate level of verification |
| Development (Preliminary) | • Convert element requirements and timelines to computer software configuration items, hardware configuration items, and operations<br>• Develop concepts for conducting and transitioning between tests |
| Development (Preliminary Design Review) | • Analyze design's ability to be tested and to have its requirements verified at configuration item and element levels<br>• Develop preliminary test plans for verifying configuration items and element |
| Development (Critical Design Review) | • Develop final test plans for verifying configuration items and element |

requirement. Demonstration operates a component and qualitatively verifies its functions. Analysis is verification using models or other mathematical methods to show requirements are satisfied. Inspection consists of reviewing designs or documents to ensure they meet requirements.

Most of the mission operations element's requirements pertain to control and processing decisions and are verified by a formal test. Test inputs are generated and read into the software or hardware, and the output is captured and compared to expected results. A certain class of requirements involves a spectrum of conditions, such as specified orbit characteristics, that can't be tested to meet all possible combinations of specified inclinations or altitudes. In these cases, verification by analysis is appropriate. That is, designers should show analytically that the equations won't produce unpredictable results.

Another way to verify wide-ranging requirements is to define a specific test agreed to by designers of the system and element. This test should have a well-defined set of input data as well as expected processing results. Usually, complex algorithms meet these requirements, which correspond to limits on system or element performance.

An area often overlooked is validating algorithms after they're developed but before they're implemented in software. Element designers should analyze them and do prototyping if necessary to prove that the methods satisfy requirements. Examples of methods that require validation are orbit determination and spacecraft power models.

*Designing for Testability.* As the element design begins to mature, designers and engineers must answer issues that appear only in a test environment, such as the ability to set a test date and time into the future from the current clock date and time, the ability to stop a test in the evening and resume the next morning, and the ability to transition an element from a test to operations. As the requirements are analyzed within the CIs and eventually specified as executable components, designers also have to examine how to verify them. This is especially true for requirements satisfied by complex algorithms. In some instances, multiple algorithms may process data sequentially or with control and feedback. Verifying that we've met the CI requirements is most difficult in these cases because we must theoretically show that the lead-off algorithm needs verifying first, before any subsequent algorithms. Unfortunately, this approach consumes a lot of time. Therefore, CI designers must configure their executables with test-only ports for input and output data, so they can verify multiple algorithms one at a time. If we don't understand this effort as the design matures, verification will cause cost and schedule slips.

Element designers create test plans for the element and its CIs as the CI development matures toward PDR. They should base these test plans in part on the mission operations concepts and articulate the test scenarios needed to verify each of the selected requirements. At this point in program development, element and CI designers must understand how many different kinds of tests and what efforts to set up, capture, and analyze data we need to verify certain requirements.

*When to Monitor Verification.* Because verifying an element's requirements is a costly attribute of its development, verification is always key in all of the subphases of definition and development. It is also key because verification occurs not only after the design is complete but also after all of the code and hardware are built. To find that we can't verify the design at this late stage puts a great strain on the development schedule for delivery of the final product. As mission operations managers we must pay close attention to the requirements that may be difficult to satisfy. One example could be timing or timeline requirements for spacecraft acquisitions. Another example is spacecraft health or safety requirements such as avoiding depleting the batteries in power-constraining situations.

*Monitor Verification.* We monitor this attribute to ensure the element designers understand verification for all system and element requirements. We also ensure they understand how the element will verify the requirements, first at a high level and, ultimately, down to individual algorithms. We must check the requirements specification, test plans, and test procedures for clear, specific details of verification—precise statements about how designers generate test input data and then capture and process test output data. For example, they should either provide or explicitly reference all mathematical methods required to support verification or requirements. Finally, we make sure test-unique requirements are placed, as needed, on the CI design to allow testing.

## 6.2.2 Ensuring Completeness and Maturation

Definition and development is a holistic process, constructed in layers. The layers represent two views at the same time. First, they represent the level of uncertainty in design and requirements—uncertainty that must be peeled away as the system and element designs evolve. Secondly, the layers represent the subphases of definition and development. The most efficient (cost effective) way to develop a space-ground system is to completely peel away each successive layer, addressing all attributes of the design while handling uncertainties in the design and requirements at each layer. As mission operations managers, our role is to monitor definition and development of the system and element so (1) requirements and concepts cover all the ground element's operations, and (2) key concepts and subsystems are mature in the early phases. At the conclusion of each of the steps discussed in Sec. 6.1.3, the program should be able to document and present a consistent level of completeness and maturity in the evolving system and element designs.

**Maturation of Requirements, Methods, and Timeline.** Too often, however, a program's maturity and completeness are measured by a schedule of calendar events instead of an understanding of concepts, requirements, and design. In this mode, a program becomes more focused on ensuring that a set of requirements is documented, rather than ensuring that the program exhibits proper understanding behind the requirements and design. To avoid this situation, we must be aware of the symptoms. For instance, suppose an audit of the requirements on the element reveals that an upper-level requirement is repeated verbatim in the next-lower layer of requirements without analysis or an operations concept to describe how the element meets the upper-tier requirement. In this case, the element design is not maturing. In fact, the element designers are merely "binning" requirements, not analyzing and designing the element. Another symptom manifests itself in methods and timeline analysis. The element design isn't maturing if the early stages of a methods development can't describe all of the inputs to the method or how a method operates on this input data. Also, if timelines allocated to an element or its tasks can't be proved or disproved by analysis or prototyping, the element design is not maturing. If any of these symptoms are present in the element design and the program still remains on schedule, the program is headed for a major failure later in the development phase.

**Design Completeness.** Another error that element and CI designers tend to make in developing an element is working locally or examining in detail a particular design attribute, while completely ignoring other attributes. For example, the CI designers may become too focused on how the software executables of a computer software configuration item are grouped and constructed, while completely ignoring the operations' concept of these processing tasks. We must ensure this fine-focused approach is not taking place at any of the definition and development steps, and especially not as the program approaches the preliminary and critical design reviews. The symptoms of this situation are difficult to detect until the actual design reviews—when all attributes of an element undergo extensive review. Early warning signals of an impending problem surface in the areas of timing and timeline analysis, verification, and processing concepts. Suppose we can't answer basic questions on starting and triggering processes, CPU requirements for executing a task, and ways to verify requirements. Yet, detailed methods are already under way. If so, the element's design attributes aren't progressing equally.

### 6.2.3  System Reviews—Concepts, Requirements, and Design

**Review Mechanisms.** As we discussed in Sec. 6.1.3, concurrent definition and development isn't free of risk. We must ensure proper system-level review and approval of the preliminary mission operations concepts and element specification during definition, as well as of the refined mission operations concepts during preliminary development. This is necessary to prevent a program from missing key concepts and definition in the most formative stages when the parallel activity offers the highest risk. We must carefully review the element's operations concepts and requirements specifications at the early stages because they set the tone for all operations and the design architecture. People who do this review must be expert in conducting operations, system definition and interface specifications, software and hardware maintenance, and computer and data processing technologies.

If at all possible a program should avoid formal reviews lasting two to four days because so much data is presented in such a short time that the opportunity to critique and examine issues is virtually eliminated. Instead, we should organize reviews as requirements and design are developed. The element's concepts to support the system's conceptual requirements are the central theme of the review. We also discuss each attribute of the design. This part of the review is essential to an element's maturation because it shows that the element designers understand all of the issues of a given design before development proceeds to the next subphase.

**Review of Requirements Maturation.** Also discussed at the reviews are the requirements allocated to the element or the function within the element, depending on the review phase. This is not intended to be a simple restatement of the requirements from the upper tiers; it is to be a discussion of what the requirements mean at the design level being reviewed and of how they will be satisfied. We keep reviewing requirements until the program's preliminary design review, when requirements have become executables. For requirements we consider key, the review must discuss how we've analyzed the concept or design to prove that it satisfies the requirements. When software algorithms are to satisfy requirements, the review should discuss them in enough detail to assure reviewers they'll do so.

Finally, the review must show that the design activities are addressing the uncertainties we're working or have newly identified. The review of these to-be-resolved (TBR) and to-be-determined (TBD) issues shouldn't simply state that we have to work an issue by a given date; it should include what specific steps will close the TBD/TBR, such as analytical analysis, prototyping, or interviews with operators.

**Role of the Mission Operations Manager.** For such a review to be successful, mission operations managers need to be an integral part of definition and development. As operations managers, we need to review the design internally, as the subphase evolves, to ensure the designers can answer the kind of questions that surface in the external review. Further, we must make sure the element-design organization treats the review as constructive, not something that has to be done because it is on the schedule or something they can prepare for in a couple of weeks. We need to ensure that the element designers make available enough materials and that enough time and resources are committed for the review. The element designers can't provide the documentation to the reviewers just ahead of the review dates. Because the review should be constructive, we must be able to secure a commitment from the external review groups to participate for months, not for a few weeks here and there. Finally, mission operations managers and participants need to view these reviews as a learning opportunity and a way to alter a wayward concept or design.

## 6.3    Scenarios and Subsystems to Monitor

The key to efficient and operable ground mission operations is understanding first all required operations and maintenance activities and second all necessary effort in definition and development. As mission operations managers, we must make sure the operations and maintenance activities are analyzed broadly and deeply so appropriate definition and development tasks are carried out. Failure to identify how much effort definition and development will take has two results. The first is that the work isn't done and thus mission operations aren't fully defined. The second is that the need for the work is identified after the contracts have been negotiated, so we'll have to absorb the work within the dollars of the current contract or change the contract.

In this section we use the principles discussed in Sec. 6.1 and 6.2 to show how element designers should analyze mission operations scenarios with regard to definition and development tasks as well as operations. Whenever possible, they should do this analysis before starting work on definition and development. If limitations on knowledge and time make an early start impossible, they should carry out the analysis during the early parts of definition and development while maturing the operations concept. This timing will ensure a properly focused effort. We provide only a few examples here. Mission operations managers should make sure all mission and maintenance activities are reviewed and key ones analyzed as illustrated in the following subsections.

This section illustrates the conceptual and analysis issues for scenarios that system and element designers must address to ensure the element design is complete at the end of definition and development. Finally, these scenarios show how broad and deep details must be to gauge the design's maturity. This section also illustrates the level of engineering effort required to fully understand the mission operations element's responsibilities in maturing scenarios involving other elements—in this case, the spacecraft element.

We'll discuss these spacecraft scenarios:

- Acquisition
- Spacecraft bus and payload calibrations
- Spacecraft safemode operations
- State of health operations
- Command and control
- Power

For each example, the discussion is broken into three parts. The first part discusses the attributes that make the scenario a key concern and require monitoring by the mission operations manager. The second focuses on what is required to support the definition phase, indicating the kind of analysis element designers must do, as well as the level of detail required at the end of the definition phase. The third part discusses the issues and analysis for the development phase, following the same format as the definition phase.

## 6.3.1    Spacecraft Acquisition

The spacecraft acquisition scenario is important to monitor because of key concerns from three design attributes. First is the critical timing and interaction between the ground and space subsystems. Second is the need to establish and obtain approval of operator tasks. Third is the need to establish displays, alarms, and procedures for normal and contingency situations.

In the definition phase, system designers establish the spacecraft acquisition events and timeline and allocate them to both the spacecraft and mission operations elements. The acquisition events start with the pre-pass checks and end with the communications link established from the ground or relay spacecraft to the mission spacecraft. System and element designers carry out analysis to demonstrate that the parts of the timeline allocated to each element are reasonable. Determining reasonable timeline allocations requires that each element establish the high-level processing and operator tasks needed to support spacecraft acquisition. The analysis specifies the different types of links (widebeam and narrowbeam), antenna scan strategies, and equipment configurations needed for acquisition. The definition phase can't be considered complete until details in four areas clearly establish element responsibilities. The first area is understanding the spacecraft's performance characteristics, such as antenna slew rates and the time needed to configure the equipment supporting acquisition. The second area involves key constraints, such as line-of-sight limitations between the ground and spacecraft, or antenna travel limitations relative to the spacecraft body. The third area is understanding the spacecraft's autonomous processing requirements for acquisition, such as the criteria for going to autotrack, or loss thresholds for losing a lock signal.

The final area pertains to messages, data, or other information, such as oral or written directives passed between the spacecraft and people in the mission operations element.

In the development phase, element designers analyze the spacecraft acquisition requirements and timelines and allocate them to the software and hardware processing tasks and operator tasks by way of the mission operations concept. They analyze to show that the timeline allocated to each task is reasonable. Preliminary alarms and procedures for operator displays are defined for normal and possible contingency acquisition. In the latter part of the development phase, which supports the preliminary and critical design reviews, further analysis confirms that the element tasks, as broken down into capabilities and executables, can meet their timelines. Element designers assign the operator tasks defined for acquisition to operator positions and then analyze the operator positions to ensure operators aren't overly tasked. They also prototype operator displays and alarms to demonstrate operability.

### 6.3.2    Spacecraft Bus and Payload Calibrations

As the mission operations managers, we monitor the scenario for calibrating the spacecraft bus and payload because of key concerns from two design attributes. First, calibration methods affect algorithm complexity. Second, flight operators must interact and coordinate with mission and spacecraft engineers and the processing tasks that support this interaction.

In the definition phase, system designers define the requirements for calibrating the spacecraft subsystems. The requirements cover spacecraft attitudes, maneuvers, equipment configurations, geometric constraints, and frequency of occurrence or other trigger mechanisms. System and element designers determine whether the calibrations need to be part of the mission plan or can wait for the activity plan. Trades are done between the spacecraft processing the calibration data onboard and the mission operations element processing the data with subsequent uplink of updated parameters. Also, designers should consider how the results of the processing will be distributed to spacecraft engineering. One more dimension in the trade is whether the ground processing needs to be online as part of mission operations, or whether spacecraft engineering can do it off-line. The definition phase is complete when these trades have been resolved and each element fully understands its role in calibration and data processing.

In the development phase, element designers develop ways to incorporate calibration activities into the activity plans. If the activity is to be planned, they must cover calculating the placement and duration of calibration activities, as well as criteria for including them in the activity plan. If the activity is to be scheduled, the methods need to cover calculating the specific attitudes, maneuvers, and equipment commands. The attitudes can be relative to the Sun, stars, or other objects. Maneuvers can be specified by required rates or a required number of star crossings. Also, in the development phase, designers define the interaction between flight operators, mission engineers, and spacecraft engineers in processing, reviewing, approving, and updating onboard parameters and ground databases.

### 6.3.3   Spacecraft Safemode Operations

This scenario covers entry into, operations within, and recovery from a safemode. It focuses on spacecraft operations when the spacecraft isn't in any imminent danger but is unable to support mission-payload operations. We monitor the scenario to ensure proper requirements analysis and maturity in support of the spacecraft's safe operations.

In the definition phase, system designers specify the events that cause entry into a safemode, such as a constraint being violated or the onboard mission load area being empty because a new load wasn't uplinked. They define the spacecraft activities that occur in the safemode, such as sunbathing to maintain power levels and planned acquisitions. These activities allow us to contact the spacecraft with the least effect on the mission plan and provide chances for loading. The various levels of safing are established from the upper level, which uplinks a set of activities to the lower level for autonomous command execution. Finally, designers establish the safing period—the number of days the spacecraft can operate in this state before requiring a new load from the ground. Because these loads can span many days of operations, scenarios must be developed for maintaining (generation and uplink) the "out-day" safing information. The definition phase is complete when the interface specifications between the spacecraft and mission operations elements are defined well enough so the ground element knows how to operate the spacecraft in an extended safemode state. Also, the interface specifications must articulate the procedures and uplink commands necessary to recover the spacecraft from its various safe mode levels to normal mission operations.

In the development phase, element designers define ways to support planning, scheduling, and loading of safing activities. They also refine the procedures for recovery from the various levels of operator roles and establish any displays needed to fully support the recovery. Finally, they carry out analysis to verify that the safing scenarios and activities result in the spacecraft's safe and efficient operation.

### 6.3.4   Spacecraft's State of Health Operations

The operations scenario to maintain a spacecraft's state of health ranges from immediate to near-term to long-term. The scenario covers areas in which mission operators' actions are critical to avoid long-term loss of spacecraft capabilities. Immediate state of health covers areas such as constraint violations, telemetry out of range, and equipment malfunction. Near-term state of health includes battery voltage approaching minimum limits and equipment temperatures nearing upper or lower limits. Long-term state of health includes propellant levels and cumulative equipment cycling limits.

We monitor this scenario as we do the safing scenario to ensure proper analysis and maturity of requirements, especially in the area of safe operations of the spacecraft during contingency or anomalous conditions. We also monitor the scenario to ensure that element designers address operability of displays, alarms, and procedures for immediate spacecraft situations. Finally, we assess how data is made available to the spacecraft engineers without affecting ongoing mission operations.

In the definition phase, system designers analyze the spacecraft design to determine how it will handle state of health over the immediate, near, and long terms. They analyze each part of the design related to state of health, so they can determine the level of monitoring and protection assigned to the spacecraft's software and hardware and to the ground's software and operators. They also identify responses to each design factor for state of health. Responses range from onboard autonomous commanding, to operator real-time commanding, to generating and uplinking replacement loads. Finally, designers determine data needed by spacecraft engineers to assess the spacecraft's state of health over the long term. The definition phase is complete when the interface specifications reflect all responses and their required timelines.

In the development phase, element designers design the displays required by an operator to monitor the spacecraft's state of health. Because displays are critical to spacecraft safe operations, they're designed for operability, including the ability to navigate efficiently through the displays and to access data quickly and assess the spacecraft's state of health. Designers must establish alarm scenarios, including the various levels of alarms and corresponding audio and visual displays. They determine how much an operator can override or reset alarms and develop operator procedures, including online executables that respond to state of health problems. They analyze displays, alarms, and procedures for operability. This analysis includes both prototype and formal evaluations involving operators and spacecraft engineers. They develop methods to support developing and uplinking replacement loads that respond to spacecraft's state of health problems (for example, defining and uplinking a sunbathing activity if power levels drop too low). They develop ways to assess long-term state of health and to access data, so spacecraft engineering's need for frequent or large amounts of data won't interfere with normal operations.

### 6.3.5   Spacecraft Command and Control

This scenario involves two fundamentally different types of spacecraft characteristics—spacecraft execution of commands previously uplinked from the ground and spacecraft execution of real-time commands from the ground.

**Spacecraft Execution of Previously Uplinked Commands.** This includes discussions on loading parts of the spacecraft's memory for execution by the spacecraft's onboard software, strategies the ground element will use for timely updates to the load, and the ground processing tasks necessary to prepare future loads. The loads generated must take into account the most recent knowledge of spacecraft state.

Spacecraft loading is one of the most difficult and complicated interfaces to work out between spacecraft and ground engineering. As mission operations managers, we monitor this scenario because we want spacecraft elements to operate as well as possible.

In the definition phase—through interface exchange meetings with spacecraft engineering—element designers determine exactly how the spacecraft software executes out of its own memory. To do so, they must know exactly how the onboard computer interprets the instructions and the specific timing of command execution. They analyze strategies to keep the spacecraft loaded with data for the mission and state of health operations. Considerations include the spacecraft's ability to accept updates to loads, the risk to mission success should an update not occur, the available onboard memory, and the ability to include the latest tasking information in the updates. Element designers need to get system approval for the loading strategy they choose because many factors influence this strategy, including sending and receipt of messages involving other elements.

The element designers analyze the ground processing tasks necessary to produce the load and allocate them to a processing timeline. They go through a series of scenario trades that cover several processing paradigms. We may load the spacecraft ad hoc or at discrete fixed points in either time or the spacecraft orbit. The processing tasks may execute in parallel or in series, depending on the interdependence or independence of the tasks and the ground processing capacities. This scenario depends on the currency of tasking requests for onboard execution of the load influence the load. Designers must develop scenarios for operator interaction, review, and approval of the load preparation and uplink.

Element designers must also develop scenarios for simulating the spacecraft execution of command loads to test whether the ground loads are correct and to verify that the ground response to certain telemetry signals is appropriate.

In the development phase, element designers understand and analyze the software complexities associated with the ground processing timeline. They finish the detailed interfaces and the spacecraft software. The CI designers finish their algorithms for load generation, including command sequencing strategies and mechanisms to trap and resolve violations of command constraints.

**Spacecraft Execution of Real-Time Commands.** This scenario differs from the scenario for executing previously loaded commands in that the real-time commands are carried out as soon as the spacecraft receives them from the ground. We use this kind of commanding for less sophisticated spacecraft that have neither onboard memory nor extensive capacity for onboard processing. We also use it for sophisticated spacecraft when the spacecraft's state of health operations are at risk or when a ground operator needs to act faster than generating a load update allows. This scenario yields a simpler interface than loaded commands because the spacecraft directly executes the real-time commands (for example, configure equipment or point antennas). These real-time commands aren't interlaced with commands generated by the spacecraft. For spacecraft with the less sophisticated onboard processing of real-time commands, there is no need to update parameters in the spacecraft's software, which would require a well coordinated effort between spacecraft and mission operations engineering.

In the definition phase, the element designers develop the scenarios for uplinking the real-time commands They trade generating the commands in real-time versus having the commands pregenerated and stored, but with the ability to select and uplink them in real time. The former approach requires fast processing to generate the appropriate commands in the spacecraft's format in real time, whereas the latter approach requires a very efficient ability to search and retrieve information from a database. Like the other commanding capability, real-time commanding requires simulation, as well as scenarios for operator interaction, review, and approval. But these areas are more critical in real-time commanding because of the very nature of the relationship between command uplink and execution.

In the development phase, element designers define the preliminary and final designs for spacecraft fault detection and isolation, command lookup, and subsequent uplink. As part of the development effort, the element and CI designers conduct studies that examine whether we can use "smart" software, such as rule-based-inference engines, to generate commands. They also explore the responsiveness and database complexities associated with using pregenerated commands that are retrieved based on real-time events.

### 6.3.6　Spacecraft Power

As mission operations managers, we monitor this scenario when the spacecraft's power capacity limits satisfying the mission objective. We monitor the scenario for two design attributes: the algorithm complexity that accompanies developing methods to support best operation of the spacecraft and the processing scenarios required to describe the roles of and relationships among the mission operations tasks for planning, scheduling, and determining spacecraft state.

Because there is a large range of complexity in modeling the subsystem, system and element designers use the definition phase to explore and decide how simple or sophisticated models need to be to satisfy the mission objectives. System designers determine the solar-array performance, onboard subsystems and payload power levels, and battery capacities. They also determine how much power-gathering activities, such as sunbathing, are needed to maintain adequate battery state-of-charge and voltage for mission activities. Finally, they establish the battery's operating requirements, including minimum allowable voltage, necessity for full-charge periods, and periodic battery reconditioning. With this information, the system and element designers determine, by way of analysis, if power limits the mission objective. Power may become a significant issue in meeting the mission objective only when efficiencies degrade in the battery or solar array. This information is critical in determining the ultimate complexity of the ground element design needed to counter this limitation.

In the development phase, if power is a resource we must allocate to spacecraft activities, element designers develop and validate the supporting methods for planning and scheduling. They determine a strategy for allocating power to activities that maximize the spacecraft's power margin while flexibly responding to the dynamics of planning and scheduling. Designers also develop ways to handle power as a spacecraft constraint. They include the processing required to recognize and compensate for a low state-of-charge in a battery before the spacecraft acts on its own. They do trades to determine how much mission planning, activity planning, and real-time command and control contribute to satisfying the scenario for spacecraft power.

## References

Larson, Wiley J. and James R. Wertz. 1992. *Space Mission Analysis and Design*. 2nd Edition. Netherlands: Kluwer Publishing.

McMinneman and Palmer. 1984. *Essential Requirements Analysis*. Englewood Cliffs, NJ: Prentice-Hall Inc.

# Chapter 7

# Activity Planning

Dave E. Kaslow, *Lockheed Martin*
Jeffrey K. Shupp, *Lockheed Martin*

One of the most visible pieces of a mission operations element is its activity plan. An *activity* represents a time reservation within which we satisfy a maintenance or mission objective. Activity planning has many facets we must explore to develop a plan that meets the needs of the mission objective in the most effective manner. This chapter presents the process for developing activity planning within mission operations from definition through preliminary development. The chapter focuses mainly on developing concepts and requirements for the system and the mission operations element in simple and complex planning environments. This chapter also addresses the aspects of a concept or requirement that require resolution at the system level rather than within mission operations. We present alternate operations concepts, along with factors that influence the selection of an operations concept through requirements and design trades. We also discuss how to develop the methods for software tasks and procedures for operator tasks that support the activity plan.

Figure 7.1 breaks out the three sections in this chapter and includes several common terms:

*Mission objectives*. The mission objectives determine the data's content, quantity, and quality. The system tasking organizations provide these objectives to mission operations.

*Maintenance objectives*. The maintenance objectives determine what we need to do to make sure the space and ground equipment remains calibrated, aligned, and

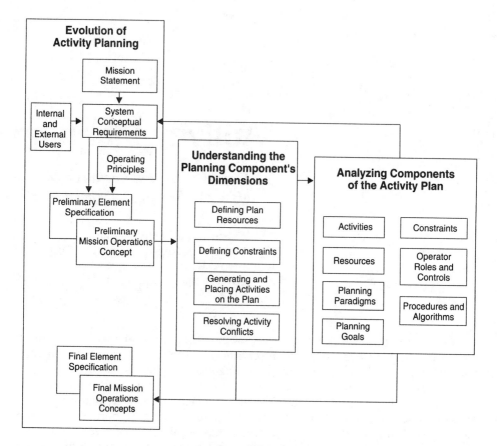

**Fig. 7.1.** **Perspectives on Activity Planning.** Applying the methods presented in Chap. 6 results in cost-effective definition and development and efficient operations for activity planning.

in good state of health throughout its operational life. Spacecraft and ground system engineering provide the maintenance objectives to mission operations.

*Constraints.* Constraints define the operational boundaries mission operations must observe in meeting the mission and maintenance objectives. One class of constraints defines the safe operating conditions for space and ground elements. These constraints are determined by spacecraft and ground system engineering. The second class of constraints, defined by the mission tasking organization, determines the geometric and temporal conditions that the mission objectives must satisfy.

*Resources.* Resources are mission components that directly contribute to generating, transmitting, and producing mission data. Examples are ground receivers, spacecraft sensors, or communications-link bandwidths. We consider a resource in the activity plan when it is in limited supply and therefore needs to be prudently or optimally allocated to help satisfy the mission objective. Spacecraft and ground-system engineering determine which components are planning resources.

*Activity.* Constraints are enforced within or between activities. Resources are assigned to activities. We satisfy the mission and maintenance objectives when corresponding activities are planned and carried out in a timely and effective manner.

*Plan.* The effectiveness of the activity plan depends on its ability to satisfy the mission and maintenance objectives, while ensuring that the spacecraft and ground elements are operated safely, efficiently, and effectively. All activity plans have the same three planning components: activities, resources, and constraints. The ultimate effectiveness of activity planning depends strongly on how designers of the system, mission operations element, and configuration items define and develop methods and procedures for handling the interaction of these components.

## 7.1 Evolution of Activity Planning

Activity planning evolves within the mission operations element, as overviewed in Fig. 7.1 and detailed in Fig. 7.2. Our discussion derives from the definition and development process described in Chap. 6, beginning with a mission statement and continuing through concepts for the mission and for operations of configuration items (CI).

Use this section as a guide for defining and developing activity planning. It shows how activity planning matures from the mission statement to the high-level operations concept. This presentation is at a high level, but it covers the breadth of activity planning. Actual definition and development should go into much greater detail. Make sure the plan components, conceptual requirements, operating principles, and high-level concept for operations are based on all the items in Fig. 7.2 and that the issues in the following paragraphs are resolved.

**Mission Statement.** As shown in Fig. 6.1, 6.2, and 6.3, the system mission statement is the basis for defining and developing the system. Correspondingly, as shown in Fig. 7.1 and 7.2, a mission statement is used to start developing the activity plan. The mission statement is "The activity plan consists of the activities necessary to support normal operations and planned maintenance of the mission operations element. Activities are arranged on a timeline to satisfy the mission and maintenance objectives, while ensuring that the spacecraft and ground adhere to constraints and operate effectively."

**Conceptual Requirements.** As shown in Fig. 6.3, we use conceptual requirements to start developing preliminary mission operations concepts and requirements. We derive these conceptual requirements from the mission statement, and they mature as the system concepts and requirements mature. We list

**Mission Statement**
- Place activities on a timeline to support operations
- Satisfy mission and maintenance objectives
- Operate effectively

**Plan Components**
- Activities
- Resources
- Constraints

**Conceptual Requirements**
- Define activities
- Approve activities
- Place activities in a plan
- Develop N days on plan
- Meet objectives
- Satisfy constraints
- Compete activities
- Efficiently allocate resources
- Negotiate resources
- Display plan
- Modify plan
- Approve plan
- Publish plan

**Operating Principles**
- Establish plan duration
- Resolve activity conflicts
- Update or replace plan
- Provide details for the plan

**High Level Operations Concept**
- Gather planning inputs
- Generate activities
- Select activity for planning and place on timeline
- Resolve conflicts

**Fig. 7.2.      Defining and Developing the Activity Plan.** The plan components, operating principles, and conceptual requirements must be defined before we can develop a preliminary operations concept. This process is iterative. As we add more detail to our operations concept, we modify components, requirements, and principles to increase mission effectiveness.

the conceptual requirements for our mission statement in Fig. 7.2 and explain them in the following paragraphs.

The mission statement is reflected, and augmented, in several of the mission's conceptual requirements. Engineers and designers include requirements to place activities in a plan so mission and activity constraints are satisfied while meeting the mission and maintenance objectives. They include other requirements to make sure the plan efficiently allocates resources needed to satisfy the mission and maintenance objectives. Requirements are added to allow negotiation on resources for which there is competition.

As we discussed in Chap. 6, there are both internal and external users of the mission operations element's products (in this case, the activity plan). The mission's conceptual requirements capture the needs that are critical to mission operations and the system. For the internal users of the activity plan, we include requirements that give operators an overview of space and ground operations. Many times the plan is displayed so operators can see the spacecraft's upcoming activities at a glance. This display can be graphical or in text, be arranged by space-craft revolution or time, and be a workstation display or a wall projection. All of these characteristics would be included at a high level in the display's conceptual requirements. Other requirements supporting the internal users are to provide a way to support the organization's authorization of the activities to be planned and for an operator to be able to modify and approve a plan. One of the requirements supporting the external users is publishing the plan periodically to elements that use the plan in order to establish their data-processing timelines, such as exploiting data from the mission downlink.

**Operating Principles.** We must get agreement on several operating principles before we can define a mission operations concept that supports the conceptual requirements. We list the operating principles in Fig. 7.2 and present them in detail in the following paragraphs.

The duration of the activity plan, which typically covers three to fourteen days into the future, will drive many other mission operations characteristics, such as the accuracy of the ephemeris and the desired amount of operator insight. Another principle to be decided is whether the plan can carry conflicts. A plan carries a con-flict if we assign time and resources to an activity in such a way that it violates a constraint with another activity or if the resources are supporting another activity. If the only way to resolve the constraint violation degrades one or more activities sharing the conflict, it may be better to inform an operator and resolve the conflict later. The plan may then contain unresolved conflicts, but the part of the plan— usually one to two days—that supports near-term scheduling and commanding must always be valid—with no unresolved conflicts.

Another principle we must determine is how often to revisit the plan based on new information and how one plan transitions into another. While we're prepar-ing a new candidate, the operational plan and its updates are in effect. Finally, we need to consider the plan's level of detail. A decision on this principle depends on how operations uses the plan. In general, the plan requires more detail when we use it to operate the spacecraft in an automated fashion. Conversely, it requires less detail if it's used as a basis for a detailed schedule or simply to guide the operators. Stated in another way, we must specify many details about the activities within the plan before translating them into spacecraft commands. These details include activity start and stop times, equipment configurations, and spacecraft orienta-tions. If this plan is used to operate the spacecraft in an automated fashion, the planning function must develop the details; otherwise, an operator or the schedul-ing function can specify them.

**Preliminary Concept for Mission Operations.** With the information from the mission's conceptual requirements and from the operating principles, CI designers can begin to develop the preliminary concept for activity planning. In Fig. 7.2, we list the four high-level tasks needed to produce an activity plan. In Fig. 7.3 we show how those tasks interact.

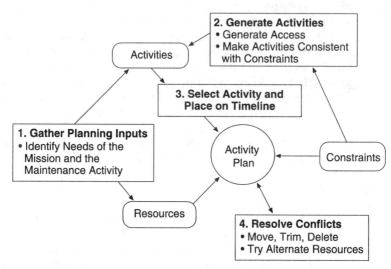

Fig. 7.3.    **Tasks Required for Activity Planning.** Activity planning consists of placing spacecraft and ground activities on a timeline and assigning planning resources to that activity while satisfying resource constraints.

The first task involves gathering all inputs that define the mission and maintenance objectives which the plan must satisfy. The next task is generating candidate activities based on activity objectives and constraints. Constraints can be geometric, time, subsystem limits, or inter-activity exclusions. The third task is selecting a candidate activity and placing it in the plan, while checking constraints and proper use of resources. Then, if we can't place the activity as defined, the fourth task involves exploring resolution strategies. We use strategies such as modifying (trimming, moving, or deleting) the activity or its competing activities already on the plan. Efficient strategies mean we won't need a lot of runtime for exhaustive enumeration, but these strategies also increase the complexity of the planning algorithms. We repeat the last two tasks until the activity meets mission and maintenance objectives, until we can't explore other placement options because of constraints, or until we've reached the time allotted for the search. Finally, we include opportunities for operators to monitor, influence, and approve the activity plan as it develops.

With this preliminary high-level concept for activity planning, configuration-item designers use analysis to develop and prototype ways of carrying out activity planning. Trades at the CI level to support a method or algorithm may influence the mission operations element's, or even the system's, definition of the requirements.

# 7.2 Understanding the Planning Components' Dimensions

As we illustrate in Fig. 7.1, there are four aspects of generating an activity plan. Each aspect has several ways of viewing the planning components, and each way influences how we generate and use the plan. In other words, there are dimensions to defining components of the activity plan. We highlight these dimensions for each of the four aspects in Fig. 7.4 and then detail each aspect in the following paragraphs.

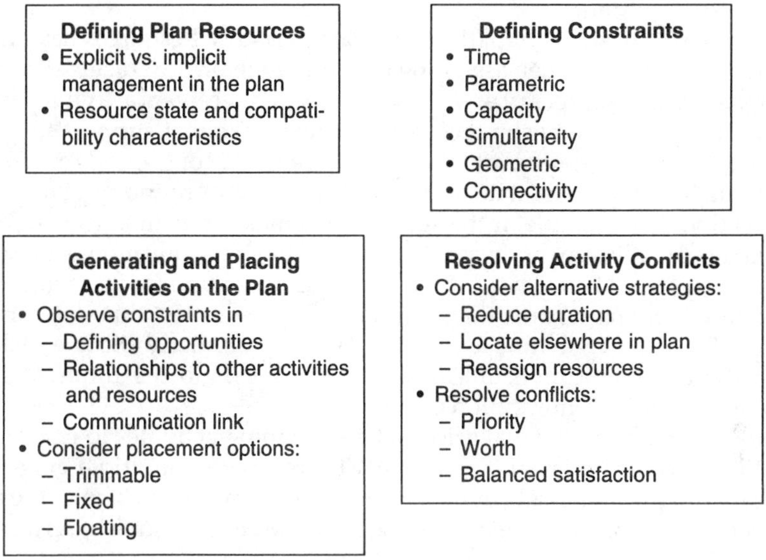

**Defining Plan Resources**
- Explicit vs. implicit management in the plan
- Resource state and compatibility characteristics

**Defining Constraints**
- Time
- Parametric
- Capacity
- Simultaneity
- Geometric
- Connectivity

**Generating and Placing Activities on the Plan**
- Observe constraints in
  - Defining opportunities
  - Relationships to other activities and resources
  - Communication link
- Consider placement options:
  - Trimmable
  - Fixed
  - Floating

**Resolving Activity Conflicts**
- Consider alternative strategies:
  - Reduce duration
  - Locate elsewhere in plan
  - Reassign resources
- Resolve conflicts:
  - Priority
  - Worth
  - Balanced satisfaction

**Fig. 7.4.** **Dimensions of the Activity-Planning Components.** We must analyze different approaches to defining resources and constraints, generating and placing activities on a timeline, and resolving conflicts.

## 7.2.1 Defining Plan Resources

As stated in the introduction, a plan resource is a system component that is in limited supply and, therefore, needs to be allocated to help satisfy the mission objective. We may define planning resources as physical assets, such as ground receivers, mission and relay spacecraft, and ground telemetry, tracking, and command (TT&C) hardware. They may also be the expendables of an asset, such as the power on a spacecraft. They may be paths or the bandwidth of a communication stream.

Space- and ground-system engineers must identify the components of space and ground elements we'll treat as resources and also identify, develop, and validate the concepts and methods for allocation.

**Explicit and Implicit Resource Management.** We use one of two ways to manage resources in a plan. The first is to explicitly assign the resource to particular activities at specific times. We use this method if assigning activities depends on knowing which resource will be needed. For example, we assign activities to a specific ground receiver if that receiver is the only one able to support the required downlink. The second way to manage resources is implicitly treating them as constraints that mission and maintenance activities must observe and placing them on the plan. In this context, the number of resources available—not any specific resource—is important.

**Resource State and Compatibility Characteristics.** Resources have characteristics we must observe when assigning activities to them. Characteristics about the state of the mission spacecraft and relay and ground receivers are vitally important to the protocols for establishing communication links that support the activities. The amount of time a spacecraft takes to acquire a relay or ground receiver may be different depending on the state of the spacecraft or the ground receiver at the time of acquisition. For example, if the spacecraft is making its first acquisition after a scheduled orbit adjustment, its position at the time of acquisition is not very well known. Therefore, we may need to use a unique procedure for this acquisition that searches a wider spatial area than normal. We must reflect this state properly in allocating the acquisition time to the plan. We must similarly allocate each component of the communication link from the spacecraft to the ultimate source or destination of the communications data.

Another characteristic that arises when assigning activities to resources is the compatibility of activities and resources. There may be multiple generations of spacecraft and ground assets in the system, with some assets being at least partly incompatible. In this case, we may not meet, or we may diminish, certain mission objectives if we assign the activity to particular resources.

## 7.2.2 Defining Constraints

In the following paragraphs, we discuss the various constraint categories, as listed in Fig. 7.4, and their application to activity planning. As mission operations managers, we should make sure the space and ground elements are thoroughly analyzed for constraints with the results documented in the element specifications and interface-control documents.

**Time Constraints.** Constraints take on many forms in a plan. One form is the relationships of activities to other activities, to resources, or to defined points in time or orbit. For example, we may prohibit a certain spacecraft activity until the spacecraft has exited the orbital umbra by at least five minutes and separate it from subsequent activities by at least fifteen minutes. Time constraints can be either prohibitive (the activity must not be closer than a defined period) or inclusive (the activity must not be farther away than a defined period). A special case of the inclusive constraint is known as the co-existence constraint, which requires that a certain activity be completely within another activity, with defined margins. Spacecraft activities normally require a coexistent contact between the spacecraft and ground receiver, plus additional time for acquiring and dropping the signal before and after the activity.

**Parametric Constraints.** These constraints require a parameter to stay above or below a particular value throughout the plan. For example, a spacecraft constraint may require the subsystem's temperature to stay below a specified value at all times. For this kind of constraint, we have to use a mathematical model to translate planned activities from a time placement and duration into the parameter being constrained. Sometimes the parameter being constrained is time, which requires no parametric translation. Limits on the spacecraft subsystem's duty cycle fall into this special case of parametric constraint. Other examples of parametrically defined constraints are power, defined by a minimum state of charge, and momentum limits, defined by a not-to-exceed torque during an orbit revolution.

**Capacity Constraints.** These constraints specify a number of occurrences of an activity that are required or are not to be exceeded over a defined time interval. For example, we may constrain the spacecraft by limiting its acquisitions during any single orbit revolution. Other examples include the rate at which tasking requests can be accepted and the rate at which mission data can be processed.

**Simultaneity Constraints.** A special case of the capacity constraint applies when the time interval being examined is set to zero. It's known as a simultaneity constraint. We usually apply it to resources expressed as a number of resources per unit of time, as opposed to those explicitly assigned to activities on a timeline. Examples include the number of ground hardware units available in TT&C to support concurrent acquisition and the number of wideband downlink streams that can be processed at one time.

**Geometric Constraints.** These constraints require specific spacecraft or payload orientations to satisfy the activity. An example would be activities that require the spacecraft's aperture to traverse specific star patterns in order to determine the spacecraft's attitude and position. Other examples include restrictions on the movement of spacecraft and ground-receiver antennas, line-of-sight restrictions induced by the horizon for ground receivers, and specific attitudes required of the spacecraft to collect the most solar energy or to transfer the most heat through thermal radiation.

**Connectivity Constraints.** Connectivity constraints try to satisfy several activities or resource conditions at the same time. An example is a constraint on the communications link. In a communications link, the spacecraft must have an established contact with a ground receiver. Sometimes, it directly contacts the ground. Other times it must use a relay spacecraft, such as TDRSS. In the latter cases, the relay spacecraft must first contact the ground receiver and then contact the mission spacecraft. So we must establish a link between the ground receivers and the TT&C

hardware in the mission operations element. While ensuring all communications links are in place, we check the plan to make sure the link resources aren't being used by another spacecraft, in maintenance, reserved, or otherwise restricted for use at the planned contact times.

**When to Place Constraints in the Plan.** We apply all constraints whenever we're trying to place an activity in the plan, but we start with those that deal with the specific activity. For example, geometric constraints usually apply to one specific activity and therefore fall into this category. The second set of constraints we apply are those that affect neighboring activities or resources in a local area of the plan. Temporal, connectivity, and simultaneity constraints fall into this category. Finally, we check constraints that deal with a large timeframe in the plan. Parametric and general-capacity constraints fall into this category. We use this strategy so we need less computing capacity to produce a plan free of constraint violations. In general, parametric and capacity constraints require more sophisticated methods than the other kinds.

### 7.2.3    Generating and Placing Activities in the Plan

Engineers must thoroughly analyze and identify all activities in the space and ground elements and document them in the element specifications as they decide which ones to include in the activity plan.

**Activities that Make up a Plan.** Normal mission operations consist of many activities. These activities include establishing communication with the spacecraft: acquiring the spacecraft, configuring the spacecraft to send or receive data, uplinking commands, downlinking mission and maintenance data, and droplinking. Other activities satisfy the mission and maintenance objectives: commanding the mission payload, thermal and momentum dumping, adjusting the orbit, collecting power, and calibrating and aligning the spacecraft bus and payload. Still others process the collected mission and maintenance data on the ground: analyzing and distributing mission data and analyzing calibration, alignment, and maintenance data. Finally, normal operations include preventive-maintenance activities, such as running diagnostics and replacing worn parts on a scheduled basis for special-purpose hardware in the TT&C subsystem or the system ground receivers.

**Activity Placement.** Each of the activities that goes into a normal operations mission plan has certain characteristics that determine where it can be satisfied on the timeline. Typically, the activity is described as a start and end time or, equivalently, a start time and duration. Activities usually have several constraint classes associated with them, some for defining opportunities, others for defining the relationships of the activities to each other and to the resources assigned to support them. Some activities require a command and telemetry link to the mission operations element's TT&C hardware. Other activities require an additional link with a high data rate to a data-processing element, while still others don't need a communications link. Finally, engineers assign activities a set of importance factors that they use when activities compete for a particular resource.

**Trimmable Activities.** Depending on an activity's nature, we may have some options for when they occur in the plan. Activities that directly support the mission objectives must occur when the spacecraft's position in the orbit allows an access that satisfies these objectives. If several accesses will satisfy the mission objective, we may place the activity on the plan at the best access point, or we can move it to another access point so we can satisfy other objectives. Access in this context means the spacecraft meets geometric constraints imposed by the mission. Satisfying the mission objective means meeting its specific characteristics. These characteristics include time of day, month, or year; the quality or quantity of the data collected; and any unique spacecraft-objective geometries. Because of this characteristic, some mission activities aren't movable. In other words, any change to the start or end time of the mission activity will cause it to fail some or all of its mission objective. But most mission activities can accept a shortened duration without failing the entire mission objective. We refer to these activities as *trimmable.*

**Fixed Activities.** Maintenance activities may or may not be similarly restricted. Certain calibrations and alignments for the spacecraft bus or payload require specific Sun- or star-to-spacecraft geometries or, as in the case of adjusting a spacecraft orbit, require the activity to occur in a very specific part of the orbit, such as perigee. We consider these activities immovable. Further, because of the nature of data collection for maintenance activities, they're not satisfied unless we can place their entire duration on the plan. For example, certain calibrations for a spacecraft subsystem may require a series of physical component settings with data collection at each setting. If we don't collect data at all of the settings, the calibration is useless. Because these activities are immovable and untrimmable, they're called *fixed activities.*

**Floating Activities.** Other maintenance activities may simply require spacecraft to meet special geometries, as is the case when pointing radiators to deep space to dump thermal energy, or pointing solar panels towards the Sun for power collection. These less restrictive constraints allow activities to be satisfied over some contiguous interval, so we consider the activities movable and call them *floating activities.* Like the fixed activities, floating activities still require a certain amount of time to be satisfied.

### 7.2.4   Resolving Activity Conflicts

In this subsection, we investigate how to place activities on a plan when they conflict with other activities or violate constraints. As operations managers, we must make sure conflict-resolution strategies are in the specification for mission operations on the ground. System managers have to approve issues of priority, worth, and balanced satisfaction. We should also ensure that mission engineers identify, develop, and verify methods for resolving constraint violations.

**Alternative Strategies for Placement.** Once we've determined opportunities for placing activities, we begin putting them into the plan against specific start and end times and resources, while watching for constraint violations and conflicts. Constraint violations differ from conflicts. A *constraint violation* is a warning that two or more activities' placements or resource assignments exceed a constraint limit. In this case, we should try to change where we place the offending activities. A *conflict* indicates that the constraint limit is exceeded and we can't change placements to clear the violation. Our alternatives for placement depend on whether the activities are fixed, floating, or trimmable. We may also have some latitude in resource assignments to clear a constraint violation, especially for communication links. The spacecraft may be able to use several different paths to satisfy the activity. Different arrangements of the link assignments may free up a resource for a particular spacecraft which has a limited number of paths, while still fully satisfying another spacecraft's needs.

**Resolving Conflicts.** Our plan has conflicts whenever we've unsuccessfully exhausted every alternative to place activities so they're free of constraint violations. We may allow the plan to carry the conflict until an operator manually deletes or modifies the activity allocation in the plan to clear the constraint violations. Or we may use software to eliminate the conflicts automatically. To remove conflicts when no more placement options are available, we eliminate or trim activities of least importance to the system until the plan meets all constraints.

**Resolution by Priority and Worth.** Engineers can identify the least important activities in many ways. One way is to define a priority scheme. If two activities are in conflict, the activity with the higher priority gets the resource and time interval. All of the lower-priority activities in conflict with it are trimmed, moved to another time interval, or deleted. A second way is to define a worth to each activity which may vary with time over the course of the activity. Then, an activity of higher worth wins out when the combined worth of the competing activities is lower but loses out when that combined worth is higher.

**Resolution by Balanced Satisfaction.** If engineers can't assign worth to all system activities, they have to use other schemes to resolve conflicts and another type of metric to resolve ties for competing resources. One concept, called balanced satisfaction, involves finding a level of satisfaction for the group of competing activities that best satisfies the overall activity plan. They can use the duration of the competing activities as a metric for balanced satisfaction. Other characteristics can determine tie breakers in a conflict, along with or instead of duration. They include whether the activity is overdue for execution and how many future opportunities exist to carry it out.

## 7.3   Analyzing Components of the Activity Plan

The hardest goal to achieve in activity planning is defining inputs, methods, and outputs that offer the most effective design. An effective design is one that satisfies the requirements and fulfills the concepts without unnecessary or extra design that adds cost and schedule risk in development and requires more effort in maintenance and operations. We can achieve this goal by analyzing concepts and requirements for the system and for mission operations. Figure 7.1 lists aspects of activity planning we must analyze to define requirements, operations concepts, operator procedures, and software methods.

We must make sure that the definition and development plans identify analysis tasks for activity planning and that the analysis is carried out in a timely, correct, and thorough manner. We also ensure the activity-planning concepts and specifications reflect the results of the analysis. Finally, we interact with all of the external and internal organizations to ensure the activity planning meets their needs.

**External Organizations.** The mission operations element is the hub of the space-ground system. Here we transform mission requests into spacecraft activities. These activities become commands that are uplinked and executed by the spacecraft, with the resulting mission data routed to the data-processing center on the ground. Because the activity plan coordinates these activities, many organizations external to mission operations influence the definition and development of activity planning. These organizations employ people in tasking, product exploitation, spacecraft engineering, ground-system engineering, and, of course, operations.

**Internal Organizations.** The designers of configuration items within activity planning work closely with external organizations. The CI designers develop the requirements and operations concepts for activity planning; they also develop and validate the software methods used to generate the activity plan. A second group who work closely with external organizations are the operations engineers. They define and analyze operator tasks, define and analyze loading on operator positions, and prototype displays and operator interactions.

We address the following aspects of activity planning

- Activities
- Constraints
- Resources
- Operator role and control
- Planning paradigms
- Planning procedures and algorithms
- Planning goals

### 7.3.1  Activities

Table 7.1 shows the issues we must consider and resolve to define and place activities in the activity plan.

**Mission Activities.** At the system level, the tasking organization and planning engineers for the system and mission operations work together to develop conceptual requirements that govern the planning and placement of mission activities.

**Table 7.1.**     **Activities.** We must define and analyze activities to develop placement strategies.

| Process Step | Issues to Address |
|---|---|
| Identify System Conceptual Requirements | • Mission-access definition, quality, and constraints<br>• Management of input tasking information<br>• Maintenance activity triggers and relationships to mission |
| Develop Element Concepts and Methods | • Role of planning versus scheduling for access accuracy<br>• Options for placing maintenance activities<br>• Automated versus manual planning of activities |

They resolve processing issues such as defining access, satisfying mission objectives, defining quality of access, and establishing geometric constraints for the activity. They also resolve issues related to managing the input-tasking information, including timelines, frequency, and throughput volume.

**Determining the Mission Activity's Access.** In response to the conceptual requirements for mission access, the CI designers develop ways to calculate the access interval, quality, and duration needed to satisfy the mission objective at the required level of performance. The *access interval* is the time in orbit that the spacecraft is in view of the mission subject. The quality is a measure of the mission data that can be collected—for example, the minimum ground distance or stellar angular resolution for collecting optical data. The required performance is the accuracy of the access calculations and the maximum allowed computer execution time, if timelines are important. Conceptual trades may help resolve issues concerning required performance for calculating mission access. For example, the accuracy of equations should never exceed the accuracy of the input data. Designers determine the accuracies of the calculations by the context in which they are used. If the access information is also used by scheduling, we may need increased accuracy, which in turn requires more exact and more complex equations.

Other trades define how planning and scheduling support the mission activity. The main purpose of planning is to allocate the system resource to support the mission and maintenance objectives, with data at a definable level of accuracy. But activity planning may be more or less involved in supporting the scheduling. At one extreme, it may act as a mission-data filter for scheduling. This concept requires more exact and complex equations to ensure that acceptable accesses are not incorrectly filtered out during planning. At the other extreme, planning employs much simpler, maybe even empirically constructed, equations. In this context, scheduling must recalculate all of the mission accesses based on more accurate data. This approach requires significant validation to ensure that the simplified equations are within the error of the remaining input data for planning and to ensure that the more accurate methods meet the timelines in scheduling.

**Maintenance Activities.** Engineers in system or mission operations planning join spacecraft and ground-systems engineers to develop requirements for planning and placing maintenance activities. They resolve processing issues, such as what maintenance activities need to be planned and what activities can wait until scheduling. They also discuss and resolve how frequently particular maintenance activities must occur, what spacecraft and ground circumstances trigger the need for these activities, and the relationships of these activities to the mission activities.

**Planning Maintenance Activities and Resolving Conflicts.** The CI designers for planning develop concepts and methods to identify and resolve conflicts when maintenance activities are added to the activity plan. Design trades for maintenance activities may help resolve some of the issues. For example, as discussed in Sec. 7.2.3, some maintenance activities can occur at various times and still meet the maintenance objective. We can consider this freedom of placement a simplification or an added complexity. As operators, we'd find it much simpler to do a maintenance activity over some interval in the defined planning period. Complexity arises when the placement choices, computed from the possible maintenance accesses, all compete with other activities on the plan. Also, certain maintenance activities may occur infrequently in operations; yet, finding placement options for them can be difficult. These activities are good candidates for operator-controlled planning, instead of planning for software to carry them out automatically. In this case, we have to trade the added operator workload against the added software complexity needed to minimize that workload.

### 7.3.2 Constraints

Table 7.2 shows what we must consider and resolve to define and use constraints in the activity plan.

**Table 7.2. Constraints.** We must analyze constraints to determine the best way to model and incorporate them in the activity plan.

| Process Step | Issues to Address |
|---|---|
| Identify System Conceptual Requirements | • Handling of hazardous versus non-hazardous constraints<br>• Planning of activities using simpler constraints |
| Develop Element Concepts and Methods | • Parametric modeling of constraints versus time budgets<br>• Highly accurate methods |

**Identifying Constraints.** Spacecraft- and ground-system engineers work with designers of the mission operations element to develop constraints on the spacecraft and ground equipment. They discuss the types of constraints and whether these constraints affect the spacecraft's or ground's immediate or long-term state of health. They also determine which commands are hazardous. For example, any commands that activate spacecraft thrusters are considered hazardous—so much so that they can't be uplinked to the spacecraft until the spacecraft has cleared its launch vehicle.

**Approach to Satisfying Constraints.** Engineers and designers can do many design trades in determining how to deal with constraints. Generally, the simpler the criteria for planning an activity to meet its constraints, the better the design. For example, the spacecraft may have certain equipment-warmup cycles that depend on the time since the equipment last operated. There are two ways to meet this constraint. The first way is to encapsulate the dependencies in the activity-planning constraints and check to see if enough time has elapsed before each use of the equipment. A much simpler way is to plan to operate the equipment regularly for a short period, whether an activity needs it or not, in order to keep the warmup time at a minimum. In this way, planning doesn't need to know the details of the equipment states or how much time has elapsed since the last time the equipment was used. Planning also doesn't need to react to situations in which planned use of the equipment has been canceled, thus changing the time dependency.

**Approach to Modeling Parametric Constraints.** Engineers and designers must be extremely careful in specifying parametric constraints. Because of their nature, parametric constraints tend to apply to large portions (e.g., 24 hours) of the planned activities and require equations to translate the activities' start and end times into units of the parameter being measured. A battery's state of charge is an example of a parametric constraint. Again, as for the warmup constraint discussed above, they must trade complexities in the design to arrive at the most effective way to meet the constraint. For this example, a more exact approach would model all of the equipment characteristics in terms of their effect on the battery's state of charge, and then would model how the equipment components support each activity in the plan. Many times, however, the equations that model the constraint require more accurate data than we can get from planning. On the other hand, if they model each activity based on historical performance to have a constant rate of change of battery charge, this rate times the activity duration gives the change in battery state of charge. This approach dramatically reduces the complexity of the design needed to meet the constraints, while still keeping operations or future spacecraft needs very flexible.

**Accuracy of Constraint Methods.** Both of these examples of constraint trades need approval by the system designers and element designers of the resources imposing the constraints. To support that approval, CI designers develop and validate methods to meet these constraints. They must analyze carefully to offer a design with the best balance between algorithmic complexity and conservatism. A highly accurate method produces the most conservative solution but with an associated increase in algorithmic complexity. Certain factors can easily destroy the perceived accuracy of the planning data and make the method useless. Such factors include the probability of not executing what was planned, variation in the parameters within the constraint equations between the predicted and post-activity values, and the willingness of operators to operate the spacecraft very near the constraint limits.

### 7.3.3   Resources

Together, the system and element designers within activity planning, as well as the spacecraft- and ground-system engineers, determine which system components are activity planning resources. They must consider whether the component is available enough to fully support the mission objective. Available means able to carry out its intended tasks. Downtime for maintenance or recovery to operations-support levels are part of this availability.

Sometimes, we can analyze best by prototyping the parts of the system to measure the level of use. A trade that can reduce the complexity of resource planning examines whether we must explicitly model the resource in the plan and specify activity assignments, or whether we can implicitly handle it through a set of capacity constraints. With this analysis and trade decisions in hand, CI designers develop and validate ways to allocate the resource.

### 7.3.4   Operator Role and Control

**Operator Role in the Planning Process.** Suppose activities are few, objectives are simple, and competition among activities for the system resources is light. In this case, an operator can construct a plan manually by using a timeline tool that checks objectives, constraints, and resources. The operator manually changes an activity to resolve conflicts. But if activities are numerous, objectives are complex, or competition is heavy, we need a more automatic planning algorithm. In this case, element designers develop a suite of algorithms to construct a plan, with those algorithms carrying out the same reasoning and actions the operator would use in constructing the plan manually. The operator's role in planning also depends on his or her skill and knowledge of how to satisfy the mission and maintenance objectives.

In either instance, the operator needs to have certain tools, controls, and procedures to construct a plan. Controls allow the operator to influence the plan in areas such as activity placement, conflict resolution and alternative solutions, stability and responsiveness, analysis and resulting changes, and approval. For example, the operator constructs a plan by selecting an activity and then determining the start time and duration that satisfy the activity's objective and constraints. Automated tasks are usually employed to help the operator select times within the access interval of the activity. If the activity also requires a link resource from the system, we must place it on the plan's timeline so it satisfies system constraints. Several placement options may be available to satisfy the activity, but some may conflict with existing activities or result in contention for resources. The operator has to modify either the new activity or one or more existing activities in order to clear the constraint violation or resource contention.

Automated algorithms will always have an absolute metric with which to gauge the quality of one set of options over another. The problem is that activity planning, as part of mission operations, deals with metrics we often can't express absolutely. So the mission operator must have the control to manipulate the plan directly or indirectly to get operational approval. The control mechanisms available to an operator depend somewhat on the processing architecture. For example, in a batch architecture, operator controls are part of process execution. But in a transactional architecture, event triggers govern operator controls.

**Plan-Visualizing Tools.** These tools help an operator place an activity on the plan. The tool typically displays activities on the timeline from multiple perspectives. One view is of the activities relative to the resource links, including mission spacecraft, relay spacecraft, ground receivers, and ground hardware. Another focuses on specific resources. For example, the display may show activities from all mission spacecraft that use a specific ground receiver. Other tools help the operator determine how well an activity satisfies its objective. Displays and messages appear when an activity's placement violates a constraint relative to other activities or when resources contend. The display should also show the extent of the violation and which other activities or resources are involved. The visualizing tool is usually a series of specialized displays using commercial, off-the-shelf (COTS) products and mission-specific software. The COTS products provide constellation views, including lines of sight from spacecraft to spacecraft and from spacecraft to ground receiver, as well as activity placements. The mission-specific code is usually reserved for mission-specific calculations and displays of the activity objective, activity constraints, and system-constraint parameters.

### 7.3.5   Planning Paradigms

Table 7.3 illustrates the range of possibilities for three planning issues.

**Table 7.3.   Planning Paradigms.** We must establish early on and then mature certain paradigms that determine the operation's approach to planning.

| Process Step | Issues to Address |
|---|---|
| Identify System Conceptual Requirements | • Creating the plan<br>  – Build a new plan versus update the old plan<br>• Operator involvement<br>  – Create the plan manually versus automatically |
| Develop Element Concepts and Methods | • Processing approach<br>  – By batch versus by transaction |

The operational look and feel of activity planning depend on decisions about processing paradigms in these three areas. They influence the operator's role in developing the activity plan, the software algorithms that assist or automate the operator's planning choices, and the approaches for ad hoc changes to the plan.

**Develop a New Plan or Change an Old One.** A plan that is built by updating a current plan offers a very stable environment for operators. Changes occur only when they add measurable value to the mission and maintenance objective. If inputs to the plan are constantly changing, or if we're optimizing over long periods, building a plan by starting from scratch yields better solutions. To choose the correct paradigm, we must analyze to get the proper mixture of best performance and stability.

**Determine the Operator's Role.** Operator-directed placement with automated constraint checking provides operations with the most control over the mission. However, if operators can rely on sophisticated automated processing to generate the activity plan, they're better able to oversee and analyze it. Also, if activity-planning inputs are too complex, trying several solutions may be cumbersome and time-consuming.

**Choose Batch or Transactional Processing.** *Batch planning* uses a progression of processes to place items into a plan. These processes have a defined task duration, with persistent data being passed from one task to the next. External updates to and feedback within them are limited, and their algorithms tend to be complex and monolithic. Because batch-processing functions have a strict process flow, they are inherently controllable. On the other hand, in *transactional planning*, processes converse with one another to arrive at a plan. Transactional processing uses continuous processes, easily incorporates updates, and normally includes feedback. For transactional processing, algorithms tend to be simpler individually but more difficult to coordinate as a group. It also requires controls, so responses to updates don't cause unwarranted changes to the already planned activities.

### 7.3.6 Procedures and Algorithms

Table 7.4 shows the issues we must consider and resolve for the planning procedures and algorithms that generate and update the activity plan.

**Define Plan-Transition Time.** We have to update the activity plan regularly as planning needs change or time passes. The plan must always be operational, so we apply updates to a candidate version of the plan. Transition of the candidate to the operational plan must be seamless. That is, the parts of both plans up to and just beyond the transition time must be the same. How far into the plan the transition time is from the current time depends on the amount of the plan we've already committed to activity schedules and the loads already uplinked to the spacecraft. If we generate a day's worth of activity schedule at a time, the normal transition time is 24 hours into the future. But if we're commanding the spacecraft in real time and generating the schedule an hour at a time, our transition time can be just two hours into the future. The closer the plan transition time is to current time, the more responsive the activity plan can be to near-term changes.

**Coordinate Plan Update and Extension.** Updates to the plan occur in two parts of the candidate plan. The first part includes updates from the transition time to the end of the current plan, usually three to fourteen days ahead. We must coordinate updates to an existing plan with all the ground components that use the plan data. Additionally, updates to an existing plan shouldn't greatly perturb the current plan. The second part to update is the new section, usually a day long. The

**Table 7.4.    Planning Procedures and Algorithms.** Planning procedures and algorithms require thorough analysis, development, and validation.

| Process Step | Issues to Address |
|---|---|
| Identify System Conceptual Requirements | • Define plan transition times<br>• Coordinate plan update and extension<br>• Trade plan stability versus responsiveness of change |
| Develop Element Concepts and Methods | • Consider time to generate the plan<br>  – Shift changes for plan personnel<br>  – Input/output and feedback timelines<br>  – Time to generate and carry out plan<br>  – Currency of plan relative to inputs<br>• Respond to dynamics<br>  – Unexpected processing results<br>  – New and updated inputs<br>  – Updated information on the spacecraft's state<br>• Consider complexities introduced by<br>  – Spacecraft and mission objectives<br>  – Resource limitations and tight constraints<br>  – Accuracy and responsiveness |

extension is a new part of the plan and must integrate smoothly into the end of the current plan, so we have to assign resources in the current plan to allow this smooth integration.

**Operator Shift and Inputs/Outputs.** The development cycle for our activity plan should run end to end during a single shift, if possible. If it does, a single operator can monitor the creation and transition of the candidate plan into operation, rather than having to coordinate that task with other operators. CI designers also must consider the timeline for inputs from users because updates to the activity plan can't be approved until we know all of these inputs—from inside or outside the mission operations element. Updates also depend on the timing of feedback that occurs because we've processed data from completed mission or maintenance activities. Finally, they also have to allow time for getting output to users.

**Generating and Carrying Out the Plan.** As stated earlier, the plan-transition time depends on the time needed to generate an activity schedule based on the activity plan, coupled with the time for generating and uplinking spacecraft commands to carry out mission or maintenance activities. Also, the processing for updates to the plan must occur within the time from the latest inputs being available and the earliest results produced. CI designers must analyze carefully to ensure there's enough time to process the inputs and produce a new candidate plan ready for transition to operations.

**Plan Currency.** How current must our plan be? Mission inputs often have continuous or regular updates. Trying to reflect all changes as they occur demands too much of our computing resources because most changes don't influence the overall activity plan and the ones that do, occur at unpredictable moments in the planning day. Engineers and designers need to analyze the system and mission operations to balance stability against responsiveness to change. Then, we must properly set the plan's currency.

**Respond to Dynamics.** Planning is dynamic because we must be able to respond to unexpected results from current mission and maintenance activities, to incorporate new or updated inputs from users, and to allow for increased accuracy of the predicted orbital geometries and our knowledge of the states of the spacecraft subsystems.

**Complexities of Plan Development.** The complexity of procedures and algorithms used to develop our plan reflect the complexity of the mission and maintenance environment—spacecraft and mission objectives, resource allocations and tight constraints, and accuracy and responsiveness.

**Complexities of Spacecraft and Mission Objectives.** A system that could cause complexity might be a spacecraft with multiple mission payloads that share common components and spacecraft resources. In this instance, we plan activities to satisfy mission objectives while efficiently allocating the spacecraft's resources, such as power, to the payloads. We must also observe intra- and inter-payload constraints, such as deployment restrictions and ability to do concurrent processing. Complexity may also arise if the spacecraft can process data faster than the downlink and ground station can. In this case, we must plan the mission and maintenance activities to satisfy these capacity limits. Other spacecraft limitations, such as memory, power, or uplink and downlink rates, may introduce complexities into the planning procedures and algorithms. Finally, the mission objective itself may impose complexities. If an activity has few opportunities to satisfy mission objectives, we'll need more complex placement options and conflict-resolution strategies.

**Complexities of Resource Limitations and Tight Constraints.** Conflict resolution to satisfy a mission objective can become even more complex in a resource-poor environment. In this case, activity planning has to best use the limited resource to satisfy the best subset of activities. Algorithms involving complex mathematical techniques, known collectively as operations-research techniques, may be required to best satisfy the activity in a reasonable computation time. A highly constrained environment can result in the same algorithm complexities as a resource-poor one because the constraints create so few options to explore. Also, testing the plan against these constraints at each combination of activity assignments requires a lot of processing.

**Complexities of Plan Accuracy and Responsiveness.** If planning must employ detailed models to stay within given parameters, or if an accurate solution requires a complex equation, planning becomes more complex. As stated earlier in our discussion of constraint considerations, engineers and designers must analyze carefully to ensure that highly accurate specifications are warranted. Complexities also arise when a deadline is so short that only the most efficient strategies to search for placement options can produce a new plan in time. This complexity is opposite from the complexity that is part of trying to find the most accurate solution.

**Reducing Complexity.** Some of the design trades that the CI designers in activity planning might use to deal with these complexities include giving some of the most complex decisions to the operator. Of course, they must take care not to overload operators with tasks that keep them from monitoring the rest of activity planning. Also, complexities can be reduced or eliminated if the system and mission operations element can agree to conservatively allocate activities to resources, satisfy constraints, and optimize mission objectives. Finally, designers can reduce the algorithm complexities if they take a converging, interactive, iterative approach rather than trying to examine everything at the same time.

### 7.3.7  Planning Goals

**Best Satisfy Objectives.** Activity planning—and procedures supporting it—must reflect the desired "goodness" of the plan. The best plan satisfies all mission and maintenance objectives, observes all system and activity constraints, and efficiently uses resources. It also produces a very good answer in minutes to hours rather than an absolutely best answer in hours to days. If the system limits our plan, designers must do trades to find the best way possible to meet mission objectives—based on the priority or worth of each activity. To determine an activity's priority or worth, we look at its importance among other competing activities, how often it must access the mission objective, and whether it must continue over a shorter or longer term.

**Provide a Flexible Plan.** A plan that leaves room for changes is an important goal. Developing a plan that uses all of the resources and drives constraints to their limits doesn't allow for additional tasking without large changes. For example, if we plan to use all available power for a given day, and a new activity is needed, we have to delete a lot of the plan to meet the new activity's power requirements. It's more operationally sound if we plan to do less than the spacecraft's absolute limits, so this kind of situation can draw on planned reserves.

# Chapter 8

# Conducting Space Mission Operations

Eileen Dukes, *Lockheed Martin*

In the previous chapters we discussed the functions required for mission operations. In this chapter we put the functions together to show how the operations flow. We discuss details of activity planning and development and mission control—concentrating on the uplink and downlink tasks, but also showing how the other functions, such as planning and analysis, feed into the flow. This chapter gives the person who has never been directly involved in operations a feel for the day-to-day activities as well as a sense for how we implement an operations timeline. We also provide recommendations for making operations more efficient and hence more cost-effective.

One of the first challenges you'll face as a mission operations manager (MOM) is matching the operations style to the mission requirements. There are two extreme perspectives on spacecraft operations. One extreme maintains that people skilled in operations should do all operations, with little or no regard for the particular spacecraft being operated. The other extreme maintains that design engineers who built the spacecraft should be the ones to operate it. As always, both extremes have their pluses and minuses, and most successful operations mix operations experts and spacecraft experts. Figure 8.1 shows some of the considerations to determine the correct mixture. System maturity is also a consideration. When a system and the spacecraft are new, the balance tends to tilt more towards the design experts. As the system matures, the spacecraft or family of spacecraft becomes well understood and the balance tends to shift toward operators.

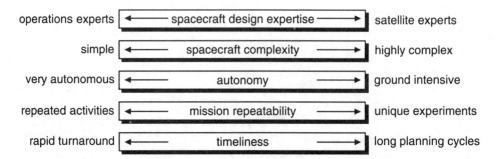

**Fig. 8.1.** **Factors Which Affect the Balance Between Operations and Spacecraft Expertise.** Use these factors to select the appropriate operations style for the mission.

Systems engineers or subsystems experts can provide the expertise. Systems engineers have a broader understanding of the spacecraft and how the subsystems interact. Subsystems experts have a narrow but deep view, usually restricted to their subsystem. Using systems engineers is usually more efficient because you can cover a broad range of spacecraft knowledge with fewer people. But if the mission is very unique or complex you'll need some subsystem experts. A good compromise is to supplement systems engineers with subsystem experts only for the operationally complex subsystems, such as command and data handling, attitude control, and the payload.

Finding the correct balance is not always easy. Not having enough spacecraft expertise readily available can result in more down time and sometimes in loss of mission. Conversely, a staff with insufficient operations experience will tend to be slow and sometimes can be more interested in using the spacecraft as an engineering testbed than in meeting the mission objectives. Also, design engineers tend to get bored with, and may be too expensive for, routine operations. We must balance these abilities to get the best ratio for the given mission.

*Routine operations* occur when the processes are well established, which happens when the procedures are fairly static, people are fully trained, timelines are usually met, and products are being regularly delivered. It doesn't mean that nothing unusual ever happens; if that were the case, we wouldn't need skilled operators. Normally, we consider routine operations less risky because we understand and practice them. But as repetitiveness becomes monotony, details can get overlooked. As the old saying goes, "familiarity breeds contempt." One of our challenges is to keep the operations job interesting and operators alert as the mission matures.

On the other hand, we consider *special operations* more risky because they're unique and often critical. They require special procedures, special training, and possibly, rehearsals or dry runs. These operations include one-time events, such as launch or a peculiar calibration, and may also include operations during or following an anomaly.

One way we can find the correct balance is through evolving operations. Although procedures can't change continually, operations need to adapt smoothly and quickly to changes. When a system and spacecraft are new, more engineers familiar with the design are needed to characterize the new spacecraft, calibrate it, and work out the procedures. These engineers then apply their knowledge to updated processes and procedures. Eventually, more operators are trained to replace the engineers. The operators understand how to do the mission and begin to streamline and improve the operations.

We need to encourage operators to improve their productivity by using computers to automate their tasks as they become standardized and repetitive. Increased automation of repetitive tasks can lead to a gradual reduction in staff. Also, the current explosion in computer technology can make a system outdated in five years and obsolete in ten years. Gradually incorporating the new technology allows for upgrades without disruptive change.

Let's consider an extreme example that illustrates this point:

Consider the operator whose job is to sort through line-printer outputs for the past 24 hours, extract the values for eight of the spacecraft's engineering measurements, and add these values to a cumulative plot. Suppose it takes an hour per measurement so it comprises the entire eight-hour shift. This job is a prime candidate for automation. As a first step, this person could research spreadsheet or graph programs which would eliminate the manual plotting. These programs usually also include basic statistical functions which could enhance the task. At the same time, by adding to his or her personal tools, the operator has a new challenge and a chance for personal growth.

As the next step, we can investigate an automated link between the telemetry processing and the plotting tool, rather than printouts. Again, the operator can improve by learning more about telemetry processing in order to define the requirements and participate in building an automated trending interface.

Once the task is automated, what used to require eight hours may now require only one hour. The operator now has seven hours to analyze trends more deeply, help others automate similar tasks, or reduce overall staffing. In exchange, the operator has had a chance to exercise some engineering skills by defining the requirements, doing trade studies, perhaps designing the implementation, testing, and incorporating new methods into procedures.

People can be reluctant to work themselves out of a job if that is how they perceive it. But most people will take the chance to grow by exercising skills or acquiring new skills. We can reduce staff through attrition and provide growth opportunities by promoting from within as higher-up or more skilled positions open up. Promoting from within rewards initiative and continuous improvement and, very importantly, keeps program expertise high.

Often, operations before launch are designed rigidly, with level staffing through the years. But this approach leads to high mission costs and doesn't allow us to take full advantage of the staff's ability to learn. Providing a structure for operations that allows it to evolve can reduce long-term mission costs.

An example of this evolution is the payload control for the Extreme Ultraviolet Explorer (EUVE). After two years of flight, they were able to transition to an automated monitoring system for payload control. The EUVE science-operations center built Artificial Intelligence (AI) software to "mimic the monitoring responsibilities of the human science payload controllers. The AI software perpetually monitors the science payload... When a rule is violated by a data point in the telemetry stream, calls are made to external processes that either rectify the problem or sound an alarm. If a problem occurs during an unstaffed shift, the software autonomously pages an anomaly response coordinator." [*AI Magazine*, 1994] This automation has allowed them to reduce costs by reducing their shift-coverage requirements.

# 8.1    A Day in the Life of Operators

Mission planning consists of collecting requests to use the system and spacecraft, balancing these requests against available resources and capabilities, and creating the activity plan. Requests come from two major groups: the product users or science teams and the analysts who operate and maintain the spacecraft and payload. Capabilities include those of the spacecraft, the payload, ground system, and mission operations. Resources include all of the people, hardware, software, and telecommunications links both inside and outside the project.

During the development phase, we produce a mission operations concept, which contains the general sequence of events taking into account planned capabilities and resources. The concept includes the purpose of the mission, the desired data return, and goals. Refer to Chap. 4 for a more detailed discussion on developing a mission operations concept.

## 8.1.1    Mission Plan

During operations, the mission operations concept documents the project's policies and generally structures mission operations. We must then translate this high-level document into a mission plan and establish an appropriate duration for the plan, taking into account the response time required by users, development cycles for activities, planning cycles for all the groups involved, and the fidelity of planning inputs such as accuracy of ephemeris prediction. For a dynamic mission, the plan covers two weeks to two months. For an interplanetary mission, it covers several months to a year. The planning horizon needs to be long enough to allow all of the mission operations functions to do their own scheduling, but not so long that it's always obsolete.

The mission plan collects information from

- Spacecraft-bus analysts
- Payload analysts
- Navigation

- Ground systems
- Telecommunications-resource scheduling
- User community

Figure 8.2 illustrates these inputs and their corresponding outputs.

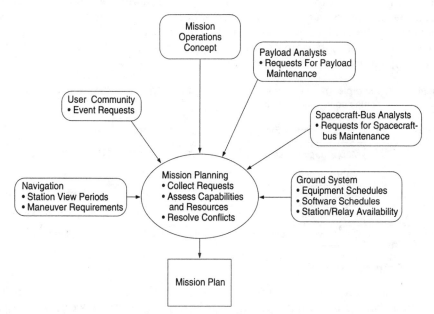

**Fig. 8.2.**     **Information Required for Mission Planning.** Mission planners must balance resources, users' request, spacecraft analysts' request, and mission objectives when creating a mission plan.

    Ideally, mission planning is independent of both the users or requesters and the spacecraft analysts. Scientists, given a free reign, will tend to drive up mission operations costs by trying to do everything without regard to resources. The spacecraft team may tend to use the spacecraft as an engineering test bed. Mission planners need to be independent and balance the spacecraft needs against operations capabilities, resources, and user requests to meet the mission objectives and stay within the mission operations budget.

    A simplistic mission plan for a mission with two-week activity plans might look like Table 8.1. The primary events for the activity plans are identified along with any special requirements or constraints. For this mission, mapping is the primary activity and is the main activity in each command load except for Activity Plan 5. Activity Plan 1 and 2 also contain calibrations of the gyroscopes related to attitude control and system-level pointing. The special consideration for Activity Plan 2 shows that it spans the Christmas holiday. It will act as a guide when we translate the mission plan into an activity plan, so staffing can be minimal on Christmas day. Because the pointing calibration is a special activity, it should definitely not be scheduled on Christmas. The special consideration on Activity Plan 3 requires more tracking coverage than usual to provide an accurate solution quickly after the orbit-trim maneuver. It notifies the scheduler of ground-system resources to schedule more station time during that plan. Spacecraft configuration or celestial geometry may influence events. In the example shown, battery reconditioning can be done only in Activity Plan 5 because it requires full Sun (i.e., no occultation). Finally, mapping can't occur during battery reconditioning because the power requirements are incompatible.

**Table 8.1.**     **Example of a Mission Plan.** This table is typical of the information and detail contained in a mission plan.

| Activity Plan | Primary Events | Geometry | Special Consideration |
|---|---|---|---|
| 1 | Mapping, gyro calibration | >20 min. occulted | Station 4 maintenance |
| 2 | Mapping, pointing calibration | >30 min. occulted | Christmas holiday |
| 3 | Mapping, orbit-trim maneuver | >20 min. occulted | Extra tracking required |
| 4 | Mapping, special test | <20 min. occulted | Timing update for test |
| 5 | Battery reconditioning | No occultation | No mapping |

### 8.1.2    Activity Plan

At some point, we must translate the mission plan into an activity plan. This translation can occur when the inputs, such as targets, maintenance schedules, and orbital geometry, are accurate and not likely to change. Usually, for a dynamic mission, this point would be a day to a week before plan activation. For a science mission, such as an interplanetary mission, it's usually six to eight weeks before plan activation.

The inputs for the activity plan are the same as for the mission plan but more detailed. Of course, changes may create new inputs. These changes might include a broken antenna at one of the stations which will require two months to fix, out-of-tolerance pointing that requires calibration, or a delay in changing telemetry flight software because the telemetry-system upgrade is behind schedule.

The *activity plan* provides the next level of detail, down to the minute; it passes directly to command development for implementation. Typically, it's a time-ordered listing of events with approximate durations allocated against resources.

This plan can be on paper or in an electronic file. Software often helps produce the activity plan, with many of the inputs also produced and delivered electronically.

Planning software would use a database of activity definitions, including required resources and duration estimates. Other inputs would include an *orbit-propagation-geometry file*, which contains spacecraft position and timing information, and a *station-allocation file*, which defines the station and times of coverage allocated to the spacecraft. (Note: we use "station" broadly to include ground stations and relay spacecraft.) The software could also include algorithms for resolving conflicts and may provide the operator with selectable options.

A simplistic activity plan might look like Fig. 8.3, which shows a ten-hour excerpt of the gyro calibration. It's now scheduled for two hours on day 1 of this activity plan. A special play-back of the calibration shows up at this level, along with the special station coverage it requires.

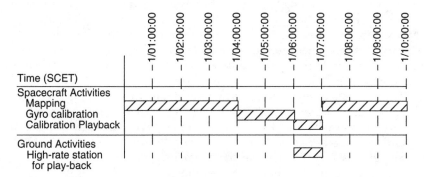

**Fig. 8.3.**     **Example of an Activity Plan.** Figure shows times in spacecraft-event times (SCET), the planned activities, and their estimated duration.

### 8.1.3   Shift Operations

Not all positions are shift positions, and different members of the team may have different shift schedules. Typically, positions for people who send commands to and receive telemetry from the spacecraft are staffed 24 hours a day. But confining operations to a single shift can greatly reduce operations costs. A high degree of spacecraft autonomy, onboard data storage, and favorable orbital geometry can combine to enable single-shift operations. A few support positions, such as computer system administrators, may also be staffed 24 hours a day depending on the mission's timeline requirements. Alternatively, support people may be on-call through telephones or pagers if required response times aren't too short. Other positions, such as management, long-range planning, analysis, and ground software are typically on a normal work schedule of 8:00 a.m. to 5:00 p.m., five days per week.

Mission requirements strongly influence coverage and shift requirements. A spacecraft in low-Earth orbit which requires short contacts every 90 minutes tends to drive 24-hour operations, whereas a geosynchronous communications spacecraft with long, easily scheduled contacts may fit into a regular work day.

A typical day in operations usually begins between 6:00 a.m. and 8:00 a.m. local time with a briefing from the overnight crew to the incoming day crew. This *hand-over briefing* includes spacecraft status, any commanding performed, anomalies that occurred, ground status such as station outages, and any other significant events. Usually, the briefing is one-on-one: each position gives a customized, detailed briefing to the counterpart on the next shift.

Actual shift schedules vary widely. Some programs operate on a traditional three-shift rotation, in which each shift spans 8.5 to 9 hours and people rotate between the shifts weekly. Others operate on a two-shift rotation, with 12-hour shifts for four days followed by four days off. People who are expected to work this way need to know in advance that it will be a shift position and what the shift rotation will look like.

Many people prefer the 12-hour days because they get four-day weekends every week. But this schedule's drawbacks include the long day and the decrease in alertness after eight hours, especially when those hours are in the early morning. Adjustments can be difficult for the night shift because they tend to revert to mostly daytime schedules on their days off and never really adjust to a nighttime schedule. This contributes to further fatigue and loss of alertness by the fourth day. Four days off can also be a drawback in a dynamic environment, leading to discontinuity and increased reliance on the quality of the shift-handover briefing when the person returns. Usually, in these arrangements, people work days or nights for two weeks to a month and then swap after their days off.

On the other hand, some people prefer the rotating, three-shift approach with rotations either weekly or in multiples of weeks. This schedule has the advantage of shorter work days and a more normal work week. However, shift changes usually are separated by only two days, which can be a short turnaround.

When designing the shift schedule, keep in mind natural circadian principles, which dictate that the rotation always be clockwise—rotating from days, to evenings, to nights. Also consider employee preferences and company policy. For instance, the night shift may be much more palatable if the company pays more for it. For a more detailed discussion of circadian principles and shift-work schedules, check "Wide Awake at 3 A.M.," by Richard M. Coleman. A poor shift schedule can cause low employee morale, decreased productivity, and mistakes that affect the mission. Conversely, a well-constructed shift policy can result in higher productivity, better health, and reduced lost time.

### 8.1.4  The Status Meeting

The *status meeting* is the main forum for communications on the operations team. It's almost always a meeting because of the improved communication face-to-face. Other groups not co-located, such as the user community, may be connected by telephone—or by video link, as it becomes more available. There's no substitute for the periodic interaction of all involved parties; still, for efficiency, meetings should be kept to a minimum. This status meeting is often held daily with a maximum interval of weekly. Twice a week, usually on Monday and Thursday, is a good compromise for a mature operation. The frequency of the status meeting may change with mission phase or during special operations. Even a system that has weekly meetings will often have at least abbreviated meetings daily during launch or other critical operations. For example, the Air Force typically conducts the status meeting as a 7:30 a.m. "stand-up" meeting, believing that, if everyone has to stand up, the meeting will run efficiently. Ideally, the status meeting lasts a half hour but no more than one hour.

The status meeting includes representatives from all of the operations areas including the control team, ground hardware, ground software, communications, operations, management, payload operations, mission planning, spacecraft engineering, navigation, and any other interested parties. A typical agenda includes reviews by all major elements of their current status, their activity plan, and the mission plan, as appropriate. The master activity plan is presented by mission planning and approved. For daily meetings, you may designate one day per week to review the mission plans, which shouldn't require daily review. But be careful not to input requirements at the meeting, except maybe as a heads-up. The team should come into the status meeting with a proposed plan—working it real-time will bog the meeting down.

Review the plan for the spacecraft and the ground plan, including software upgrades, hardware maintenance, and planned outages. Although schedules are coordinated at the lower levels, you may not recognize many conflicts until you have the complete picture.

### 8.1.5   Developing Command Loads

The *command load* flows directly from the activity plan and is often referred to as the "standard" sequence or standard commanding. Sometimes, you'll need commanding that wasn't included in the activity plan—called *non-standard commanding*. These commands can also be stored commands, although they're usually stored in an area of onboard memory different from that for the standard command loads. Non-standard commands can also be *real-time* or immediate-execution commands: those which execute when the computer or hardware receives them. In general practice, the operations concept always states there will be little or no unplanned or non-standard commands, but they're inevitably required. You'll need to define a disciplined system to handle non-standard commands and avoid errors, while also responding quickly when anomalies require the commands.

After the activity plan has been reviewed and approved, you'll begin developing the command load to carry it out. If the timeline is very tight, initial development may start in parallel with the status meeting. The command-load-development group carries out the plan. This group or function is variously called planning and scheduling, planning and analysis, sequence design, or uplink design. Follow the timeline for command-load development discussed in Chap. 7.

The predominant method of commanding is through *stored command loads*, which are sequences of commands with time tags. These commands reside in an onboard computer for execution as their time occurs. They're designed to repeat functions while changing only parameters and command details. The structure and function of these loads are extensively validated before launch. The dynamic parameters, such as time and turn angle, are input to execute the specific activity plan. Depending on the commanding complexity and the computer's memory space, stored command loads may cover from several hours to several weeks. Other factors, such as the accuracy of the orbit propagation and prediction, also influence the duration.

The inputs for developing the command load are very similar to those for mission planning (see Fig. 8.4), although they're now more detailed and should be electronic. Highly automated development leads to speed and accuracy because electronic interfaces, once validated, are repeatable and reliable.

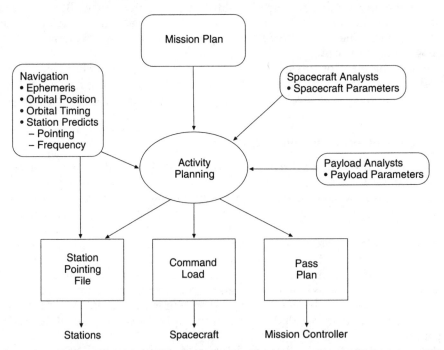

**Fig. 8.4.    Information Required for and Products from Activity Planning.** Inputs are similar to those for the mission plan, but they require more detail.

Timelines for missions with a daily activity plan are the most strenuous. Extended planning cycles have similar development processes with an extended duration. Extended timelines also allow you to develop command loads through several cycles. Daily development more likely will have a single cycle flow. Multiple cycles allow you to repeatedly refine the product. Often, the scientists and investigators tend to provide late inputs. In some cases the final plan bears little or no resemblance to the initial cycle. Try to limit this tendency; it wastes effort reviewing products that change drastically, and the final product has fewer reviews than would be indicated. In this case, either there are too many cycles or the process starts too early.

Sometimes, you can't avoid late changes because a spacecraft change or a new discovery demands them. But late changes shouldn't be the norm. Ideally, command loads change only when dynamic parameters, such as orbit determination and timing, become more accurate.

Figure 8.5 shows a typical flow for command development. First, expand the plan into its component events using specific timing and parameters. The output of this expansion is an *activity-event file* that contains precise timing information, commands, command parameters, and station information. It may also include spacecraft-status information which shows how commands affect the spacecraft's state and allows state propagation and checking. The activity-event file should be in both electronic and humanly readable formats, though these aren't necessarily separate.

Next, review and check the activity-event file and validate the commands. Check for resource conflicts that weren't visible at the plan level, violations of mission rules, limits, command timing, and configurations of the spacecraft and ground equipment. Include groups who handle the payload, spacecraft, ground element, and planning in your product review.

You can do this review manually or through automated software. The trend is towards more automated checking with software that can encompass rules defining timing, configuration, and limits. More sophisticated software can propagate states and model dynamics to verify spacecraft maneuvering and pointing, thermal characteristics to check temperature limits, power levels to verify power margins, and payload characteristics.

In an ideal world, you'd find no errors during the review, but the world isn't ideal. If you discover errors, you must identify them to the rest of the project and propose corrections. Sometimes, corrections are simple, such as a mistyped parameter. Other times, the error can be very complicated and require a change to the plan.

An example of a major change is a desired calibration we can't do because it violates a power limit. You could delay the calibration until solar occultation is lower, add commands to power off equipment not being used during the calibration, or abbreviate the calibration. For your first option, mission planning must replan the event or, possibly, add an event to a later activity plan when the

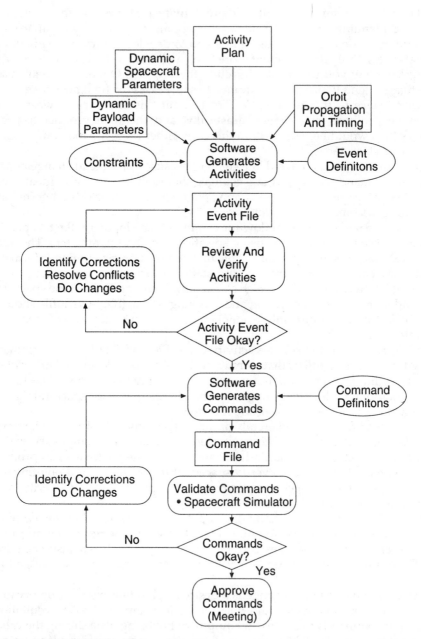

**Fig. 8.5.**    **Developing Command Loads.** This process generates a command load ready for up-linking to the spacecraft.

resources are available to move the calibration. Your second option involves all of the spacecraft's subsystems, including the payload, but wouldn't require replanning. It would require adding commands to reconfigure the equipment. Your third option mainly affects the requester, who needs to determine if an abbreviated calibration would suffice; it may require changes to input parameters, or adding or deleting commands.

After the activity-event file has successfully passed all reviews, it's translated into low-level, binary computer instructions. This step requires the command definitions, usually contained in a command database, to translate the command name to the proper computer instructions. It also usually includes translating the time-tags from a standard time format (UTC or Julian) to the spacecraft clock's time. Time-tags tell the computer when to carry out a command. The translation software also assigns memory locations to the commands in the file. Depending on the complexity of the sequencer or onboard computer, this allocation may be simple or very complex. It may require the command-translation software to maintain a map of the computer memory and propagate the state from event to event. A less reliable method would require the operator to input the desired memory addresses and manually maintain a memory map.

The command-translation software also adds appropriate headers and trailers, which contain such information as identification codes, message type, destination, message size, and the error-detection codes required by the command decoder or the flight software.

Ancillary products may appear at this point in the flow if they weren't already produced with the activity event file. Flight controllers or ground stations use these products, which are sequences of events or predicts. Examples are the contact-support plan (CSP), integrated sequence of events (ISOE), keyword files, and station control files.

The output of command translation is a *command file* that is ready for uplink to the spacecraft. But before uplinking it, you must validate it. Because command validation is extremely important, we discuss it separately below. Once the commands are validated, you seek command approval. Through *command approval*, you make sure all of the development steps have occurred properly, commands have no remaining errors, all of the required coordination has taken place, and all affected groups approve the command upload. You can get command approval in two main ways: a command-approval meeting or authorized persons electronically entering an approval code. Whether through meeting and signing a piece of paper or by electronic means, command approval releases the command file to the mission controllers for uplink to the spacecraft. During operations development, the mission operations managers need to decide what level of authority is required for command approval and who can approve violation overrides and waivers. Document any violations or waivers, along with the reason for approval.

## 8.1.6    Command Validation

*Command validation* consists of the reviews and simulations that ensure the command load is free of errors. *Command errors* are any commands sent to the spacecraft that result in an action other than what was intended. Some command errors are harmless, but others are catastrophic to the mission. When you try to reduce mission operations costs, emphasize catching errors that affect, limit, or are catastrophic to the mission. Don't sweat the small stuff. The cost to catch 100% of the errors is very high, so you must determine which errors to catch and which ones to let slide. As a result, you can reduce the amount of review and validation for the commands you designate "no impact."

A Magellan experience illustrates the problem with judging the severity of command errors. We determined that memory readout commands were harmless and that an erroneous command would result only in our not getting the area of memory we expected. We'd need to re-send a correct command, but the impact would be a delay and a little more work. However, in one case, because of a ground-software limitation, we decided to store a memory buffer for the attitude and articulation control subsystem (AACS) in the command and data subsystem (CDS) memory. We'd read it out periodically by transferring the memory contents back to the AACS memory and doing the standard memory readout. Unfortunately, nobody recognized that the dynamic parameters for the gyro-bias estimate were also in this memory buffer. When the old contents stored in the CDS computer were transferred back to the AACS memory, old values for gyro bias were written back into the AACS and used for attitude propagation. Using the old values resulted in the spacecraft drifting from Earth point, which called for several days of recovery operations. We learned two lessons from this incident. First, avoid becoming too complacent about harmless commands because we may be using the command in a different way. Second, never read data back into an active area of memory.

During design and development, we must determine what type of command validation we need and how much is affordable for this mission. The mission operations concept should address when we'll validate, who will do it, whether it will be online or off-line, and whether it will be automated or done manually. This design then becomes part of the mission plan, which defines in greater detail how and when we'll validate commands.

Often, most of command validation occurs before flight using a simulator and the spacecraft. All of the individual commands and many of the stored command files are validated on the spacecraft during ground test. The files are tested with a representative set of values or parameters; we can't test all values because of time and budget constraints.

The most common way of validating commands after launch is by simulation. We can do this simulation on a real-time simulator of the type used during development. It usually consists of a flight computer, at least a breadboard version of the spacecraft, a version of the flight software, some or all of the interface hardware, and software models of some or all of the remaining hardware. We also model spacecraft dynamics. This type of simulator provides the most extensive validation, but it's also expensive. It may also be time consuming because it executes at the same rate as the actual spacecraft, which makes it impractical for running all commands in large loads.

As spacecraft are becoming smaller and less expensive, highly accurate simulators using flight hardware are less likely for development and are rarely justifiable for operations alone. We can use software simulators that run the flight software, potentially on a different platform, with software models of all the hardware. The disadvantage of this type of simulator is less accurate timing and interfaces. But one of the great advantages is that these simulations can run on a faster computer than the flight computer, which enables them to run faster than real-time.

We may also combine methods. An example would be a software simulation of the flight system for commanding that runs very fast and checks command structure, ranges, and memory management. Large command loads could run as a matter of course. In addition, the real-time simulator could run all or part of the critical or unique command loads.

In addition to simulation, or along with it, spacecraft configuration and state propagation are important to command validation. Commands or activities may have a prerequisite hardware configuration or precursor commands. Relationships between activities or command loads are very important, but they're also one of the hardest things to keep track of. For instance, activity A requires that box A be powered—its normal state. But an anomaly occurs before activity A executes. Then activity B is uplinked and executed, which results in box A being powered off. Activity A won't execute properly and is a command error. We must track the spacecraft's state very closely and propagate it forward to make sure we meet expectations and requirements.

## 8.2　Uplink Process

While the planners and schedulers are building the command load, the control team is working on the pass plan, which includes the contact-support plan. The controllers need to know how long this particular load is going to take to uplink and whether to schedule it over a single pass or multiple passes. A *pass* (sometimes called a contact) is the period between acquiring and losing the spacecraft's signal. For a low-Earth orbiter using remote-tracking stations, a pass usually varies from 4 to 12 minutes. A pass can last several hours for a geosynchronous spacecraft. TDRSS or a relay spacecraft allow a pass of several hours. Elements of the *pass plan* are:

- When does this command load get uplinked?

- Timing constraints—absolute time codes or untimed?
- Over what station?
- How long does it take to uplink it?
- What configuration does the uplink require?
- What is the uplink rate?
- When does the station rise?
- What is the earliest acquisition? Usually, the spacecraft must be at least five degrees off the horizon. For some stations, it's higher than that because of surrounding mountains.

## 8.2.1   Preparing for the Pass

An *uplink* is a contiguous stream of data which is transmitted from the ground to the spacecraft while the ground transmitter is active and maintains the spacecraft's receiver in-lock. It includes all of the header information required by the spacecraft receiver and decoder, as well as commands or sequences of commands. If the transmitter is active, then off, then active within a single pass, it has done two uplinks. An *uplink window* is the time during which we may send the uplink. We can specify it as no-earlier-than or no-later-than, or we can bound its transmission with earliest and latest times. The uplink window occurs over a station pass or may span several passes. We may also schedule several uplinks over a single pass.

How we determine the uplink window for a given command load or command can be fairly complicated. Command loads are usually time-tagged with a spacecraft clock time at which time it should start executing. Based on the available memory in the onboard computer, we can uplink this command load hours or days ahead of its execution time. If memory space is limited, the timing constraints may be quite tight, so we may have to uplink it after the current command load completes but before the new execution time. Immediate-execution commands start when the spacecraft receives them, so the time of uplink completely controls the execution timing. Between stored commands and immediate commands, many other command types exist which vary with each spacecraft's command architecture. A command handbook should fully document the types of command loads and the requirements and constraints for their use. We can also place the command requirements in software which can automatically determine the type of command load and the applicable rules. Then, by accessing an electronic version of the pass schedule, the software can provide possible command windows to the operator.

*Uplinking* or *commanding* is the process of transmitting commands from a ground computer through a series of ground equipment and transmitters to the spacecraft. The US Air Force uses command centers at the Onizuka Air Force Station and Falcon Air Force Base, with transmission through its network of Remote Tracking Stations (RTS) at eight locations worldwide. NASA's Earth-orbiting spacecraft have various command centers, including the Goddard Space Flight

Center, with transmission over the Satellite Tracking and Data Network (STDN)—a world-wide network of ground antennas. Interplanetary spacecraft are controlled mainly from a command center at the Jet Propulsion Laboratory (JPL) with transmission through the Deep Space Network (DSN), which has large (up to 70 m) antennas at three sites: Goldstone, California; Madrid, Spain; and Canberra, Australia. Spacecraft relays are also used for various programs. The most widely known of these relay spacecraft is NASA's Tracking and Data Relay Satellite System (TDRSS), which is a constellation of three geosynchronous spacecraft. These spacecraft are controlled by a command center at White Sands, New Mexico. A relay system such as TDRSS provides the advantage of nearly continuous uplink and downlink for any spacecraft in LEO. Science missions that produce a lot of data, such as the Hubble Space Telescope, use TDRSS extensively. Chapter 12 describes various tracking networks.

The plan that describes the command uplink carries various names, such as contact-support plan or an integrated sequence of events. This plan lays out all of the particulars needed for this particular pass or uplink, such as the configuration of the station and its associated ground equipment, the command rate, telemetry and data rates, time for acquiring the signal, and time for losing the signal. The plan can be a computer-executed series of transactions or it may be a paper listing of the required events, which operators do manually. For simplicity, we'll refer to this plan as the *pass plan*.

The pass plan actually begins with prepass operations, well before the spacecraft appears. During this period, we configure the ground system and brief everyone involved in the pass. We calibrate the station and do loop-back tests. A *loop-back test* consists of sending a command from the console through the ground-commanding system to the antenna. Instead of actually sending the signal, we loop it back to the controller. Also at this time, the station predictions are loaded or activated at the station. As discussed earlier in Sec. 3.5, these predictions include antenna-pointing angles, frequencies, and timing information.

The *mission controller* is the senior person on the real-time control team and is responsible for carrying out the pass plan. The primary interface to the stations is the *range* or *resources controller* who coordinates the communications configuration and monitors and reports the station readiness. For LEO spacecraft using the remote-tracking station, a typical prepass operation begins eight minutes before acquiring the signal, as depicted in Fig. 8.6. A typical prepass timeline for a DSN pass is 1.5 hours. The ground-commanding configuration includes such items as the transmitter rate and power, crypto and decoder units, spoof-checking enable, and ranging-signal enable, as described in Chap. 11.

### 8.2.2 Executing the Pass

A station's *view period* is the time (as defined by the geometry) that the spacecraft is visible from the desired antenna. Chapter 10 shows how to calculate the

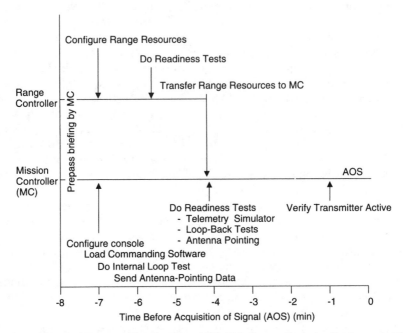

**Fig. 8.6.** **A Typical Prepass Timeline for a LEO.** This figures illustrates the types and timing of activities to prepare for a pass.

geometry. A *pass* may encompass all or part of the view period and is the time that the antenna and associated ground resources are dedicated to that spacecraft.

*Acquisition of signal* occurs whenever the ground antenna's receiver locks onto the spacecraft's electromagnetic transmission (or carrier signal). Ground equipment then processes the signal into a usable data stream, as described later. The telemetry is *in-lock* whenever spacecraft telemetry is being processed and displayed. We usually reserve the initial period of telemetry for spacecraft status and health checking, as shown in Fig. 8.7. We examine this telemetry to determine the state of the spacecraft and verify that it's in the expected state based on the last pass and any known commands which may have executed while it was out of contact.

In a typical LEO pass, the status-and-health period may last 30–90 seconds depending on the spacecraft's complexity and the level of automation. In a highly automated system, software does all checking based on a set of rules expressed as conditions on the data values. In practice, given the processing ability and technology of modern computers, we can reduce the status-and-health check to a "status OK" message, as discussed for payload monitoring in the EUVE. However, defining the rules for measurement processing may be difficult, and autonomous state prediction is imperfect. Also, a good operator develops a sense about what nomi-

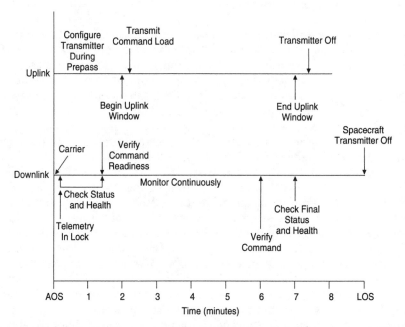

**Fig. 8.7.** **A Timeline of Typical Activities for a Pass by a Low-Earth Orbiter.** Acquisition of signal (AOS) and loss of signal (LOS) define the pass duration.

nal values should be and may often catch a problem in its early stages, before any rule base would have indicated it.

Mission controllers use a console that consists of video-display terminals, a telephone handset or headset tied into the various voice networks, notebooks containing procedures, spacecraft data, a logbook, and anything else they might need close at hand. Typically, each operations position has its own console with the appropriate procedures and data.

During the pass, mission controllers check the status-and-health, continuously monitor telemetry, analyze that telemetry in real time, transmit any command loads or commands scheduled for this pass or uplink window, and verify the receipt and execution of transmitted commands. They also coordinate with all other positions and lead anomaly investigations. If they detect any problems during the pass, such as violation of a red-alarm limit during the status check, they abort the command plan (at least temporarily) and start the appropriate response defined by procedures. Chapter 16 fully discusses how to handle anomalies.

In addition to the mission controllers and range controllers, several other people may be involved in the pass. A *deputy mission controller* works much like the mission controller and shares the workload during busy passes or critical events. He or she would also be the second approval for command enable, if required. Another control-team position is the *spacecraft analyst*, who more carefully monitors the spacecraft telemetry and verifies functional commands. Of all the real-time positions, this person knows the most about how the spacecraft works. Depending on the level of spacecraft expertise required for this mission, this person may be either a systems engineer who helped design the spacecraft or an operator who has had extra training on it. Besides the spacecraft analyst, subsystem analysts may monitor telemetry and verify functional commands, especially during critical events.

Once the status-and-health check is successful, commanding may begin. Commands that are going to be sent during this pass are promoted to the command system or included in the command data-base designated for this pass. Requiring a positive action to make the command available is part of a system of checks and balances which reduces the chance of selecting and transmitting an erroneous command. Many operations also require the buddy system, in which at least two people must verify that the selected command is correct before sending it. Operational procedures or the commanding software ensure we use this system.

The command database may be set up so all commands or selected commands are available for each pass. Some commands are restricted so we can't select them in real-time. *Restricted commands* are considered dangerous—inadvertent execution could be catastrophic to the mission. For example, a flag or bit might indicate that the command is restricted and that we must remove or reset this flag during command development and approve it during the command-approval meeting. Examples of restricted commands are igniting solid rocket motors, deploying appendages, and enabling and firing pyrotechnics.

Even though the transmitter was configured and checked before the pass, it remains in a standby state. As part of sending the command, we must apply high power to the transmitter. Applying high power is often referred to as *going active* on the command. The transmitter is active during the uplink. In spacecraft that use a coherent signal or an active ranging signal, a continuously uplinked carrier signal keeps the spacecraft receiver in-lock and provides a frequency reference for the downlink signal. The command transmission is then modulated onto the carrier signal. We check commands at various points along the transmission path to ensure the correct signal goes to the spacecraft. These checks include parity checks, sequence ID, and packet timing. If a check fails, we abort the transmission and retransmit the uplink, noting the time and duration of the transmission in a paper or electronic log.

The pass plan can have one or more uplinks which may require us to verify telemetry between them. A command that changes the state of the spacecraft may require verification before sending a subsequent command requiring that state. The commanding procedure usually specifies a minimum set of verifications for all commands, as well as additional or specific requirements for certain commands. For non-routine events or commanding, the spacecraft subsystem or payload analysts may also monitor the telemetry in real time and tell the mission controller which commands to transmit.

_____

An example of this real-time interaction would be battery reconditioning. A battery has been disconnected from the main bus, discharged, and then connected to a charge circuit. An expert on the spacecraft battery monitors the telemetry on the battery's temperature, voltage, and charge current to decide whether or not the battery should be reconnected to the main bus during this pass. After examining the battery telemetry and deciding the battery is recharged and ready to be reconnected, the battery analyst would use proper protocol to notify the mission controller over a voice net. It might go something like "Power to mission controller. Battery 1 has completed recharge and is ready to be connected. Power is 'go' for transmit of B1CNT, battery 1 connect." And the mission controller might reply "Copy. Power is 'go' for B1CNT. It is in the queue and will go active in 30 seconds."

_____

An example of multiple uplinks with verifications is a memory load to write-protected memory that requires the write-protects to be disabled or the on-board computer will reject the memory load and register a fault or alarm. In this case, the commands to disable the write-protects should be sent and positively verified by telemetry before uplinking the memory load.

Command verification takes two basic forms: command acceptance or authentication and functional verification. _Command acceptance_ consists of telemetry from the command decoder or on-board computer which registers receipt of the command through counters, an acknowledge bit, or simply the lack of command-reject errors. The spacecraft design includes command-receipt logic and error detection.

_Functional command verification_ consists of verifying that the intended action of the uplinked command was successful. In the write-protect example, a telemetry measurement containing the status of the write-protects (ENABLE/DISABLE) would be a functional check of the state. There may not be any direct measurements, such as a state, to do the functional verification, so we may need several measurements or indirect measurements. For example, a command goes out to turn on a heater, but the heater has no status bit. An increase in the battery-discharge current of a predicted amount would indirectly verify that the heater did turn on. An increase in the temperature of the item being heated would be a correlating measurement.

Somewhat the inverse of command verification is _fault detection_. During the real-time pass, this detection mainly takes the form of checking alarm limits. Rules for the expected state or value of the telemetry measurement are defined and the current value of the measurement is compared to these rules. If a violation occurs, an alarm notifies the controller. Typically, we use yellow and red alarms. Yellow alarm limits are set to warn us that we should look at something more closely. Usually, they indicate that the spacecraft analysts should take a closer look. Red alarm limits are set for action—usually an immediate action by the controller.

The display and formatting of the telemetry at the console is very important, including such human factors as colors, the size of the font, and the amount of information. Using color on the measurement values to show status—green for nominal, yellow for exceeding the yellow alarm limit, and red for exceeding the red alarm limit—is a common technique which helps the controller process and understand the information. If color is not available, we can use boldface letters, reverse video, and audio alarms. We need to balance overwhelming the controller with information against providing enough information to do what's required. Different controller positions require different displays. Real-time plotting of analog measurements may show trends. Simple schematics may help the controller understand how the measurement works. New, state-of-the-art graphics can include full three-dimensional depictions of the spacecraft with point and click interfaces to bring up subsystem details. But resist the urge to make the display interface too flashy if it harms ease of operation and clarity.

After all commands specified in the pass plan have been sent and verified, we do a final health-and-status check. We also note the spacecraft's state. We can't always functionally verify each command, especially in the case of a stored command uplink which may execute over days or weeks, but we should verify command receipt.

*Loss of signal* occurs whenever the spacecraft is no longer visible to the ground station. If a spacecraft is geosynchronous or uses a relay spacecraft, loss of signal may occur at a specified time.

### 8.2.3    Post-Pass

After loss of signal, deconfigure ground resources and release shared resources from support. Complete required paperwork immediately while the pass is still fresh in your memory. Fill out and submit discrepancy reports or failure reports on any anomalies for the spacecraft or the ground system. At this point, downlink post-processing begins. Process collected ranging data and make it available to the navigation people. Store spacecraft telemetry and make it available to the spacecraft analysts for trending and analysis. Route payload or science data to the appropriate processing location and make it available to the payload analysts.

## 8.3    Downlink Process

*Downlink* refers to all data originated at the spacecraft and transmitted to Earth. We describe below the various types of data and what they do.

The spacecraft's main mission is to provide science, mission, or payload data, including everything from the spacecraft images you see on the evening weather report to measurements of the solar wind. For communications spacecraft, the data is the continuous stream of telephone or picture data.

Often, operators don't see mission data because it goes directly to users. For some Earth-orbiting spacecraft, such as communications spacecraft, the mission data travels on a distinct communications link that is dedicated to the mission data and independent of the spacecraft bus. In other cases, we interleave the payload and engineering data before transmitting it all to the ground and downlink it over the same communications network. Ground processing then strips it out.

*Engineering* or *housekeeping data* comes from the subsystems of the spacecraft bus. It includes such data as temperature, voltage, and attitude. During spacecraft design, the design engineers determine what measurements are important to determine the health, status, and performance of their piece of hardware. (See Chap. 15 for a discussion of spacecraft telemetry.) All of these measurements get collected and traded off against the available space in the data stream to make up the engineering data. Data from the computer shows how the software is working or provides information, such as use of memory, that the software computes.

As onboard computers get more sophisticated, the engineering-telemetry stream is becoming more programmable. There may be more sensors than there is room in the telemetry stream, which may lead to selecting multiple telemetry formats based on mission phase, activity, and hardware configuration. Increased flexibility onboard requires more flexibility on the ground, so we can determine the format in use and process it correctly (see decommutation).

*Tracking* or *ranging data* provides the position and velocity for the spacecraft's orbit or trajectory. We can collect this data in various ways. The most basic way is by determining where the ground antenna points when it contacts the spacecraft and the delta from where we predicted it would be. *Active ranging* provides more data about velocity. It consists of adding a signal to the uplink, called ranging tones, at a known frequency and measuring the difference in the returned signal. *Doppler ranging* also measures the frequency shift of the downlink signal. Doppler ranging can be totally passive, in which frequency shift is measured relative to a predicted, spacecraft-produced frequency from an onboard crystal oscillator. A more accurate method involves a coherent signal—where the spacecraft's downlink frequency is a fixed multiple of the uplink frequency. By measuring the phase difference between the uplink and downlink signals, we can calculate a distance based on the doppler shift. For interplanetary spacecraft, we can get even more accuracy by using more than one ground station simultaneously to do very-long-baseline interferometry (VLBI). Earth-orbiting spacecraft often do their own position estimates based on position information from the global positioning system (GPS).

*Ancillary data* consists of any other information the user needs to process or interpret the mission data. It may include other spacecraft measurements such as the payload's operating temperature or an attitude estimate, or other ground-processed information such as the orbital position. If we process the mission and engineering data separately, we must develop a way to deliver the ancillary data to the user. In the case where data are interleaved, we can strip out the appropriate spacecraft measurements and payload data during the original processing.

### 8.3.1 Processing Telemetry

The first step in processing the downlink data is *synchronizing frames*. A repeating pattern in the serial stream marks the boundary between major frames. Once the processing software finds this frame boundary, it can locate the data slots in the stream as defined by a decommutation map.

*Decommutation* converts the analog stream to a digital stream of data words. In essence, it's the exact reverse of commutation, which takes place onboard the spacecraft. Also, at this point, interleaved payload data may be separated from the engineering data if it requires different processing.

Once the data stream is decommutated, it may be stored or displayed in its raw format, but it usually undergoes *decalibration*. This process applies calibration parameters to the data words. These calibration parameters can include binary to hex conversions; polynomials which define a mapping from a binary to analog conversion; or a definition of a state value, such as on or off, depending on how we're going to display or further process the data. To execute the pass, the decalibrated engineering telemetry appears on the operator's console. For a detailed description of the hardware and software needed for telemetry processing, refer to Chap. 13.

Once the telemetry is processed, we can store it in a database, display it, or further process it. The flexibility of displays has increased greatly over the years with the advent of window systems and affordable color displays. Such displays as functional schematics with the appropriate data interspersed are now possible. Real-time plots have virtually replaced the old strip-chart recorders, providing a more flexible software equivalent of strip charts (and the pens don't clog).

Improved computers also give operators and analysts more freedom to customize their displays. Some people like to look at the measurement displayed in its engineering units; others prefer hexadecimal. We must control the decommutation and decalibration data to ensure its integrity but leave the display to the discretion of the person who has to look at it.

### 8.3.2 Archiving and Retrieval

One requirement we should define during development is how much engineering data we'll store—and for how long. The data can be stored online, where it's immediately accessible, or in long-term media, such as tape or optical disk. The requirement for data storage may drive the type of hardware selected. In general, we must strike a balance between the need for long-term trending and the cost of storing so much information. Usually we keep engineering data online 30–45 days and then archive it until the end of the mission.

Large database systems that can store and process a lot of data have helped lower mission operations costs. This capability has had to keep pace with the more complicated spacecraft and its higher downlink rates. In the past, we stored very little data online, thus requiring that we maintain tape libraries and mount tapes when we needed to access the data.

A quick guideline for determining storage requirements is to expect at least 30 days of data. Take the engineering data rate in bits per second times the average contact length times the number of contacts per day. If the data is going to be stored as processed data rather than raw data, some multiplier should account for such things as floating point values requiring more space than bits. However, storing only completely processed data limits flexibility for post-processing later on.

# 8.4    Planning and Analysis Functions

### 8.4.1    Spacecraft Planning and Analysis

Spacecraft planning and analysis consists of analyzing and assessing the spacecraft bus. It usually occurs off-line in mission operations and in parallel with real-time commanding and downlinking.

The major tasks are:

- Predicting spacecraft performance
- Assessing the health of the spacecraft by trending
- Assessing actual spacecraft performance against design goals and requirements
- Maintaining spacecraft parameters which may need changing or adjusting
- Making inputs to mission planning regarding future spacecraft activities such as calibrations
- Sometimes, validating stored commands
- Often, maintaining flight software

Chapter 15 discusses these tasks in more detail.

Typically, people who do these tasks are the spacecraft experts and are often organized by subsystem. For a new or complicated spacecraft design, the space-craft-analysis group consists of two or three systems engineers who monitor the spacecraft's overall performance and are usually the most involved in mission planning and commanding. The subsystems represented include command and data, attitude control, power, thermal, propulsion, and telecommunications. We often combine related subsystems, such as power and thermal or attitude control and propulsion, to reduce staffing. Depending on the spacecraft, the flight-software expertise may reside in the command and data subsystem, be split between command and data and attitude control, or be a separate group.

*Telemetry analysis* is the main way to assess and trend the spacecraft's health and performance. This analysis falls under either short-term or long-term trending. In this context, short-term means orbit or day; long-term trending usually occurs over weeks or months. How we determine this time scale helps drive the archive capability discussed earlier. By trending over the two different time scales, we can distinguish between event-driven consequences, such as a new attitude that changes the heating profile, and hardware aging or degradation, such as a decrease in the solar panel's current output caused when silicon solar cells degrade.

The tools used to do this trending analysis include displays similar to the real-time displays, database query and retrieval tools, plotting routines, and programs for statistical evaluation. Which tools are used varies by subsystem and with the particular measurement within the subsystem. Sometimes the telemetry data goes to ground-processing software which calculates derived parameters. For example, attitude error is the downlinked parameter from which the ground software reconstructs the spacecraft's actual attitude.

At the point where trending analysis begins looking at derived quantities, it starts to become performance analysis. *Performance analysis* consists of measuring the spacecraft's performance against the design. It often requires us to combine measurements and do more post-processing. Chapter 15 discusses this topic in more depth.

Automation of trending tasks can save a lot of money in mission operations. For example, having ground software that executes a trending script every morning and produces plots or summaries by the time analysts arrive can save the analysts many hours. Database management is one of the areas that lends itself to automation and therefore reduces costs. An analyst with a printout of numbers, a pencil, and a piece of graph paper can spend hours producing a plot that a computer can generate in minutes.

Outputs of trending go into the mission-planning cycle. Examples are requests for calibrations that may be indicated by degrading pointing performance, for more Sun-pointing time due to lower power output, or simply for a parameter update to improve performance. We need a way to flow these types of requests into the mission-planning cycle. But we can't always get our inputs into the mission plan. If the input is too late for the mission plan, we want to get it into the activity plan and eventually to the spacecraft through standard command files. However, whenever events need to occur sooner than the standard process will allow, we need a non-standard command process. We can use this same process to do corrections for anomalies, as discussed below.

Predicting the spacecraft's future performance is closely tied to trending analysis. Prediction consists of extrapolating the future performance based on current and past performance. The long-term mission plan is an input to the predictions. These predictions may also be the basis for transferring values of spacecraft parameters, such as command parameters or flight-software variables, that we need to generate activity event files.

Trending and prediction of consumable items are very important. *Consumables* include propellant, battery charge/discharge cycles, mechanical cycles on gimbals or other motors, and anything else that gets used up. For example, predicting when the spacecraft will run out of fuel usually contributes significantly to mission-life analyses.

Another way to reduce mission operations costs is to develop dual-use software during design. When you're developing software to analyze the design, considering long-term use will cost very little more. So develop software that is compatible with the operations-software environment. For instance, while designing the attitude-control subsystem, develop routine software for validating control pointing and performance that also covers analysis and planning during mission operations. That way, you won't have to recreate and validate new software for operations.

Assessing flight software is another part of analyzing the spacecraft's performance. Typically, this analysis is a by-product of other spacecraft-performance verification or anomaly resolution. In other words, you assess the results of the software execution and analyze the outputs. You may also have some direct measurements of the software's performance such as measures of CPU use or other onboard diagnostics.

Maintaining flight software consists of changing the flight code based on your assessments of the software's performance, results of anomaly investigations, or new requirements. New requirements can occur if the mission lasts for a long time because people who become familiar with the system want to enhance or expand the spacecraft's capabilities.

Maintaining flight software follows largely the same rules under which it was developed except that we can sometimes speed it up to resolve anomalies. In this case, a requestor writes an engineering change notice (ECN), or the equivalent. Anyone who's interested then receives the ECN. Remember, a change in the flight software can often affect procedures, telemetry displays, and other operations tools. For example, the change may create a new software parameter that we must add to the ground database. We may also have to change procedures to trend and update or maintain this parameter.

After the ECN is approved, we code and test the change. A patch is the most common way to update flight software. Because the onboard software is normally executing while we're updating it, we rarely recompile and reload the flight software totally unless onboard computers are redundant enough so we can reload one computer while executing out of the other one. Also, uplink rates are typically slow (<1000 bits per second), so it's very time-consuming to completely reload the flight software. Patches to the code are usually small (<100 words) and can be loaded in spare memory, verified through memory readout, and branched to or moved.

The saying, "no flight-software change is a small change" tells us it's good policy to resist changing flight software. Typically, we can't test a flight-software patch as thoroughly as we tested the flight load before launch. No matter how good the simulator or testbed is, the patch will run on the spacecraft for the first time when it's operating in space. If problems occur, it's not easy just to reboot.

Often, anomalies bring out problems with flight software. An *anomaly* is any unexpected occurrence on the spacecraft. The consequence of an anomaly ranges from the annoying, such as doing a memory readout of the wrong address, to catastrophic for the mission. Anomaly resolution has three parts: detection, investigation, and correction.

We can detect an anomaly from a telemetry alarm or from performance trending and analysis. For severe anomalies or ones that require quick response, a contingency procedure usually defines the detection criteria, including correlating evidence, and outlines the steps to be taken.

Once a problem is detected that doesn't meet the criteria for pre-established contingency procedures, an anomaly-investigation team looks at the problem, tries to determine the cause, and recommends corrective actions. This team contains members of the flight team and others, as appropriate.

After the anomaly-investigation team has determined the most probable cause of the anomaly and recommended a resolution, we have to correct it. Usually, causes fall into categories such as human error, hardware failures, procedures, ground software, flight software, or others. Corrective action can include increased training, procedural updates, or changes to ground or flight software. Chapter 16 covers anomalies in more detail.

### 8.4.2     Payload Planning and Analysis

Payload planning is very closely associated with mission planning. It involves determining what data the payload should take and when, often referred to as *payload tasking*. A dedicated payload group usually does this tasking. For remote-sensing missions, especially military ones, payload tasking may be in nearly real time. In these cases, the mission plan may just have time periods blocked out for payload activities. The activity plan will then narrow down these opportunities by time or particular sensor. Finally, at command-load development, the precise tasking occurs within the constraints specified by the activity plan. The payload group then deconflicts tasking during command-load development.

For science missions, such as the Hubble Space Telescope (HST), a science working group does the tasking and deconfliction. These working groups meet well in advance, and the tasking plan is integral to the mission plan. When Comet Shoemaker-Levy 93 was discovered, HST went through major replanning of the mission to take data related to the comet's hitting Jupiter. Planners had to bump the science that had been scheduled during that timeframe. Normally we want to stick with the mission plan but we have to be flexible enough to seize such rare opportunities.

Analyzing the payload is very similar to analyzing the spacecraft's performance. The payload, which may be one instrument or multiple instruments, is trended and analyzed as other subsystems are.

Payload analysts also use collected data to assess the payload performance. Analysts for the HST wide-field camera were able to look at the processed pictures from the collected data to determine and analyze a system problem. Analysts routinely examine product quality, including the data processing on the ground, to measure the system's performance. This analysis may also be a secondary indication of a spacecraft problem that hasn't shown up in subsystem trending.

### 8.4.3    Navigation Planning and Analysis

We must also determine the spacecraft's exact location and motion as a function of time, both for where the spacecraft has been and for predicting where it will be. The parameters of the spacecraft's location and motion are known collectively as the *spacecraft ephemeris*.

Knowing where the spacecraft has been, the *reconstructed ephemeris*, is usually required for processing payload data so we can accurately determine the payload's location when it captured the data. Reconstructed ephemeris is usually very accurate because it's based on direct measurements. Reconstruction also provides a way to predict the future by updating such variables as spacecraft drag or solar torques in the knowledge database.

Knowing where the spacecraft is going to be, the *predicted ephemeris*, is important from long-range planning all the way through to generating pointing predictions for the ground antenna. Predicted ephemeris may also be uplinked to the spacecraft to maintain its knowledge of the position of celestial bodies, such as the Earth and Sun. Normally, prediction intervals for the ephemeris will be tied to the time spans of the products they are required to support: the mission plan, the activity plan, command load duration, and pass duration. The orbit-determination software is fairly complex and can be computationally intensive, so our long-term prediction may be limited.

Routine operations include collecting and processing the tracking data collected during the downlink pass. This processing can occur after each pass, or we can collect and process some number of passes at the same time. Again, because the orbit-determination software is complex, we usually batch-process the tracking data. Using data from many passes also makes our solution more accurate. We should at least get ephemeris at the same frequency as command-load development because we need the latest and best position information for accurate payload tasking and subsequent data collection. Chapter 10 includes more detail on navigation.

Another function of navigation is orbit maintenance. As part of ephemeris prediction, we identify the need for future orbital maneuvers, also known as *orbit-trim maneuvers* (OTM). The system design will set requirements on orbit altitude, inclination, or ground track. We must monitor and predict these parameters so we can insert the appropriate OTM into the plan as early as possible.

Once we've identified the need for an OTM, we must have a way to translate that requirement into the commands the spacecraft needs to properly execute it. OTM planning is iterative, involving navigation, attitude control, and propulsion to design the maneuver. The telecommunications, thermal, and power subsystems may also be involved in determining an attitude for the maneuver. If we want or must have real-time telemetry, communications may affect the timing and attitude for the burn. The thermal and power subsystems may constrain the attitude, but even more importantly, we may need thermal modeling to predict temperatures so we can accurately model and predict propulsive performance. Tracking requirements may also increase right after the OTM to provide a solution for the new orbit.

The perception of the complexity and risk associated with OTMs varies greatly from mission to mission and largely depends on their frequency and size. Configuring and reconfiguring the spacecraft to do the OTM also adds to this complexity. For many communications spacecraft, the thrusters mount on the spacecraft faces and point into or opposite the velocity vector. Thus, we don't need to reorient the attitude, and normal payload activity isn't interrupted. For these types of programs, orbital maintenance is routine.

In the other extreme, once a spacecraft first reaches its operating position, OTMs are risky because they require a lot of spacecraft reconfiguration. The mounting of the propulsion system may require reorienting the attitude, which interrupts the payload's operation. Some spacecraft with long, flexible appendages may require us to lock the appendages in position or even restow them to a launch configuration.

Spacecraft design varies with the required frequency and difficulty of OTMs for the spacecraft. Sometimes unpredicted events, such as unusually high solar activity, may cause us to execute OTMs more often than we designed or planned.

## References

Agrawal, Brij N. 1986. *Design of Geosynchronous Spacecraft*. NJ: Prentice-Hall, Inc.

AI Magazine, Vol. 15, No. 4, 1994. AI Approach Demonstrates Low-Cost Satellite Strategy.

Coleman, Richard M. 1986. *Wide Awake at 3:00 A.M.* New York, NY: W.H. Freeman and Co.

Data System Modernization (DSM) Position Level Training Guide. 1985. Technical Training Division, Air Force Satellite Control Facility, USAF.

Wall, Stephen D. and Kenneth W. Ledbetter. 1991. *Design of Mission Operations for Scientific Remote Sensing*. London: Taylor and Francis.

# Chapter 9

# Launch and Early-Orbit (L&EO) Operations

**Michael Fatig,** *AlliedSignal Technical Services Corporation*

In this chapter we discuss the environment for launch and early-orbit (L&EO) operations, key issues regarding them, a process for developing them, and methods for reducing costs. We also present ways to deal effectively with the demands of L&EO operations and encourage developing plans for them early in the space mission's life cycle. This early planning allows us to coordinate these plans with the designs of the space and ground components for a more effective end-to-end architecture.

## 9.1 Handling the Demands of the L&EO Environment

### 9.1.1 The Environment for L&EO Operations

Understanding the demands, rigors, and special characteristics of the environment for L&EO operations is vitally important to planning and to operational success. Depending on the program's approach and the space mission's architecture, L&EO activities can range from complex and lengthy to simple and automated. Understanding the L&EO environment and characterizing operations within it allow mission operations managers to make informed decisions about key issues, which help control the cost, performance, and risks of this critical mission phase.

Tough decisions face you and the program team in developing operations for L&EO. It tends to be the shortest phase but is also most stressful, critical, and demanding on the systems and people. Far too often the mission is designed for normal operations, with special activities of L&EO operations neither understood nor assessed. The results can be disastrous. The mission design may overlook the demands and risks of L&EO operations until costly late requirements cause risk in developing systems, plans, and procedures. How much should we spend on systems and facilities for the safety and comfort of L&EO operations, when often these systems, facilities, and people won't do anything for most of the mission's lifetime? This becomes a major issue of cost versus risk.

The environment for L&EO operations is short, intense, critical, and dynamic, so the operations team must have skill and quick reactions to handle

- Increased volume
- Shortened cycle times
- Increased risk

**Increased Volume.** L&EO operations commonly require resources two to three times that of normal operations. You can expect large increases in

- Number of real-time events
- Volume of commanding
- Amount of data processing needed for engineering telemetry
- Number of people involved in operations
- Number of activities which must be scheduled, checked, and conducted

For example, in normal operations you may require one contact with the spacecraft every other orbit—driven by rates of data collection, capability for onboard storage, and requirements for routine commands. However, L&EO operations typically involve a series of critical functions done rapidly after spacecraft separation (such as deploying appendages, stabilizing attitude, powering up critical components, and stabilizing and recharging power). During this period, you may need to contact the spacecraft many times per orbit in order to respond quickly to these mission-threatening activities. The number of real-time events increases dramatically, as does the command volume and processing of engineering data.

Because L&EO operations are critical, key design engineers for the spacecraft and ground system are commonly present. It's often necessary and advisable to have these design engineers near command, control, and communications to allow rapid analysis and response to anomalies. The two to four people for normal operations can become a staff of 20 to 30 for L&EO. Design of the operations room and system should match this temporary expansion (additional displays, voice circuits, and work area).

Because L&EO demands many more activities, the volume of commanding and data processing greatly increases. Many non-routine activities are done during L&EO, including such items as clock adjustments, system configurations and mode changes, system biasing, powering up components, and frequent calibrations. These activities and associated command procedures must be scheduled, compiled into pass plans and command loads, and verified.

**Shortened Cycle Times.** Shortened cycle times result from increased volume and operational dynamics. The increased volume means doing more in the same amount of time, so you have less time to move from activity to activity. During L&EO you may have only minutes to prepare for the next real-time event, so you may need to reconfigure rapidly. Not considering this need could severely restrict L&EO operations or place them at risk. Increased volume also drives shorter cycle times by demanding more resources than normal operations would require. For example, if you're commanding at two to four times the volume of normal operations, the onboard processor of stored commands may not be sized for this increase. Generating and uplinking of stored command loads may have to occur every eight hours versus every 24 hours as planned for normal operations. The operations team may therefore have to include staff for mission planning and activity around the clock, versus day staff only for normal operations.

The dynamics of operations also shorten cycle times. Often during L&EO operations critical activities require validating one step before moving to the next. So we must plan the mission and do real-time operations not in large, contiguous blocks, but in short segments. Two or more options usually follow each short segment and depend on completion and performance of the previous step, so planning and decision making are much faster. Again, this approach significantly affects the design of the operations team and the ground system. If the team or system performance doesn't allow these shortened cycle times, we may miss critical events or inadequately handle anomalies.

**Increased Risk.** Perhaps the most significant attribute of the L&EO environment is risk: the tremendous forces on the rocket during its ascent and during operation of the on-orbit kick motor; the "firsts" associated with each activity and system; the rapid, sequential activities; single-point failures in the system design; and demands on the operations team's reactions. Major problems you may encounter in this phase include

- Off-nominal launch-vehicle performance (improper orbit)
- Damage from the forces of launch (acoustics, g-forces, vibrations, changes in atmospheric pressure, temperature changes, or outgassing)
- Improperly deploying the spacecraft bus's key appendages
- Inability to acquire stable attitude or a power positive mode

- Inability or difficulties in acquiring good rf communications with the ground element
- Inability to detect and respond to time-critical anomalies
- Improper decisions in stressful, time-critical situations
- Unexpected performance levels in on-orbit systems
- Improper procedures and operator errors
- Unplanned contingencies with quickly recurring hazards
- Problems with software for the ground or flight element

For example, an improper orbit by the launch vehicle drastically changes the timeline for contacting the spacecraft through a ground- or space-based communications network. We must rapidly redo and reschedule all operations plans, then distribute changes to many organizations, facilities, and systems.

Another example is variance in the spacecraft systems' actual versus expected performance on orbit. For example, the power subsystem may not generate as much power as expected. If undetected, under voltage may occur. At least we must redefine the operations plan for the new levels of available power. Another example is the unexpected vibration of the solar array panels as the spacecraft enters the sunlight part of the orbit experienced by the Hubble Space Telescope. This "ping" induced vibrations into the attitude-control system, causing target-pointing problems. Other examples include the unexpected charging of the batteries in the Earth Radiation Budget Satellite due to albedo as the spacecraft was in a 180° pitch configuration, or uplink signal margins that were better than expected on the Solar Anomalous and Magnetosphere Explorer mission, which caused receiver lock on a sidelobe.

Because L&EO operations are risky, we must plan for potential conditions so we can establish a team, system, and procedures to handle these conditions as they occur. To do so requires a systems approach (space, ground, operations plan, and operations team) across various elements and organizations (Fig. 9.1). Understanding and characterizing the environment for L&EO operations early in mission planning will result in a system, plan, and team that can handle its demands.

### 9.1.2    Phases of L&EO Operations

In this text, we define the *L&EO operations phase* as the period from the beginning of live spacecraft activities during launch countdown to the transition into normal operations in orbit. The *transition to normal operations* is loosely defined as the time when we've largely completed special activation and checkout, made sure the space element works properly and performs well on orbit, and started activities to meet mission objectives. For experimental spacecraft, this would be when we start collecting primary science data. For communications spacecraft, it would be the beginning of communication services to the main user. Figure 9.2 and the

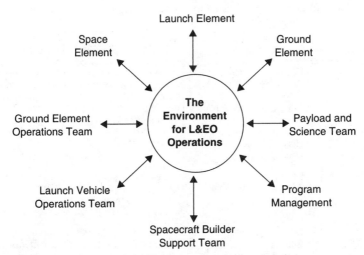

**Fig. 9.1.    The Environment for L&EO Operations.** L&EO operations demand close and rapid co-ordination of many system elements and organization.

following table define and illustrate the typical phases and subphases for L&EO operations. We define subphases and steps during early planning for L&EO operations, as discussed in Sec. 9.3.

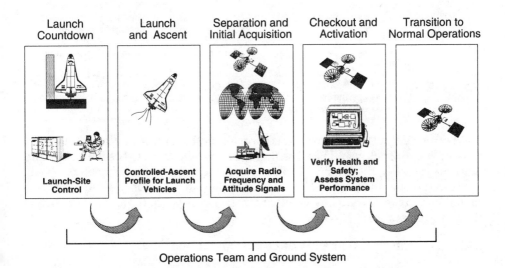

**Fig. 9.2.    The Phases of L&EO Operations.** Defines major groups of activities for further decomposition.

**Table 9.1.    Description of L&EO Operations Phase.** L&EO operations has distinct subphases with separate objectives and activities.

| Subphase | Steps | Description |
|---|---|---|
| Launch Countdown | • Power-up spacecraft<br>• Test spacecraft's functions<br>• Test ground system's readiness<br>• Determine go/no-go<br>• Configure for final launch<br>• Monitor system | Involves the final launch readiness tests of the end-to-end system (short aliveness tests) to support a final go/no-go determination by project management. Once the launch is a "go," places the systems in their final configuration for lift-off. |
| Launch and Ascent | • Lift-off<br>• Ascend through the atmosphere | Entails the pre-programmed, powered-flight profile to achieve orbit. Spacecraft operations during this phase are typically passive, with some automated operations controlled by timers. |
| Separation and Initial Acquisition | • Separate spacecraft<br>• Start activities immediately after separation<br>• Acquire initial attitude<br>• Acquire initial signal with ground<br>• Transfer to operational orbit (if required) | Involves the spacecraft's separation from the launch vehicle and the critical operations associated with acquiring an initial attitude, settling into a power positive mode, acquiring routine communications with the ground system, and deploying appendages. The separation activities are often automatically controlled by timers, separation switches, or stored commands. |
| Activation and Checkout | • Activate subsystems<br>• Activate instruments<br>• Check out system performance | Activate the spacecraft's subsystems and instruments; configure for orbital operations; adjust biasing, settling, and drift rate; do checkout. Checkout entails verifying modes and configurations and doing initial system calibrations. Verify overall system performance against predictions, adjust ground models, and modify operating plans. |
| Transition to Normal Operations | • Configure for normal operations<br>• Begin normal activities | Configure the spacecraft bus and instruments for routine operations to meet mission objectives. |

We must define L&EO operations in phases and subphases because their activities differ significantly and because this approach allows us to break this complex phase into smaller, more manageable parts. Finally, it provides convenient breakpoints in the operations plan, allowing for replanning periods, decision points, or transitioning responsibility to a new organization or team. It's a vital step in handling the demands of LEO operations.

## 9.1.3    Handling the Demands of L&EO Operations

From experience, we define the demands of L&EO early in the design for mission operations and develop ways to handle them. Thus, we can decide where best to handle these demands within the space-mission architecture and operations team. Without this foresight, the operations team must meet them under stress, often by overusing command, control, and communications. In this section, we'll look at demands on the

- Operations timeline
- Operations team
- Ground element
- Space element

**Demands on the Operations Timeline.** The timeline for L&EO operations requires much higher levels of activity, many more critical operations, shorter planning, and more data processing than for normal operations. Many missions are too eager to start, often resulting in an aggressive timeline with little room for problems. Experience tells us that, in almost all missions, we'll have trouble maintaining the schedule. Experience also teaches us that a more cautious, slower approach is more likely to start the mission on time—with lower risks and stress. Avoiding these traps produces a more efficient, error-free, and timely L&EO.

Table 9.2 describes some of the demands on the operations timeline and ways to meet them.

**Table 9.2.    The L&EO Environment's Demands on the Operations Timeline.** Operations are critical, dynamic, and complex.

| Demands On The Operations Timeline | How to Handle Demands |
|---|---|
| Complex, time-critical operations | • Plan operations at a 50% activity level; schedule backup events for critical and demanding activities; develop and test the detailed plan against the spacecraft and simulator<br>• Develop user-friendly graphics and flow charts<br>• Minimize the number of operations that depend on previous operations activities |
| Many operations depend on previous sequences | • Understand the interdependencies between operations activities and schedule slack in the timeline<br>• Document the interdependencies for use in rapid replanning<br>• Consider using a smart planning tool |
| Off-nominal space-element performance, higher anomaly rate | • Have readily available expected performance results (from test or models), resource budgets, and operational and survival limits to identify off-nominal performance and determine effects on the resource budget<br>• Have an onsite engineering team for rapid identification and response<br>• Preplan major anomalies and establish real-time monitors to recognize anomalies. Have flow charts and other aids that help you quickly determine a response.<br>• Collect as much engineering data as possible during orbital operations and frequently downlink this data for analysis as needed |

Table 9.2.    The  L&EO  Environment's  Demands  on  the  Operations  Timeline.  (Continued)
Operations are critical, dynamic, and complex.

| Demands On The Operations Timeline | How to Handle Demands |
|---|---|
| More offline data processing, often in a rapid mode | • Estimate the amount and frequency with which to collect engineering data, then size ground systems and operations to handle this data<br>• Establish a separate system and team for continuous data processing |
| Offline data processing driving the next planned real-time operations | • Make sure the data-processing system and team can process the data in the amount of time needed<br>• Add slack to the schedule and simulate the process |
| Shorter mission-planning windows | • Establish daily planning cycles, including an 8- or 12-hour shift of "no operations" in which daily replanning can occur without stress |
| Rapid replanning | • Ensure the system, people, and procedures are in place for rapid replanning at any point in the timeline<br>• Have predefined operations breakpoints and re-entry points—simulate replanning<br>• Design the operations plan in small, self-contained modules to allow operations to be easily altered and repeated<br>• Know the hazards, constraints, and interdependencies of operations activities—have these in readily accessible, concise checklists to use in rapid reviews of the hazards and constraints involved with changes to the operations plans and procedures |
| High activity level (3 to 5 times greater than normal operations) | • Estimate the volume and phasing of the operations, and ensure the mission operations system and staff are the right size<br>• Try to avoid major fluctuations in work levels and don't build an aggressive timeline (50% activity loading, schedule backup events into timeline) |

**Demands on the Operations Team.** The team is larger. It's physically dispersed between the control site, the launch site, and possibly other significant areas. The activity level, and therefore the stress level, is higher, and the team must rapidly communicate, coordinate, and make decisions. Table 9.3 identifies the major demands on the operations team and provides guidelines for handling these increased demands. You should use graphic timelines and integrated scripts with easy-to-follow steps to keep all the participants synchronized with one another and with the dynamic activities. Train and simulate to prepare teams, test communications, build a team approach, and familiarize support engineers with the operations environment. Use design and test people to cover all technical disciplines.

**Demands on the Ground Element.** The many operations activities and the large operations team place extra demands on the mission operations system. Activity levels can be three to four times normal. Not recognizing these demands will stress mission operations systems so they can't handle L&EO requirements. The result is workarounds, which add more complexity and further stress the operations team and process. To produce a capable mission operations system, we

**Table 9.3. The L&EO Environment's Demands on the Operations Team.** Requires unique skills for a demanding job.

| Demands on the Operations Team | How to Handle Demands |
|---|---|
| Large, diverse, and often physically distributed operations team | • Organize well and clearly define roles<br>• Identify the team early so you can budget, allocate resources, and retain key engineers whose skills will be critical during acquisition, start-up, and checkout<br>• Ensure the team covers all technical disciplines and includes managers and test personnel who have hands-on experience with the spacecraft<br>• Have clear, concise, and easily interpreted documentation on console for all operators<br>• Conduct all-hands briefings and simulations to make sure everyone knows all the players and to practice team communications |
| Operations 24 hours a day with long shifts | • Do team-building exercises with the operations team, developing cooperation and relationships<br>• Meet the physical needs of the operations team (provide food, coffee, and rooms for rest, relaxation, and even sleep) |
| High levels of activity and high reasoning requirements | • Ensure ready access to technical information<br>• Establish clear, concise responsibilities for monitoring each activity<br>• Establish clear, unambiguous values for identifying off-nominal performance, synchronized to the spacecraft's changing configuration and orbital variations<br>• Automate critical monitoring requirements as much as possible |
| Rapid, real-time judgments | • Ensure ready access to technical information<br>• Establish a clear process for operations management and decisions—make it fast and as unlayered as possible. |
| Many communication paths across many groups (voice circuits and face-to-face communications) | • Ensure effective communications paths, good voice protocol, and minimum talk<br>• Establish adequate facility capabilities (identified early and provided as design requirements). Assign communications paths to facilities to support coordination and face-to-face communications over voice circuits—these assignments change from day to day as the focus of the operations for that day changes from subsystem to subsystem. |
| High visibility from management, the press, and the public | • Establish a room and an information flow so people not directly in the loop will know the mission status<br>• Have a message center for directing incoming calls and outgoing information (avoids interruptions of the console team; ensures the controlled release of information to the media, management, and other interested persons) |

must identify to designers early on all L&EO operations, special requirements, and estimated volume, sizing, and phasing. Table 9.4 shows how to deal with some of these demands.

**Table 9.4.   The L&EO Environment's Demands on the Ground Element.** This environment stresses systems and calls for special configurations.

| Demands On The Ground Element | How to Handle Demands |
|---|---|
| More systems to support the larger user community and more frequent ground-to-space communications | • Establish voice circuits that allow the diverse and distributed team to communicate effectively. We often use three to six voice loops and allocate circuits to operations control, engineering discussions, management discussions, and mission-planning coordination. Voice circuits typically don't cost much; yet, inadequate voice circuits can cause serious problems during contingencies. |
| Unique systems and configurations | • Define, document, and test the configurations needed for L&EO operations<br>• Include volume and reconfiguration requirements in the definition and testing of these unique configurations |
| Backup systems are often required and must be rapidly configured when failures occur | • Define the critical operations and the possible failures that may jeopardize the mission<br>• Establish backups for these single-point failures<br>• Ensure the system design allows for rapid switching to backup systems during operations<br>• Simulate failures and switching to backup systems |
| Increased command and data volumes | • Ensure the systems have been designed and tested to the levels of performance and volume expected during L&EO operations |
| Longer periods of use | • Ensure systems are designed for longer periods of operation<br>• Include provisions in the operations timeline for reboot and reinitializing systems (refresh) |
| Quick access to data and reference information to respond to the high anomaly rates and requirements for rapid replanning | • Provide for rapid replanning, including the rapid replanning of operations for off-nominal launch times or orbit insertions<br>• Ask the operations team what references they may need and make this information available<br>• Consider online systems for rapid access to technical information |

**Demands on the Space Element.** Finally, the conditions surrounding the space element during L&EO operations are also complex, unique, and dynamic. In many ways, these conditions (see Table 9.5) drive demands on the operations timeline, team, and ground system.

**Table 9.5.   The L&EO Environment's Demands on the Space Element.** The spacecraft undergoes stress, operates in the space environment for the first time, and may face many problems.

| Demands On The Space Element | How to Handle Demands |
|---|---|
| Stressful and rapidly changing environmental conditions | • Continuously collect engineering data; downlink and process frequently; monitor critical parameters against expected, operating, and survival limits |

**Table 9.5.    The L&EO Environment's Demands on the Space Element. (Continued)** The spacecraft undergoes stress, operates in the space environment for the first time, and may face many problems.

| Demands On The Space Element | How to Handle Demands |
|---|---|
| Changing spacecraft configurations | • Develop a plan to monitor configurations based on planned operations<br>• Monitor the configuration frequently<br>• Collect engineering data for performance assessments or for each configuration |
| Higher anomaly rates | • Have engineering expertise on hand, collect engineering data, downlink and process it frequently, and ensure technical references are on hand (drawings, component specifications, simulation and test results) |
| Deviations from expected values in system performance | • Have expected performance data on hand, establish ways to compare actual to expected performance, compare performance to operating and survival limits, and modify the operating plan for off-nominal performance |
| Stress is close to design margins for many spacecraft components (batteries, propulsion systems, and thermal subsystem) | • Assess L&EO operations early in design and influence design<br>• Ensure adequate design margins<br>• Identify stressed components and lessen the effects when defining the L&EO plan<br>• Closely monitor operations |
| Outgassing | • Know outgassing requirements—protect optics and other components sensitive to contamination |
| Thermal settling and its affect on spacecraft components | • Have expected performance data on hand<br>• Know test results for cold and hot cases; understand the effect of thermal variations on component performance (oscillator frequency drifts, thruster performance, and signal noise); include these effects in the plan for L&EO operations |

You'll need to understand these conditions, characterize them early in the mission life cycle, and determine how to handle them (take the risk, add tools and capabilities to the ground system, automate, or add and train a large team). Once you've taken these steps, you can confidently enter the environment for L&EO operations and increase your chances of success.

## 9.2   Developing L&EO Operations

In this section, we present a structure for planning, testing, and simulating L&EO operations. The degree to which a mission uses the process depends on time, resources, and anticipated mission complexity. Still, it frames the activities most missions require and serves as a checklist of considerations for all missions.

The process for developing L&EO operations, shown in Fig. 9.3, consists of seven major steps designed to collect, characterize, and determine the detailed plans and procedures for L&EO operations; test the plans and procedures; and prepare the team for actual operations.

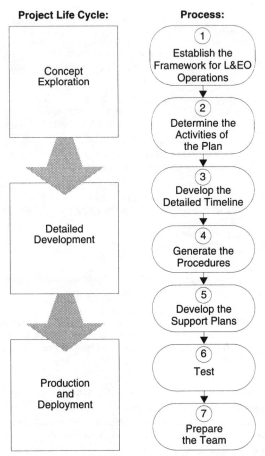

**Fig. 9.3.    The L&EO Planning Process.** Begins with concept exploration and evolves towards production and deployment. Early characterization ensures we consider L&EO demands in designing the space mission architecture.

Do steps 1 through 3 (establish a framework, determine activities, develop timeline) as early in the mission life cycle as possible. Estimate these details during concept development, so you can provide design inputs for systems development. Validate and refine these estimates as the system design matures to ensure a synchronized system and plan.

Do steps 4 and 5 (develop procedures and support plan) after detailed system design and before integration and test. In this way, you can use actual procedures and plans for integration and testing, which will then validate procedures and match testing of systems to their intended use.

Do steps 6 and 7 (test, simulate, prepare team) during integration, test, and dress rehearsal up to the launch day.

### 9.2.1    The Process Steps

*Step 1: Establish the Framework for L&EO Operations*

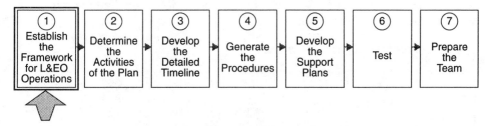

The *operations framework* is the foundation, or set of givens, that guides and bounds planning for L&EO operations. Established by the program's management and systems-engineering team, it's intended to lessen rework by properly guiding the effort with the program view. Experience has found that approaches to L&EO operations differ. Building a consensus reduces redevelopment and reapproval of the L&EO plans. The framework includes

- **Priorities.** Decide the priority of various mission components and objectives (e.g. check out prime instrument first).
- **Duration and Schedule.** Define the broad schedule or duration for L&EO activities. For example, check out the spacecraft bus within two weeks and the instrument after 30 days. Often there are technical reasons for schedules or duration, such as a target of opportunity we may miss, orbit decay, or outgassing.
- **Redundancy Philosophy.** Decide whether to check out redundant capabilities as part of the checkout plan, or to forgo it until a failure occurs. Redundancy checkout adds time, complexity, and risk, but it may be necessary if the spacecraft is designed for automatic recovery from failure.
- **Mode-Checkout Philosophy.** Check out all modes before normal operations or as we come to them in the mission plan.
- **Activity Level.** Establish guidelines regarding the pace and level of activity. For example, a project may elect not to pay for more staff and resources needed for an around-the-clock operation.

- **Resources.** Issue guidelines regarding use of resources (such as tracking sites, radars, additional facilities, or engineering specialists) to avoid costs.
- **Organizational Elements.** Define the organizations involved and a point of contact.
- **Approval Process.** Briefly describe the approval process, if different from other project practices. Because L&EO often involves more than one organizational element, its approval process (typically a joint board or committee) is often different from normal practice.

By specifying these program-level conditions before doing detailed planning, you can avoid problems and rework.

*Step 2: Determine the Activities of the Plan*

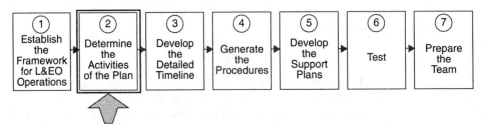

After you establish a framework, collect the activities for L&EO operations, determine attributes (actual or estimates), review and approve the list, and group logically. These actions provide the foundation for all further planning.

At this point in the program, it's useful to define a common structure for the information we want to collect and generate. As mission complexity increases, so does the amount of information we need for planning and execution. Defining a structure provides a common language for all participants and, more importantly, provides a framework for managing information (moving away from documentation, providing online access to technical information). Figure 9.4 describes a five-level structure used in many missions. Figure 9.5 shows how you can use this structure.

Assess each spacecraft subsystem and instrument and define all activities for initializing, checking out, and activating the subsystem or instrument. Include the

- Activity name
- Purpose or objective
- Key success criteria
- Major constraints or interdependencies

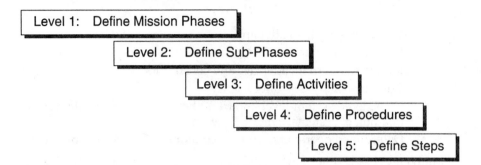

**Fig. 9.4.     The Five Levels for Defining a Structured Operations Activity.** With operations complexity increasing, we need good, structured management of operations information.

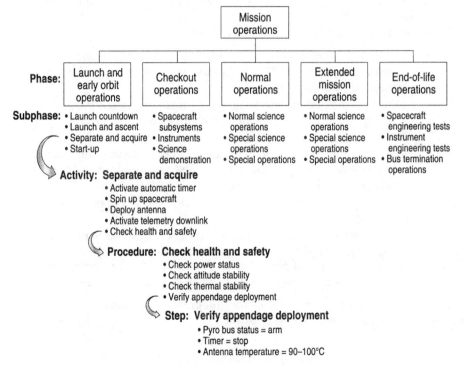

**Fig. 9.5.     An Example of the Five-Level Structure for Operations Information.** This structure breaks down the L&EO phase into manageable blocks of information.

Compile this information into an integrated list and review it to make sure you've

- Completely defined all items
- Followed the framework
- Identified interdependencies and constraints
- Combined activities for efficiency
- Considered system-level impact of the integrated subsystem plans (loading of systems, overall duration, and data volume)
- Determined operations won't need more resources than planned

Group checkout activities into logical units to define subphases. Define operations phases and subphases that logically bound how you collect, integrate, plan, and conduct mission operations activities. In this way, you'll have manageable tasks that contain logical breakpoints (or milestones) for tracking the status of development, implementation, and replanning as needed. With a complete, approved list of activities in hand, you can start grouping them into mission-specific subphases.

To define the subphases,

- Locate logical transition periods for operations—places where there is a major shift in the activity level, objective, spacecraft configuration, or ground-support posture
- Minimize the dependencies across groups—plan each group separately and reduce the effect of changes rippling throughout the plan
- Identify points where you need major acceptance or qualification criteria—breaking up operations at these points allows entry into contingency paths should the qualifications not be met
- Identify major contractual or organizational shifts in responsibilities (if applicable)
- Avoid segments that are extremely large or contiguous in time to avoid long, complex plans and procedures; delegate groups to different teams; and provide convenient points for re-entry and schedule changes if you need to replan during operations

Compile the results of Step 2 into the Integrated Checkout Plan, (Fig. 9.6), which is distributed, reviewed, and approved.

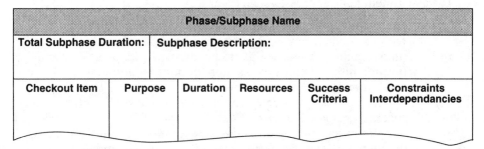

| Phase/Subphase Name | | | | | |
|---|---|---|---|---|---|
| **Total Subphase Duration:** | **Subphase Description:** | | | | |
| **Checkout Item** | **Purpose** | **Duration** | **Resources** | **Success Criteria** | **Constraints Interdependancies** |
| | | | | | |

**Fig. 9.6.**     **Format of the Integrated Checkout Plan.** The integrated checkout plan forces early thinking about the major L&EO activities and their attributes. It helps us build consensus and identify early any factors that affect system design.

## Step 3: Develop the Detailed Timeline

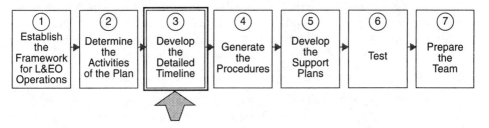

Collect all detailed information you need to schedule the activities into an integrated timeline. Use the approved integrated checkout plan to schedule and assess each checkout item within a Timeline Assessment Table, shown in Fig. 9.7. Use this table to develop and modify the timeline through successive revisions, including rapid changes during the mission. The table becomes the database for scheduling and carrying out each L&EO activity.

| | | Execution Option: | | | | | | | | |
|---|---|---|---|---|---|---|---|---|---|---|
| **Ref. No.** | **Checkout Item** | **Real time** | **Stored Command** | **Stored Command with Real-time Visibility** | **No. Cmds** | **Freq.** | **Commands per Stored Command Load** | **Duration** | **Time Format** | **Interdependancies and Constraints** |
| | | | | | | | | | | |

**Fig. 9.7.**     **Timeline Assessment Table.** This table is the database for scheduling and executing each L&EO activity.

The assessment table addresses such items as:

- **Execution Options.** For most missions, we can control the activity with real-time commands, commands stored onboard without real-time visibility, or commands stored onboard with real-time visibility. (Future missions may give us more options for automation.) But it's still important to define how we'll carry out the activity. Schedule critical activities that require decisions which can't be automatically checked or predicted for real-time execution. Schedule activities that don't require real-time decisions for execution by a stored command, do them outside the real-time event (in the blind), and verify them at the next scheduled event. Some critical activities require real-time, go/no-go options but need stored commands because we don't want to risk losing the real-time rf link with the spacecraft. Schedule these activities for execution by onboard stored commands during a real-time event.

- **Command Volume.** Estimate the command volume, including number of commands for a single occurrence, the number of occurrences, and the total number of commands. Assess total command volumes for all checkout items to determine if the L&EO plan exceeds the systems' capacities for stored or real-time commands.

- **Telemetry Format.** Various activities often require different telemetry formats. Identify these activities so you can plan them properly, plan telemetry formats, and avoid many format changes.

- **Interdependencies and Constraints.** Identify prerequisites, post-requisites, and constraints for an activity to allow proper planning of activities relative to each other and relative to the support plans (Step 5). Also, assess constraints on how to handle the constraint. Three options are software, procedural, or manual checks.

If you're using an expert planning system, take information from the Assessment Table to develop the rule base. But maintain the table even after you build the rule base so you can use it in unplanned situations or as a tool to understand and validate the rule base.

**Building the Integrated Timeline for L&EO Operations.** Use the information in your Timeline Assessment Table to develop an integrated timeline. The timeline graphically illustrates the activities over universal time and mission elapsed time. Tailor the format of the timeline to each mission, but consider including

- All available network coverage based on the planned orbit (ground or space-based telemetry and command systems). As activities are scheduled, shade the scheduled network resources.

- Telemetry format and changes

- Stored-command-load periods and volume
- Spacecraft activities with some graphical way to distinguish the method of execution (real-time, automatic, or stored command)
- Payload activities
- Record cycles and dumps for onboard data storage
- Relevant orbital events (such as eclipse, South Atlantic Anomaly entry, or exit)
- Major ground-based events, such as planning meetings, generating commands, data processing, or any other activity that is important and directly related to integrating and carrying out the L&EO plan

While developing the timeline,

- Ensure you have enough time to acquire the signal and check health and safety for real-time activities
- Schedule backup events for critical operations and events that have small time margins
- Consider shift handovers and the operations team's abilities. If you have only one expert resource, you can't schedule around-the-clock activities that depend on this person.
- Consider a 50% loading rate (50% busy and 50% free for contingencies). In most cases you'll use the contingency time or need a break!
- Identify convenient breakpoints in the timeline for replanning and re-entry
- Allow parallel activities, if needed, to expedite the timeline. Scheduling several activities to occur simultaneously through stored commands or with real-time is possible, but you must consider shared resources such as power, telemetry format, attitude position, and people.

Once you've completed the Assessment Table and timeline, get them reviewed and approved. It may also be useful at this point to develop time-phased loading plots. For example a time-phased plot of the power consumption and battery depth-of-discharge (DOD) may be helpful for power-limited missions. You could plot command volume per stored load to show the size and uplink time of each stored command load. These time-phased plots allow the operations team and engineers to understand system use, stress, and capacity for changes.

When designing the operations timeline, consider the ability to change it rapidly during the mission. There are few off-the-shelf tools, but you can adapt spreadsheet programs with plotting capabilities, or better yet, project-management tools with timeline and resource abilities. With a tool such as this, you can quickly change the timeline to keep it accurate. A well-thought-out timeline graphic becomes one of the most useful tools in presenting an often complex plan for L&EO operations to the large and diverse operations team. Figure 9.8 shows part of the timeline from NASA's Small Explorer SAMPEX mission.

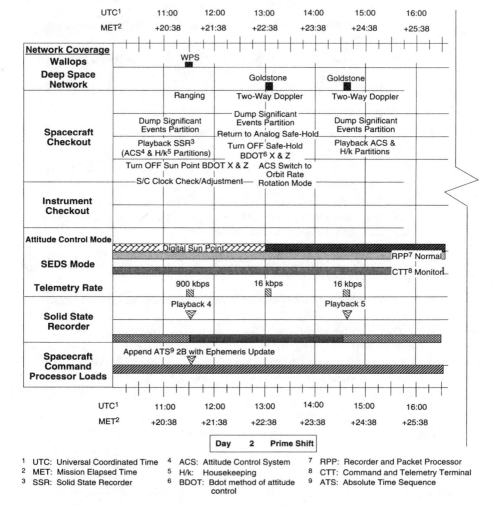

**Fig. 9.8.    L&EO Operations Timeline.** An example of the L&EO timeline for NASA's SAMPEX mission. [NASA, 1992]

*Step 4: Generate The Procedures*

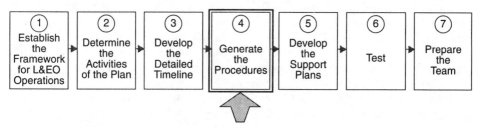

Review the Assessment Table and timeline and identify all procedures required to support the operations. Define procedures in logical units that conduct discrete events, so you'll have flexibility during the mission. Include

- A launch-configuration table
- Configuration-state tables
- Stored command loads
- Flight-software tables
- Real-time command and control
- Contingencies

**Launch-Configuration Table.** This defines the spacecraft's configuration for launch. It meets the launch constraints (what can and can't be turned on) and the needs for separation and acquisition on orbit. Set the configuration for minimum commanding once the spacecraft is in orbit and minimum power-discharge levels during ascent. Include in your launch-configuration table

- Component, switch, or software item
- Telemetry indicator
- Desired launch site
- Command to produce the desired state
- Any necessary remarks

Usually, the integration and testing team finally configures the spacecraft at the launch site. The launch-configuration table becomes an agreement between this team and the operations team on the spacecraft's state for launch.

**Configuration-State Tables.** Throughout L&EO the spacecraft, its subsystems, and its payloads cycle through many modes, so the configuration changes dynamically. By defining configuration states for the various phases, you can use these state definitions for power management, command management, and health and safety checks. With a good set of state definitions, synchronized to the L&EO timeline, the operations team can closely monitor changing configurations.

**Stored Command Loads.** We recommend defining and generating stored command loads early in mission design. Consider several factors when building these loads. First, segment them into logical units that fit within the system capacity. Second, segment and schedule them for uplink based on uplink opportunities, including backup opportunities. Third, make sure they contain any safing or reconfiguration commands necessary to safe the spacecraft if the next load isn't uplinked in time. Finally, generate, error check, and test them, remembering they often use the stored command processor in ways other than expected for normal operations (larger loads, unusual timing, and sequencing).

**Flight-Software Tables.** With the increasing use of flight software to control spacecraft operations, you may need to generate, load, and modify from a few to several hundred tables during L&EO operations. Managing, testing, uplinking, and verifying these tables may require responsive systems. Early on you'll need to construct and test the tables needed to support the L&EO timeline.

**Procedures for Real-Time Command and Control.** For L&EO operations, generate real-time command procedures for all activities designated to execute under real-time command and control. As with stored command loads, you must segment the activities into procedures. Procedures should accomplish a given activity, yet not be too lengthy or inflexible. They should have logical break points that allow for return and re-entry, and they should include the interactive steps of commanding and telemetry verification into a single logic flow. Finally, for more complex procedures, we recommend flowcharts. They help you develop the procedural logic and are good training tools.

**Procedures for Contingencies.** Contingency procedures are perhaps the most difficult and time consuming to develop. First, define what contingencies may occur that require a preplanned response to avoid serious problems on orbit. To do so, you can assess spacecraft Failure Mode and Effects Analyses, spacecraft integration and test anomalies, and experience with similar components or designs on previous missions (Fig. 9.9). These may guide you to where hazards exist and what failures are most common, but the best method is a good brainstorming session with key spacecraft and ground-system engineers. They can help you work through the L&EO timeline and assess what could go wrong, how serious it is, whether a quick response is essential (or can you take time to better assess it), and how to respond to it. The result will be a set of contingency paths off the nominal timeline for L&EO operations and the procedures to support them. You'll also identify points of re-entry into the timeline.

Procedure development may vary with the unique design of the spacecraft, ground system, or timeline for L&EO operations. Still, you should develop them early, use them to integrate and test the spacecraft and ground system, make sure they're well controlled and documented, and map them to the timeline for L&EO operations.

You may also need to vary the procedures for ground testing. Ground-related constraints may keep you from completing procedures in test (such as thruster and pyro firings). And you may need to change some procedures for flight, especially to avoid problems in orbit.

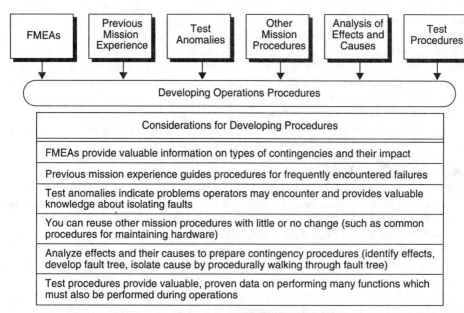

| Considerations for Developing Procedures |
| --- |
| FMEAs provide valuable information on types of contingencies and their impact |
| Previous mission experience guides procedures for frequently encountered failures |
| Test anomalies indicate problems operators may encounter and provides valuable knowledge about isolating faults |
| You can reuse other mission procedures with little or no change (such as common procedures for maintaining hardware) |
| Analyze effects and their causes to prepare contingency procedures (identify effects, develop fault tree, isolate cause by procedurally walking through fault tree) |
| Test procedures provide valuable, proven data on performing many functions which must also be performed during operations |

**Fig. 9.9.**    **Considerations for Developing Procedures.** Considerations beyond the activities timeline develop better procedures, especially contingency procedures. (FMEA = failure mode and effects analysis)

## *Step 5: Develop The Support Plans*

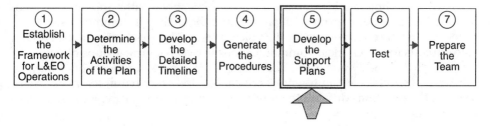

Support plans may include

- Data processing
- Display
- Facility and voice
- Mission planning
- Operations management
- Status and announcement

**Data-Processing Plan.** Two activities in the L&EO timeline may be interdependent, so we must analyze results of the first step before the second step. It then becomes important to retrieve the data, process it, analyze it, and decide whether to proceed or not. For instances where this is critical, or the timing is tight, you may want a data-processing plan that shows how to handle specific data sets. You should also query each of the many support specialists for data-processing needs and collect results in the plan. In most instances, data processing in support of L&EO operations is more significant than during normal operations. It's best to know in advance rather than to find out needs during operations and be unable to meet them because of constraints on systems or staffing.

**Display Plan.** You may need a display plan if the number of available displays doesn't allow everyone control over a display device. It defines the available displays and maps them to the timeline. Given the large support staff, a dynamic timeline, and many unique needs during L&EO, you may need special and rapidly changing displays.

**Facility and Voice Plan.** L&EO operations often requires support from design engineers for the spacecraft and mission operations system, program managers, and people who do integration and testing. Further, special L&EO facilities are often involved in the operation, so you need to determine who is critically needed, as well as when and where they're needed. The results go into a facility plan. Figure 9.10 lists the facilities used in launching NASA's Cosmic Background Explorer. Note that it used special launch facilities physically distributed across the United States. It became important to decide who would be at each facility, as well as what and how much each could do. Define your operations team, support team, and facility requirements (space, capabilities) early to ensure you've defined interfaces and designed to meet needs.

After assigning facilities, assess the communications paths among these people. Define the main groups of people and how they'll communicate; then, generate a set of requirements for the voice system. The goal is to provide enough voice circuits for quick, clear communications between logical groups of people in support of the L&EO activities. Too many people on a single channel may be restrictive, especially when timing is critical. On the other hand, too many channels makes operations difficult to monitor. One way to design a voice system is to allocate loops for

- Management
- Spacecraft engineering
- Ground-element engineering
- Real-time operations
- Navigation

An additional "all-hands" loop may provide status, direction, and control of the overall timeline. This loop keeps everyone synchronized across the physically

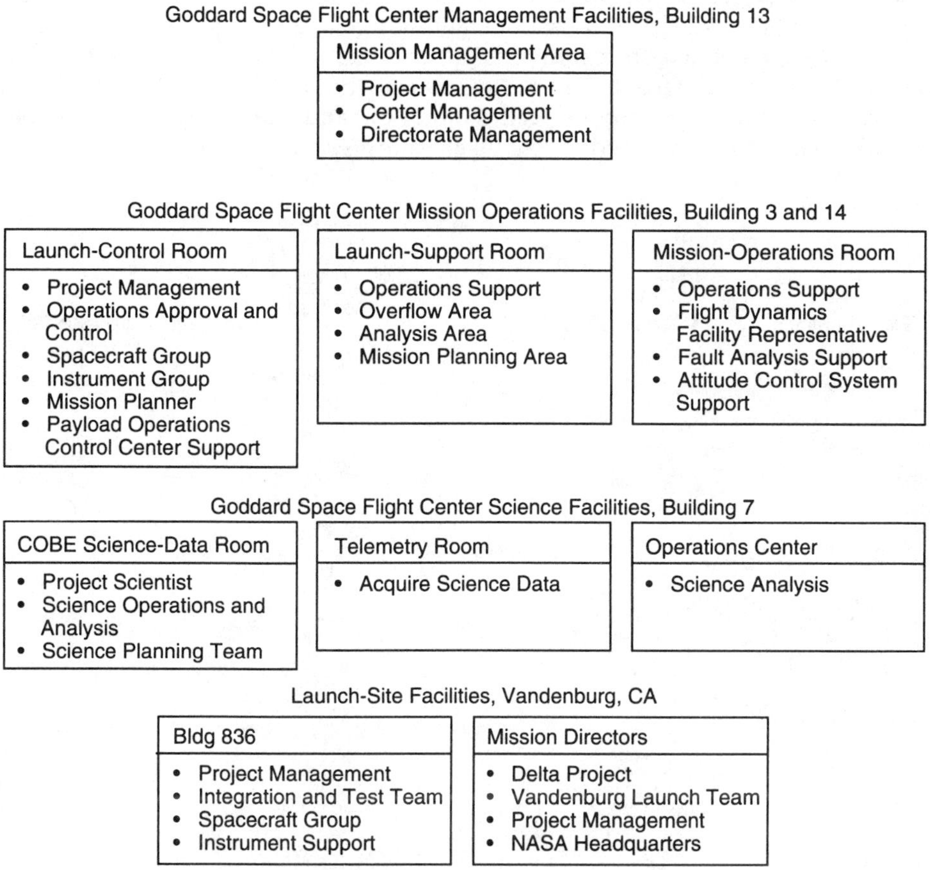

Goddard Space Flight Center Management Facilities, Building 13

| Mission Management Area |
| --- |
| • Project Management<br>• Center Management<br>• Directorate Management |

Goddard Space Flight Center Mission Operations Facilities, Building 3 and 14

| Launch-Control Room | Launch-Support Room | Mission-Operations Room |
| --- | --- | --- |
| • Project Management<br>• Operations Approval and<br>  Control<br>• Spacecraft Group<br>• Instrument Group<br>• Mission Planner<br>• Payload Operations<br>  Control Center Support | • Operations Support<br>• Overflow Area<br>• Analysis Area<br>• Mission Planning Area | • Operations Support<br>• Flight Dynamics<br>  Facility Representative<br>• Fault Analysis Support<br>• Attitude Control System<br>  Support |

Goddard Space Flight Center Science Facilities, Building 7

| COBE Science-Data Room | Telemetry Room | Operations Center |
| --- | --- | --- |
| • Project Scientist<br>• Science Operations and<br>  Analysis<br>• Science Planning Team | • Acquire Science Data | • Science Analysis |

Launch-Site Facilities, Vandenburg, CA

| Bldg 836 | Mission Directors |
| --- | --- |
| • Project Management<br>• Integration and Test Team<br>• Spacecraft Group<br>• Instrument Support | • Delta Project<br>• Vandenburg Launch Team<br>• Project Management<br>• NASA Headquarters |

**Fig. 9.10. L&EO Facilities for Cosmic Background Explorer (COBE) at Goddard Space Flight Center (GSFC).** Many diverse facilities are involved in L&EO operations, so we must plan for people and communications. [COBE, 1989]

distributed operations. Actual voice-loop configurations depend on the number of people and the locations.

**Mission-Planning Plan.** This plan defines daily planning and decision making. Mission planning during L&EO operations is typically more frequent—with shorter planning periods. Due to the high rate of anomalies and "firsts," plans often require daily refinement. Consider replanning daily: assess results of the current day and review the next day's plans for refinements. During L&EO operations, you may also have a different operations-management team.

**Operations-Management Plan.** This plan defines the roles, responsibilities, and decision process for the launch countdown and orbital-operations phase.

Launch-countdown decisions usually entail the launch-vehicle team and scheduled go/no-go decisions. Orbital-operations management provides go/no-go decisions for major activities and decisions to solve anomalies. Figure 9.11 shows a typical operations-management process. Due to the large team, diverse locations, and critical nature of operations, well-thought-out and defined operations reporting and management will reduce confusion during actual operation.

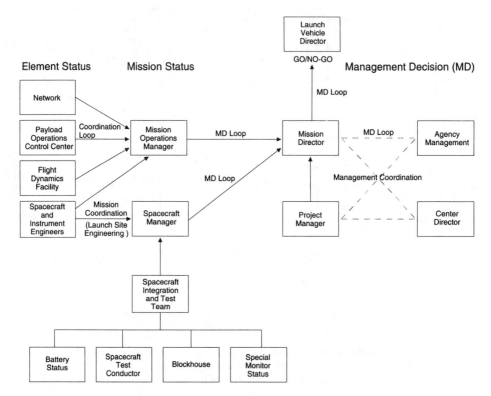

**Fig. 9.11.    Operations Management for L&EO Operations.** Element information usually flows up to mission status, with management decisions based on the status.

**Status and Announcement Plan.** This plan discusses how to release announcements regarding mission status. Typically, many people are interested in this status. We've found that, unless we satisfy these interests, we'll receive many phone calls and visitors to the control complex. Having a defined status and announcement plan will remove this job from the control-center team, thereby reducing the complexity and confusion. Status and announcement plans may consist of press releases, press briefings, phone-message systems, or postings.

Develop all of your support plans consistent with the timeline. For example, the status and announcement plan may schedule a press briefing just after a major event. If so, you can schedule daily planning meetings at a time of the day when there is gap in real-time coverage, thereby avoiding conflicts with real-time operations. The complexity and need for each of these plans varies from mission to mission, so you must develop only what you need.

*Step 6: Test*

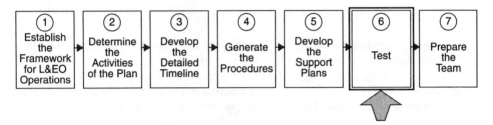

The purpose of testing is to validate the systems, procedures, timelines, and people as launch ready. We recommend you

- Test with simulators first
- Validate the timeline and procedures with the actual systems and spacecraft
- Provide realistic test environments (loading, phasing)—L&EO operations tend to stress the system and people, so you'll need to properly load and phase the activities to replicate the worst-case scenarios
- Have operators conduct the tests, with developers in support
- Combine (as much as possible) validation and readiness testing for L&EO with integration and test of the ground and space elements
- Plan for retest of modifications and improvements
- Include tests late in the preparation for launch, including launch-site or launch-pad testing
- Test anomalies by using simulators to inject faults
- Define criteria for success

Develop test plans early in the program so people can build unique tools and interfaces and include the tests in integration and test schedules for the space and ground elements.

*Step 7: Prepare the Team*

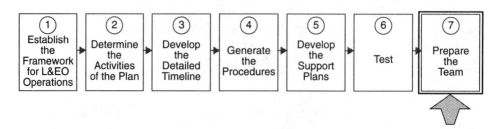

Make sure all staff are aware of their roles, of the integrated team and timeline, and of how to follow along (including tracking the changes that may occur daily). Because the operations team for L&EO often consists of developers, some people won't know console operations and systems. Special training and more simulation may be necessary for them. To ensure people are prepared and aware,

- **Develop an Integrated Console Script.** This script integrates all activities and keeps everyone synchronized. It orchestrates and controls the activities of the ground, space, launch-vehicle, and operations teams according to the timeline.

- **Train People Who Aren't Operators.** Train in console procedures, voice procedures, and system operations. Console operations in a time-critical environment can be difficult and confusing. The team should be comfortable enough to focus their attention fully on operations (versus how to work the console and systems).

- **Brief All Hands.** Gather all hands (or as many people as you can) and brief them. Summarize all aspects of the L&EO plan including the timeline and support plans. Hand out all launch-ready materials.

- **Simulate All or Parts of the L&EO Operations.** Integrate the actual procedures, people, and systems as much as possible. Also inject faults into the system with a simulator or by voice. Simulate actual loading and timing. Have a team monitor the simulation and debrief to go over the results.

### 9.2.2    Development Team, Products, and Tools for L&EO Operations

**Development Team for L&EO Operations.** Developing timelines and procedures for L&EO operations requires the support and involvement of many groups across the mission system (Fig. 9.12). The operations team typically takes the lead in developing plans for L&EO operations. A mission operations working group or similar forum is used to develop and review the timelines, plans, and procedures outlined in the previous section. The team should meet early in mission development to define L&EO operations in a checkout plan and draft timeline. These early drafts help define the capabilities needed in simulators, interfaces, systems, consoles, and voice circuits. In this way the needs of operations are built into the system, versus having an L&EO plan that fits operations to a fixed design.

You'll also need to coordinate with the launch-vehicle team in order to get the typical inputs shown in Fig. 9.12.

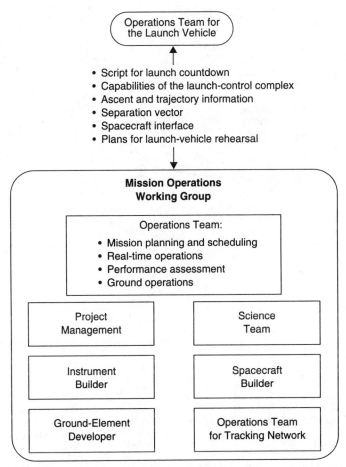

**Fig. 9.12.    Development Team for L&EO.** Integrated team represents all elements. It's established early in mission development and it coordinates all processes and products of L&EO planning.

**Products for L&EO Operations.** Capture the results of L&EO development in an L&EO Operations Handbook, which is a quick-reference guide to L&EO operations. Distribute it to all members of the operations and support teams for L&EO,

thereby ensuring everyone operates from the same script, timeline, and procedures. The handbook contains all key products and is often indexed and tabbed for quick reference. Figure 9.13 shows typical contents of an L&EO Operations Handbook.

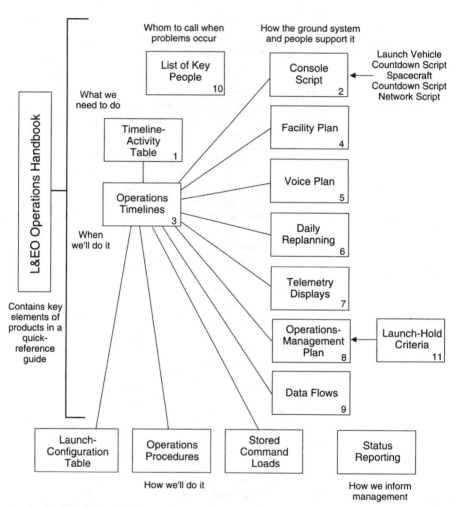

**Fig. 9.13.    L&EO Operations Handbook.** Provides key information in an easy reference guide to the large, diverse operations team for L&EO.

**Configuration Control.** As with any critical information, we need to control timelines, plans, and procedures for L&EO operations. Depending on the project, the level of configuration management may vary. But it's important to control changes once testing and validation is complete, as well as to label clearly the version (draft, review, in test, approved, or validated), revision numbers and dates.

**Tools for L&EO Operations.** The commercially available tools to develop and carry out L&EO operations are limited. Depending on the complexity of the development and operations task, you may find a timeline-development tool beneficial. For example, planning for the mission to service the Hubble Space Telescope (HST) was a complex task. It required many activities executed in close coordination with Shuttle-crew activities and within short operations periods. Further complicating the planning was the real possibility of a launch delay or slip to later days in the launch window. Consequently, an expert planning tool was developed which contained ·a rule base for scheduling activities against the timeline. It also allowed rapid replanning to recompute the timeline given a launch delay. This tool proved invaluable during operations. In contrast, NASA's Small Explorer mission used a simple graphics-drawing tool to develop the timeline, as its operations were simple, with few external variables. Assess the complexity and number of external variables before deciding whether to develop a tool. In some cases, a tool will save many hours of manual replanning. In other cases, the planning may be too simple to justify development costs.

### 9.2.3   Special Planning Multi-Mission Operations and Human Spaceflight

**Multi-Mission Operations.** With the trend towards smaller missions and multiple spacecraft launches, we may need to plan for operations over multiple missions. Multi-mission planning involves six steps designed to overlay the operations of each mission on the common mission operations elements, assess the effect of multi-mission loading and phasing, and develop strategies for handling the resultant multi-mission plans. Figure 9.14 illustrates this process. Multi-mission planning is useful when launching a single mission using an existing mission system or team, or when launching multiple missions at the same time.

First, define *operations cases*—distinct mission phases where the operations may vary significantly from the previous case. Characterize and assess data volume, operator tasks, and system use for loading. Include the total volume and the phasing of activities over time according to the launch and orbit characteristics. Define areas of concern: where loading is beyond planned systems, team capacities, or cost ceilings. With this knowledge early in mission development, define various strategies to handle each area, such as automating, migrating to onboard, taking risks, adding systems or staff, and so forth. You can then use the most cost-effective strategy.

**Implications of Human Flight.** L&EO operations for flight crews require complex coordination of spacecraft and crew activities, as well as a new, larger set of responses to anomalies. Finally, crew safety requires more constraints. While human flight adds options for launching the mission, it also adds complexity.

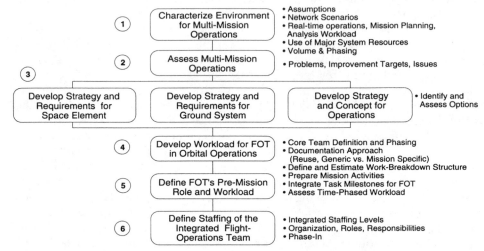

**Fig. 9.14.    Assessing Multi-Mission Operations.** Handling the demands of operations for multiple missions requires special planning. (FOT = flight-operations team) [NASA, 1993]

## 9.3   Conducting L&EO Operations

Conducting L&EO operations is the team's most stressful yet most exciting activity. It culminates all the work of many people over several years. It's a technical and emotional activity that taxes systems and people while determining the success or failure of the mission. We've outlined many characteristics of the environment for L&EO operations (Sec. 9.2.1). We've defined the unique demands on the spacecraft, ground system, and operations team. But during actual L&EO operations, these demands are simultaneous, and your planning, testing, and preparation begin to pay off. Here's what you can expect.

    1. **Operations Team.** People will be anxious, stressed, and possibly tired (depending on the pace of activities leading up to launch). Conflict is a real possibility, but you can lessen it with discipline, a high state of readiness, good operations products that reduce confusion, minimal voice traffic, frequent status messages to keep all informed, and calm reactions. Selecting the operations team and shift assignments, skill matching the team to each other and to the timeline, and developing a timeline with a moderate activity level reduce stress and conflict. Many NASA missions have successfully used a two-shift arrange-

ment. An "A" team, staffed with the most skilled people and experts, operates one 12-hour shift, and conducts all of the critical activities and major checkouts. A "B" team operates the second 12-hour shift. They monitor health and safety, collect and process data, and replan. The B shift is therefore a period to catch up, replan, and re-synchronize while the A team rests.

2. **Ground Element.** Nearly all parts of the ground element are online. The ground system during this mission phase is the largest and most complex of the mission, having many people, systems, and interfaces. Data volumes, system use, and system-performance needs are at a peak. Expect problems with an interface or a system and be prepared for switching to backup systems or altering the operations plans. Reduce problems by testing and simulating before launch; having a well defined way to report, discuss, and resolve anomalies; often checking a system's readiness before using it; configuring systems well in advance; and freezing the configuration.

3. **Space Element.** The space element undergoes extreme environmental conditions, rapid changes, unusual and new configurations, and the first operation of all systems in space. As with the ground system, expect failures and problems. If the team can stay rational and controlled, avoiding quick reactions and hasty conclusions, L&EO operations will be smooth. You can decrease risk and ensure a smooth activation despite variances if you have good reference material handy; know the expected orbital signatures and values of the systems; have good, quick processing systems on the ground for assessing health and safety; and have defined a process for discussing and resolving anomalies.

Although L&EO is the most challenging phase, it's also the most rewarding. Enter it with optimism, control, confidence, and a readiness to deviate from the plans. After completing L&EO operations, write a lessons-learned report and give the results to the next mission for process improvement.

## 9.4    Reducing the Cost of L&EO Operations

We've discussed ways of succeeding in L&EO operations, but our recommendations and methods have costs, and not all are necessary for each mission. Your budget may restrict development or operations, so you may need to drop the best approach for a less costly approach, and then further trade development costs against operations costs. As mentioned before, deciding to spend money for a mission phase which can be as short as several days is difficult. Tools and capabilities unique to L&EO will largely sit idle after this short phase. Still, the L&EO phase is often the most critical, most stressful, and most likely to have mission-critical problems.

To reduce the cost of planning for and conducting L&EO operations, you need to determine which of the following options apply to your mission.

1. **Plan early and influence designs for the spacecraft and mission operations.** Characterize the loading and performance requirements of L&EO and determine how to phase activities. Make educated guesses if you have to. Your guesses will at least establish a consistent baseline across the system so the ground elements, spacecraft system, and operations teams are developing to the same activity levels. Go through at least Step 3 of the planning process in the concept phase, influencing design with the results. Good early planning will avoid late and changing requirements, which are not only costly but stressful.

2. **Use Common Systems for Mission Operations and for Integration and Testing.** Using the same system for integration and testing and operations can save development money. Often we build separate systems and transfer the information, files, databases, and knowledge to operations. Having the same system avoids translations and transfers, lessens risks, reduces validation requirements, and lowers costs.

3. **Share Overflow Facilities and Systems.** As we've mentioned throughout this chapter, L&EO operations demand abnormal capabilities. Designing mission-unique systems to handle the demands of L&EO is too expensive because much of the system's capability will be under-used after L&EO operations. Identifying resources from an existing mission or operation can alleviate the need to develop unique facilities or to suffer with a team and system impaired because it's sized for normal operations.

4. **Base Integration and Testing on Operations Plans and Procedures.** Combine the integration, testing, and operations test plans; use operations procedures in the test environment; and capture the systems' signatures, values, and behaviors during integration and testing. These actions will help reduce costs (avoids duplicate tests and procedures) and risk (ensures the systems are tested as they'll be used).

5. **Design Spacecraft With Ample Margins.** Include margins for the demands of launch and early-orbit operations. Ample margins remove the complexities and risks of planning and operating a limited resource.

6. **Review Plans Frequently and at Critical Points in the Process.** Reviewing plans and procedures frequently and with everyone involved will ensure the planning doesn't go down a path unacceptable to others or inconsistent with systems design.

7. **Automate Spacecraft and Monitor Critical Hazards Onboard.** Avoid as much as possible ground control of time-critical, sequence-dependent tasks during the initial early orbits. This will reduce complexity and the expense of adding redundant capabilities to the ground system.

8. **Use People from Integration and Testing.** With the increased demands of L&EO operations, you need more hands from the integration and testing teams. In fact, these people don't require much training because they must know the systems well and develop trouble shooting skills before launch. Using them on the operation team provides valuable practical skills.

9. **Use Ground Support Equipment Developed to Support Integration and Testing.** L&EO operations may often require quick processing of instrument data to assess instrument performance. If so, the ground support equipment developed to integrate and test the instrument may provide a quick, proven capability. This equipment can be at the mission-control center, thereby reducing complexity and improving responsiveness.

10. **Reduce Interdependencies.** Design the systems and the timeline to avoid dependencies between activities. If interdependencies must occur, schedule ample time to verify prerequisites.

11. **Automate.** Automate as much as possible the time-critical events, especially the scenarios for separation and early-orbit activation. It's often difficult and expensive to get ground-station coverage at the point when the launch vehicle separates. Further, reliably acquiring a signal just after separation is unlikely.

12. **Place the Team Close Together.** This reduces interface costs, decreases complexity, and promotes the teamwork needed to respond to the large number of possible anomalies.

13. **Reduce Numbers of People.** Doing so decreases system and facility requirements, complexity and confusion, and training costs.

# References

National Aeronautics and Space Administration. May, 1989. "Cosmix Background Explorer Launch and Early Orbit Operations Development Plan." NASA Goddard Space Flight Center. Greenbelt, MD.

National Aeronautics and Space Administration. July, 1992. "SAMPEX Flight Operations Plan." NASA Goddard Space Flight Center, Greenbelt, MD.

National Aeronautics and Space Administration. September, 1993. "Small Explorer Multi-Mission Assessment." NASA Goddard Space Flight Center. Greenbelt, MD.

# Chapter 10

# Space Navigation and Maneuvering[*]

Daryl G. Boden, *United States Air Force Academy*
Paul Graziani, *Analytical Graphics, Inc.*

In this chapter we provide the tools needed to calculate some basic parameters for orbit trajectory and mission geometry. We use these parameters to plan for and analyze the navigation function, as described in Sec. 3.5. We also help you understand basic orbital theory, define key terms, and provide basic equations that allow you to calculate many orbital parameters by hand. We begin by explaining terms used to describe orbits and giving you equations for calculating orbital parameters. The next four sections describe aspects of Earth-orbiting spacecraft: orbit perturbations, orbit maneuvering, launch-window calculations, and orbit maintenance. We next describe interplanetary trajectories and a way to estimate velocity budgets for interplanetary trajectories. Finally, we describe parameters of orbit geometry for spacecraft in Earth orbit. Several textbooks are available in the areas of satellite orbits and celestial mechanics. Some of the most popular are Bate, Mueller, and White [1971], Battin [1987], Danby [1962], Escobal [1965], Kaplan [1976], and Roy [1978].

---

[*] Sections 10.1 through 10.5 have been adapted with permission from Larson and Wertz [1992].

# 10.1 Keplerian Orbits

Explaining the motion of celestial bodies, especially the planets, has challenged observers for many centuries. The early Greeks attempted to describe the motion of celestial bodies about the Earth in terms of circular motion. In 1543, Nicolaus Copernicus proposed a heliocentric (Sun-centered) system with the planets following circular orbits. Finally, with the help of Tycho Brahe's observational data, Johannes Kepler described elliptical planetary orbits about the Sun. Later, Isaac Newton mathematically solved this system based on an inverse-square gravitational force.

Kepler spent several years reconciling the differences between Tycho Brahe's careful observations of the planets and their predicted motion based on previous theories. Having found that the data matched a geometric solution of elliptical orbits, he published his first two laws of planetary motion in 1609 and his third law in 1619. Kepler's three laws of planetary motion (which also apply to spacecraft orbiting the Earth) are:

**First Law:** The orbit of each planet is an ellipse, with the Sun at one focus.

**Second Law:** The line joining the planet to the Sun sweeps out equal areas in equal times.

**Third Law:** The square of the period of a planet is proportional to the cube of its mean distance from the Sun.

## 10.1.1　Equations of Motion for Satellites

Figure 10.1 depicts the key parameters of an elliptical orbit. The *eccentricity, e,* of the ellipse (not shown in the figure) is equal to $c/a$; it measures the deviation of the ellipse from a circle.

Isaac Newton explained mathematically why the planets (and satellites) follow elliptical orbits. Newton's Second Law of Motion, applied to a system with constant mass, combined with his Law of Universal Gravitation, provides the mathematical basis for analyzing satellite orbits. Newton's law of gravitation states that any two bodies attract each other with a force proportional to the product of their masses and inversely proportional to the square of the distance between them. The equation for the magnitude of the force due to gravity, $F$, is

$$F = -GMm/r^2$$
$$\equiv -\mu m/r^2 \tag{10.1}$$

where $G$ is the universal constant of gravitation, $M$ is the mass of the Earth, $m$ is the mass of the satellite, $r$ is the distance from the center of the Earth to the satellite, and $\mu \equiv GM$. The Earth's gravitational constant, $\mu$, is 398,600.5 $km^3 sec^{-2}$.

Combining Newton's second law with his law of gravitation, we get an equation for the satellite's acceleration vector:

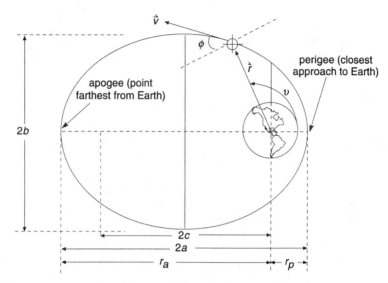

**Fig. 10.1.** **Geometry of an Ellipse and Orbital Parameters.** $\phi$ = flight-path angle, the angle between the velocity vector and a line perpendicular to the position vector; $\vec{r}$ = position vector of satellite relative to the center of the Earth; $\vec{v}$ = velocity vector of satellite relative to
the center of the Earth; a = semi-major axis of the ellipse; b = semi-minor axis of the ellipse; c = the distance from the center of the orbit to one of the focii; $v$ = the polar angle
of the ellipse, measured in the direction of motion from perigee to the position vector

$$\ddot{\vec{r}} + (\mu r^{-3}) \vec{r} = \vec{0} \tag{10.2}$$

This equation, called the *2-body equation of motion,* is the relative equation of
motion of a satellite's position vector as the satellite orbits the Earth. In deriving it,
we assumed gravity is the only force, the Earth is spherically symmetrical, the
Earth's mass is much greater than the satellite's mass, and the Earth and satellite
are the only two bodies in the system.

A solution to the two-body equation of motion for a satellite orbiting the Earth
is the *polar equation of a conic section.* It gives the magnitude of the position vector
in terms of the satellite's location in the orbit,

$$r = a (1 - e^2) / (1 + e \cos v) \tag{10.3}$$

where a is the semi-major axis, e is the eccentricity, and v is the polar angle or true
anomaly.

A *conic section* is a curve formed by passing a plane through a right circular
cone. As Fig. 10.2 shows, the angular orientation of the plane relative to the cone
determines whether the conic section is a *circle, ellipse, parabola,* or *hyperbola.* We can

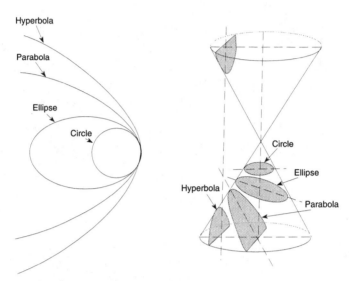

**Fig. 10.2.**    Satellite Orbits Can Be Any of Four Conic Sections: a Circle, an Ellipse, a Parabola, or a Hyperbola.

define all conic sections in terms of the eccentricity, $e$, in Eq. (10.3) above. The type
of conic section is also related to the semi-major axis, $a$, and the energy, $\varepsilon$. Table 10.1
shows how energy, eccentricity, and semi-major axis relate to the type of conic
section.

**Table 10.1.**    **Conic Sections.** See text for discussion.

| Conic | Energy, $\varepsilon$ | Semi-Major Axis, $a$ | Eccentricity, $e$ |
|-------|----------|---------------------|-------------------|
| circle | < 0 | = radius | 0 |
| ellipse | < 0 | > 0 | 0 < e < 1 |
| parabola | 0 | $\infty$ | 1 |
| hyperbola | > 0 | < 0 | > 1 |

### 10.1.2    Constants of Motion

Using the two-body equation of motion, we can derive several constants of
motion of a satellite orbit. The first is

$$\varepsilon = V^2/2 - \mu/r = -\mu/(2a) \tag{10.4}$$

where $\varepsilon$ is the total *specific mechanical energy*, or mechanical energy per unit mass, for the system. It's also the sum of the kinetic energy per unit mass and potential energy per unit mass. Equation (10.4) is referred to as the *energy equation*. Because the forces in the system are conservative, the energy is a constant. The term for potential energy, $-\mu/r$, defines the potential energy to be zero at infinity and negative at any radius less than infinity. Using this definition, we find the specific mechanical energy of elliptical orbits will always be negative. As the energy increases (approaches zero), the ellipse gets larger, and the elliptical trajectory approaches a parabolic trajectory. From the energy Eq. (10.4), we find that the satellite moves fastest at the orbit's perigee and slowest at apogee.

We also know that for a circle the semi-major axis equals the radius, which is constant. Using the energy equation, we discover the velocity of a satellite in a circular orbit is

$$V_{cs} = (\mu/r)^{1/2} \tag{10.5}$$

or, for orbit around the Earth,

$$\cong 7.905\ 366\ (R_E/r)^{1/2}$$

$$\cong 631.3481\ r^{-1/2}$$

where $V_{cs}$ is the circular velocity in km/sec, $R_E$ is the radius of the Earth, and $r$ is the orbit radius in km.

From Table 10.1, the energy of a parabolic trajectory is zero. A parabolic trajectory is one with the minimum energy needed to escape the Earth's gravitational attraction. Thus, we can calculate the velocity required to escape from the Earth at any distance, $r$, by setting energy equal to zero in Eq. (10.4) and solving for velocity:

$$V_{esc} = (2\mu/r)^{1/2} \tag{10.6}$$

$$\cong 11.179\ 88\ (R_E/r)^{1/2}$$

$$\cong 892.8611\ r^{-1/2}$$

where $V_{esc}$ is the escape velocity in km/sec, and $r$ is in km.

Another quantity associated with a satellite orbit is the *specific angular momentum*, $h$, which is the satellite's total angular momentum divided by its mass. We can find it from the cross product of the position and velocity vectors.

$$\vec{h} = \vec{r} \times \vec{v} \tag{10.7}$$

We see that, from Kepler's second law, the angular momentum is constant in magnitude and direction for the two-body problem. Therefore, the plane of the orbit defined by the position and velocity vectors must remain fixed in inertial space.

### 10.1.3    Classical Orbital Elements

When solving the two-body equations of motion, we need six constants of integration (initial conditions). Theoretically, the three components of position and velocity at any time could be found in terms of the position and velocity at any other time. Alternatively, we can completely describe the orbit with five constants and one quantity which varies with time. These quantities, called *classical orbital elements*, are defined below and are shown in Fig. 10.3. The coordinate frame shown in the figure is the geocentric inertial frame,[*] or GCI. Its origin is at the center of the Earth, with the X axis in the equatorial plane and pointing to the vernal equinox. Also, the Z axis is parallel to the Earth's spin axis (the North Pole), and the Y axis completes the right-hand set in the equatorial plane.

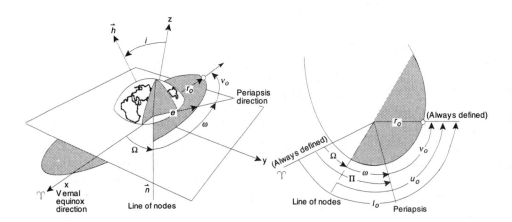

**Fig. 10.3.**    **Definition of the Keplerian Orbital Elements of a Satellite in an Elliptic Orbit.** Elements are defined relative to the geocentric inertial coordinate frame.

The classical orbital elements are:

- $a$:    *semi-major axis*: describes the size of the ellipse (see Fig. 10.1).
- $e$:    *eccentricity*: describes the shape of the ellipse (see Fig. 10.1).
- $i$:    *inclination*: the angle between the angular momentum vector and the unit vector in the Z-direction.
- $\Omega$:    *right ascension of ascending node*: the angle from the Vernal Equinox to

---

[*] A *sufficiently inertial coordinate frame* is a coordinate frame that can be considered to be non-accelerating for the particular application. The GCI frame is sufficiently inertial for Earth-orbiting satellites but is inadequate for interplanetary travel because of its rotational acceleration around the Sun.

the ascending node. The *ascending node* is the point where the satellite passes through the equatorial plane moving from south to north. Right ascension is measured as a right-handed rotation about the pole, Z.

$\omega$:  *argument of perigee:* the angle from the ascending node to the eccentricity vector measured in the direction of the satellite's motion. The *eccentricity vector* points from the center of the Earth to perigee with a magnitude equal to the eccentricity of the orbit.

$v$:  *true anomaly:* the angle from the eccentricity vector to the satellite position vector, measured in the direction of satellite motion. Alternately, we could use *time since perigee passage, T.*

Given these definitions, we can solve for the elements if we know the satellite's position and velocity vectors. Equations (10.4) and (10.7) allow us to solve for the energy and angular momentum vector. An equation for the *nodal vector, n,* in the direction of the ascending node is

$$\vec{n} = \vec{Z} \times \vec{h} \tag{10.8}$$

We can calculate the eccentricity vector from the following equation:

$$\vec{e} = (1/\mu) \{ (v^2 - \mu/r)\vec{r} - (\vec{r} \cdot \vec{v})\vec{v} \} \tag{10.9}$$

Table 10.2 lists equations to derive the classical orbital elements and related parameters for an elliptical orbit.

Equatorial ($i = 0$) and circular ($e = 0$) orbits demand alternate orbital elements (Fig. 10.3) to solve the equations in Table 10.2. For equatorial orbits, a single angle, $\Pi$, can replace the right ascension of ascending node and argument of perigee. Called the *longitude of perigee,* this angle is the algebraic sum of $\Omega$ and $\omega$. As $i$ approaches 0, $\Pi$ approaches the angle from the X-axis to perigee. For circular orbits ($e = 0$), a single angle, $u \equiv \omega + v$, can replace the argument of perigee and true anomaly. This angle is the *argument of latitude;* when $e = 0$, it equals the angle from the nodal vector to the satellite's position vector. Finally, if the orbit is both circular and equatorial, a single angle, $l$, or *true longitude,* specifies the angle between the X-axis and the satellite's position vector.

## 10.1.4   Satellite Ground Tracks

A satellite's ground track is the trace of the points formed by the intersection of the satellite's position vector with the Earth's surface. In this section we will evaluate ground tracks using a flat map of the Earth.

Although ground tracks are generated from the satellite's orbital elements, we can gain insight by determining the orbital elements from a given ground track.

**Table 10.2.    Classical Orbital Elements.** For the right ascension of ascending node, argument of perigee, and true anomaly, if the quantities in parentheses are positive, use the angle calculated. If the quantities are negative, use 360° minus the angle calculated.

| Symbol | Name | Equation | Check |
|--------|------|----------|-------|
| $a$ | semi-major axis | $a = -\mu/(2\mathcal{E}) = (r_a + r_p)/2$ | |
| $e$ | eccentricity | $e = \lvert e \rvert = 1 - r_p/a = r_a/a - 1$ | |
| $i$ | inclination | $i = \cos^{-1}(h_z/h)$ | |
| $\Omega$ | right ascension of ascending node | $\Omega = \cos^{-1}(n_x/n)$ | $(n_Y > 0)$ |
| $\omega$ | argument of perigee | $\omega = \cos^{-1}[(\vec{n}\cdot\vec{e})/(n\cdot e)]$ | $(e_z > 0)$ |
| $v$ | true anomaly | $v = \cos^{-1}[(\vec{e}\cdot\vec{r})/(e\cdot r)]$ | $(\vec{r}\cdot\vec{v} > 0)$ |
| $r_p$ | radius of perigee | $r_p = a(1 - e)$ | |
| $r_a$ | radius of apogee | $r_a = a(1 + e)$ | |
| $P$ | period | $P = 2\pi (a^3/\mu)^{1/2}$ $\cong 84.489\,(a/R_E)^{3/2}$ min $\cong 0.000\,165\,87 a^{3/2}$ min, $a$ in km | |
| $\omega_0$ | orbit frequency | $\omega_0 = (\mu/a^3)^{1/2}$ $\cong 631.348\,16\ a^{-3/2}$ rad/sec, $a$ in km | |

Figure 10.4 shows ground tracks for satellites with different orbital altitudes and, therefore, different orbital periods. The time it takes for the Earth to rotate through the difference in longitude between two successive ascending nodes equals the orbital period. For *direct* orbits, in which the satellite moves eastward, we measure the change positive to the East. For *retrograde* orbits, in which the satellite moves westward, positive is measured to the West.[*] With these definitions in mind, the period, $P$, in minutes is

$$P = 4\,(360° - \Delta L) \quad \text{direct orbit} \tag{10.10}$$

$$P = 4\,(\Delta L - 360°) \quad \text{retrograde orbit}$$

where $\Delta L$ is the longitudinal change in degrees that the satellite goes through between successive ascending nodes. The difference in longitude between two successive ascending nodes for a direct orbit will always be less than 360° and, in fact, will be negative for orbits at altitudes higher than geosynchronous altitude. For retrograde orbits, the difference in longitude between two successive ascending nodes (positive change is measured to the West) is always greater than 360°.

---

[*] This convenient empirical definition doesn't apply for nearly polar orbits. More formally, a prograde or direct orbit has i < 90°. A retrograde orbit has i > 90°. A polar orbit has i = 90°.

Once we know the period, we can determine the semi-major axis by using the equation for the period of an elliptical orbit:

$$a = [(P/2\pi)^2 \mu]^{1/3} \tag{10.11}$$

$$\cong 331.249\ 15\ P^{2/3}\ \text{km}$$

where the period is in minutes.

Figure 10.4 shows one revolution each for the ground tracks of several orbits with an increasing semi-major axis. The period of a *geosynchronous* orbit, $E$, is 1436 minutes, matching the Earth's rotational motion.

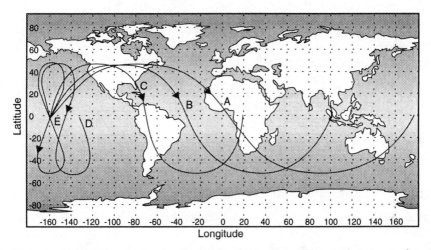

**Fig. 10.4.**      **Orbital Ground Tracks of Circular Orbits with Different Periods.** (A) $\Delta L = 335°$, $P = 100$ min; (B) $\Delta L = 260°$, $P = 400$ min; (C) $\Delta L = 180°$, $P = 720$ min; (D)$\Delta L = 28°$, $P = 1328$ min; and (E) $\Delta L = 0°$, $P = 1436$ min. [Sellers, 1994]

We can determine the orbit's inclination by the ground track's maximum latitude. For direct orbits, the inclination is equal to the ground track's maximum latitude, and for retrograde orbits, the inclination is equal to 180° minus the ground track's maximum latitude.

The orbit is circular if a ground track is symmetrical about both the equator and a line of longitude extending down from the ground track's maximum latitude. Figure 10.4 shows several different circular orbits:

Figure 10.5 shows examples of ground tracks for the following orbits:

A: Shuttle parking orbit, $a = 6700$ km, $e = 0, i = 28.4°$;

B: Low-altitude retrograde, $a = 6700$ km, $e = 0, i = 98.0°$;

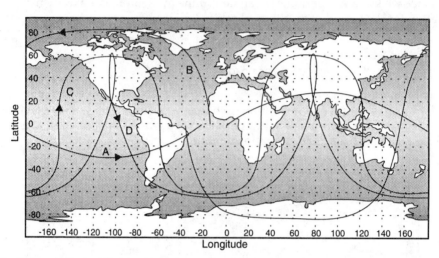

**Fig. 10.5.** **Typical Ground Tracks.** (A) Shuttle Parking, (B) Low-Altitude Retrograde, (C) GPS, and (D) Molniya Orbits. See text for orbital elements. [Sellers, 1994]

C: GPS orbit, $a = 26,600$ km, $e = 0$, $i = 60.0°$; and

D: Molniya orbits, $a = 26,600$ km, $e = 0.75$, $i = 63.4°$, $\omega = 270°$.

## 10.1.5    Time-of-Flight in an Elliptical Orbit

In analyzing Brahe's observational data, Kepler was able to solve the problem of relating position in the orbit to the elapsed time, $t - t_0$, or conversely, how long it takes to go from one point in an orbit to another. To do so, Kepler introduced the quantity $M$, called the *mean anomaly* and expressed as an angle, which is the fraction of an orbit period that has elapsed since perigee. The mean anomaly equals the true anomaly for a circular orbit. By definition,

$$M - M_0 \equiv n\,(t - t_0) \tag{10.12}$$

where $M_0$ is the mean anomaly at time $t_0$, and $n$ is the *mean motion*, or average angular velocity, determined from the semi-major axis of the orbit:

$$n \equiv (\mu\,/\,a^3)^{1/2} \tag{10.13}$$

$$\cong 36{,}173.585\ a^{-3/2}\ \text{deg/sec}$$

$$\cong 8{,}681{,}660.4\ a^{-3/2}\ \text{rev/day}$$

$$\cong 3.125\ 297\ 7 \times 10^9\ a^{-3/2}\ \text{deg/day}$$

where $a$ is in km.

This solution will give the average position and velocity, but satellite orbits are elliptical, with a radius constantly varying in orbit. Because the satellite's velocity depends on this varying radius, it changes as well. To resolve this problem, we can define an intermediate variable called *eccentric anomaly*, $E$, for elliptical orbits. Table 10.3 lists the equations necessary to relate time-of-flight to orbital position.

**Table 10.3.**    **Time-of-Flight in an Elliptical Orbit.** All angular quantities are in radians.

| Variable | Name | Equation |
|---|---|---|
| $n$ | mean motion | $n = (\mu / a^3)^{1/2}$ <br> $\approx 631.348\ 16\ a^{-3/2}$ rad/sec       ($a$ in km) |
| $E$ | eccentric anomaly | $\cos E = (e + \cos v) / (1 + e \cos v)$ |
| $M$ | mean anomaly | $M = E - e \sin (E)$            ($M$ in rad) <br> $M = M_0 + n (t - t_0)$      ($M$ in rad) |
| $t - t_0$ | time-of-flight | $t - t_0 = (M - M_0) / n$      ($t - t_0$ in sec) |
| $v$ | true anomaly | $v \approx M + 2e \sin M + 1.25 e^2 \sin(2M)$ |

As an example, let's find the time it takes a satellite to go from perigee to an angle 90° from perigee, for an orbit with a semi-major axis of 7000 km and an eccentricity of 0.1. For this example,

| | | | |
|---|---|---|---|
| $v_0$ | $= E_0 = M_0 = 0.0$ rad | $t_0$ | $= 0.0$ sec |
| $v$ | $= 1.5708$ rad | $E$ | $= 1.4706$ rad |
| $M$ | $= 1.3711$ rad | $n$ | $= 0.001\ 08$ rad/sec |
| $t$ | $= 1271.88$ sec | | |

Finding the position in an orbit after a specified period is more complex. For this problem, we calculate the mean anomaly, $M$, using time-of-flight and the mean motion, using Eq. (10.12). Next, we determine the true anomaly, $v$, using the series expansion shown in Table 10.3, a good approximation for small eccentricity (the error is of the order $e^3$). If we need greater accuracy, we must solve the equation in Table 10.3, relating mean anomaly to eccentric anomaly. Because this is a transcendental function, we must iterate to find the eccentric anomaly, after which we can calculate the true anomaly directly.

## 10.1.6    Orbit Determination

Up to this point, we've assumed we know the satellite's position and velocity in inertial space—the classical orbital elements. But we often cannot directly observe the satellite's inertial position and velocity. Instead, we commonly receive data from radar, telemetry, optics, or the Global Positioning System (GPS). Radar and telemetry data consist of range, azimuth, elevation, and possibly the rates of change of one or more of these quantities, relative to a site attached to the rotating

Earth. GPS provides the range and range-rate relative to a set of satellites. Optical data consist of right ascension and declination relative to the celestial sphere. In any case, we must combine and convert this data to inertial position and velocity before determining the orbital elements. Bate, Mueller, and White [1971] and Escobal [1965] cover methods for combining data, so we won't cover them here.

The type of data we use for orbit determination depends on the orbit selected, accuracy requirements, and weight restrictions on the payload. Because radar and optical systems collect data passively, they require no additional payload weight, but they are also the least accurate methods of orbit determination. Conversely, GPS data is more accurate but it requires additional payload weight. We can also use it for semi-autonomous orbit determination because it requires no ground support.

## 10.2  Orbit Perturbations

The Keplerian orbit discussed above provides an excellent reference, but other forces act on the satellite to perturb it away from the nominal orbit. We can classify these *perturbations*, or variations in the orbital elements, based on how they affect the Keplerian elements.

Figure 10.6 illustrates a typical variation in one of the orbital elements because of a perturbing force. *Secular variations* represent a linear variation in the element. *Short-period variations* are periodic in the element with a period less than or equal to the orbital period. *Long-period variations* are those with a period greater than the orbital period. Because secular variations have long-term effects on orbit prediction (the orbital elements affected continue to increase or decrease), we'll discuss them in detail. If the spacecraft mission demands that we precisely determine the orbit, we must include the periodic variations as well. Battin [1987], Danby [1962], and Escobal [1965] describe methods of determining and predicting orbits for non-Keplerian motion.

When perturbing forces are considered, the classical orbital elements vary with time. To predict the orbit we must determine this time variation using techniques of either special or general perturbations. *Special perturbations* employ direct numerical integration of the equations of motion. Most common is Cowell's method, in which the accelerations are integrated directly to obtain velocity and again to obtain position.

*General perturbations* analytically solve some aspects of the motion of a satellite subjected to perturbing forces. For example, the polar equation of a conic applies to the two-body equations of motion. Unfortunately, most perturbing forces yield not to a direct analytical solution but to series expansions and approximations. Because the orbital elements are nearly constant, general perturbation techniques usually solve directly for the orbital elements rather than for the inertial position and velocity. They are more difficult and approximate, but they allow us to under-

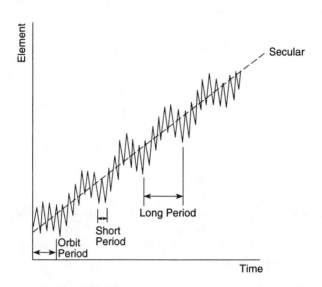

**Fig. 10.6.**　　**Secular and Periodic Variations of an Orbital Element.** Secular variations represent linear variations in the element, short-period variations have a period less than the orbital period, and long-period variations have a period longer than the orbital period.

stand better how the perturbations affect a large class of orbits. We can also get solutions much faster than with special perturbations.

The primary forces which perturb a satellite orbit arise from third bodies such as the Sun and the Moon, the nonspherical mass distribution of the Earth, atmospheric drag, and solar-radiation pressure. Each of these is described below.

### 10.2.1　　Third-Body Perturbations

The gravitational forces of the Sun and the Moon cause periodic variations in all of the orbital elements, but only the right ascension of ascending node, argument of perigee, and mean anomaly experience secular variations. These secular variations arise from a gyroscopic precession of the orbit about the ecliptic pole. The secular variation in mean anomaly is much smaller than the mean motion and has little effect on the orbit; however, the secular variations in right ascension of the ascending node and argument of perigee are important, especially for high-altitude orbits.

For nearly circular orbits, $e^2$ is near zero, and the resulting error is of the order $e^2$. In this case, the equations for the secular rates of change resulting from the Sun and Moon are right ascension of ascending node:

$$\dot{\Omega}_{MOON} = -0.003\,38\,(\cos i)/n \qquad (10.14)$$

$$\dot{\Omega}_{SUN} = -0.001\,54\,(\cos i)/n \qquad (10.15)$$

argument of perigee:

$$\dot{\omega}_{MOON} = 0.001\,69\,(4 - 5\sin^2 i)/n \qquad (10.16)$$

$$\dot{\omega}_{SUN} = 0.000\,77\,(4 - 5\sin^2 i)/n \qquad (10.17)$$

where $i$ is the orbit inclination, $n$ is the number of orbit revolutions per day, and $\dot{\Omega}$ and $\dot{\omega}$ are in deg/day. These equations are only approximate; they neglect the variation caused by the changing orientation of the orbital plane with respect to both the Moon's orbital plane and the ecliptic plane.

## 10.2.2    Perturbations Because of a Nonspherical Earth

When developing the two-body equations of motion, we assumed the Earth was a spherically symmetrical, homogeneous mass. In fact, the Earth is neither homogeneous nor spherical. The most dominant features are a bulge at the equator, a slight pear shape, and flattening at the poles. For a potential function of the Earth, $\Phi$, we can find a satellite's acceleration by taking the gradient of the potential function. One widely used form of the geopotential function is

$$\Phi = (\mu/r)\,[1 - \sum J_n\,(R_E/r)^n P_n\,(\sin L)\,] \qquad (10.18)$$

where $\mu \equiv GM$ is the Earth's gravitational constant, $R_E$ is the Earth's equatorial radius, $P_n$ are Legendre polynomials, $L$ is the geocentric latitude, and $J_n$ are the dimensionless geopotential coefficients, of which the first several are

$J_2 = 0.00108263$
$J_3 = -0.00000254$
$J_4 = -0.00000161$

This form of the geopotential function depends on latitude, and the geopotential coefficients, $J_n$, are called the *zonal coefficients*. Other, more general expressions for the geopotential include sectoral and tesseral terms in the expansion. The *sectoral terms* divide the Earth into slices and depend only on longitude. The *tesseral terms* in the expansion depend on both longitude and latitude. They divide the Earth into a checkerboard pattern of regions that alternately add to and subtract from the two-body potential.

The potential generated by the non-spherical Earth causes periodic variations in all of the orbital elements. But the dominant effects are secular variations in right ascension of ascending node and argument of perigee because of the Earth's oblateness, represented by the $J_2$ term in the geopotential expansion. The rates of change of $\Omega$ and $\omega$ due to $J_2$ are

$$\dot{\Omega}_{J_2} = -1.5 \, n \, J_2 \, (R_E/a)^2 \, (\cos i) \, (1 - e^2)^{-2} \qquad (10.19)$$

$$\cong -2.06474 \times 10^{14} a^{-7/2} (\cos i) \, (1 - e^2)^{-2}$$

$$\dot{\omega}_{J_2} = 0.75 n \, J_2 \, (R_E/a)^2 \, (4 - 5 \sin^2 i) \, (1 - e^2)^{-2} \qquad (10.20)$$

$$\cong 1.03237 \times 10^{14} a^{-7/2} (4 - 5 \sin^2 i) \, (1 - e^2)^{-2}$$

where $n$ is the mean motion in deg/day, $R_E$ is the Earth's equatorial radius, $a$ is the semi-major axis in km, $e$ is the eccentricity, $i$ is the inclination, and $\dot{\Omega}$ and $\dot{\omega}$ are in deg/day. Table 10.4 compares the rates of change of right ascension of ascending node and argument of perigee resulting from the Earth's oblateness, the Sun, and the Moon.

**Table 10.4.** **Secular Variations in Right Ascension of the Ascending Node and Argument of Perigee.** For spacecraft in GEO and below, the $J_2$ perturbations dominate; for spacecraft above GEO, the Sun and Moon perturbations dominate.

| Orbit | Effect of $J_2$ (Eqs. 10.19, 10.20) (deg/day) | Effect of Moon (Eqs. 10.14, 10.16) (deg/day) | Effect of Sun (Eqs. 10.15, 10.17) (deg/day) |
|---|---|---|---|
| Shuttle | $a = 6700$ km, $e = 0.0$, $i = 28°$ | | |
| $\Delta\Omega$ | −7.35 | −0.00019 | −0.00008 |
| $\Delta\omega$ | 12.05 | 0.00242 | 0.00110 |
| GPS | $a = 26{,}600$ km, $e = 0.0$, $i = 60.0°$ | | |
| $\Delta\Omega$ | −0.033 | −0.00085 | −0.00038 |
| $\Delta\omega$ | 0.008 | 0.00021 | 0.00010 |
| Molniya | $a = 26{,}600$ km, $e = 0.75$, $i = 63.4°$ | | |
| $\Delta\Omega$ | −0.30 | −0.00076 | −0.00034 |
| $\Delta\omega$ | 0.00 | 0.00000 | 0.00000 |
| Geosynchronous | $a = 42{,}160$ km, $e = 0$, $i = 0°$ | | |
| $\Delta\Omega$ | −0.013 | −0.00338 | −0.00154 |
| $\Delta\omega$ | 0.025 | 0.00676 | 0.00307 |

*Molniya orbits* are highly eccentric ($e \cong 0.75$) with approximately 12-hour periods (2 revolutions/day). The orbital inclination is chosen so the rate of change of the argument of perigee, Eq. (10.20), is zero. This condition occurs at inclinations of 63.4° and 116.6°. For these orbits, the argument of perigee is typically placed in the southern hemisphere, so the spacecraft remains above the northern hemisphere near apogee for approximately 11 hours/orbit. The perigee altitude is chosen to meet the mission constraints. Typical perigee altitudes vary from 200 to 1000 km. We can calculate the eccentricity and apogee altitude using the semi-major axis and perigee.

In a *Sun-synchronous orbit*, the spacecraft's orbital plane remains approximately fixed with respect to the Sun because we match the secular variation in the right ascension of ascending node [Eq. (10.19)] to the Earth's rate of rotation around the Sun. A nodal precession rate of 0.9856 deg/day will match the Earth's rate of average rotation about the Sun. Because this rotation is positive, Sun-synchronous orbits must be retrograde. For a given semi-major axis, $a$, and eccentricity, we can use Eq. (10.19) to find the inclination that will keep the orbit Sun-synchronous.

### 10.2.3    Perturbations From Atmospheric Drag

The principal nongravitational force acting on spacecraft in low-Earth orbit is atmospheric drag. Drag acts in a direction opposite that of the velocity vector and removes energy from the orbit. This reduction of energy causes the orbit to get smaller, leading to further increases in drag. Eventually, the orbit's altitude becomes so small that the spacecraft reenters the atmosphere.

The equation for acceleration due to drag on a spacecraft is

$$a_D = -(1/2)\rho\,(C_D A/m)V^2 \tag{10.21}$$

where $\rho$ is the atmospheric density, $C_D$ is the coefficient of drag $\approx 2.2$, $A$ is the spacecraft's cross-sectional area, $m$ is the satellite's mass, and $V$ is the spacecraft's velocity with respect to the atmosphere.

We can approximate the changes in semi-major axis and eccentricity per revolution, and the lifetime of a spacecraft in a circular orbit, using the following equations:

$$\Delta a_{rev} = -2\pi\,(C_D A/m)a^2\,\rho_p \exp(-c)\,[I_0 + 2eI_1] \tag{10.22}$$

$$\Delta e_{rev} = -2\pi\,(C_D A/m)a\,\rho_p \exp(-c)\,[I_1 + e/2\,(I_0 + I_2)] \tag{10.23}$$

where $\rho_p$ is the atmospheric density at perigee, $c \equiv ae/H$, $H$ is the density scale height, and $I_i$ are Modified Bessel Functions[*] of order $i$ and argument $c$. The term $m/(C_D A)$, or *ballistic coefficient*, is modelled as a constant for most spacecraft.

For circular orbits, we can use the above equations to derive the much simpler expressions:

$$\Delta a_{rev} = -2\pi (C_D A/m)\rho a^2 \tag{10.24}$$

$$\Delta P_{rev} = -6\pi^2 (C_D A/m)\rho a^2 / V \tag{10.25}$$

$$\Delta V_{rev} = \pi (C_D A/m)\rho a V \tag{10.26}$$

$$\Delta e_{rev} = 0 \tag{10.27}$$

where $P$ is the orbit period, and $V$ is the spacecraft's velocity.

A rough estimate of the satellite's lifetime, $L$, due to drag can be computed from

$$L \approx -H/\Delta a_{rev} \tag{10.28}$$

where, as above, $H$ is the atmospheric density scale height.

### 10.2.4    Perturbations from Solar Radiation

Solar-radiation pressure causes periodic variations in all of the orbital elements. Its effect is strongest for spacecraft with low ballistic coefficients—light vehicles with large frontal areas such as Echo. The magnitude of the acceleration in m/sec$^2$ arising from solar-radiation pressure at the Earth is

$$a_R \approx -4.5 \times 10^{-6} (A/m) \tag{10.29}$$

where $A$ is the effective cross-sectional area of the spacecraft exposed to the Sun in m$^2$, and $m$ is the spacecraft's mass in kg. For satellites below 800 km altitude, acceleration from atmospheric drag is greater than that from solar-radiation pressure; above 800 km, acceleration from solar-radiation pressure is greater.

# 10.3  Orbit Maneuvering

At some point during the lifetime of most spacecraft, we must change one or more of the orbital elements. For example, we may need to transfer from an initial parking orbit to the final mission orbit, rendezvous with or intercept another spacecraft, or correct the orbital elements to adjust for the perturbations discussed

---

[*] Tables of values for $I_i$ can be found in many standard mathematical tables.

in the previous section. Most frequently, we must change the orbit's altitude, plane, or both. To change a spacecraft's orbit, we have to change the velocity vector in magnitude or direction. Most propulsion systems operate for only a short time compared to the orbital period, so we can treat the maneuver as an impulsive change in the velocity while the position remains fixed. For this reason, any maneuver changing a spacecraft's orbit must occur at a point where the old orbit intersects the new orbit. If the two orbits don't intersect, we have to use an intermediate orbit that intersects both. In this case, the total maneuver will require at least two propulsive burns.

In general, the change in the velocity vector to go from one orbit to another is given by

$$\Delta \overline{V} = \overline{V}_{NEED} - \overline{V}_{CURRENT} \tag{10.30}$$

We can find the current and needed velocity vectors from the orbital elements, keeping in mind that the position vector doesn't change much during impulsive burns.

### 10.3.1   Coplanar Orbit Transfers

The most common type of in-plane maneuver changes the orbit's size and energy, usually from a low-altitude parking orbit to a higher-altitude mission orbit such as a geosynchronous orbit. Because the initial and final orbit don't intersect (see Fig. 10.7), the maneuver requires a transfer orbit. Figure 10.7 represents a Hohmann[*] Transfer Orbit. In this case, the transfer orbit's ellipse is tangent to both the initial and final circular orbits at the transfer orbit's perigee and apogee, respectively. The orbits are tangential, so the velocity vectors are collinear, and the Hohmann Transfer represents the most fuel-efficient transfer between two circular, coplanar orbits. When transferring from a smaller orbit to a larger orbit, the change in velocity is applied in the direction of motion; when transferring from a larger orbit to a smaller, the change of velocity is opposite to the direction of motion.

The total change in velocity required for the transfer is the sum of the velocity changes at perigee and apogee of the transfer ellipse. Because the velocity vectors are collinear, the velocity changes are just the differences in magnitudes of the velocities in each orbit. We can find these differences from the energy equation if we know the size of each orbit. If we know the initial and final orbits ($r_A$ and $r_B$), we can calculate the semi-major axis of the transfer ellipse, $a_{tx}$, and the total velocity change (the sum of the velocity changes required at points $A$ and $B$) using the

---

[*] Walter Hohmann, a German engineer and architect, wrote *The Attainability of Celestial Bodies* [1925], consisting of a mathematical discussion of the conditions for leaving and returning to Earth.

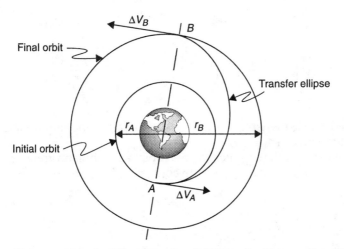

**Fig. 10.7.**     **Hohmann-Transfer Ellipse Showing Orbit Transfer Between Two Circular, Coplanar Orbits.**

following algorithm. An example illustrates this technique: transferring from an initial circular orbit of 6567 km to a final circular orbit of 42,160 km.

| STEP | EQUATIONS | | EXAMPLE |
|---|---|---|---|
| 1. | $a_{tx}$ | $= (r_A + r_B)/2$ | $= 24{,}364$ km |
| 2. | $V_{iA}$ | $= (\mu/r_A)^{1/2} = 631.3481(r_A)^{-1/2}$ | $= 7.79$ km/sec |
| 3. | $V_{fB}$ | $= (\mu/r_B)^{1/2}$ | |
| | | $= 631.3481(r_B)^{-1/2}$ | $= 3.08$ km/sec |
| 4. | $V_{txA}$ | $= [\mu(2/r_A - 1/a_{tx})]^{1/2}$ | |
| | | $= 631.3481\,[(2/r_A - 1/a_{tx})]^{1/2}$ | $= 10.25$ km/sec |
| 5. | $V_{txB}$ | $= [\mu(2/r_B - 1/a_{tx})]^{1/2}$ | |
| | | $= 631.3481\,[(2/r_B - 1/a_{tx})]^{1/2}$ | $= 1.59$ km/sec |
| 6. | $\Delta V_A$ | $= \mid V_{txA} - V_{iA} \mid$ | $= 2.46$ km/sec |
| 7. | $\Delta V_B$ | $= \mid V_{fB} - V_{txB} \mid$ | $= 1.49$ km/sec |
| 8. | $\Delta V_{TOTAL} = \Delta V_A + \Delta V_B$ | | $= 3.95$ km/sec |
| 9. | Time of transfer $= P/2$ | | $= 5$ hrs 15 mins |

Or we can write the total $\Delta V$ required for a two-burn transfer between circular orbits at altitude $r_A$ and $r_B$ as

$$\Delta v_{total} = \Delta v_A + \Delta v_B \tag{10.31}$$

$$= \sqrt{\mu} \left[ \left| \left( \frac{2}{r_A} - \frac{1}{a_{tx}} \right)^{\frac{1}{2}} - \left( \frac{1}{r_A} \right)^{\frac{1}{2}} \right| + \left| \left( \frac{2}{r_B} - \frac{1}{a_{tx}} \right)^{\frac{1}{2}} - \left( \frac{1}{r_B} \right)^{\frac{1}{2}} \right| \right] \tag{10.32}$$

where $\sqrt{\mu} = 631.3481$ when $\Delta V$ is in km/sec and all of the semi-major axes are in km. As in step 1, $a_{tx} = (r_A + r_B)/2$.

The above expression applies to any coplanar Hohmann transfer. In the case of small transfers ($r_A$ close to $r_B$), we can conveniently approximate it in two forms:

$$\Delta V \approx V_{iA} - V_{fB} \tag{10.33}$$

$$\Delta V \approx 0.5 \, (\Delta r/r) \, V_{A/B} \tag{10.34}$$

where

$$\Delta r \equiv r_B - r_A \tag{10.35}$$

and

$$r \approx r_A \approx r_B \qquad V_{A/B} \approx V_{iA} \approx V_{fB} \tag{10.36}$$

The two small burns are of nearly equal magnitude.

The result in Eq. (10.33) is more unusual than it might at first seem. Assume that a spacecraft is in a circular orbit with velocity $V_{iA}$. In two burns we *increase* the velocity by an amount $\Delta V$. The result is that the spacecraft is higher and traveling *slower* than originally by the amount $\Delta V$. An example will clarify this result. Consider a spacecraft in a circular orbit at 400 km such that $r_A = 6778$ km and $V_{iA} = 7700$ m/s. We'll apply a total $\Delta V$ of 20 m/s (= 0.26% of $V_{iA}$) in two burns of 10 m/s each. From Eq. (10.34) the total $\Delta r$ will be 0.52% of 6778 km or 35 km. Thus, the final orbit will be circular at an altitude of 6813 km. Immediately following the first burn of 10 m/s, the spacecraft will be at perigee of the transfer orbit with a velocity of 7710 m/s. When the spacecraft reaches apogee at 6813 km, it will have slowed according to Kepler's second law by 0.52% to 7670 m/s. We then apply the second burn of 10 m/s to circularize the orbit at 7680 m/s which is 20 m/s slower than its original velocity. We've added energy to the spacecraft, which has raised the orbit and resulted in a lower kinetic energy but enough more potential energy to make up for the reduced speed and the added $\Delta V$.

Sometimes, we may need to transfer a spacecraft between orbits faster than the Hohmann transfer will allow. Figure (10.8) shows a faster transfer called the *One-Tangent-Burn*. In this instance the transfer orbit is tangential to the initial orbit. It intersects the final orbit at an angle equal to the flight-path angle of the transfer orbit at the point of intersection. An infinite number of transfer orbits are tangential to the initial orbit and intersect the final orbit at some angle. Thus, we may choose the transfer orbit by specifying the size of the transfer orbit, the angular change of the transfer, or the time required to complete the transfer. We can then define the transfer orbit and calculate the required velocities.

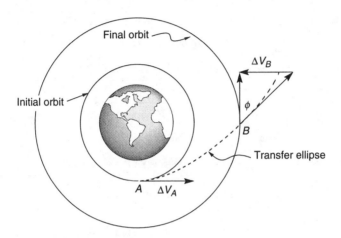

**Fig. 10.8.**      **Transfer Orbit Using One-Tangent Burn between Two Circular, Coplanar Orbits.**

For example, we may specify the size of the transfer orbit, choosing any semi-major axis that is greater than the semi-major axis of the Hohmann-transfer ellipse. Once we know the semi-major axis of the ellipse ($a_{tx}$), we can use the equations in Table 10.5 calculate the eccentricity, the angular distance traveled in the transfer, the velocity change required for the transfer, and the time required to complete the transfer using the equations in Table 10.5.

Table 10.6 compares the total required velocity change and time-of-flight for a Hohmann transfer and a one-tangent burn transfer from a low-altitude parking orbit to geosynchronous orbit.

Another option for changing the size of the orbit is to use a constant-low-thrust burn, which results in a *spiral transfer*. We can approximate the velocity change for this type of orbit transfer by

$$\Delta V = |V_2 - V_1| \tag{10.37}$$

where the velocities are the circular velocities of the two orbits. Following the previous example, the total velocity change required to go from low-Earth orbit to

**Table 10.5.**    **Computations for an Orbit Transfer Using a One-Tangent Burn.** For example, see Battin [1987].

| Quantity | Equation |
|---|---|
| Eccentricity | $e = 1 - r_A / a_{tx}$ |
| True anomaly at second burn | $v = \cos^{-1}[(a(1 - e^2) / r_B - 1) / e]$ |
| Flight-path angle at second burn | $\phi = \tan^{-1}[e \sin v / (1 + e \cos v)]$ |
| Initial velocity | $V_{iA} = 631.3481 \, r_A^{-1/2}$ |
| Velocity on transfer orbit at initial orbit | $V_{txA} = 631.3481 \, [2/ r_A - 1/ a_{tx}]^{1/2}$ |
| Initial velocity change | $\Delta V_A = |V_{txA} - V_{iA}|$ |
| Final velocity | $V_{fB} = 631.3481 \, r_B^{-1/2}$ |
| Velocity on transfer orbit at final orbit | $V_{txB} = 631.3481 [2/ r_B - 1/ a_{tx}]^{1/2}$ |
| Final velocity change | $\Delta V_B = [V_{fB}^2 + V_{txB}^2 - 2 V_{fB} V_{txB} \cos\phi]^{1/2}$ |
| Total velocity change | $\Delta V_T = \Delta V_A + \Delta V_B$ |
| Eccentric anomaly | $E = \tan^{-1}[(1 - e^2)^{1/2} \sin v / (e + \cos v)]$ |
| Time-of-flight | $TOF = 0.001583913 \, a^{3/2} (E - e \sin E)$, $E$ in rads |

**Table 10.6.**    **Comparison of Coplanar Orbit Transfers from Low-Earth Orbit to Geosynchronous Orbit.**

| Variable | Hohmann Transfer | One-Tangent-Burn Transfer |
|---|---|---|
| $r_A$ | 6570 km | 6570 km |
| $r_B$ | 42,200 km | 42,200 km |
| $a_{tx}$ | 24,385 km | 28,633 km |
| $\Delta V_T$ | 3.935 km/s | 4.699 km/s |
| $TOF$ | 5.256 hr | 3.457 hr |

geosynchronous is 4.71 km/s using a spiral transfer. We get this value by subtracting the results of step 3 from the results of step 2 in the above example.

## 10.3.2   Changes in the Orbital Plane

To change the orientation of the satellite's orbital plane, typically the inclination, we must change the direction of the velocity vector. This maneuver requires a component of $\Delta V$ to be perpendicular to the orbital plane and, therefore, perpendicular to the initial velocity vector. If the size of the orbit remains constant, the maneuver is called a *simple plane change* (Fig. 10.9a). We can find the required change in velocity by using the law of cosines. For the case in which $V_f$ is equal to $V_i$ this expression reduces to

$$\Delta V = 2V_i \sin (\theta/2) \tag{10.38}$$

where $V_i$ is the velocity before and after the burn, and $\theta$ is the required angle change.

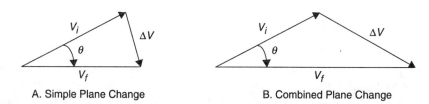

A. Simple Plane Change　　　　　　　　　　　　B. Combined Plane Change

**Fig. 10.9.** **Vector Representation of Simple and Combined Changes in Orbital Plane.** For the simple plane change, the initial and final velocities are equal in magnitude.

For example, the change in velocity required to transfer from a low-altitude ($h = 185$ km), inclined ($i = 28°$) orbit to an equatorial orbit ($i = 0$) at the same altitude is:

$$r = 6563 \text{ km} \quad V_i = 7.79 \text{ km/sec} \quad \Delta V = 3.77 \text{ km/sec}$$

From Eq. (10.38) we see that if the angular change is equal to 60°, the required change in velocity is equal to current velocity. Plane changes are very expensive in terms of the required change in velocity and resulting fuel consumption. To hold down cost, we should change the plane at a point where the spacecraft's velocity is lowest: at apogee for an elliptical orbit. In some cases, it may even be cheaper to boost the spacecraft into a higher orbit, change the orbit plane at apogee, and return the spacecraft to its original orbit.

Typically, orbital transfers require changes in the orbit's size and plane, such as transferring from an inclined parking orbit at low altitude to a zero-inclination orbit at geosynchronous altitude. We can do this transfer in two steps: a Hohmann transfer to change the size of the orbit and a simple plane change to make the orbit equatorial. A more efficient method (less total change in velocity) would be to combine the plane change with the tangential burn at apogee of the transfer orbit (Fig. 10.9B). As we must change the velocity vector's magnitude and direction, we can find the required change in velocity using the law of cosines:

$$\Delta V = (V_i^2 + V_f^2 - 2V_i V_f \cos \theta)^{1/2} \tag{10.39}$$

where $V_i$ is the initial velocity, $V_f$ is the final velocity, and $\theta$ is the required angle change.

For example, we find the total change in velocity to transfer from a Shuttle parking orbit to a geosynchronous, equatorial orbit as follows:

| | | | |
|---|---|---|---|
| $r_i$ | = 6563 km | $r_f$ | = 42,159 km |
| $i_i$ | = 28° | $i_f$ | = 0° |
| $V_i$ | = 7.79 km/sec | $V_f$ | = 3.08 km/sec |
| $\Delta V_A$ | = 2.46 km/sec | | |
| $\Delta V_B$ | = 1.83 km/sec | | |
| $\Delta V_{TOTAL}$ | = 4.29 km/sec | | |

Completing a Hohmann transfer followed by a simple plane change would require a velocity change of 5.44 km/sec, so the Hohmann transfer with a combined plane change at apogee of the transfer orbit represents a savings of 1.15 km/sec. As you can see from Eq. (10.39), we can combine a small plane change ($\theta \geq 0$) with an energy change for almost no additional cost in $\Delta V$ or propellant. So, in practice, geosynchronous transfer uses a small plane change at perigee and most of the plane change at apogee.

Another option is to complete the maneuver using three burns. The first burn is a coplanar maneuver placing the spacecraft into a transfer orbit with an apogee much higher than the final orbit. A combined plane change follows when the spacecraft reaches apogee of the transfer orbit. This maneuver places the spacecraft in a second transfer orbit which is coplanar with the final orbit and has a perigee altitude equal to the altitude of the final orbit. Finally, when the spacecraft reaches perigee of the second transfer orbit, another coplanar maneuver places it into the final orbit. This three-burn maneuver may save fuel, but the fuel savings comes at the expense of the total time required to complete it.

### 10.3.3 Orbit Rendezvous

Orbital transfer becomes more complicated when the objective is to rendezvous with or intercept another object in space: both the interceptor and target must arrive at the rendezvous point at the same time. This precision demands a phasing orbit to do the maneuver. A *phasing orbit* is any orbit which results in the interceptor achieving the desired geometry relative to the target to start a Hohmann transfer. If the initial and final orbits are circular, coplanar, and of different sizes, the phasing orbit is simply the interceptor's initial orbit (Fig. 10.10). The interceptor remains in the initial orbit until the relative motion between the interceptor and target results in the desired geometry. At that point, we would inject the interceptor into a Hohmann transfer orbit. The equation to solve for the wait time in the initial orbit is

$$\text{Wait Time} = (\phi_i - \phi_f + 2k\pi)/(\omega_{int} - \omega_{tgt}) \qquad (10.40)$$

where $\phi_f$ is the phase angle (angular separation of target and interceptor) needed for rendezvous; $\phi_i$ is the initial phase angle; $k$ is the number of rendezvous opportunities (for the first opportunity, $k = 0$); $\omega_{int}$ is the interceptor's angular velocity; and $\omega_{tgt}$ is the target's angular velocity. We calculate the lead angle, $\alpha_L$, by multiplying $\omega_{tgt}$ by the time of flight for the Hohmann transfer; $\phi_f$ is 180° minus $\alpha_L$.

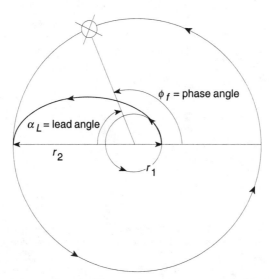

**Fig. 10.10.**      **Geometry Depicting Rendezvous between Two Circular, Coplanar Orbits.** The phase angle is the angular separation between the target and interceptor at the start of the rendezvous, and the lead angle is the distance the target travels from the start until rendezvous occurs.

The total time to rendezvous is equal to the wait time from Eq. (10.40) plus the time-of-flight of the Hohmann transfer orbit.

The denominator in Eq. (10.40) represents the relative motion between the interceptor and target. As the size of the interceptor's orbit approaches the size of the target's orbit, the relative motion approaches zero, and the wait time approaches infinity. If the two orbits are exactly the same size, the interceptor must enter a new phasing orbit to rendezvous with the target (Fig. 10.11). For this situation, the rendezvous occurs at the point where the interceptor enters the phasing orbit. The period of the phasing orbit is equal to the time it takes the target to get to the rendezvous point. Once we know the period, we can calculate the semi-major axis. The two orbits are tangential at their point of intersection, so the change in velocity is again the difference in magnitudes of the two velocities at the point where the two orbits intersect. Because we know the size of the two orbits, and therefore, the energies, we can use the energy Eq. (10.4) to solve for the current and needed velocity.

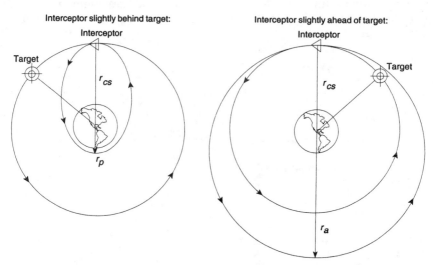

**Fig. 10.11.**     **Rendezvous from Same Orbit Showing the Target Leading and Trailing the Interceptor.** The period of the phasing orbit is equal to the time it takes the target to travel to the rendezvous point.

Spacecraft in circular orbits must often adjust their relative phasing in the orbit. We can do so by making the spacecraft drift relative to its initial position. The *drift rate* in deg/orbit for spacecraft in Earth orbit is

$$drift\ rate = 1080\ \Delta V/V \tag{10.41}$$

where $V$ is the orbit's nominal velocity, and $\Delta V$ is the velocity change required to start or stop the drift.

## 10.4  Launch Windows

Similar to the rendezvous problem is the launch-window problem, or determining the appropriate time to launch from the Earth's surface into the desired orbital plane. Because the orbital plane is fixed in inertial space, the launch window is the time when the launch site on the Earth's surface rotates through the orbital plane. As Fig. 10.12 shows, the time of the launch depends on the launch site's latitude and longitude and the spacecraft orbit's inclination and right ascension of ascending node.

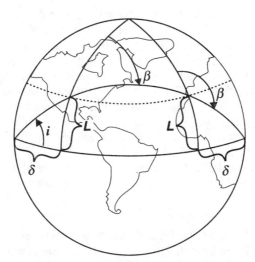

**Fig. 10.12.** **Launch Window Geometry for Launches near the Ascending Node and Descending Node.** The angles shown are the orbit's inclination ($i$), launch-site latitude ($L$), and launch azimuth ($\beta$). $\delta$ is the angle measured from the nearest node to the launch-site longitude.

For a launch window to exist, the launch site must pass through the orbital plane, placing restrictions on the orbital inclinations, $i$, possible from a given launch latitude, $L$:

- No launch windows exist if $L > i$ for direct orbit or $L > 180° − i$ for retrograde orbits
- One launch window exists if $L = i$ or $L = 180° − i$
- Two launch windows exist if $L < i$ or $L < 180° − i$

The *launch azimuth*, $\beta$, is the angle measured clockwise from north to the velocity vector. If a launch window exists, the launch azimuth required to achieve an inclination, $i$, from a given launch latitude, $L$, is

$$\beta = \beta_I \pm \gamma \approx \beta_I \tag{10.42a}$$

where

$$\sin \beta_I = \cos i \ / \ \cos L \tag{10.42b}$$

and

$$\tan \gamma = \frac{V_L \cos \beta_I}{V_0 - V_{eq} \cos i} \approx \left( \frac{V_L}{V_0} \right) \cos \beta_I \tag{10.42c}$$

where $V_L$ is the launch site's inertial velocity as given by Eq. (10.46) below, $V_{eq} =$ 464.5 m/s is the velocity of the Earth's rotation at the equator, and $V_0 \approx 7.8$ km/s is the velocity of the satellite immediately after launch. $\beta_I$ is the inertial launch azimuth, and $\gamma$ is a small correction to account for the velocity contributed by the Earth's rotation. For launches to low-Earth orbit, $\gamma$ ranges from 0 for a due-east launch to 3.0° for launch to a polar orbit. The approximation for $\gamma$ in Eq. (11.42c) is good to better than 0.1° for low-Earth orbits. For launches near the ascending node, $\beta$ is in the first or fourth quadrant, and the plus sign applies in Eq. (11.42a). For launches near the descending node, $\beta$ is the second or third quadrant, and the minus sign applies in Eq. (11.42a).

Let $\delta$, shown in Fig. 10.12, be the angle in the equatorial plane from the nearest node to the longitude of the launch site. We can determine $\delta$ from

$$\cos \delta = \cos \beta / \sin i \tag{10.43}$$

where $\delta$ is positive for direct orbits and negative for retrograde orbits. Finally, the *local sidereal time, LST*, of launch is the angle at the time of launch from the vernal equinox to the longitude of the launch site:

$$LST = \Omega + \delta \qquad \text{(launch at ascending node)}$$
$$= \Omega + 180° - \delta \qquad \text{(launch at descending node)} \tag{10.44}$$

where $\Omega$ is the resulting orbit's right ascension of the ascending node.

Having calculated the launch azimuth required to achieve the desired orbit, we can now calculate the velocity needed to accelerate the payload from rest at the launch site to the required burnout velocity. To do so, we would use topocentric-horizon coordinates with velocity components $V_S$, $V_E$, $V_Z$:

$$V_S = -V_{bo} \cos \phi \cos \beta_b$$
$$V_E = V_{bo} \cos \phi \sin \beta_b - V_L$$
$$V_Z = V_{bo} \sin \phi \tag{10.45}$$

where $V_{bo}$ is the velocity at burnout (usually equal to the circular-orbit velocity at the prescribed altitude), $\phi$ is the flight-path angle at burnout, $\beta_b$ is the launch azimuth at burnout, and $V_L$ is the velocity of the launch site on the Earth at a given latitude, $L$, as given by

$$V_L = (464.5 \text{ m/sec}) \cos L \tag{10.46}$$

Equation (10.45) doesn't include losses in the launch vehicle's velocity because of atmospheric drag and gravity—approximately 1500 m/sec for a typical launch vehicle. Also, in Eq. (10.45) we assume the azimuths at launch and at burnout are the same. Changes in the launch vehicle's latitude and longitude during powered flight will introduce small errors into calculating the burnout conditions. We can calculate the velocity required at burnout from the energy equation if we know the orbit's semi-major axis and radius of burnout.

## 10.5  Orbit Maintenance

Once in their mission orbits, many spacecraft need no orbital adjustments. But mission requirements may demand that we maneuver the spacecraft to correct the orbital elements when perturbing forces have changed them. Two particular cases of note are spacecraft with repeating ground tracks and geosynchronous-equatorial spacecraft.

Using two-body equations of motion, we can show that a spacecraft will have a repeating ground track if it has exactly an integer number of revolutions per integer number of days. Its period must therefore be

$$P = (m \text{ sidereal days})/(K \text{ revolutions}) \tag{10.47}$$

where $m$ and $k$ are integers, and 1 sidereal day = 1436.068 min. For example, a spacecraft orbiting the Earth exactly 16 times per day will have a period of 89.75 min and a semi-major axis of 6640 km.

Next we would modify the spacecraft's period to account for the drift in the orbital plane caused by the Earth's oblateness ($J_2$). We can calculate the rate of change of the right ascension of ascending node, $\Delta\Omega$ because of $J_2$ from the two-body orbital elements. In this case the new period is

$$P_{New} = P_{Two\text{-}body} + \Delta\Omega/\omega_{Earth} \tag{10.48}$$

Because the nodal drift is based on the two-body orbital elements, we must iterate to find the new orbital period and semi-major axis. Continuing with the previous example, assume a perigee altitude of 120 km and an inclination of 45°. In this case, we find the compensated period is 88.20 min and the new semi-major axis is 6563 km.

Table 10.7 shows examples of spacecraft placed in orbits with repeating ground tracks.

**Table 10.7.    Examples of Spacecraft in Orbits with Repeating Ground Tracks.**

| Satellite | Semi-Major Axis (km) | Revs | Days |
|-----------|----------------------|------|------|
| SEASAT | 7168.3 | 43 | 3 |
| LANDSAT 4/5 | 7077.8 | 233 | 16 |
| GEOSAT | 7173.6 | 244 | 17 |

The Earth's oblateness also causes the direction of perigee to rotate around the orbit. If the orbit isn't circular, and the mission limits the altitude over specific targets, we must control the location of perigee. One possibility is to select the orbital inclination to be at the critical inclination (63.4° for a direct orbit and 116.4° for a retrograde orbit), so the location of perigee is fixed. If other constraints make this selection impossible, we must maintain the orbit through orbital maneuvers. We can change the location of perigee by changing the flight-path angle by an angle $\theta$. Only the direction of the velocity vector is changing, so we can find the change in velocity from the equation for a simple plane change:

$$\Delta V = 2V \sin \theta/2 \qquad (10.49)$$

A final consideration for a low-altitude orbit with repeating ground tracks is the change in the semi-major axis and eccentricity due to atmospheric drag. Drag causes the orbit to become smaller. As the orbit becomes smaller, the period also reduces, causing the ground track to appear to shift eastward. If some tolerance is specified, such as a maximum distance between the actual and desired ground track, the satellite must periodically maneuver to maintain the desired orbit.

We can use Eq. (10.22) to calculate the change in semi-major axis per revolution of the orbit. Given the change in the orbit's size, we can also determine the change in the period:

$$\Delta P = 3\pi \, \Delta a/(na) \qquad (10.50)$$

If constraints exist for the orbit's period or semi-major axis, we can use Eqs. (10.22) and (10.50) to keep track of the period and semi-major axis until the orbit needs to be corrected. Applying a tangential velocity change at perigee will adjust the semi-major axis when required. Again, we can find the current and needed velocities from the energy Eq. (10.4), because we know the size, and therefore the energy, of the two orbits.

Geosynchronous-equatorial orbits also require orbital maintenance. Spacecraft in these orbits drift when perturbations occur from the oblate Earth and from third-body interactions with the Sun and Moon. Matching the period of a geostationary orbit with the Earth's rotational velocity results in a resonance with the $J_{22}$ term in the geopotential. This resonance term results in a transverse acceleration—an acceleration in the orbital plane—which causes the spacecraft to drift in longitude (*East-West drift*). The Sun and the Moon cause out-of-plane accelerations, which make the spacecraft drift in latitude (*North-South drift*).

North-South stationkeeping is necessary when mission requirements limit the drift in latitude or inclination. If not corrected, the orbit's inclination varies between 0° and 15° with a period of approximately 55 years. The approximate equations to solve for the worst-case change in velocity are

$$\Delta V_{Moon} = 102.67 \cos \alpha \sin \alpha \quad \text{(m/sec per year)} \qquad (10.51)$$

$$\approx 36.93 \text{ m/sec per year, for } i = 0$$

$$\Delta V_{Sun} = 40.17 \cos \gamma \sin \gamma \quad \text{(m/sec per year))} \tag{10.52}$$

$$\approx 14.45 \text{ m/sec per year, for } i = 0$$

where $\alpha$ is the angle between the orbital plane and the Moon's orbit, and $\gamma$ is the angle between the orbital plane and ecliptic.

The transverse acceleration caused by resonance with the $J_{22}$ term results in periodic motion about either of two stable longitudes: 75° and 255° East. If a spacecraft is placed at any other longitude, it will tend to orbit the closest of these two longitudes, resulting in East-West drift of up to 180° with periods of up to 900 days. Suppose a mission for a geostationary spacecraft specifies a required longitude, $l_D$. The change in velocity required to compensate for the drift and maintain the spacecraft near the specified longitude is

$$\Delta V = 1.715 \left| \sin \left( 2 \left( l_D - l_S \right) \right) \right| \tag{10.53}$$

where $l_D$ is the desired longitude, $l_S$ is the closest stable longitude, and $\Delta V$ is in m/sec per year.

For example, we can find the velocity change required for one year if the desired longitude is 60° west:

$$l_D = -60° \qquad\qquad l_S = 255°$$

$$\Delta V = 1.715 \text{ m/s per year}$$

After the spacecraft's mission is complete, several optioı·з exist, depending on the orbit. We may allow low-altitude orbits to decay and reenter the atmosphere or use a velocity change to speed up the process. We may also boost spacecraft at all altitudes into benign orbits to reduce the probability of collision with active payloads, especially at synchronous altitudes. Because coplanar velocity changes are more efficient than plane changes, we would normally apply tangential changes in velocity. Their magnitude would depend on the difference in energy of the two orbits. For example, the velocity change required to deorbit (drop perigee altitude to 0 km) a satellite in a circular orbit at radius, $r$, and velocity, $V$, is

$$\Delta V_{deorbit} \approx V \left( 1 - \sqrt{\frac{2R_E}{R_E + r}} \right) \tag{10.54}$$

We don't have to reduce perigee altitude to 0 km. If we choose a less conservative deorbit altitude, $H_{deorbit}$, we can determine the deorbit $\Delta V$ from Eq. (10.55) by replacing $R_E$ with $(R_E + H_{deorbit})$. Thus, choosing a 50-km deorbit altitude would reduce the FireSat $\Delta V_{deorbit}$ to 183 m/s. Note that only perigee is reduced in the deorbit burn. Reducing perigee to 100 to 150 km could result in several orbits over which apogee is reduced before the spacecraft reenters, which might not allow adequate control of the deorbit conditions.

# 10.6 Interplanetary Trajectories

Many missions of interest require the spacecraft to escape the Earth's influence and travel to other objects (planets, moons, comets, and asteroids) in the Solar System. The spacecraft may pass by the object, enter into an orbit around it, or land on its surface. By adding bodies to the problem, we've greatly increased its complexity. Thus, we can't solve the problem analytically. However, by splitting the trajectory up into parts, we can find an approximate solution that estimates the velocity budget and transfer time for a spacecraft traveling between two bodies in the solar system. We call this method of solution a patched-conic approximation. It divides the interplanetary trajectory into three segments: the hyperbolic escape from the first body, the elliptical transfer orbit about the Sun, and the hyperbolic arrival at the second body. We solve each segment of the trajectory with the previously defined two-body equations of motion and match the conditions from one segment of the trajectory to the next.

For Earth-orbiting spacecraft, we used the Earth-centered inertial (ECI) coordinate frame to describe and solve the equations. The ECI frame is inertial enough to solve Earth-orbiting trajectories, but we must define a new frame for interplanetary problems: the heliocentric, ecliptic coordinate frame shown in Fig. 10.13. By definition, the center of the frame is at the center of the Sun, and the fundamental plane is the plane of the ecliptic—the Earth's orbital plane. We choose the principal direction, $I$, as the vector pointing from the Sun to the vernal Equinox; $K$, as the direction perpendicular to the ecliptic plane; and $J$ in the ecliptic plane to complete our right-handed system.

## 10.6.1    Patched-Conic Approximation

The only forces we consider in this problem are the gravitational forces of the Earth, the Sun, and the target planet acting on the spacecraft. The spacecraft's equations of motion are defined by combining Newton's second law and his law of gravitation to get

$$\sum \vec{F} = m\ddot{\vec{r}} = \vec{F}_{\text{gravity Sun}} + \vec{F}_{\text{gravity Earth}} + \vec{F}_{\text{gravity target}} \tag{10.55}$$

We can't solve this problem directly, but we can solve it if we divide it into the three separate trajectories shown in Fig. 10.14. We define the *sphere of influence* (SOI) to be the point where the spacecraft transitions from the Earth's or the target planet's influence to the Sun's influence, or vice versa. Inside the SOI we consider the two-body equations of motion relative to the planet; outside the SOI, we consider the two-body equations of motion about the Sun.

We first solve the Sun-centered, elliptical transfer orbit from the Earth to the target planet. We must make several assumptions to solve this problem. We assume the Earth and the target planet are in circular, coplanar orbits and the transfer orbit is tangential to the orbits of the Earth and the target planet. In this

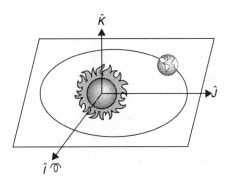

**Fig. 10.13.    Heliocentric-Ecliptic Coordinate System for Interplanetary Transfer.** Origin—center of the Sun; fundamental plane—ecliptic plane (plane of the Earth's orbit around the Sun); principal direction—vernal-equinox direction. [Sellers, 1994]

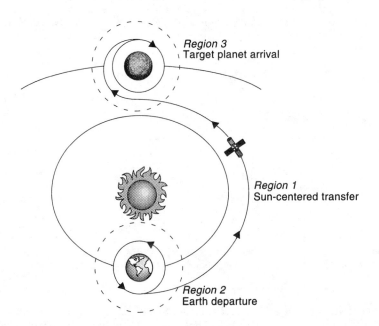

**Fig. 10.14.    Three Regions of the Patched-Conic Approximation.** We can break the trajectory for interplanetary transfer into three distinct regions in which the gravitational pull of only one body dominates the spacecraft. [Sellers, 1994]

case, the transfer orbit is a Hohmann transfer, and we use the methods described in Section 10.3.1 to solve for the velocities referenced to the Sun. We define these velocities below:

$V_{Earth}$ is the Earth's circular velocity about the Sun

$V_{transfer\ at\ Earth}$ is the velocity on the transfer orbit near the Earth

$V_{SOI-Earth}$ is the velocity of the spacecraft relative to the Earth at the SOI

$V_{target}$ is the target planet's circular velocity about the Sun

$V_{transfer\ at\ target}$ is the velocity on the transfer orbit near the target planet

$V_{SOI-target}$ is the velocity of the spacecraft relative to the target planet at the SOI

Table 10.8 lists the orbit radius, circular velocity, and gravitational parameter for the planets in the solar system and the gravitational parameter for the Sun.

**Table 10.8.**     **Parameters of the Solar System.** We use these parameters to find the various velocities on the interplanetary trajectory.

| Planet/Sun | Mean Radius (km) | Circular Velocity (km/sec) | Gravitational Parameter (km³/sec²) |
|---|---|---|---|
| Sun | -- | -- | $1.327 \times 10^{11}$ |
| Mercury | $57.9 \times 10^6$ | 47.87 | $2.232 \times 10^4$ |
| Venus | $10.81 \times 10^7$ | 35.04 | $3.257 \times 10^5$ |
| Earth | $14.95 \times 10^7$ | 29.79 | $3.986 \times 10^5$ |
| Mars | $22.78 \times 10^7$ | 24.14 | $4.305 \times 10^4$ |
| Jupiter | $77.8 \times 10^7$ | 13.06 | $1.268 \times 10^8$ |
| Saturn | $14.26 \times 10^8$ | 9.65 | $3.795 \times 10^7$ |
| Uranus | $28.68 \times 10^8$ | 6.80 | $5.820 \times 10^6$ |
| Neptune | $44.94 \times 10^8$ | 5.49 | $6.896 \times 10^6$ |
| Pluto | $58.96 \times 10^8$ | 4.74 | $3.587 \times 10^5$ |

The velocities of the spacecraft relative to the Earth and the target planet are called the *hyperbolic excess velocities*. They're equal to the difference between the planet's velocity and the spacecraft's velocity on the transfer orbit. We match these velocities with the spacecraft's velocity relative to the planets at the SOI to define the hyperbolic escape and arrival trajectories relative to the planets. If we assume the potential energy of the orbit relative to the planet is zero at the SOI, the specific mechanical energies of the hyperbolic trajectories are defined by the kinetic energies:

$$\varepsilon_H = V_{SOI}^2 / 2 \tag{10.56}$$

Because we assume two-body motion, the energy at the SOI defines the size of the orbit and the semi-major axis (Eq. (10.4)). Once we know the energy of the orbit, we can solve for the velocity at any other point in the orbit, relative to the planet, from the energy equation:

$$V = \sqrt{2\left(\varepsilon_H + \frac{\mu_{planet}}{r}\right)} \tag{10.57}$$

This value defines the burn-out velocity for departure. If the spacecraft is already in a parking òrbit, we can calculate the velocity change needed to start the transfer (see Fig. 10.15.).

$$V_{bo} = \sqrt{2\left(\varepsilon_H + \frac{\mu_{planet}}{r_{bo}}\right)} \tag{10.58}$$

$$\Delta V = \left| V_{bo} - V_{parking} \right| \tag{10.59}$$

Similarly, we can calculate the velocity at any point along the hyperbolic arrival trajectory relative to the target planet (see Fig. 10.16):

$$\varepsilon_H = V_{SOI}^2 / 2 \tag{10.60}$$

$$V_{retro} = \sqrt{2\left(\varepsilon_H + \frac{\mu_{planet}}{r_{retro}}\right)} \tag{10.61}$$

$$\Delta V = \left| V_{parking} - V_{retro} \right| \tag{10.62}$$

We apply the velocity change only at the target planet if we want the target planet to capture our spacecraft. If we don't decrease the velocity, the spacecraft continues on the hyperbolic trajectory and exits the target planet's SOI with the same hyperbolic velocity it had when it arrived at the planet's SOI. The two velocity changes approximate the velocity budget necessary to complete an interplanetary Hohmann transfer.

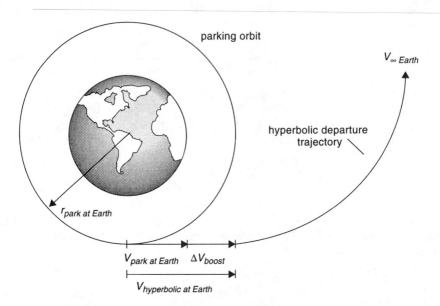

**Fig. 10.15.**    **Hyperbolic Escape from Earth.** This problem uses an Earth-centered perspective and requires the spacecraft to increase its velocity by an amount $\Delta V_{boost}$. [Sellers, 1994]

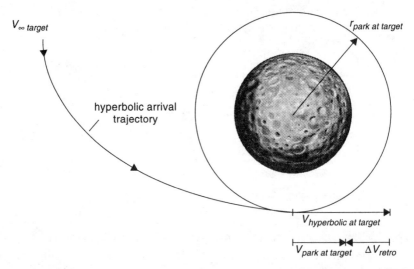

**Fig. 10.16.**    **Hyperbolic Arrival at a Target Planet.** This problem focuses on the target planet and requires the spacecraft to slow down by an amount $\Delta V_{retro}$ to drop into orbit. [Sellers, 1994]

The time of flight (TOF) to complete the transfer is about equal to the TOF of the heliocentric, elliptical transfer orbit. Because we assume a Hohmann transfer, the TOF is one half the period of the elliptical transfer orbit.

$$TOF = \pi \sqrt{\frac{(r_{Earth} + r_{target})^3}{8\mu_{Sun}}} \qquad (10.63)$$

Another critical step in solving the interplanetary transfer is ensuring proper phasing so the spacecraft meets the target planet when it arrives at the planet's orbit. We solve this problem as we did the rendezvous problem described in Sec. 10.3.3. Table 10.9 lists the steps for estimating the velocity budget and the time of flight for interplanetary, Hohmann transfers.

**Table 10.9.** **Interplanetary Hohmann Transfer.** We assume circular, coplanar orbits for the planets and solve the problem using the patched-conic method.

| Step | Where |
| --- | --- |
| 1. Determine velocity of Earth and target planet | Table 10.8 |
| 2. Determine velocities on transfer orbit at Earth and target planet | Eq. (10.4), use $\mu$ of the Sun |
| 3. Find $V_{SOI}$ at Earth and target planet | $V_{SOI} = |V_{transfer\ orbit} - V_{planet}|$ |
| 4. Determine hyperbolic escape/arrival trajectory | Eq. (10.56) |
| 5. Calculate $\Delta V_{bo}$ and $\Delta V_{retro}$ | Eq. (10.58) and (10.59) |
| 6. Calculate time-of-flight | Eq. (10.63) |

## 10.6.2   Swingby Trajectories

In the previous section we saw how to get from one planet to another using an interplanetary, Hohmann-transfer orbit. But suppose we choose not to retro-fire at the target planet to enter orbit or land on the planet. In this case, the spacecraft will continue along the hyperbolic trajectory relative to the target planet and exit the SOI with a relative velocity the same as when it entered the SOI, but the direction of the velocity vector will change. (Figure 10.17)

We see in the figure that we can increase or decrease the spacecraft's velocity relative to the Sun by passing behind or in front of the planet relative to the Sun. This maneuver is called a gravity-assist trajectory because it uses the planet's gravitational acceleration to change the spacecraft's velocity relative to the Sun. We can consider this velocity change "free" because it doesn't require any onboard propellant to complete the maneuver. The amount of velocity change available is a function of the arrival velocity and the turning angle, $\delta$, which is in turn a function of the closest approach distance to the planet. The magnitude of the arrival velocity

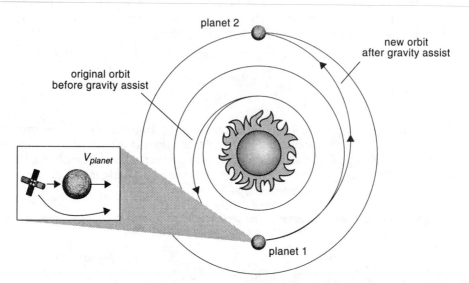

**Fig. 10.17.    Gravity Assist.** During a gravity assist, a planet pulls the spacecraft, changing its velocity with respect to the Sun and thus altering its orbit around the Sun. The planet's orbit also changes, but very little. [Sellers, 1994]

is determined by the transfer orbit and can't be changed, so the only design parameter is the closest approach distance. We calculate the turning angle from the following equations.

$$e = 1 + \frac{r_\rho V_h^2}{\mu_{planet}}$$  (10.64)

where $r_\rho$ equals distance of closest approach to planet.

$$\delta = 2\sin^{-1}\left(\frac{1}{e}\right)$$  (10.65)

If the spacecraft passes in front of the planet, its velocity relative to the Sun decreases and it drops into a smaller orbit. But if the spacecraft passes behind the planet, its velocity increases relative to the Sun. (Fig. 10.18)

Gravity-assist trajectories reduce the propellant requirements for interplanetary travel by changing the spacecraft's velocity relative to the Sun without using propulsive burns. The Galileo spacecraft used three gravity assists (one from Venus and two from Earth) to travel from Earth to Jupiter.

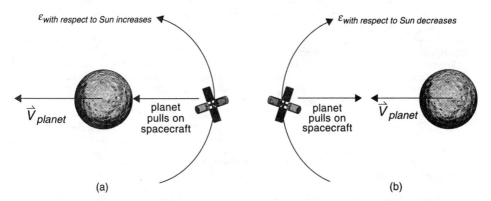

**Fig. 10.18.**    **Gravity Assist with the Spacecraft Passing Behind and in Front of a Planet.** (a) During a gravity-assist maneuver, a spacecraft's energy will increase with respect to the Sun if it passes behind the planet. (b) During a gravity-assist maneuver, a spacecraft's energy will decrease with respect to the Sun if it passes in front of the planet. [Sellers, 1994]

## 10.7 Mission Geometry

### 10.7.1   Coordinate Systems

To discuss mission geometry, we must define several coordinate systems. Different coordinate systems are best for different geometries. We specify a *coordinate system* by defining an origin and the orientation of a set of linearly independent axes. Coordinates are quantities, such as distances and angles, which allow us to specify a location within a coordinate system. We may use different types of coordinates within a single coordinate system.

Table 10.10 shows several coordinate frames used to calculate mission geometries.

### 10.7.2   Parameters of Mission Geometry

Many software packages are available to calculate parameters of mission geometry. We describe several packages in the next section. But often, we may want a quick method to approximate mission-geometry parameters such as elevation angle, coverage area, and maximum time in view. Figure 10.19 shows the relationship between selected parameters, the spacecraft's orbit, and the Earth.

First, we find the Earth's angular radius, $\rho$, from the orbit altitude, $H$

$$\sin\rho = \frac{R_E}{R_E + H} \tag{10.66}$$

**Table 10.10.    Coordinate Frames Used to Calculate Mission Geometries.** We specify coordinate frames by listing origin, fundamental plane (plane containing X and Y unit vectors), and the directions of the X and Z axis.

| Coordinate Frame | Origin | Fundamental Plane | X-axis | Z-axis | Comment |
|---|---|---|---|---|---|
| Earth-Centered Inertial (ECI) | Center of Earth | Equatorial plane | Vernal Equinox | Earth spin axis | X and Z axis are specified on a particular date. The most common are Jan. 1, 1950 or 2000 |
| Earth-Centered Fixed (ECF) | Center of Earth | Equatorial plane | Greenwich Meridian | North Pole | Coordinate frame rotates with the Earth |
| Topocentric Horizon (SEZ) | Point on or near surface of Earth | Local horizon | Local South | Local Vertical | Used to express vectors relative to station coordinates |
| Vehicle Velocity, Local Horizon (VVLH) | Center of spacecraft | Spacecraft local horizon | Direction of velocity vector projected into plane of local horizon | Nadir | Used for Earth observations |

Next, we must specify either the elevation angle, $\varepsilon$, the nadir angle, $\eta$, or the Earth-central angle, $\lambda$. If we specify the elevation angle based on some constraint on minimum elevation, we solve for the nadir angle, the Earth-central angle, and the distance from the spacecraft to the target using

$$\sin\eta \;=\; \sin\rho\cos\varepsilon \tag{10.67}$$

$$\lambda \;=\; 90° - \eta - \varepsilon \tag{10.68}$$

$$D \;=\; R_E\,(\sin\lambda/\sin\eta) \tag{10.69}$$

Or we can specify the nadir angle and then solve for the elevation angle (solving Eq. (10.67) for $\varepsilon$), central angle, and distance. We may also want to find the elevation angle and nadir angle for a given subsatellite point and target on the Earth's surface. For this problem, we start with the coordinates (latitude and longitude) of the subsatellite point and the target, as well as the altitude of the orbit. We again solve for the Earth's angular radius using Eq. (10.63). If the latitude and longitude of the subsatellite and target are $(\delta_T, L_T)$ and $(\delta_S, L_S)$ respectively, we solve for the Earth-central angle, nadir angle, elevation angle, and distance from the spacecraft to the target:

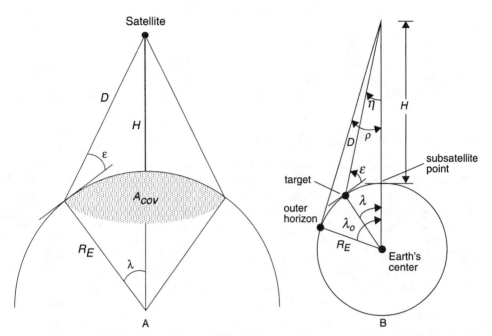

**Fig. 10.19.**     **Mission Geometry.** Figure A shows the coverage area for a given geometry, and B depicts the angular relationship between the spacecraft, the target, and the center of the Earth.

$$\cos\lambda = \sin\delta_S \ \sin\delta_T + \cos\delta_S \ \cos\delta_T \ \cos |L_S - L_T| \qquad (10.70)$$

$$\tan\eta = \sin\rho \ \sin\lambda / (1 - \sin\rho \ \cos\lambda) \qquad (10.71)$$

$$\cos\varepsilon = \sin\eta / \sin\rho \qquad (10.72)$$

$$D = R_E \ (\sin\lambda / \sin\eta) \qquad (10.73)$$

Note that $\lambda$, $\eta$, and $\varepsilon$ must always sum to 90°. Also, $R_E$ is the Earth's radius and equals 6378.14 km.

Next, we calculate the coverage area, $A_{cov}$, and the maximum viewing time, $t_{view}$, using the Earth-central angle and the orbit period. The coverage area is the part of the Earth's surface in view of the spacecraft sensor(s), with limits specified by either the nadir angle or the elevation angle. Maximum viewing time is the maximum length of time that a spacecraft is visible from a point on the surface of the Earth at a specified minimum elevation angle, assuming the spacecraft passes directly over the point on the Earth's surface.

$$A_{cov} = (2\pi R_e^2)\,(1 - \cos\lambda)$$

$$t_{view} = P\lambda_{max}/180°$$

where $P$ is the orbit period and $\lambda_{max}$ is the Earth-central angle, which corresponds to the minimum elevation angle.

Please read Larson and Wertz, Chap. 5, for a more complete discussion of mission geometry.

### 10.7.3    Mission-Geometry Software

Many software packages can help with various aspects of mission planning. These packages divide into commercial off-the-shelf (COTS), government off-the-shelf (GOTS), and custom-developed packages. COTS packages are characterized by a broad customer base which funds the support and further development of the product. GOTS packages are usually funded by one particular office of the government and then made available to other government offices and contractors. Custom-developed code is funded by one program and used only on that program.

Custom development is very expensive. At today's industry-standard rates of between \$50 and \$100 per line of code, a 100,000-line program would cost \$5–\$10 million to develop. Maintaining this package would cost four times that, or \$20–\$40 million, spread out over the project's life cycle. Of course, the program gets exactly what it needs.

GOTS packages have similar costs but they leverage that cost out among a broader set of users. This approach amplifies the benefit of the software but still leaves all the initial development and maintenance cost on one entity's shoulders. Eventually, the funding organization may not be able to bear it. Also, support for these packages usually doesn't exist or is inadequate.

COTS development can cost much less than government-developed code because commercial providers don't have to observe rigid regulations. Also, commercial providers are very sensitive to efficiency, whereas government developers aren't. Further, these reduced costs are spread out over a very wide customer base, each customer receiving the full benefit and yet only a tiny fraction of the costs. Also, if the commercial provider is successful, the product will be continually advanced, bringing more benefit to all the users. Although this software will be much cheaper, it may not meet 100% of the customer's requirements. But sometimes the customers get functions they may not have recognized they needed.

### References

Bate, Roger R., Donald D. Mueller, and Jerry E. White. 1971. *Fundamentals of Astrodynamics*. New York, NY: Dover Publications.

Battin, Richard H. 1987. *An Introduction to the Mathematics and Methods of Astrodynamics*. New York, NY: AIAA Education Series.

Danby, J. M. A. 1962. *Fundamentals of Celestial Mechanics*. New York, NY: Macmillan.

Escobal, Pedro R. 1965. *Methods of Orbit Determination*. Malabar, FL: Robert E. Krieger Publishing Co.

Kaplan, Marshall H. 1976. *Modern Spacecraft Dynamics and Control*. New York, NY: Wiley and Sons.

King-Hele, D. 1964. *Theory of Satellite Orbits in an Atmosphere*. London: Butterworths Mathematical Texts.

Larson, Wiley J. and James R. Wertz. 1992. *Space Mission Analysis and Design*. Second Edition. Netherlands: Kluwer Academic Publishers.

Pocha, J. J. 1987. *An Introduction to Mission Design for Geostationary Satellites*. Boston, MA: D. Reidel Publishing Company.

Roy, A. E. 1978. *Orbital Motion*. Bristol and Philadelphia: Adam Hilger.

Sellers, Jerry Jon. 1994. *Understanding Space: An Introduction to Astronautics*. New York, NY: McGraw Hill.

Wiesel, William E. 1989. *Spaceflight Dynamics*. New York, NY: McGraw-Hill Book Company.

# Chapter 11

# Communications Architecture

Gary M. Comparetto, *The MITRE Corporation*
Richard S. Davies, *Stanford Telecommunications, Inc.*

The *communications architecture* is the arrangement of components which transport and deliver data. We use communications systems to transmit and receive data to and from the spacecraft. In some cases, the communications system on the spacecraft is the payload, relaying information from one location to another.

Designing the communications architecture is complicated. It's often done ad hoc by various people who are technically competent in a wide range of disciplines. The whole process can intimidate a mission operations manager (MOM) on a particular program and, if unstructured, can result in a poorly documented, weakly justified end product that becomes more difficult to change as it ages. The inertia builds up, so to speak, and critical questions often appear heretical. Hence, the design is outdated by the scheduled operational date!

One school of thought contends that trying to design a large, complicated communications architecture is a lot like trying to invent new products, so it requires unstructured trial and error. Structure, they say, inhibits original thought and creativity and is much too cumbersome to incorporate dynamic design iterations and changes. Although some of these arguments may be valid, past experience in designing large communications architectures calls for a process by which we document architecture options, technically justify them, and review them periodically to ensure they remain cost-effective while meeting the stated system requirements.

Thus, in this chapter, we adopt a top-down, structured approach to designing communications architecture. As Fig. 11.1 shows, it has three main phases. In Phase 1, defining and evaluating requirements, the technical community identifies driving requirements for the communications system.

In Phase 2, which considers design options, we evaluate communications designs in terms of complexity, power, size, and other factors. The result is a set of building blocks, consistent with the requirements derived in Phase 1, that we can use to construct alternative architectures.

In Phase 3, forming the architecture, we construct alternatives based on the design characteristics and options from Phase 2 and the requirements from Phase 1. Phase 3 is iterative, so we often must revisit Phases 1 and 2 to determine whether each alternative can meet program requirements while remaining technically feasible.

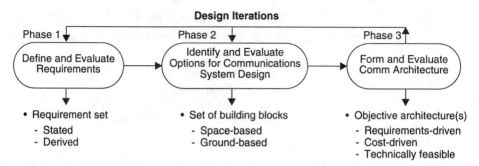

**Fig. 11.1.   Designing Communications Architectures.** Phase 1 of communications architecture design results in requirements that guide us in developing building blocks from which we can form alternative architectures.

Other factors may be beyond the control of communications-architecture engineers. Examples include existing assets, heritage, and politics, which may interfere with this structured approach. But even an imperfect process will be more likely to produce a well documented, technically justified architecture that cost-effectively meets requirements.

## 11.1 Defining and Evaluating Requirements

Figure 11.2 shows the key characteristics of requirements definition, during which we identify performance requirements, operational requirements, and system constraints. We then analyze them within program constraints such as existing assets, the technology baseline, program schedule, and allowable system cost. The following paragraphs overview the more common performance requirements, operational requirements, system constraints, and program constraints for communications-architecture designs.

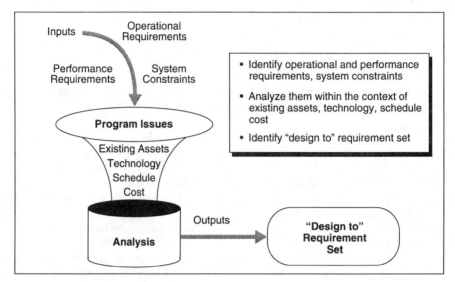

**Fig. 11.2.   Defining Requirements.** In this phase, we use performance and operational requirements plus system constraints to develop "design to" requirements.

### 11.1.1    Performance Requirements

The key performance requirements for satellite-communications (SATCOM)
systems are the data rate, bit-error-rate (BER), end-to-end (E/E) delay, and link
availability. Military SATCOM systems also require anti-jam (A/J) capability and
must operate through nuclear scintillation.

*Data rate* will largely dictate the design of two of the space/ground link's main
features—the high-power amplifier (HPA) in the ground terminal and the terminal antennas in the spacecraft and the ground terminal. The data rate is
proportional to the quantity of information per unit time transferred between the
spacecraft and ground station. The higher the data-rate requirement, the larger the
transmitter power (HPA power) or antenna size required to close the communications link. In general, the HPA's power requirement is a linear function of the data-
rate requirement, whereas the antenna-size requirement will vary as the square
root of the data-rate requirement. For example, if the data-rate requirement dou-
bles for a given space/ground link and the antenna sizes can't be changed, the
HPA power needed to support the new data-rate requirement would be double
the original design. But, if the HPA power can't change, the required antenna size
(transmit or receive) would need to increase by a factor of 1.414.

Often, we don't have the data-rate requirement for a particular system, so we
must derive it using other system requirements. For example, a system may have
to support message transfer at a rate of $x$ messages per second. Assuming the mes-

sages were each composed of $y$ bytes of information, with each byte represented by eight bits of data, the required data rate would be

$$R_d = x \text{ (messages/sec)} \cdot y \text{ (bytes/message)} \cdot 8 \text{ (bits/byte)} \qquad (11.1)$$

$$= 8xy \text{ bits/sec or bps}$$

So, for a required message rate of 25 messages/second and a message size of 80 bytes/message, the resulting data-rate requirement would be 16,000 bps or 16 kbps.

Finally, the concept of link duty factor affects the data rate. The *link duty factor* is the fraction of time that the communications link is actively supporting the data-rate requirement. For example, a duty factor of 20% for a ground terminal means the terminal is transmitting data only 20% of the time. A duty factor less than 100% could occur because the link must support multiple users or because something obscures the satellite's line-of-sight for part of its orbit. Whatever the reason, a link duty factor of less than 100% means we must transmit at rates higher than the data-rate requirement to maintain throughput. For example, if the required data rate is 20 kbps and the link has a duty cycle of 10%, the effective data-rate requirement would be 200 kbps to compensate for the link's not transmitting 90% of the time— 200 kbps for 10% of the time and 0 kbps for 90% of the time results in a throughput of 20 kbps.

*Bit-error-rate (BER)* is another key requirement that drives communications-architecture design. The BER is somewhat of a misnomer because the term really doesn't represent a rate but a probability of bit error. The BER is defined as the probability that a bit will be demodulated or decoded (if applicable) incorrectly. So, a BER of $10^{-5}$ would mean that, on average, one bit in 100,000 will be demodulated or decoded in error.

The BER is a function of the received signal-to-noise ratio (SNR). Depending on how we modulate and code, as well as the characteristics of the transmission channel, the relationship between the BER and the received SNR can be complex. The expression below provides the BER, or probability of bit error ($P_b$), as a function of the received SNR for binary-phase-shift key (BPSK) modulation if we use coherent demodulation (including phase information) but don't use forward-error-correction coding (FECC):

$$P_b = (1/2) \, erfc\left(\sqrt{\frac{SNR}{2}}\right) \qquad (11.2)$$

where $erfc(x)$ is the complementary error function, defined as

$$erfc(x) = \frac{2}{\sqrt{\pi}} \int_x^\infty e^{-t^2} dt \qquad (11.3)$$

where

$$x = \sqrt{\frac{SNR}{2}}$$

Several excellent references address this topic, including Sklar [1988], Ha [1986], Gagliardi [1984], Proakis [1983], Simon, et. al. [1985], and Spilker [1977]. The actual derivation of the BER equations is complex and involves probability, random variables, and stochastic processes. Papoulis [1965] has served as a benchmark reference in these areas for years. Finally, Fig. 11.3 shows the relationship between the probability of a bit error, or BER, and the received $E_b/N_o$, or SNR, assuming BPSK modulation with and without using several FECC techniques. As shown, the BER can vary significantly with only small variations in the received $E_b/N_o$.

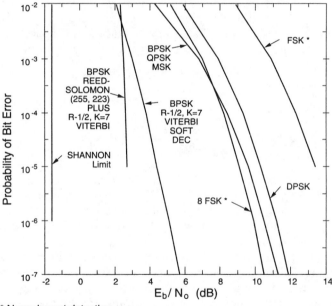

* Noncoherent detection

**Fig. 11.3.**     **Bit Error Probability as a Function of $E_b/N_o$ or Equivalently, SNR.** We can approach the theoretical performance (Shannon) limit by using error-correction coding. [Larson and Wertz, 1992].

*End-to-end (E/E) delay* in a communications architecture having multiple nodes can be thought of as a composite delay made up of separate delays in transmission, propagation, queuing, and processing. The *transmission delay* is simply the time it takes to transmit a block or packet of data. We compute it by dividing the data block size by the transmission data rate. For example, the transmission delay for an 80-byte block of data transmitted at a data rate of 9.6 kbps would be

$$\tau_{trans} = (80 \text{ bytes}) (8 \text{ bits/byte}) / 9.6 \text{ kbps} \tag{11.4}$$

$$= 66.7 \times 10^{-3} \text{ seconds} = 66.7 \text{ ms}$$

*Propagation delay*, also referred to as the speed-of-light delay, represents the amount of time the transmitted signal takes to traverse the path between the source emitter and the destination receiver. In calculating the propagation delay, we assumed the transmitted signal (an electromagnetic wave) travels at the speed of light, $c$.

As an example, assume a spacecraft is in a geostationary orbit. This orbit is common for communications satellites (COMSATs) because the spacecraft appears to be stationary from Earth. The altitude for a geostationary spacecraft is ~35,785 km. For the case in which the ground terminal is directly under the spacecraft, the source-to-destination distance is equal to the spacecraft's altitude. Consequently, we calculate the uplink propagation delay as follows:

$$\tau_{prop} = \frac{(\text{Source-to-destination distance})}{c} \tag{11.5}$$

$$\frac{35,785 \times 10^3 \text{ km}}{3 \times 10^8 \text{ m/s}} = 0.12 \text{ s}$$

The one-way, source-to-destination, propagation delay consists of uplink and downlink components and is 0.24 ($2 \times 0.12$) seconds. That's why we commonly refer to the propagation delay of a geostationary spacecraft as a quarter-second delay "per hop," where a hop represents the composite uplink and downlink paths from source to destination.

The *queuing delay* applies only to a communications architecture that employs packet switching. Queuing arises when a packet arrives at a processing node (satellite-based or ground-based) when that node is busy processing other packets. The queuing delay is the amount of time that transpires before the packet is processed and depends on several parameters, including the packet's mean arrival rate, the service time for each packet, and the use time—the fraction of time that the server is busy. Typically, the queuing delay is relatively flat and small (i.e., several milliseconds) for widely ranging use factors. Stallings [1985] fully discusses queuing delays along with today's common data-switching methods, including packet switching, message switching, and circuit switching.

The *processing delay* is a catch-all term that captures the delays associated with communications equipment. For example, in a typical spacecraft, the received signal undergoes frequency translation, signal amplification, and antenna routing before it's retransmitted. The processing delay represents the composite delay the signal experiences as it goes through the spacecraft or ground terminal. Typically, it's quite small and can be ignored but for complex processing in a space- or ground-based terminal, the delay may be significant. In these cases, we should consider the delay in designing the communications-system architecture.

*Link availability* in a communications system depends on the environment's affect on the transmission channel. The communications link is typically designed so the received SNR results in the required BER for a benign environment (no rain or other ill effects in the transmission channel). Link margin is then added to compensate for other expected operating conditions. For example, we'd add a link margin equal to the worst rain attenuation we anticipate on a given link to compensate for the rain-induced power fades.

The link margin for rain can be significant and directly depends on the requirements for link availability. Figure 11.4 shows the expected rain attenuation as a function of frequency and elevation angles for link availabilities of 98.0% and 99.5%. We generated the data shown in this figure using the Crane model—a set of tables and equations based on observed climatic data used to estimate rain attenuation [Crane, 1980]—and assuming the climate of the United States' northeastern section. The link availability for a specified rain attenuation will be higher in the Midwest and lower in the Southeast, compared to those shown in the figure. Note that the Crane rain model attempts to empirically model average weather conditions over a period of years; it doesn't account for higher or lower rainfall in a particular year.

As shown in Fig. 11.4, rain attenuation increases dramatically for space-to-ground frequencies over 10 GHz. Figure 11.4 also shows that rain attenuation increases rapidly as the antenna's elevation angle decreases below 20°. When operating at frequencies above 10 GHz, a good rule of thumb is to limit the spacecraft's minimum elevation angle to 20°, especially in high rainfall areas. But increasing the minimum elevation angle dramatically reduces the coverage area, which may mean the communications architecture won't meet other system requirements. We must balance all key performance and operational requirements.

*Jamming* can generally be described as the presence of unwanted signal components at the receiver of a friendly terminal—intentional (as in electronic warfare) or not (as in radio-frequency interference, or RFI). Intentional jamming is typically much more challenging to design against in a communications architecture.

*Anti-jam* (A/J) capability refers to a system's ability to overcome the harmful effects of a jammer. A requirement for A/J will mainly affect our design of the communications waveform (the choice of modulation format, coding, interleaving, and spread-spectrum technique) and the terminals (the antenna's design and the amplifier's power rating). Jammer types include

- Wideband (or broadband) noise jammer—The jamming signal's power is transmitted evenly over the communications signal's total bandwidth

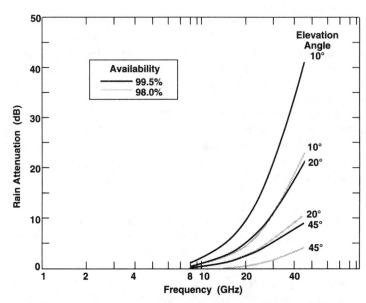

**Fig. 11.4.   Rain Attenuation Predicted by Crane Model for Climate Typical of the Northeastern United States.** Ground-station altitude = 0 km, latitude = 40°. For other cases, see Ippolito [1986]. [Larson and Wertz, 1992]

- Partial-band noise jammer—Similar to wideband noise jammers except that only part of the signal bandwidth is jammed
- Pulse jammer—A high-energy signal repeatedly transmits over a very narrow bandwidth
- Tone (multi-tone) jammer—A high-energy signal continually transmits over single (multiple) frequency tone(s)
- Repeat-back (or frequency-follower) jammer—The jammer intercepts the transmitted signal, determines the transmitted frequency, then re-transmits interference at this particular frequency. It must finish this process before the friendly signal changes to a new frequency.

The main goal for each of these jammers is to degrade the received BER at the friendly terminal. Various design techniques have been developed over the years to combat jamming:

- Spread-spectrum processing, which decreases the jammer signal's affect on the received friendly signal (e.g., frequency hopping [FH] and direct-sequence or pseudo-random noise [PRN])
- Antenna nulling, in which we detect a jammer's presence and sharply decrease the antenna's gain in its direction

- Forward-error-correction coding (FECC), which compensates for the additional error jamming causes

Torrieri [1992] and Simon [1985] treat jamming and jamming countermeasures in more detail.

Jamming isn't the only source of disturbance to the communication signals. For over thirty years, we've known that spacecraft signals suffer propagation disturbances in the natural ionosphere [Yeh, 1959]. Many reports state that these signals undergo even greater disturbances following high-altitude nuclear detonations (NUDETs) [King, 1980]. In both a natural or artificially disturbed (resulting from a NUDET) ionosphere, random fluctuations of electron density cause radio waves to be scattered as they propagate through such regions, producing random variations in the received signal's amplitude and phase. Such signal variations are called *scintillations* or fading. Middlestead [1987] discusses the extensive research on nuclear detonations in the ionosphere and their affect on signal propagation.

Clearly, these studies show that requiring a communications system to operate through a nuclear-disturbed region (a *nuclear-scintillation* requirement) will strongly affect choices on almost every characteristic of its design. An example would be modulation, coding, interleaving, and spread-spectrum choices in a signal-waveform design. We may have to employ satellite-to-satellite crosslinks in order to preclude transmissions through a nuclear-disturbed region. Or we may choose the spacecraft's orbits so they're less susceptible to the physical and communications-channel effects of NUDETs detonated at particular altitudes. The link margin may have to increase drastically to overcome the fading effects of scintillation, so the signal will have enough power at the receive terminal (received SNR) to meet the required BER. A nuclear-scintillation requirement will always dramatically affect our choices in developing the communications architecture; thus, we should validate it before moving the design forward.

## 11.1.2    Operational Requirements

The key operational requirements areas for satellite communications systems are security, standardization, backward compatibility, access, spacecraft orbit, spacecraft mobility, user-terminal characteristics, and channel characteristics.

Communications-system architectures have two main security components: communications security (COMSEC) and transmission security (TRANSEC). *COMSEC* means disguising the actual transmitted data and typically involves data encryption. *TRANSEC* means disguising the transmission method and normally involves generating security keys and variables that support spread-spectrum techniques. We can separate the communications-architecture design from TRANSEC if we focus our attention on the spread-spectrum technique rather than how we generate the security keys. We take this approach here. The following paragraphs briefly overview common techniques for COMSEC and TRANSEC.

We typically encrypt blocks or data streams. With *block encryption*, the plaintext (data to be encrypted) is segmented into blocks of fixed size and each block is then separately encrypted. With *data-stream* (or simply stream) *encryption*, blocks don't have a fixed size and each plain-text bit is separately encrypted. Block encryption is more common; the two most widely used techniques are the data-encryption standard and public-key encryption.

The data-encryption standard uses a 56-bit key and operates on 64-bit blocks. The same key is used to encrypt and decrypt the data. The algorithm employs a technique described as *product ciphering*, in which we repeatedly do simple encryption in tandem, resulting in an overall cipher system that is more powerful than any individual part. In effect, the complexity of the product of the encryption components is greater than their sum.

*Public-key encryption* techniques differ in that we use two different keys, one for encryption and one for decryption. In public-key systems, the encryption key can be publicly revealed without compromising the system's security. The encryption and decryption keys are developed in pairs and only the decryption key is kept secret. Deriving such sets of keys depends on complicated mathematical functions but involves the relatively simple premise that certain mathematics are easier to compute in one direction than in the reverse direction. For example, it's easier for most of us to calculate the cube of six in our heads than to calculate the cube root of 216! Sklar [1988] is one of the few texts that completely discuss data-encryption standard and public-key systems, as well as other encryption techniques.

Within this chapter, having to provide COMSEC in a communications architecture will alter the communications-terminal design by requiring more equipment to do the encryption. More equipment means increased terminal weight and cost. Processing delays may increase so the system can't meet the requirement for end-to-end delay. And the options for data routing may be limited if we must completely decrypt the full data packet at each intermediary communications node.

As stated, we'll focus our discussion of TRANSEC on the spread-spectrum techniques used to achieve transmission security rather than on generating security keys. The two main spread-spectrum techniques are frequency hopping (FH) and direct sequence (DS) or pseudo-random noise (PRN).

In an FH system, the frequency of the transmitted signal hops across a wide bandwidth (relative to the data bandwidth) at a rate fast enough to keep a frequency-follower jammer from intercepting the signal, determining the frequency, and retransmitting an interfering signal before the next hop. FH systems can use an effective spread bandwidth in the GHz range, with hop rates in the khps (kilo hops/sec) range.

In a DS or PRN system, the transmitted data are modulated with a very high-rate PRN code. The result is that the transmitted signal occupies a much wider instantaneous bandwidth than is actually needed to transmit the given data but at a lower average energy across the transmitted frequency spectrum. In this way, the transmitted signal appears as background noise to the unintended receiver. The spread-bandwidth limit for DS systems is about 100–300 MHz. Dixon [1976] and volume one of Simon [1985] are comprehensive references on the topics of spread-spectrum systems using FH and DS.

Standardization supposedly

- Maximizes a system's operational effectiveness and minimizes its life-cycle costs
- Enhances the system's performance, interoperability, maintainability, portability, reliability, and availability

An additional "ility" sometimes given in number two above is scalability, which is a system's ability to get bigger easily without having to scrap the current system.

Standardizing means we must

- Limit excursions from current design techniques and do detailed technical justifications if we vary them
- Emphasize existing standards (government and commercial) and processes
- Emphasize commercial, off-the-shelf (COTS) equipment and non-developmental items (NDI)

We should certainly consider standardizing when we develop an architecture, but we must be sure it doesn't become the primary driver. If it does, we may not be able to meet other stated performance and operational requirements.

Standardizing can cost us more downstream if we have to extend an architecture/infrastructure into the future. In most cases, to reduce life-cycle costs, we have to accrue certain costs up front by incorporating new equipment, processes, and techniques that will save money later. The up-front development and acquisition costs can be substantial, and often planners will rule against an apparently expensive architecture that will have lower life-cycle costs than the alternatives.

In most cases, communications architectures aren't built from scratch. Instead, we extend an existing infrastructure of some sort into the future. If this is the case, and depending on the size of the infrastructure, all types of ground and space assets may not have reached their useful lifetime. Requiring the users to procure new equipment so it will conform with the new communications architecture would cost too much. Thus, we have to develop a plan that addresses using existing equipment, capabilities, and facilities throughout their lifetimes. In other words, we have to look at backward compatibility.

Backward compatibility can significantly burden the architecture designer and often precludes the best solution. For example, it's almost always possible to develop a new communications architecture that meets all stated requirements (except for, perhaps, cost!) using a clean sheet of paper, but we may not be able to meet all performance requirements if we must integrate old equipment into the new architecture. One way to address backward compatibility is to develop a solid plan that phases in the newly developed architecture to match major program milestones for assets in the existing infrastructure.

If we had unlimited money and resources, we'd provide access to all assets and allow everyone to use them as they saw fit. However, in the real world, there are typically more users than assets, so we must somehow limit access. One common way to do so is through demand-assigned multiple access (DAMA). DAMA allocates access of a channel by geographically dispersed communications terminals according to demand. In a DAMA scheme, the user isn't guaranteed use of a given resource, but if it's designed well, we'll seldom encounter an asset-busy signal. In a sense, the phone system employs DAMA concepts. Normally, one simply picks up the receiver and dials a number to access it. But during periods of very high demand (e.g., Mother's Day!), we may not get through because of the high demand and lack of available resources (phone circuits).

The government is quite serious about using DAMA on voice circuits. Two military standards, MIL-STD-188-182 ("Air Force DAMA") and MIL-STD-188-183 ("Navy DAMA"), are under development. They describe the DAMA waveforms for 5 kHz and 25 kHz ultra-high-frequency (UHF) channels (0.3 to 3 GHz), respectively. The government has mandated compliance with these standards by the end of fiscal year 1996. There's also an effort under way to develop a DAMA-waveform standard for super-high-frequency (SHF) applications (3 to 30 GHz). A working group formed in July, 1992, to establish an SHF standard for DAMA, with representatives from the Air Force, Army, Navy, and other government agencies, as well as support contractors. The bottom line is that communications architectures required to provide access through UHF or SHF voice channels will be driven by the above military standards. Thus, we have to be sure the architecture meets these standards.

Spacecraft orbits include three key parameters that will influence communications-architecture design [Gagliardi, 1984]:

1. *Coverage area*, $A_{cov}$—The part of the Earth's surface that can receive the spacecraft's transmission assuming the ground terminals are operating at an elevation angle larger than a prescribed minimum. (Note: The elevation angle of a ground terminal is the angle defined between a line tangent to the Earth at the ground terminal's location and the pointing direction to the spacecraft).

2. *Slant range*, $S$—The line-of-sight distance from a fixed point on the Earth to the spacecraft

3. *Viewing time*, $\tau_{view}$—The length of time that a spacecraft is visible from a ground terminal assuming a prescribed elevation angle

The relationships between the coverage area, elevation angle, slant range, and viewing time are discussed in Sec. 10.7.

Characteristics of the spacecraft's orbit will directly affect the design of the communications architecture. We may choose higher spacecraft altitudes to increase the coverage area per spacecraft and viewing times for a given minimum elevation angle. However, the cost to place a spacecraft into orbit increases dramatically as the altitude increases and may outweigh the potential benefits of increased coverage area and viewing time. Instead, we may need to place more spacecraft into lower orbits, with each one having a smaller coverage area and viewing time relative to the geostationary spacecraft. But together, as a composite constellation, they'll be able to meet the requirements at lower cost for the overall system.

*Spacecraft mobility* is the ability to reposition a spacecraft while it's in orbit. The main way to actively reposition a spacecraft is through thrusting engines. In fact, the spacecraft uses the same abilities to initially achieve its final orbital location. A spacecraft must reposition after orbit insertion to

- Compensate for the effects of natural drag and other orbital perturbations
- Move into a different orbital location in order to meet updated mission requirements (e.g., move a launch-sensing spacecraft over a specific area of operations)
- Avoid collisions with other orbiting spacecraft and space debris
- Move an in-orbit spare to operational position
- Survive by avoiding direct-collision weapons, enhanced radiation belts (natural or NUDET-induced), high-energy laser or radio-frequency (rf) weapons, etc.

Mobility requirements for a spacecraft can significantly affect our design, typically translating into increased spacecraft size to support the storage of thruster fuel. Increased spacecraft size always translates into high cost because the cost to deliver a spacecraft into a particular orbit depends strongly on the launch weight. Thus, we have to substantiate the mobility requirements for a given spacecraft system and address anticipated spacecraft operations throughout end-of-life (EOL). (See Chap. 10)

*User terminals* are either fixed, mobile, or transportable. Fixed users typically employ large, stationary ground terminals. They aren't easily relocated and represent a large amount of the communications system's overall cost. However, they can put through a lot of data because of the high achievable gains of their antenna designs.

Mobile users typically employ small terminals that can communicate while moving and that a user can easily transport. Examples of this type of terminal include cellular phones, airplanes, and car radios. Note that larger mobile terminals do exist to support maritime communications between ships (mobile, but relatively slow) and shore stations.

Transportable users represent the middle ground. That is, they employ medium-size terminals that are too large to support communication links while moving but are small enough to be packaged straightforwardly and relocated to another ground terminal. Transportable terminals are common in the military, where theaters of operation move quickly during a conflict, and the communications system must keep up with the advancing columns.

Each user category has associated characteristics which affect the communications-architecture design. For example, fixed terminals normally employ large antennas that support requirements for high communications throughput (a plus) but are limited in their ability to track spacecraft in low-Earth orbit (a negative). As another example, mobile users, because of their small size and mobility, normally employ omni-directional antennas which are simple to design and manufacture (a plus) but which can't generate any appreciable signal gain (a negative). For omni-directional antennas at ground terminals, the spacecraft must compensate for this lack of gain by increasing the size of its transmit antenna, increasing the output transmit power level, or decreasing the slant range between the spacecraft and the mobile user (low-Earth orbits versus geostationary orbits). In summary, we must consider the characteristics of the user's terminal when forming options for the space-based architecture.

*Channel characteristics* directly affect the communications-system architecture in almost every aspect—from the modulation format, to the antenna design, to the number and location of ground terminals and spacecraft. The *channel* is the medium through which we communicate. For conversations, the channel is air. For local area networks, it's twisted cable. For satellite communications, it's a combination of atmosphere and space. Channel characteristics can be natural or made by humans. Typically, a channel comprised of natural phenomena is referred to as a *benign channel* (although communications in a benign channel can be quite challenging depending on the natural phenomena!). In contrast, a channel comprised of human-made phenomena (e.g., nuclear detonations and electronic jamming) is referred to as a *stressed channel*. We've discussed the main stressed-channel characteristics—A/J capability and nuclear scintillation—in Sec. 11.1.1, so we'll focus on the benign channel and its natural phenomena, including rain, atmospheric absorption, sand or dust, and foliage.

*Rain attenuation* has long been recognized as a principal cause of unwanted signal loss in satellite communications (SATCOM) systems operating from 3 to 300 GHz. Signal attenuation of tens of dB is possible depending on the specific characteristics of the rain event. So we have to account for signal attenuation from rain in developing the communications architecture to ensure the communications link works properly. Because we discussed rain in Sec. 11.1.1 in terms of link availability we won't discuss it further.

The one-way attenuation due to atmospheric absorption depends on operating frequency, the Earth's surface temperature, humidity, and the Earth-station antenna elevation angle. The International Radio Consultative Committee (CCIR) of the International Telecommunications Union has calculated theoretical one-way attenuation for a United States standard atmosphere. This calculation is for July at 45° North latitude and for frequencies between 7 and 50 GHz. From this data,

Schwab [1980] has approximated the vertical (zenith) path attenuation (in dB) with the following polynomial expression:

$$L_{atm}(f, zenith) = e^{\left(k_1 f + k_2 f^2 + k_3 f^3 + k_4 f^4\right)} - 1 \qquad (11.6)$$

where $f$ is the operating frequency in GHz, zenith refers to a vertical (90° elevation) path, and the coefficients ($k_1, k_2, k_3,$ and $k_4$) are given in Table 11.1. Values of atmospheric attenuation for elevation angles ($\varepsilon$) other than the antenna's zenith pointing angle are given by

$$L_{atm}(f, \varepsilon) = \frac{L_{atm}(f, zenith)}{\sin(\varepsilon)} \qquad (11.7)$$

**Table 11.1.** **Coefficient Values for the Schwab Polynomial Approximation.** We can calculate the atmospheric absorption through Eq. (11.6) for the given frequency ranges.

| Coefficients | 7 GHz $\leq f \leq$ 22 GHz | 22 GHz $< f \leq$ 50GHz |
|:---:|:---:|:---:|
| $k_1$ | −0.00617564 | 0.13030320 |
| $k_2$ | 0.00432368 | −0.00816987 |
| $k_3$ | −0.00044445 | 0.00015648 |
| $k_4$ | 0.00001358 | −0.00000074 |

Using Eqs. (11.6) and (11.7), we generated the atmospheric absorption values in Table 11.2 for several elevation angles and for frequencies ranging from 20 to 50 GHz.

**Table 11.2.** **Atmospheric Absorption (in dB) as a Function of Frequency and Antenna Elevation Angle ($\varepsilon$).** The atmospheric absorption generally increases with increasing frequency and decreasing elevation angle.

| $f$ (GHz) | Elevation Angle, $\varepsilon$ (deg) | | | |
|:---:|:---:|:---:|:---:|:---:|
| | 90° | 30° | 20° | 10° |
| 20 | 0.25 | 0.50 | 0.73 | 1.44 |
| 30 | 0.20 | 0.40 | 0.58 | 1.15 |
| 40 | 0.30 | 0.60 | 0.88 | 1.73 |
| 44 | 0.60 | 1.20 | 1.75 | 3.46 |
| 50 | 1.79 | 3.58 | 5.23 | 10.31 |

Most studies to date concerning the attenuation of SATCOM signals from *dust* have assumed the dust results from a nuclear detonation. However, experience from the Desert Storm operations has shown we must consider signal attenuation from local dust storms in designing communications architectures. SATCOM signals incident on dust particles in the atmosphere undergo some absorption and scattering, depending on the particles' size, shape, and complex dielectric constant as well as the signal's wavelength (or frequency). A thorough discussion of dust attenuation is beyond the level of this text. Refer to Comparetto [1993] for an overview of this topic. Comparetto shows the signal attenuation from dust is about 0.5 dB/km at 45 GHz and 0.2 dB/km at 20 GHz. Clearly, a requirement to operate through dust in the benign channel may significantly reduce the communication link's performance, depending on the operating frequency, and may require a lot of link margin in the communications architecture to overcome this effect.

Of all the signal-propagation effects in a benign channel, attenuation from foliage is the most difficult to quantify accurately. Widely varying types and density of foliage makes estimating attenuation equally variable. Most studies of propagation through foliage have been concerned with horizontal propagation (parallel to the ground) and have considered only frequencies below 1250 MHz (see, for example, Tamir [1977], Horwitz [1979], and Tewari [1990]). In this region, the leaves are small compared to the wavelength and the forest is treated as a homogeneous layer above the Earth. But we can't assume homogeneity in the higher frequency regimes (EHF or 30-300 GHz), where the leaves and the spaces between the leaves are large compared to the signal wavelength.

As one might imagine, the inconsistency in characterizing foliage from one study to another has made empirical models inconsistent. Comparetto [1993] broadly overviews various empirical models and data on foliage attenuation generated over a wide frequency range. That study determined foliage attenuation was difficult to quantify accurately. But it also showed that foliage attenuation could severely degrade, and potentially preclude, satellite communications regardless of the available link margin. The signal attenuation was 3 to 4 dB *per meter* of foliage in the 44/20 GHz operating regime but much less severe at frequencies below 1 GHz. Comparetto concluded that communications through any length of foliage should be discouraged at high frequencies (above 1 GHz) and that operational adjustments and procedures, not link-margin design, would be more likely to ensure communications through foliage. Thus, we might consider alternate frequencies or not using satellite communications if we have to operate through foliage.

### 11.1.3  System Constraints

Many aspects of the communications-architecture design can be influenced by system constraints, such as

- Maximum antenna size at the ground terminal or spacecraft (based on mobility and maneuverability requirements, costs, limits on real-estate, and requirements for launch-vehicle integration)
- Maximum transmit power (based on requirements for cooling the high-power amplifier, available technology, and environmental limitations)
- Maximum spacecraft weight (based on launch costs and the launch vehicle's lifting capacity)
- Restrictions on basing ground stations at overseas locations (based on survivability concerns, operational cost, and length of use)
- Limitations concerning transmission media (based on existing infrastructure using a particular media)

These constraints are only samples of the types that could affect the communications-system architecture. In the early phases of design, we must identify all system constraints so we can bound the architecture's trade space. By doing so, we won't consider non-viable alternatives and will shorten the process for designing a communications architecture.

## 11.1.4   Program Constraints

Cost, schedule, and the state of current and projected technology also affect our design. For example, the best ways to meet performance and operational requirements might end up being untenable when we consider cost. One example of a system for which cost is becoming increasingly important is the Milstar COMSAT system. It was designed to operate through severe nuclear and electronic jamming environments. Planners emphasized meeting these and other operational and performance requirements without rigidly constraining costs. The result is a military-communications system that meets its operational and performance requirements very well—but at a high cost. The cost of this system is undergoing intense scrutiny because of cuts in defense spending. In a sense, the ground rules were changed in mid-stream by shifting emphasis from meeting requirements to minimizing cost. As a result, Milstar is probably suffering unfairly.

Schedules also drive our design. For example, stipulating short- or near-term schedules for reaching initial operational capability (IOC) on a particular system may drive the system designers to use existing and proven technology to reduce risk. As a result, the system may not be able to meet certain performance requirements without future upgrades to the system. This approach can lead to short-term solutions for long-term problems, thereby fueling the "stove pipe" method commonly found in communications-system architectures. Make sure short-term solutions don't preclude including enhanced technology as the system matures.

As mentioned earlier, a strong constraint on communications architectures is the requirement to incorporate an existing infrastructure. Of course, this constraint depends on the size of the infrastructure, it always drives a number of design deci-

sions during development. An example is the Air Force's Satellite Control
Network (AFSCN). Efforts are under way to develop a satellite-control architecture for the DoD that responds to Air Force, Army, and Navy requirements. This,
in itself, is a challenging task given the services' diversity. Worse yet, each service
does some degree of spacecraft control and therefore has some infrastructure to
accomplish their missions. In some cases, this existing infrastructure is substantial.
For example, the AFSCN has been operating for nearly 30 years and represents a
yearly commitment of more than $500M. The common-user segment of the
AFSCN includes nine remote tracking stations, 16 space/ground antennas, two
major operational control nodes, and various dedicated and support facilities and
assets. Obviously, all future spacecraft-control architectures must address the current AFSCN infrastructure if they're to be considered fiscally sound and
technically doable. Although using this infrastructure may save money, it may
also keep the communications architecture from meeting operational and performance requirements.

Finally, the proposed lifetime of the communications-system architecture
drives both cost and design decisions in its development. For example, if the proposed lifetime of a particular spacecraft exceeds the projected lifetime of the
communications system's components, we may need to use redundant components to meet requirements for end-of-life (EOL) reliability. In many spacecraft
systems today, we use redundancy for traveling-wave tube amplifiers, onboard
power supplies, modems, multiplexers, and frequency up/down converters to
meet the lifetime requirements. Costs therefore go up because of the number of
units and the cost of launching their greater weight and volume. Also, the specified
lifetime will affect the design choices made in terms of projected technology. For
example, if the system lifetime includes several generations of the communications
architecture's key components (e.g., computers), the architecture must be upgrad-
able without requiring us to redesign it overall.

## 11.2 Evaluating Design Options

Figure 11.5 shows the key characteristics of this phase, in which we evaluate
the design characteristics of the ground terminal and spacecraft payload, together
with the performance and operational requirements identified in the require-
ments-definition phase. To do so, we must consider the system constraints (which
typically include system complexity, power, size, and other factors) while identifying options for the communications architecture. We keep the most favorable
options as the architecture building blocks we'll use to develop the architecture.
The following paragraphs overview the more common options related to the
spacecraft and ground terminal.

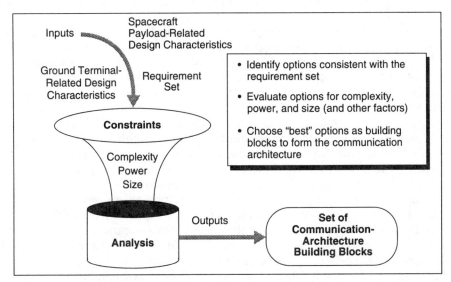

**Fig. 11.5.    Evaluating Design Options.** In the design option phase, satellite payload and ground terminal based design characteristics are combined with the Phase 1 requirement set in order to develop a set of building blocks for the architecture.

## 11.2.1    Spacecraft

Most communication satellites today use repeater transponders (more commonly referred to as *bent-pipe transponders*). In a *bent-pipe configuration*, the spacecraft receives the uplink signal, translates the frequency of the received signal to the appropriate downlink frequency, amplifies the signal, and then retransmits the signal on the downlink. The bent-pipe transponder incorporates proven technology and is highly reliable and straightforward to use. However, bent-pipe transponders do have several key disadvantages when the spacecraft receives multiple signals at the same time.

First, bent-pipe transponders are inherently non-linear. That is, the amount of amplification varies depending on the input-signal level. Normally, we want to operate the transponder in the linear region, where a step change in input-signal level will result in a corresponding step change in output-signal level. We do so to avoid generating intermodulation (IM) products and to reduce signal-suppression effects. The IM products interfere with the uplink signals and, essentially, rob part of the transponder's available transmit power.

A spacecraft's transponder, when operating in the non-linear regime, will tend to amplify stronger signals more than weaker signals, in a sense suppressing the transponder's output of the weaker signals. We call this effect *signal suppression*. To make sure the transponder operates linearly, we back off its output power about 3 to 6 dB from the maximum potential output power, thus reducing the transmitter's power output to roughly one-half to one-fourth of the maximum (saturated) power. Comparetto [1989 and 1990] further discusses IM products and signal-suppression effects in non-linear transponders.

An alternative to the bent-pipe transponder involves onboard processing using a *regenerative transponder*. In this case, the signal is demodulated on board the spacecraft and then routed to the appropriate downlink modulator/transmitter or antenna beam.

When ground terminals at the source and destination are within line-of-sight (LOS) of the same spacecraft, communications is straightforward, as shown in Fig. 11.6A. But if they're not, we need a data relay to support *source-to-destination connectivity*. The two main ways to relay data in satellite communications are ground-terminal relay (Fig. 11.6B) and satellite crosslinks (Fig. 11.6C).

Ground-terminal relay involves a double hop and an increased delay which can affect certain types of communications (e.g., the round-trip propagation delay is about one second, which is quite noticeable in voice communications). Ground terminals may also drive up costs, depending on their design and location. In many cases, we need an overseas ground terminal, which requires the host nation's approval before it can operate. Despite these drawbacks, ground-terminal relays are a proven and reliable way to support connectivity beyond line-of-sight in satellite-communications architectures.

To date, very few satellite crosslinks have been incorporated in communications architectures. The reasons vary but typically include

- Lack of technological maturity
- Limited availability of the required space-qualified hardware
- Difficulty of space-based antenna tracking, especially for low-Earth orbits
- Increased weight to the spacecraft payload as a result of the satellite crosslink system. (This weight increases launch cost and may exceed the launch vehicle's ability to insert the spacecraft into orbit.)
- Favorable economics for the ground terminal relay to date

However, satellite crosslinks are becoming increasingly popular as their technologies mature and their costs are becoming more competitive with ground-terminal relays. For example, they'll provide source-to-destination connectivity in the IRIDIUM system under development by Motorola. IRIDIUM is a space-based, mobile, satellite-communications system that will employ 66 spacecraft in low-Earth orbit. The constellation consists of six orbital planes, with 11 spacecraft per plane. Each spacecraft will operate four 25-Mbps satellite-satellite crosslinks in the 22.55 to 23.55 GHz frequency band. The expected operational date for IRIDIUM is in 1998, but regulatory and licensing issues may delay it. We'll see whether or not Motorola will be able to overcome this architecture's significant technical challenges at a reasonable cost. Comparetto [1993 and 1994] discusses the IRIDIUM

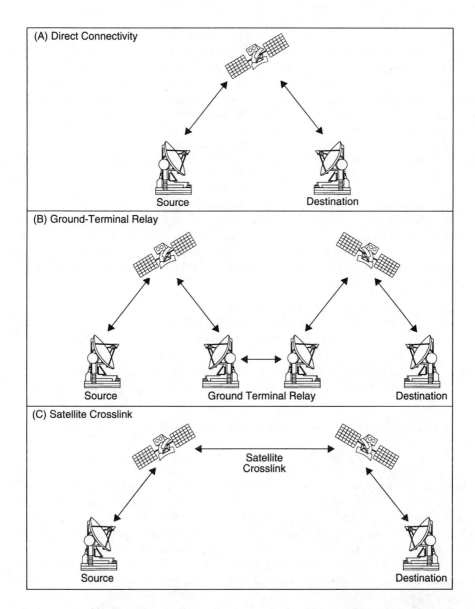

**Fig. 11.6.    Various Configurations of Sources and Destination for Connectivity.** Direct transmission, ground-terminal relay, or satellite crosslinks can satisfy source-to-destination
connectivity.

system, as well as two competing designs referred to as Odyssey (by TRW, Inc.) and Globalstar (by Loral/QUALCOMM Inc.).

Requiring a ground terminal to support continuous spacecraft-to-ground connectivity drives the choice of orbits in a communications architecture. The only way truly to support continuous connectivity is through geostationary orbits. In a geostationary orbit, the inclination is nominally zero degrees (positioned at the equator), and the period is one sidereal day (23 hours, 56 minutes, 4 seconds). It is expensive to insert a spacecraft into a geostationary orbit, and the launch cost increases dramatically with increasing weight. One other way to support "quasi" continuous connectivity between the ground and a spacecraft is through a network of ground stations, by which at least one ground terminal is in view of the spacecraft at any given time. We use the term "quasi" because the individual spacecraft-to-ground contacts are periodically broken. We'd need a geographically dispersed global network for truly continuous coverage.

If continuous connectivity isn't necessary, we may be able to employ a single ground station on the continental United States (CONUS) and support spacecraft-to-ground connectivity whenever the spacecraft is in view of the ground station as it completes it orbital journey. This approach would save a lot of money over the continuous-coverage case because

- We don't need overseas ground stations, thereby reducing operating cost and eliminating host-nation approval
- Launch cost per spacecraft decreases (no need to use a geostationary orbit)
- Flexibility increase for choosing the launch vehicle because more launch vehicles can insert a spacecraft into a sub-geostationary orbit

In summary, we must carefully validate a requirement for continuous spacecraft-to-ground connectivity because it will increase the cost and complexity of the SATCOM's architecture.

A key to operating both types of transponders discussed earlier is their amplifier. Three categories of transponder amplifiers are the traveling-wave tube amplifier (TWTA), the solid-state amplifier, and the klystron. Of the three, TWTAs are the most extensively developed. Their theory is also well understood, and they've been used successfully in all types of space missions. For these reasons, TWTAs have emerged as the universal choice in power amplifiers for Earth stations and satellites operating at 30 GHz and beyond.

The *klystron* is used mainly in ground-terminal designs, which require high transmission powers (kilowatts) at high frequencies (>30 GHz). They are large and typically require cooling to reduce the system noise temperature that results from random electron motion. For these reasons, they're seldom used onboard a satellite.

In recent years, research and development has accelerated on *solid-state amplifiers* (SSA). The two most common types of SSAs include Gallium Arsenide, field-effect transistor (GASFET) amplifiers and impact-avalanche transit time (IMPATT) diode amplifiers. Solid-state amplifiers aren't as efficient as TWTAs or klystrons; power-combining techniques are typically required to attain output power of 10–20 watts for operating frequencies up to 44 GHz. But solid-state amplifiers offer significant advantages over TWTAs and klystrons, including decreased size and weight, ruggedness, compactness, and longer life. Of course, all these advantages translate into cost savings for the system. We expect solid-state amplifiers to become the amplifier of choice for satellite applications as their output power, reliability, and component availability increase over time.

### 11.2.2 Design Options for Ground Terminals

Some missions may require multiple uplinks or downlinks, especially for communications-system architectures that integrate a number of spacecraft and ground stations into a single network. In such systems, it costs less to share the limited amount of satellite link capacity among users. Figure 11.7 shows three basic techniques for sharing link capacity: frequency-division multiple access (FDMA), time-division multiple access (TDMA), and code-division multiple access (CDMA).

In *FDMA systems*, a set of Earth stations transmit uplink carriers to be relayed simultaneously by the spacecraft to various downlink Earth stations. Each uplink carrier is assigned a frequency band within the available rf bandwidth of the spacecraft. In a bent-pipe transponder, the entire rf frequency spectrum appearing at the spacecraft input is frequency-translated to form the downlink. The destination ground station receives the source ground's station signal by tuning to the proper band in the downlink spectrum. FDMA represents the simplest way to achieve multiple access. The required system technology and hardware are readily available in today's communication market [Gagliardi, 1984].

In *TDMA systems*, a single time slot in each time frame is assigned to a single input channel. A digitized input signal is sampled and stored in buffer memory. These samples then transmit as short bursts within the assigned time slots. The bit rate during the burst is high, therefore requiring a high peak transmitter power. At the receiver, the samples are sorted, stored, and then read out at the original rate. These samples are then converted to an analog signal, if required, and smoothed to obtain a replica of the original input signal. If the spacecraft uses a multiple-beam antenna, we may use a switching matrix on the spacecraft with TDMA to route each time-slot burst to the desired downlink antenna's beam. NASA's Advanced Communications and Technology Satellite (ACTS) system uses this technique, which is known as Satellite Switched (SS) -TDMA [Naderi, 1988]. The FLTSATCOM EHF package (FEP) is a system that employs FDMA on the uplink and TDMA on the downlink [McElroy, 1988]. This method requires onboard processing of the signal in the spacecraft.

In *CDMA systems*, the data is first phase-modulated by a carrier (normally using binary phase-shift keying—BPSK or quadrature phase-shift keying—QPSK

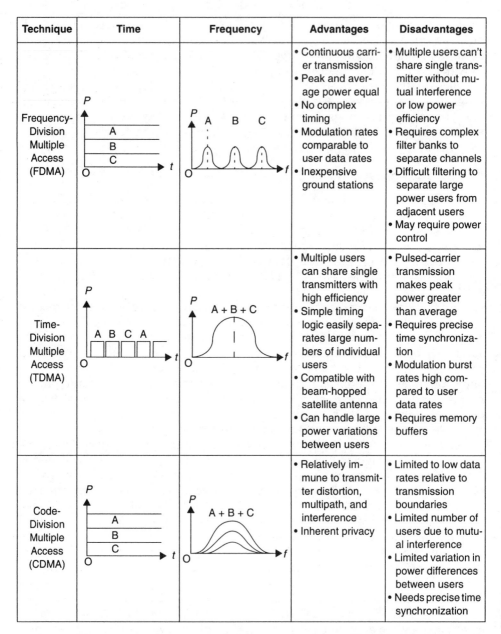

| Technique | Time | Frequency | Advantages | Disadvantages |
|---|---|---|---|---|
| Frequency-Division Multiple Access (FDMA) | | | • Continuous carrier transmission<br>• Peak and average power equal<br>• No complex timing<br>• Modulation rates comparable to user data rates<br>• Inexpensive ground stations | • Multiple users can't share single transmitter without mutual interference or low power efficiency<br>• Requires complex filter banks to separate channels<br>• Difficult filtering to separate large power users from adjacent users<br>• May require power control |
| Time-Division Multiple Access (TDMA) | | | • Multiple users can share single transmitters with high efficiency<br>• Simple timing logic easily separates large numbers of individual users<br>• Compatible with beam-hopped satellite antenna<br>• Can handle large power variations between users | • Pulsed-carrier transmission makes peak power greater than average<br>• Requires precise time synchronization<br>• Modulation burst rates high compared to user data rates<br>• Requires memory buffers |
| Code-Division Multiple Access (CDMA) | | | • Relatively immune to transmitter distortion, multipath, and interference<br>• Inherent privacy | • Limited to low data rates relative to transmission boundaries<br>• Limited number of users due to mutual interference<br>• Limited variation in power differences between users<br>• Needs precise time synchronization |

**Fig. 11.7.    Comparison of FDMA, TDMA, and CDMA Multi-Access Techniques.** Multiple-access techniques allow different users to share the same transmission channel. The plots represent power (*P*) as a function of time (*t*) or frequency (*f*) for multiple users. [Larson and Wertz, 1992]

techniques), with the resulting carrier then biphase-modulated with a pseudo-random noise (PRN) code. The data rate is much lower than the PRN code, or chip rate (Note: The acronym sometimes used for pseudo-random noise is PN versus PRN). Thus, many code bits (or chips) are transmitted per data bit.

The receiver employs a PRN-code generator which replicates the PRN code of the desired transmitted signal. The PRN codes are designed to have low cross-correlation properties so multiple signals can simultaneously transmit at the same frequency and experience very little mutual interference. The received signals, when mixed with the locally generated PRN code, appear as noise to the receiver except when the locally generated PRN code is identical to the one used on the desired transmit signal and they are perfectly aligned, or synchronized, in time. When this happens, the output of the mixer is a carrier containing only the narrow-band data modulation, and the PRN-code modulation is completely removed. In a sense the desired signal is despread. Similarly, because of the low cross-correlation properties of the PRN codes used, the signals from the undesired users appear at the output of the mixer as low-energy, wide-bandwidth (about twice the PRN-code rate) noise. Therefore, only a low-energy portion of the undesired signal will pass through the band pass filter for the narrow-band output.

The number of simultaneous users that a CDMA system can handle is limited by the low-energy noise the undesired user's signals generate. Assuming all CDMA users' code rates and data rates are the same and their carrier powers at the receiver input are equal, the question becomes: How many users can the system support? Equation (11.8) estimates the maximum number of users, $N_{users}$, the system can handle as a function of the energy-per-bit to noise power density ratio, $E_b/N_o$ (note: $E_b/N_o$ equals the received SNR)[*]:

$$N_{users} = 1 + R_c \left( \frac{1}{(C/N_o)_{Req}} - \frac{1}{(C/N_o)_{Act}} \right) \tag{11.8}$$

where $R_c$ is the PRN-code chip rate[†], $R_d$ is the data rate ($R_c$ is much greater than $R_d$), $(C/N_o)_{Req}$ is the required carrier-to-noise power density ratio, and $(C/N_o)_{Act}$ is the actual carrier-to-noise power density ratio. The ratios are defined by:

$$(C/N_o)_{Req} \equiv (E_b/N_o)_{Req} + 10 \log (R_d) \tag{11.9}$$

$$(C/N_o)_{Act} \equiv (E_b/N_o)_{Act} + 10 \log (R_d) \tag{11.10}$$

---

[*] Equation (11.8) was derived assuming an operating point at which the received $E_b/N_o$ drops below the threshold needed to achieve a particular BER.

[†] Chip rate is commonly used in a spread spectrum system and refers to the hop rate or pseudo-random noise code rate.

Note that $C/N_o$ and $E_b/N_o$ are expressed in power ratios, not in dB[*], in Eqs. (11.8) and (11.9). Typical values for chip rate ($R_c$) are from 1 to 100 M chips per second (Mcps).

The Global Positioning System (GPS) is an example of a system employing CDMA. Each spacecraft transmits a PRN code with a different time phase. The PRN chip rate, $R_c$, is 1.023 Mcps; the data rate, $R_d$, is 50 bps; $(E_b/N_o)_{Req} = 10$ dB = 10 (not expressed in dB); and $(C/N_o)_{Act} = 38.6$ dB-Hz = 7,244 Hz (not expressed in dB). Substituting into Eq. (11.8), we find the system can support $N_{users} = 1,906$.

CDMA is usually less bandwidth-efficient than FDMA or TDMA. But it's less susceptible to interference, including multipath caused by reflections from buildings or other objects, which makes CDMA especially interesting for satellite-communication systems with mobile terminals. For more information on multiple-access techniques, see Gagliardi [1984], Sklar [1988], or Ha [1986].

In recent years, lasers generating narrow-band energy at optical frequencies have provided an attractive alternative to microwave-frequency transmission. Unfortunately, clouds and rain seriously attenuate optical links, so these optical links have limited application in satellite-Earth communications. However, an optical link is well suited for crosslink communications between satellites. Inter-satellite links have been proposed using optical links with capacities above 300 Mbps.

Optical crosslinks are superior to microwave crosslinks for high data rates because they can support these rates using relatively small antenna diameters and system weights, as shown in Figs. 11.8 and 11.9. On the other hand, the narrow optical beams typical of laser communications (beamwidth is inversely proportional to operating frequency) are difficult to acquire and point accurately, requiring complex and sometimes heavy pointing mechanisms. Figure 11.9 compares rf and laser crosslinks, showing that rf links are usually better for data rates less than about ~100 Mbps because of their lower mass and power. But development of more efficient lasers with lighter and steerable optics may some day make lower-rate optical links attractive.

One application of optical links between spacecraft and Earth is the blue-green laser link being developed by ARPA and the US Navy for submarine communications [Weiner, 1980]. The laser frequency of $6 \times 10^{14}$ Hz (equivalent to a wavelength of 0.5 μm) was chosen for its ability to penetrate sea water. Even so, the water loss can range from 5 to 50 dB or more, depending on the actual depth of the submarine. In addition, loss due to cloud scattering is 4 to 14 dB. These losses are overcome by using low data rates of 10 to 100 bps, advanced coding techniques, and high-gain, narrow-beam optics.

The carrier frequency affects the ground-terminal antenna's achievable size and beamwidth. In turn, these factors indirectly affect spacecraft size, mass, and

---

[*] To express a number, $x$, that is in dB as a power ratio, simply make $y=10^{(x/10)}$. So, the number 40 dB expressed as a power ratio would be $10^{(40/10)}$ or 10,000.

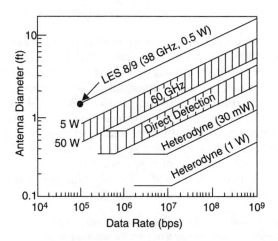

**Fig. 11.8.     Antenna Diameter and Data Rate for Several rf and LASERCOM Systems.** Optical systems (direct-detection and heterodyne) require smaller antenna diameters compared to rf crosslinks. [Larson and Wertz, 1992].

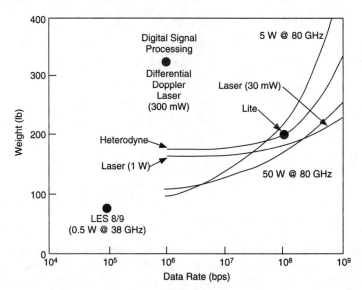

**Fig. 11.9.     A Comparison of Crosslink Package Weights.** At data rates greater than about 100 Mbps, an optical system provides the lightest package [Larson and Wertz, 1992].

complexity. The carrier frequency also varies how much attenuation we can anticipate due to rain and other benign- and stressed-channel characteristics (see Sec.
11.1.2).

Regulations constrain our choice of operational frequency, the usable data
bandwidth, and the allowable power flux density that impinges on the Earth's surface. Table 11.3 lists the frequency bands allocated to space communications. These
assignments originated with the International Telecommunications Union (ITU)
and the World Administrative Radio Conference (WARC). They are administered
in the United States by the Federal Communications Commission (FCC) for commercial users and by the Interdepartmental Radio Advisory Committee (IRAC) for
military users. The system designer must apply for and receive from the appropriate agency permission to operate at a specified frequency, orbit, and ground-
terminal location. This procedure is often time-consuming. For an excellent summary of this complex subject, see Morgan and Gordon [1989].

**Table 11.3. The International Telecommunications Union's Limits on Frequency Bands and Flux Densities.** Power-density limits are for elevation angle > 25°. They are about 10 dB less for lower angles. [Larson and Wertz, 1992]

| Frequency Band | Frequency Range (GHz) | | Service | Limit on Power Flux Density for Downlinks (dBW/m$^2$) |
| | Uplink | Downlink | | |
| --- | --- | --- | --- | --- |
| UHF | 0.2–0.45 | 0.2–0.45 | Military | -- |
| L | 1.635–1.66 | 1.535–1.56 | Maritime/Nav | −144/4 kHz |
| S | 2.65–2.69 | 2.5–2.54 | Broadcast | −137/4 kHz |
| C | 5.9–6.4 | 3.7–4.2 | Domestic Comsat | −142/4 kHz |
| X | 7.9–8.4 | 7.25–7.75 | Military Comsat | −142/4 kHz* |
| $K_u$ | 14.0–14.5 | 12.5–12.75 | Domestic Comsat | −138/4 kHz |
| $K_a$ | 27.5–31.0 | 17.7–19.7 | Domestic Comsat | −105/1 MHz |
| SHF/EHF | 43.5–45.5 | 19.7–20.7 | Military Comsat | -- |
| V | 60 | | Satellite Crosslinks | -- |

\* No limit in exclusively military band of 7.70–7.75 GHz

One criterion for allocating frequency bands is possible interference between
links. Extensive analysis is required when applying for a frequency band and orbit
to avoid interference with, or by, existing services, such as terrestrial microwave
links and ground-based radar operations. Especially significant are the antenna's
sidelobe levels and the dynamic range of powers over which the system must
operate. For ground-station antennas operating in the 4–6 GHz and 12–14 GHz

bands, the FCC specifies the maximum sidelobe gain as 32–25 log $\phi$ (in dBi) for $1° \le \phi \le 48°$ and –10 dBi for $48° \le \phi \le 180°$. Here, dBi is dB relative to an isotropic radiator (0 dB gain), and $\phi$ is the angle in degrees off the axis of the main antenna beam. Also shown in Table 11.3 are the maximum allowed power flux densities radiated by the spacecraft onto the Earth. These limits, set by the ITU, are necessary to avoid interference to existing terrestrial services, such as microwave relay links.

Two geostationary spacecraft in approximately the same orbit location servicing the same ground area may share the same frequency band by (1) separating adjacent spacecraft by some minimum angle (typically 2°) that is larger than the ground station's beamwidth; or (2) polarizing the transmitting and receiving carriers orthogonally, which allows two carriers to be received at the same frequency without significant mutual interference. Right-hand and left-hand circular polarization are orthogonal, as are horizontal and vertical linear polarization. Commercial systems use these frequency-sharing techniques extensively [Morgan and Gordon, 1989].

Two broad types of communications networks exist within the context of a communications system architecture: switched communications networks and broadcast communications networks. In a *switched communications network*, data transfers from source to destination through a series of intermediate nodes. These nodes aren't concerned with the content of the data; rather, they switch data from node to node until it reaches its destination. *Broadcast communications networks* have no intermediate switching nodes. At each node a transmitter/receiver communicates over a medium shared by other nodes. A transmission from any one node is broadcast to and received by all other nodes within the transmission footprint. Most SATCOM networks employ broadcast-communication techniques in that the transmitted data are receivable by any ground terminal within the spacecraft's footprint; however, future SATCOM systems employing satellite/satellite crosslinks will lend themselves to switched-network techniques that could enhance the architecture's network performance.

Two common types of switched communications networks include circuit-switched networks and packet-switched networks. In a *circuit-switched network*, a dedicated communications path is established between the source and destination node through a number of intermediate nodes in the network. Essentially, the path is a connected sequence of physical links between nodes. On each link, a logical channel is dedicated to the connection. Data generated by the source node transmits along the dedicated path continuously. No routing or switching delay accrues along the path (other than the initial circuit-setup delay) because the path is fixed. The most common example of a circuit-switched network is the telephone network.

In a *packet-switched network*, it's not necessary to dedicate transmission capacity along a dedicated path within the network. Rather, data goes out in a sequence of small bundles, called packets. Each packet is independently routed through the network from node to node, traveling from source to destination. At each node, the entire packet is received, briefly stored, and then transmitted to the next node. Message switching is a form of packet-switching in which the transmitted packet consists of the entire message to be transmitted. One benefit of message switching is that we don't need to reconstruct the received data packets because each packet received comprises the entire message. The primary drawbacks to message switching are that the transmission delay may be quite large, depending on the message size, and may vary if the message sizes aren't uniform.

No SATCOM networks employ either packet- or message-switched techniques. However, as stated previously, a number of systems are on the drawing board (e.g., IRIDIUM and TELEDESIC) that include satellite-satellite crosslinks in a richly connected space segment. They'll require some type of network-switching technique to ensure end-to-end connectivity. For more detail on network communications, see Stallings [1985].

# 11.3  Forming the Architecture[*]

Figure 11.10 shows the key characteristics of this phase, in which everything comes together. We consider the existing infrastructure, defined requirements, and the architectures identified in the design option phase to put together potential communications architectures. Much to the chagrin of communications-system engineers, past experiences (both good and bad) and politics often affect the outcome of this phase. We must evaluate each architecture for operational, technical, schedule, and cost feasibility.

In this section, we describe the more common types of architectures and the process for designing the space/ground communications link. An example of a design iteration needed to meet a change in the stated performance requirements concludes this section.

## 11.3.1   Architecture Categories and Selection Criteria

One way of categorizing communications-system architectures is by describing the spacecraft orbits or the main way of distributing the data throughout the network. Figure 11.11 shows several examples.

The *store and forward* architecture (Fig. 11.11A) for relaying communications data through spacecraft appeared in 1960 when the US Army launched the Courier satellite [Mottley, 1960]. In this configuration, the spacecraft is positioned in a low-altitude orbit (under 1000 km), receives data, and stores it in memory until it moves within view of a receive ground station, at which time it transmits the stored data. This architecture results in a relatively low launch cost because of

---

[*] Adapted with permission from Chapter 13 of *Space Mission Analysis and Design*. [Davies, 1992].

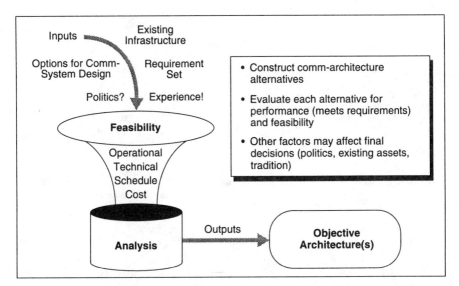

**Fig. 11.10.    Forming the Architecture.** We now integrate the results of the previous phases to generate architecture alternatives.

using low-altitude orbits. The spacecraft cost also stays relatively low because of a simpler communications payload. One example uses omni-directional (or, at least, wide-beamwidth) antennas to close the communications link. The larger beamwidth results in lower antenna gain, but the nearness of the low-altitude spacecraft decreases free-space loss to compensate for this effect. Additionally, the wide-beamwidth antenna greatly decreases the complexity required for pointing and stabilizing the antennas, and spacecraft stationkeeping is normally not required. The main disadvantage with this architecture, however, is its long access time and transmission delay, perhaps hours, which is the result of waiting for the spacecraft to pass within view of the user's ground station before delivering its data.

Virtually all communication spacecraft, as well as many others, use a geostationary orbit (Fig. 11.11B). The spacecraft enters a near-zero-degree inclination orbit at an altitude of approximately 35,785 km. The period of the orbit is exactly equal to the period of the Earth's rotation (one sidereal day or 23 hours, 56 minutes, 4 seconds), making the spacecraft appear stationary when viewed from the ground. The cost of ground stations is usually less for this architecture because it requires little or no antenna-pointing control. A geostationary network is far easier to set up, monitor, and control compared to a dynamic network containing non-geostationary spacecraft. There is no need to switch from one spacecraft to another, for the spacecraft is always in view of the ground station. The main disadvantages of this architecture include the lack of coverage available above 70° latitude and the high launch cost associated with placing a spacecraft into a geo-

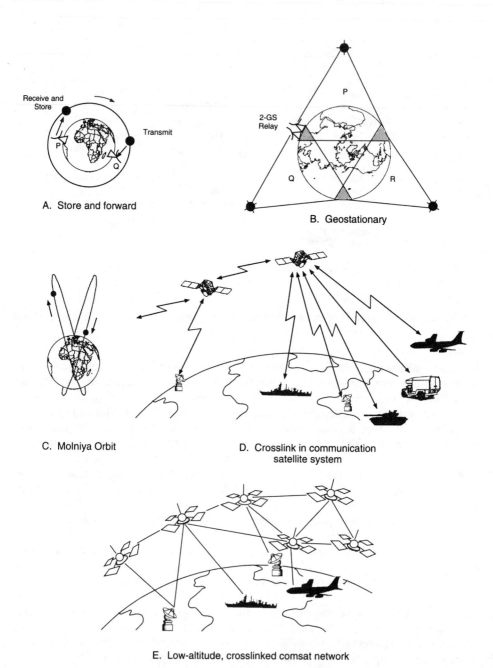

A. Store and forward

B. Geostationary

C. Molniya Orbit

D. Crosslink in communication
   satellite system

E. Low-altitude, crosslinked comsat network

**Fig. 11.11.    Typical Communications Architectures Used to Satisfy Different Mission Requirements.** Orbit configurations or space elements categorize the communications architectures. [Larson and Wertz, 1992]

stationary orbit. Furthermore, the one-way, source-to-destination propagation delay for a geostationary spacecraft was shown earlier to be approximately 0.25 second, which can sometimes cause problems in communications-satellite systems. Problems include voice echoes in voice-communications systems and acknowledgment delays in stop-and-wait message protocols.

Examples of spacecraft systems employing geostationary orbits are INTELSAT, INMARSAT, the Galaxy satellite system, the Defense Satellite Communications Systems (DSCS), the Tracking and Data Relay Satellite System (TDRSS), and the Fleet Satellite Communications System (FLTSAT). Refer to Comparetto [1993] for a more detailed description of the INTELSAT and INMARSAT systems.

The Russian space program uses a *Molniya-orbit architecture* (Fig. 11.11C) to cover the northern polar regions. The spacecraft are in highly elliptical orbits with an apogee of 40,000 km, a perigee of 500 km, and an inclination angle of 63.4°. The apogee is over the North Pole to cover northern latitudes. The period of the orbit is 12 hours, but because it is highly elliptical, the spacecraft spends about eight hours of each period over the northern hemisphere. Two or more spacecraft orbit in different planes, phased so that at least one is always in view from all northern latitudes. Unfortunately, the Molniya orbit requires continuous changing of antenna-pointing angles at the ground station and switching links between spacecraft as they move into and out of view of the ground station. This factor makes ground terminals more complex. Furthermore, although this architecture covers the northern latitudes well, it covers the southern latitudes rather poorly.

*Satellite-satellite crosslinks* can support source-destination connectivity for those cases in which the respective ground terminals are not within view of the same spacecraft. Satellite crosslinks were previously addressed in Sec. 11.2.1. Figure 11.11D shows a configuration in which satellite crosslinks are used between geostationary spacecraft (as opposed to the low-orbiting spacecraft in Fig. 11.11E). In crosslink configurations, an intermediate spacecraft relays the data to the target spacecraft, which contrasts with doing a double hop using two adjacent ground stations, as shown in Fig. 11.11B. The crosslink configuration performs better than the double-hop configuration because it has a smaller propagation delay (the double hop's delay is about 0.5). Crosslinks also don't require landing rights from a foreign country for ground-terminal operations (for communications architectures needing global connectivity). Foreign-based ground terminals can be costly and decrease system survivability and security. The obvious disadvantage of a satellite-crosslink architecture, however, is having to develop and implement the crosslink package, which can greatly increase the system's overall complexity, risk, and cost.

The *low-altitude, crosslinked architecture* (Fig. 11.11E) is basically an excursion from geostationary crosslinks. Multiple spacecraft are placed in low-altitude (500 to 3000 km) orbits, and satellite-satellite crosslinks are used to support a richly connected network. The data would typically transfer in packets of a few hundred or thousand bits (packet switching), with each packet time-stamped and labeled with its destination. The data packets may arrive at the destination node by different paths exhibiting different propagation delays, depending on the spacecraft/ ground station geometry at the time of transmission. The receiving station must be able to sort and reassemble the packets in the correct order in order to obtain the original message.

The low-altitude, cross-linked architecture is highly survivable because it has many paths between a given source and destination. The number of spacecraft, together with their low-altitude orbits, improves immunity to jamming from the ground. That's because each spacecraft can only be seen by a relatively small segment of the Earth's surface, so many jammer terminals would be necessary to effectively disrupt network communications. Finally, the uplink transmitter power needed at the ground terminal is lower because the ground terminal and spacecraft are closer together. This closeness also decreases the probability of unauthorized reception. On the other hand, this architecture requires complex network synchronization and spacecraft control. One example of this type of architecture that has received notoriety is the IRIDIUM system discussed previously in Sec. 11.2.1.

We can categorize satellite-communications architectures by two main functions: collecting and distributing data and communicating. Weather satellites are a good example of the first function. By using sophisticated sensing equipment, they measure atmospheric conditions, photograph the atmosphere and cloud activity, and then send the data to ground stations. On the ground, the data goes through further processing and, perhaps, further distribution to interested users such as the military, local television stations, or civil-emergency agencies. We can distribute data in various ways, including direct broadcast, relay transmissions, multiple hops, and interfaces with ground-communications networks. The actual method isn't important, as long as the architecture collects or generates data and delivers it to the user.

All spacecraft communicate, but some live to communicate. An example of this type of architecture would be the INTELSAT system [Comparetto, 1993]. An international organization formed in August, 1964, to produce, own, manage, and use a global communications-satellite system—the INTELSAT system. It consists of 120 member nations and supports direct communications links among 180 countries, territories, and dependencies using more than 1,300 antennas located at over 800 Earth stations. As of October, 1993, the INTELSAT space segment included 19 active spacecraft, with the more sophisticated ones able to support up to 24,000 voice circuits each! As stated in the INTELSAT Agreements (the governing documentation for this body), the prime objective of the INTELSAT system is:

> *To provide, on a commercial basis, high quality, reliable international public telecommunications services.*

INTELSAT's only purpose in life is to support telecommunications services, which they define as telephony, telegraphy, telex, facsimile, data transmission, and radio/television. Thus, they chose a spacecraft, antennas, ground-terminal locations, and orbital configurations to produce these services.

Telemetry represents the spacecraft's health and status data. Tracking involves transferring and updating the orbital-element set and other data to maintain accurate data on the spacecraft's position and motion. Commanding is the process of transmitting specific command sequences to the spacecraft in order to accomplish certain actions (e.g., fire a thruster, move an antenna). There are basically two philosophies for doing TT&C in a communications-system architecture: keep the complexity on the ground or put it in the spacecraft.

In keeping the complexity on the ground, operators at ground stations control the mission in (near) real time by transmitting commands directly to the spacecraft. One advantage of this approach is flexibility to changing requirements because changes are easier in the ground element than in space. Other advantages are greater reliability and a less complex, lower-cost spacecraft. The disadvantages include greater vulnerability to human error and the costs associated with the ground-control element.

If the complexity is onboard the spacecraft, the spacecraft itself does TT&C using onboard data sensing and programmed decision making. This arrangement replaces ground control, is highly survivable, has fast response time (communication link delays are eliminated), excludes errors introduced by human operators, and reduces ground equipment and operations cost. However, this method is less responsive to changing or unanticipated requirements, and the spacecraft itself is more complex, more costly, and potentially less reliable. Even when using an autonomous control architecture, a ground station is usually required to collect data from the spacecraft and to back up the onboard control system. In the future, however, we expect the spacecraft to handle more functions, such as stationkeeping, to reduce dependence on control from the ground station. Chapter 12 discusses the TT&C functions in more detail.

### 11.3.2    Designing the Ground/Space Communications Link

*Modulation* is the process by which an input signal varies the characteristics of a radio frequency carrier (usually a sine wave). These characteristics include amplitude, phase, and frequency. *Coding* is the process by which we add redundancy to the transmitted data stream in order to improve power efficiency for a given modulation technique, though at the expense of bandwidth. That is, the required received signal-to-noise ratio (SNR) to achieve a given bit-error-rate (BER) is lower by an amount referred to as the coding gain for a system that employs FECC techniques.

*Demodulation* measures the received carrier's frequency and phase to estimate the transmitted data stream. *Decoding* removes the data redundancy from the received signal to detect and correct transmission errors. The most common form of FECC is convolutional encoding with Viterbi decoding. The theory behind FECC techniques is beyond the scope of this chapter, so we'll focus on common modulation techniques used in satellite-communications architectures. See Lin and Costello [1983], Proakis [1983], and Sklar [1988] for detailed discussions of FECC techniques along with the supporting theory.

Amplitude modulation, though common in terrestrial services, seldom appears in spacecraft systems because it requires larger (and more costly) transmitters. It also results in unwanted intermodulation (IM) products and signal-suppression effects in the satellite transponders (because they don't operate linearly). Both phase- and frequency-modulation techniques are common in SATCOM applications because the transponder can operate at or near saturation, resulting in maximum power efficiency.

Figure 11.12 depicts the most common modulation techniques used in space-craft systems today. *Binary phase-shift keying* (BPSK) is most common. It consists of setting the carrier phase to 0° to represent a binary 0, and to 180° to represent a binary 1. In quadrature phase-shift keying (QPSK), we examine two bits and identify one of four symbols (00, 01, 10, and 11). We then map each symbol to one of four carrier phases (0°, 90°, 180°, and 270°). Note that for QPSK, the symbol rate is equal to one half of the bit rate, which results in a decrease in the required transmission bandwidth by a factor of one half.[*]

*Frequency-shift keying* (FSK) is another common modulation scheme. In FSK we assign two separate carrier frequencies, F1 and F2, to a binary 0 and a binary 1, respectively. The separation between F1 and F2 must at least be equal to the data rate in order to avoid performance loss that would otherwise result from mutual interference. Thus, the required transmission bandwidth is at least twice the width of the spectrum generated by BPSK (a minus), but the demodulation techniques available for FSK are less complicated than those for BPSK or QPSK (a plus). In M-ary frequency shift keying (MFSK), we map M separate frequencies to the M combinations of the transmitted bit stream, taken $\log_2 M$ bits at a time. For example, in a typical 8-ary FSK system, the input bit stream is sampled 3 bits at a time ($\log_2 8$) resulting in eight possible combinations (000, 001, 010, 011, 100, 101, 110, 111). Depending on the 3-bit combination observed, we assign one of the 8 possible frequencies to the carrier (F1, F2,..., F8). For 8-ary FSK, the symbol rate is one-third the bit rate (only one symbol is transmitted for every three input bits), and the required transmission bandwidth is approximately 8/3 the bit rate. The factor of eight results from using a frequency spacing between each of the eight possible frequencies that is equal to the transmitted data rate. Less common modulation schemes include minimum shift keying (MSK), offset QPSK (OQPSK), and 8-ary PSK. Sklar [1988] describes these modulation schemes.

---

[*] One of the most common FECC techniques used today is the rate 1/2 convolutional code. The rate 1/2 means that for every data bit that enters the encoder, two coded symbols exit. As a result, the system enjoys a coding gain (a plus) at the expense of doubling the required transmission bandwidth (a minus). Rate 1/2 convolutional encoding is often used with QPSK to offset this drawback.

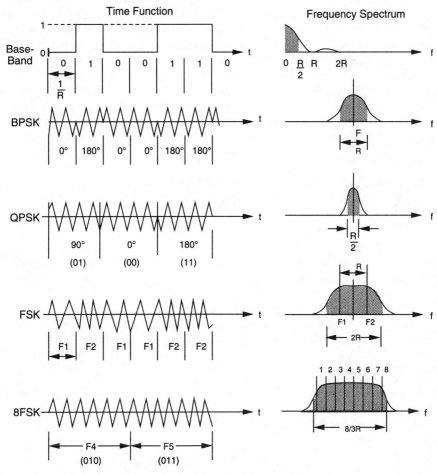

**Fig. 11.12. Modulation Types Commonly Used for Digital Signal Transmission in Satellite Communications.** R is the data rate. The shaded region is the required bandwidth. [Larson and Wertz, 1992]

To reliably demodulate a received digital data bit, the amount of received energy per bit, $E_b$, must exceed the noise power spectral density, $N_0$, by a specified amount. In other words, we need a minimum $E_b/N_0$ to achieve a given BER. Communications theorists derive analytical expressions that relate the received $E_b/N_0$ to the BER as a function of the modulation/demodulation type and FECC/decoding techniques applied. As stated earlier, this area can become quite complex for M-ary modulation schemes combined with FECC techniques. We'll simply acknowledge the relationship between modulation type and bandwidth, as well as

FECC's effect on overall system performance. These concepts are significant in developing a communications-link design, as discussed in the following section.

A link-budget equation quantifies the available received signal-to-noise ratio (SNR) on a given link. The received signal-to-noise ratio is also commonly referred to as the energy per bit to noise power density ratio, or $E_b/N_o$[*]. Once we determine the received SNR, we compare it to the SNR required to achieve a required bit-error-rate (BER) and refer to the difference as *link margin*. A negative link margin tells us the communications-link design won't support the given data rate at the required BER, whereas a positive link margin implies enough excess received SNR is available. A link margin of 0 dB tells us the received SNR is just enough to close the link, and any further, or unforeseen, link degradations will reduce link performance.

The link-budget equation in its most general form is

$$SNR_{avail} = (E_b/N_o)_{avail} = EIRP + G/T - L_{fs} - L_{other} - k - R_d \qquad (11.11)$$

where
$EIRP$ = effective isotropic radiated power of the transmit terminal
$G$    = receive antenna gain
$T$    = system noise temperature of the receive terminal[†]
$L_{fs}$ = free space signal loss
$L_{other}$ = term that accounts for the other link loss terms (antenna pointing loss, rain, atmospheric absorption and implementation)
$k$    = Boltzmann's constant (–228.6 dBW/K-Hz)
$R_d$  = data rate

Note that each of the terms identified in Eq. (11.11) above are expressed in units of dB.[‡]

We then compute the link margin, $M$, as follows:

$$M = (E_b/N_o)_{avail} - (E_b/N_o)_{req'd} \qquad (11.12)$$

where $(E_b/N_o)_{avail}$ is the SNR available at the receiver and $(E_b/N_o)_{req'd}$ is the SNR required to achieve a given BER. $(E_b/N_o)_{req'd}$ is a function of the modulation format, the presence of forward-error-correction coding (FECC), and whether the operational environment is benign or stressed.

---

[*] In satellite communications, the term "density" measures a given parameter per Hz of bandwidth. For example, the noise power density refers to the noise power (in watts) per Hz of receiver bandwidth (in Hz), resulting in units of W/Hz or, more commonly, dBW/Hz.
[†] The parameter G/T is typically referred to as the figure-of-merit for a receive terminal and is a common operational characteristic given when describing a receive terminal.
[‡] For more detail on how the link-budget equation is derived, see either Gagliardi [1984] or Sklar [1988]. Both are excellent.

Typically, for a given communications architecture, we can juggle the *EIRP*, *G/T*, and $R_d$ terms to meet all of the performance and operational requirements for the communications link. The $L_{fs}$ term is a calculable parameter, whereas we typically use the $L_{other}$ term to account for most of the link uncertainties. Each of these parameters is discussed further below.

We can compute the *EIRP* simply by adding the transmitted power to the gain of the transmitting antenna, with both values expressed in dB. Suppose we have a ground terminal employing a 1000 W high-power amplifier (HPA) with an antenna providing a gain of 15 dB. The *EIRP* would be

$$EIRP = 10 \log (1000) + 15 = 45 \text{ dBW} \tag{11.13}$$

Note that the antenna gain was already in units of dB in the above equation.

We usually know the receive *G/T* of off-the-shelf ground terminals. But we have to develop the *G/T* requirement while meeting other requirements. The receive *G/T* is simply the receive antenna's gain (not in dB) divided by the receive system's temperature (also not in dB). In calculating *G/T* for a given ground-terminal design, we commonly treat the terms individually and generate their values in units of dB. The *G/T* value is then the difference between the receive antenna's gain and receive system's temperature, with both values expressed in dB:

$$\text{Figure of Merit} = G/T = G \text{ (in dB)} - T \text{ (in dB)} \tag{11.14}$$

The maximum antenna gain, *G*, for a directional antenna is given by:

$$G = \eta \left( \frac{4\pi A}{\lambda^2} \right) \tag{11.15}$$

where $\eta$ is the antenna efficiency (typically 0.55 for parabolic-dish antennas), $\lambda$ is the carrier wavelength, and *A* is the antenna aperture area, which is $\pi D^2/4$ for a parabolic-dish antenna having a circular aperture of diameter *D*. The carrier wavelength is related to the carrier frequency by $\lambda f = c$, where *c* is the speed of light or $3 \times 10^8$ m/s. Let's express the antenna gain in terms of the carrier frequency, substitute the expression above for the antenna-aperture area, and express the antenna gain in units of dB. Eq. (11.15) then becomes

$$G = \text{Constant} + 10 \log (\eta) + 20 \log (f) + 20 \log (D) \tag{11.16}$$

where the constant is a function of the units used for the carrier frequency and antenna diameter, as shown in Table 11.4.

As you can see from Eqs. (11.15) and (11.16), the antenna gain is proportional to frequency squared (doubling the carrier frequency quadruples the resulting gain). Also, antenna gain is proportional to the antenna's diameter squared (doubling the antenna's size quadruples the resulting antenna gain). These relationships are significant drivers of a communications system architecture.

**Table 11.4. Constants for the Antenna-Gain Equation as a Function of the Units Used for Frequency and Antenna Diameter.** The antenna gain is a function of the square of the frequency and the antenna diameter.

| Constant (dB) | Units of Frequency, f | Units of Antenna Diameter, D |
|:---:|:---:|:---:|
| −159.6 | Hz | meters, m |
| 20.4 | GHz | meters, m |
| −169.9 | Hz | feet, ft |
| 10.1 | GHz | feet, ft |

As an example of using Eq. (11.16), assume an antenna diameter of 60 ft. and a carrier frequency of 7 GHz—the nominal downlink operating characteristics of an FSC-78, DSCS, fixed ground terminal). The resulting antenna gain using Eq. (11.16) would be:

$$G = 10.1 + 10 \log(.55) + 20 \log(7) + 20 \log(60) = 60 \text{ dB} \qquad (11.17)$$

Another antenna characteristic that will affect our design is the antenna beamwidth. The *antenna beamwidth* describes the angular spread of the transmitted electromagnetic wave as it traverses the path from the source (or transmitter) antenna to the destination (or receiver) antenna. The beamwidth is a first-order indication of the accuracy with which we must point the antenna toward its target. A common antenna beamwidth prescribed for antennas is $\theta_{3dB}$—the angle across which the antenna gain is within 3 dB (or 50%) of the maximum gain. Recall that Eqs. (11.16) and (11.17) provide the maximum antenna gain for a parabolic-dish antenna. Gagliardi [1984] defines the 3 dB beamwidth as

$$\theta_{3dB} \text{ (in degrees)} = \frac{(1.02)\,\lambda}{D}\left(\frac{180}{\pi}\right) = \frac{(1.02)\,c}{fD}\left(\frac{180}{\pi}\right) \qquad (11.18)$$

where $c$ is the speed of light ($3 \times 10^8$ m/s), $f$ is the carrier frequency in Hz, $D$ is the antenna diameter in meters, and the term $(180/\pi)$ is used to convert from radians to degrees. So, for the example given above, we calculate the 3 dB beamwidth using Eq. (11.18) to be ~0.14°.

The second term in the figure-of-merit expression, $G/T$, is the *system noise-temperature term*, $T$. In practice, the system noise temperature consists of three primary components: antenna noise temperature, line-loss noise temperature, and receiver noise temperature. Table 11.5 shows typical values for these three components, together with the system noise temperature (which is the sum of the three components) as a function of frequency for satellite systems using uncooled receivers. To calculate the receive $G/T$ or figure-of-merit for a given ground receiver, we simply subtract the value of the system noise temperature given in Table 11.5 from the calculated antenna gain.

For example, let's use the antenna gain of 60 dB calculated previously for a DSCS FSC-78 fixed ground terminal and a value of 27.4 dB-K for the system noise temperature from Table 11.5. (Note the downlink is 7 GHz, which is in the 2-12 GHz column.) In this case, the receive G/T is 60-27.4 or 32.6 dB/K. If this value of G/T isn't large enough to support the link-budget calculations, one alternative would be to cool the receive terminal, thereby decreasing the system noise temperature. The result would be a larger value for G/T. As you can see in Eq. (11.11), the larger G/T would increase the available or received SNR and, consequently, increase the link margin from Eq. (11.12).

**Table 11.5. Typical System Noise Temperatures in Satellite-Communication Links in Clear Weather.** The temperatures are referred to the antenna terminal. Cooling techniques can reduce the antenna noise temperature, which affects the ground terminal's figure of merit, G/T. [Larson and Wertz, 1992]

|  | Frequency (GHz) | | | | | |
|---|---|---|---|---|---|---|
|  | Downlink | | | Crosslink | Uplink | |
| **Noise Temperature** | **0.2** | **2–12** | **20** | **60** | **0.2–20** | **40** |
| Antenna Noise (K) | 150 | 25 | 100 | 20 | 290 | 290 |
| Line Loss Noise (K) | 35 | 35 | 35 | 35 | 35 | 35 |
| Receiver Noise (K) | 190 | 492 | 592 | 1728 | 970 | 1505 |
| System Noise Temp (K) | 375 | 552 | 727 | 1783 | 1295 | 1830 |
| System Noise Temp (dB-K) | 25.7 | 27.4 | 28.6 | 32.5 | 31.1 | 32.6 |

The *free-space loss*, $L_{fs}$, is defined by

$$L_{fs} = \left( \frac{4 \pi S}{\lambda} \right)^2 \qquad (11.19)$$

where $S$ is the source-to-destination range (slant range), and $\lambda$ is the carrier wavelength.

Note that the product of the carrier wavelength, $\lambda$, and the carrier frequency, $f$, is equal to the speed of light, $c$. By applying the appropriate constants to account for the desired units, we can express the free-space loss in units of dB as

$$L_{fs} = \text{Constant} + 20 \log (f) + 20 \log (S) \qquad (11.20)$$

where the constant depends on which units we use for the carrier frequency and slant range, as shown in Table 11.6.

As seen from Eqs. (11.19) and (11.20), the loss from free-space attenuation is proportional to frequency squared (doubling the carrier frequency quadruples the

Table 11.6.  **Constants for the Free-Space Loss Equation as a Function of the Units used for Frequency and Slant Range.** The free-space loss varies as a function of the square of both frequency and slant range.

| Constant (dB) | Units of Frequency, *f* | Units of Slant Range, *S* |
|:---:|:---:|:---:|
| 96.58 | GHz | Statute Miles |
| 92.45 | GHz | km |
| 97.79 | GHz | Nautical Miles, nm |
| 32.45 | MHz | km |

resulting loss). Also, loss from free-space attenuation is proportional to distance squared (doubling the slant range quadruples the resulting loss). These relationships significantly drive communications-system architectures.

Finally, as an example of using Eq. (11.20), assume a slant range of 35,785 Km (vertical distance to a geostationary satellite) and a carrier frequency of six GHz (typical uplink frequency for commercial satellites). Using Eq. (11.20), we find the free-space loss to be

$$L_{fs} = 92.45 + 20 \log (6) + 20 \log (35,875) = 199.11 \text{ dB} \qquad (11.21)$$

The *other-loss term*, $L_{other}$, is a catch-all term that represents all of the signal loss mechanisms in the benign channel, including rain, atmospheric absorption, dust, foliage, antenna pointing, and system implementation. The losses due to rain, atmospheric absorption, dust, and foliage can be quite severe depending on the carrier frequency. We discussed them in Sec. 11.1.2 under channel characteristics. Here we'll briefly discuss losses from antenna pointing and implementing the system.

The *antenna-pointing loss* accounts for inaccuracies in aligning the transmitter and receiver antennas. The loss is a function of the 3 dB beamwidth, $\theta_{3dB}$, and the antenna pointing error, $\theta_e$. For a parabolic dish antenna (see Gagliardi [1984]), this loss is

$$L_{pointing} = 10 \log \left( e^{-2.76 \left( \theta_{3dB} / \theta_e \right)^2} \right) \qquad (11.22)$$

The antenna-pointing error, $\theta_e$, depends on the antenna design and fabrication process and is typically specified for an antenna. We can calculate the 3 dB beamwidth, given the antenna diameter and operating frequency, by using Eq. (11.18). Let's continue with the example of the DSCS FSC-78 fixed ground terminal operating at a downlink frequency of 7 GHz. We previously calculate the 3 dB beamwidth to be 0.14°. Assuming an antenna-pointing error of 0.05°, the pointing loss, using Eq. (11.22), would be:

$$L_{pointing} = 10 \log\left( e^{-2.76\,(0.05/0.14)^2} \right) = -1.53 \text{ dB} \qquad (11.23)$$

The *implementation loss* is a term that accounts for most of the non-ideal qualities of communications-link equipment. These typically include uplink-waveform distortion, timing errors, frequency errors, data-detection-matched filter losses, non-ideal payload hardware (phase noise and filtering), and interference between adjacent channels. The implementation-loss term, as a whole, doesn't lend itself to analytical calculations. We can use analysis to calculate some components, such as adjacent-channel interference, but most others require testing. Thus, we usually estimate the implementation loss and then, after operationally testing the communications-link equipment, use test data to verify it.

### 11.3.3 Example of Designing a Communications-System Architecture

We can't show here a detailed example of how to design the communications-system architecture for a complicated system. In practice, teams do this work over months to years. But we can use a simplified example to develop a basic understanding of how we might develop more complicated architectures. Our task will be to design a space-to-ground communications link while addressing each phase of the communications-architecture design.

We'll assume requirements are already defined, as shown in Table 11.7. Table 11.7 includes performance requirements (P1 through P6), operational requirements (O1 through O9), system constraints (SC1 through SC3), and program constraints (PC1 and PC2). We've listed them in the order discussed previously. We'll discuss below each requirement and constraint in Table 11.7, showing how it would affect the design.

The data-rate requirement of 51.2 Mbps was derived using Eq. (11.1) and the link's being required to support a message rate of 40,000 messages/sec, with a message size of 160 bytes.

As shown in Table 11.7, we used *message error rate* (MER) instead of the BER in this example. The MER is simply the probability that a received message (or packet) will contain an error after it has been demodulated and decoded. It's a function of the message size (in bits) and the BER:

$$\text{MER} = \text{BER} * \text{Message Size (in bits)} \qquad (11.24)$$

In this example, the MER requirement is 0.002 with a message size of 160 bytes or 1280 bits. We can derive the required BER by rearranging the terms in Eq. (11.24):

$$\text{BER} = \frac{\text{MER}}{\text{Message Size}} = \frac{0.002}{1280} = 1.56 \times 10^{-6} \qquad (11.25)$$

To be conservative, we set the BER requirement at $1 \times 10^{-6}$.

**Table 11.7. Summary of Requirements for an Example Communications-System Architecture.** This table compiles the performance requirements (P1–P6), operational requirements (O1–O9), system constraints (SC1–SC3), and program constraints (PC1 and PC2) for a simple communications-system architecture.

| ID | Required Label | Description | Impact |
|---|---|---|---|
| P1 | Data Rate | 40,000 Messages/sec; (160 bytes message) | 51.2 Mbps |
| P2 | BER | MER < 0.002 | $10^{-6}$ |
| P3 | E/E Delay | Minimize | TBD |
| P4 | Link Availability | ≥ 99.5% at minimum elevation angle of 10° | Will affect rain margin; function of frequency and geographic region |
| P5 | A/J | N/A | None |
| P6 | Nuclear Scintillation | N/A | None |
| O1 | Security | N/A | None |
| O2 | Standardization | N/A | None |
| O3 | Backward Compatibility | Use DSCS FSC-78 Fixed Ground Terminal | Receive and transmit characteristics are given |
| O4 | Access | Single Access | No FDMA, TDMA, or CDMA |
| O5 | Satellite Orbital Characteristics | ≥ 33% of Earth's surface, elev angle ≥10° | Geostationary satellite |
| O6 | Satellite Mobility | No requirement | None |
| O7 | User Terminal Characteristics | Fixed | No need for satellite tracking antenna |
| O8 | Data Source | Digital | No sampling required |
| O9 | Channel Characteristics | Rain, Atmos Absorption, No Dust or Foliage | Must be accounted for in link-budget calculations |
| SC1 | Parabolic dish for satellite antenna; 0.3 m ≤ diameter ≤ 4 m | | |
| SC2 | Maximum satellite transmit power of 50 w | | |
| SC3 | Military application; downlink - Use X-band frequency | | |
| PC1 | Use currently available technology - No sophisticated engineering | | |
| PC2 | Use non-development items to maximum extent practical | | |

Continuing through Table 11.7, we must minimize E/E delay, which may affect our design but remains to be determined. The link availability is set at $\geq$ 99.5% for a minimum antenna-elevation angle of 10°. This requirement will affect the rain-loss part of the $L_{other}$ term in the link-budget equation [see Eq. (11.11)]. No A/J, nuclear scintillation, security, or standardization requirements are identified. Under backward compatibility, we must use the DSCS FSC-78 fixed ground terminal discussed in the previous section. We'll use single access, so we don't need to consider FDMA, TDMA, or CDMA techniques.

The orbital-characteristics requirement states that the spacecraft must be able to view at least 33% of the Earth's surface, assuming a minimum elevation angle of 10°. So we choose a geostationary orbit which, based on previous calculations in Sec. 11.1.2, will allow the spacecraft to illuminate ~34% of the Earth's surface. Table 11.7 lists no spacecraft-mobility requirements. The user's ground terminal is fixed (DSCS FSC-78 fixed ground terminal). The data source is digital, which implies we don't need to address analog data-sampling techniques. The channel characteristics include rain and atmospheric absorption but not dust or foliage. Once we choose the downlink frequency, we can quantify the amount of rain and atmospheric absorption.

Table 11.7 identifies three system constraints. The first two limit the size of the spacecraft antenna and the amount of spacecraft transmit power, respectively. Both will come into play during the link-budget calculations. The last one, SC3, indicates the communications link will be solely for military applications. As a result of this requirement, the fact that a DSCS FSC-78 fixed ground terminal must be used, and the allocation of frequency bands previously identified in Table 11.3, we decide to use the X-band frequency band for the space-to-ground communications link in this example. Based on the data in Table 11.3, we also choose a downlink frequency of 7.4 GHz.

Finally, the two program constraints are fairly general and won't drive our design choices much. But note that, if SC1 wasn't identified, the program constraints in Table 11.7 would probably have driven us to choose a parabolic dish antenna for the spacecraft anyway.

We've already identified most of the data we need to design the space-to-ground communications link in this simple example. But let's briefly go on to the design-option phase to generate any other design characteristics we need. We've addressed the design-option areas below in the order originally discussed in Sec. 11.2.

As previously stated in Sec. 11.1.1, the E/E delay consists of transmission, propagation, queuing, and processing delays. The transmission delay is a function of the message size and the data rate, both of which are fixed in this example, so it's also fixed. The propagation delay is a function of the source-to-destination distance and the speed of light, both of which are again fixed in this example. Thus, the propagation delay is fixed. A queuing delay occurs only in multiple user/server networks, so it doesn't apply. Finally, the delay for onboard processing is the only one we can influence in this example. Performance requirement P3 from

Table 11.3 calls for minimum E/E delay. Consequently, we decide to use a bent-pipe transponder instead of onboard processing in order to eliminate the signal-processing delay on the spacecraft. Of course, other factors may exist in a more complicated communications system that could cause us to employ onboard processing even though it might violate requirement P3.

Because of requirement O5, we choose a geostationary orbit to support continuous space/ground connectivity.

For the reasons stated above, the transponder design will be a non-regenerative or bent-pipe design. System constraint SC2 requires maximum output power to stay below 50W. Fortunately, SSAs and TWTAs at 7.4 GHz can both produce 50W of output power, so either design would be acceptable in this example. Klystrons wouldn't be an option because they're typically limited to ground applications.

We assume a single dedicated user in this example.

We choose rf over LASERCOM because of the program constraints identified in Table 11.3. One constraint requires available technology, and LASERCOM is still considered a maturing technology. Another requires us to use non-development items (NDI) as much as possible, and only a few space-qualified LASERCOM components are readily available. Finally, requirement O3 states we must use the DSCS FSC 78 ground terminal as the receiver terminal. The DSCS FSC 78 terminal is designed to receive X-band rf signals and doesn't suit LASERCOM applications.

We select an operational frequency of 7.4 GHz for the reasons stated in the paragraphs above.

We're now able to combine the stated and derived requirements, along with the identified design characteristics, into a target architecture design. In this example, the architecture simply consists of the space-to-ground communications link. Figure 11.13 summarizes the communications-link design to this point. As shown in the left hand part of Fig. 11.13, the required BER is $10^{-6}$, the required data rate is 51.2 Mbps, and the frequency is 7.4 GHz. The spacecraft is in a geostationary satellite orbit, and the slant range, $S$, between the ground terminal and the spacecraft is calculated to be 40,585 km assuming a minimum elevation angle of 10°. The free-space loss, $L_{fs}$, is calculated to be 202 dB, using Eq. (11.21) with the data in Table 11.3, and assuming an operating frequency of 7.4 GHz and the slant range calculated above.

The other loss term, $L_{other}$, consists of rain and atmospheric absorption. We estimate the rain loss to be ~2 dB, using Fig. 11.4 and assuming a required link availability of 99.5% at a minimum elevation angle of 10° and an operating frequency of 7.4 GHz. We calculate the atmospheric absorption to be ~0.3 dB, using Eqs. (11.6) and (11.7) with the data in Table 11.1. Again, the operating frequency is 7.4 GHz.

The top right part of Fig. 11.13 identifies several assumptions. The modulation chosen in this design is BPSK—consistent with the program constraints in Table 12.7, which calls for available technology and NDI where practical. BPSK modems are widely used in SATCOM and are supported by a mature technology base. We

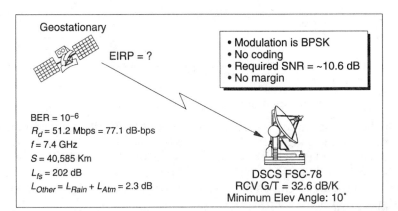

**Fig. 11.13. Communications-Link Summary.** We use the performance and operational requirements, along with system and program constraints, to develop a candidate architecture for the communications link. The only task left in this example is to solve for the required satellite EIRP.

won't use FECC techniques so we can keep the design as simple as possible. We can introduce FECC if we need greater power efficiency even though it will reduce bandwidth efficiency. The received SNR required to achieve a BER of $10^{-6}$ using BPSK modulation is ~10.6 dB. We can directly calculate this value using Eq. (11.2) or estimate it using Fig. 11.3. Finally, we put no margin into this link, so it will be a bare-minimum design.

As shown in Fig. 11.13, the only parameter left to complete the design is the spacecraft EIRP required to close the link. We calculate the EIRP by rearranging Eq. (11.11):

$$\text{EIRP} = \text{SNR}_{\text{req}} - G/T + L_{fs} + L_{other} + k + R_d \qquad (11.26)$$

$$= 10.6 - 32.6 + 202 + 2.3 - 228.6 + 77.1$$

$$= 30.8 \text{ dBW}$$

Note that the EIRP is simply the sum of the transmit power and the transmit-antenna gain (assuming the units of both terms are expressed in dB). We can now size both parameters based on the required spacecraft EIRP of 30.8 dBW. If we assume the spacecraft transponder's output power is 1W (equivalent to 0 dBW), the resulting transmit-antenna gain must be 30.8 dB. We can then use Eq. (11.16) and the data in Table 11.4 to determine the required size of the spacecraft's dish antenna. Using units of GHz for the transmit frequency and meters for the antenna diameter, we can rewrite Eq. (11.16) to solve for the required antenna diameter:

$$D = 10^{\left( \frac{G - \text{Constant} - 10 \log (\eta) - 20 \log (f)}{20} \right)} \qquad (11.27)$$

$$= 10^{\left( \frac{30.8 - 20.4 - 10 \log (0.55) - 20 \log (7.4)}{20} \right)}$$

$$= 0.6 \text{ meters}$$

Note that our assuming 1 W for the spacecraft's transmit power and the resulting diameter of the spacecraft's antenna are consistent with the first two system constraints (SC1 and SC2) in Table 11.7. Thus, the resulting design is acceptable: it meets the stated and derived requirements, is consistent with the system and program constraints, and includes the design options and characteristics identified in the design-option phase. Our work is done!

What happens if the requirement set changes after we've designed the communications architecture? Depending on the amount of change, we may have to repeat many steps in the process and revisit each phase in order to develop a new communications architecture that fully responds to the changed requirements. We'll discuss one minor change to the requirements so we can see the effect on our design example. Let's assume the message-rate requirement (P1 in Table 12.7) increases from 40,000 messages/sec to 80,000 messages/sec. The resulting data rate required to support this new message rate is 102.4 Mbps. How can we change our design, in the least painful way, to satisfy the increased data rate?

First, let's examine the relationship between data rate and the other parameters in the link-budget equation [Eq. (11.11)], which we'll rewrite to solve for the data rate[*]:

$$R_d = \text{EIRP} + G/T - L_{fs} - L_{other} - k - (E_b/N_o)_{\text{req'd}} \qquad (11.28)$$

Note that the design change requires us to double (increase by 3 dB) the achievable data rate, which is what Eq. (11.28) solves for. In examining Eq. (11.28), with an eye to doubling the achievable data rate, we see that

1.  The EIRP can be increased by 3 dB. This would require doubling the transmit power to 2 W or doubling the spacecraft antenna's gain by increasing its diameter by a factor of 1.414 (the square root of 2), or a combination of the two.
2.  The receive figure-of-merit, $G/T$, can be increased by 3 dB. The receive antenna's gain is fixed to that of the DSCS FSC-78 antenna. We'd have to use cooling to decrease the receive-system temperature by 3 dB

---

[*] We assumed a link margin of 0 in developing the above equation, so we could set $(E_b/N_o)_{avail}$ equal to $(E_b/N_o)_{req'd}$.

(thereby increasing G/T by 3 dB). This wouldn't be an easy fix in terms of cost or operations.

3. The free-space loss is essentially fixed because the slant range and operating frequency aren't negotiable

4. $L_{other}$ is also essentially fixed because the operating frequency drives it

5. Boltzmann's constant is fixed

6. The required $E_b/N_o$ can be decreased by 3 dB by introducing FECC into the design

In summary, the one option that would least affect the overall design would be to increase the spacecraft's transmit power from 1 W to 2 W. We may be able to use the same transponder if we originally chose a transponder with some margin in output power. Another option would be to increase the antenna diameter from 0.6 m to 0.85 m, but we couldn't do so if the antenna were already built. Finally, we could incorporate FECC or possibly get a BPSK modem that already incorporates it.

Obviously, designing communications architectures is complicated. It consists of various technical areas, each requiring significant expertise. That's why teams of qualified people normally design all but the simplest systems.

At the beginning of this chapter, we argued that designing communications architectures requires a structured approach to make sure the resulting options are well documented, technically justified, and cost-effective. Consistent with this philosophy, we showed you a process that included phases for defining requirements, evaluating options, and forming the candidate architecture. We described each phase and discussed key components and technical areas in terms of how they could affect the cost and performance of space/ground communications. For a complete view of communications architectures, you should also read Chap. 12 (ground systems) and Chap. 13 (data processing).

We couldn't cover every detail of a communications architecture here, so you should consult references mentioned in each section for more detail and theory.

Of course, no matter how structured we try to make it, designing communications architectures isn't a clear-cut, step-by-step process that results in a single correct answer. Often, it's iterative, and the result is a family of options. The key thing to keep in mind is that you must tie the resulting architecture options to specified and derived requirements and constraints, and you must carefully document the process with detailed technical justification. If you don't a mission operations manager can easily discard and discount the staff-years of effort on one design in favor of another product, often basing this decision more on opinion than on fact.

## References

Altshuler, E. E.1983. "The Effects of a Low-Altitude Nuclear Burst on Millimeter Wave Propagation." *Rome Air Development Center In-House Report.* RADC-TR-83-286, December.

Chan, V.W.S. 1988. "Intersatellite Optical Heterodyne Communications Systems." *The Lincoln Laboratory Journal.* 1(2):169-183.

Comparetto, G. M. 1994. "Global Mobile Satellite Communications: A Review of Three Contenders." Presented at the *1994 AIAA 15th International Communications Satellite Systems Conference.* San Diego, CA., 27 Feb – 3 March.

Comparetto, G. 1993. "A Technical Description of Several Global Mobile Satellite Communications Systems." *J. of Space Comm.,* Vol. 11, no. 2, October: 97–104.

Comparetto, G. and Maj. M. Kaura. 1993. "Using Global Mobile Satellite Communications Systems to Support Army Requirements." Presented at the *AIAA Space Programs and Technologies Conference and Exhibit.* Huntsville, AL., 21–23 September.

Comparetto, G. 1993. "On the Use of INTELSAT and INMARSAT to Support DoD Communications Requirements." Presented at *MILCOM '93,* Paper number 1.2, Boston, MA., 11–14 October 1993.

Comparetto, G. 1993. "The Impact of Dust and Foliage on Signal Attenuation in the Millimeter Wave Regime." *J. of Space Comm.,* Vol. 11, no. 1, July: 13–20.

Comparetto, G., and W. Foose. 1990. "An Evaluation of the Gaussian Approximation Technique Applied to the Multiple Input Signal Case of an Ideal Hard-Limiter." Presented at *MILCOM '90,* Monterey, CA., 30 September – 3 October, Paper # 26.4.

Comparetto, G. 1989. "Signal Suppression Effects in an Ideal Hard-limiter for the Many-Carrier Case." *Int. J. of Sat. Comm.,* Vol. 7, No. 5, December.

Comparetto, G. and D. Ayers. 1989. "An Analytic Expression for the Magnitudes of the Signal and IM Outputs of an Ideal Hard Limiter Assuming "n" Input Signals Plus Gaussian Noise." *Int. J. of Sat. Comm.,* Vol. 7, No. 1, January.

Crane, R. K. 1980. "Prediction of Attenuation by Rain." *IEEE Trans. on Commun.,* Com-28(9): 1717–1733.

Davies, R. 1992. *Space Mission Analysis and Design.* Chapter 13 "Communications Architecture." Netherlands: Kluwer Publishing.

Dixon, R. C. 1976. *Spread Spectrum Systems.* John Wiley, New York, NY.

Gagliardi, R. 1984. *Satellite Communications.* Van Nostrand Reinhold Co., New York, NY.

Ha, T. 1986. *Digital Satellite Communications.* Macmillan Publishing Co., New York, NY.

Horwitz, G. M. 1979. "Optimization of Radio Tracking Frequencies." *Trans. App. Phys.,* Vol. 27, May: 393–398.

Ippolito, L. J. 1986. *Radiowave Propagation in Satellite Communications.* New York: Van Nostrand Reinhold.

Jasik, Henry, ed. 1961. *Antenna Engineering Handbook.* New York: McGraw-Hill.

King, M. A. and P.B. Fleming. "An Overview of the Effects of Nuclear Weapons on Communications Capabilities." *Signal,* Jan 1980: 59–66.

Larson, Wiley J., and James R. Wertz. 1992. *Space Mission Analysis and Design*. Netherlands: Kluwer Publishing.

Lin, S. and J. Costello. 1983. *Error Control Coding: Fundamentals and Applications*. Prentice-Hall, Inc., Englewood Cliffs, NJ.

McElroy, D. 1988. "The FEP Communications System." *AIAA 12th International Communication Satellite Systems Conference Proceedings*, 395–402.

Middlestead, R. W., et. al. "Satellite Crosslink Communications Vulnerability in a Nuclear Environment." *IEEE Journal on Selected Areas in Communications*, col. SAC-5, no. 2, Feb 1987.

Mie, G. 1908. "A Contribution to the Optics of Turbid Media, Especially Colloidal Metallic Suspensions." *Ann. Phys.*, Vol. 25: 377–445.

Morgan, Walter L. and Gary D. Gordon. 1989. *Communications Satellite Handbook*. New York: John Wiley & Sons.

Mottley, T.P., D. H. Marx, and W. P. Teetsel. 1960. "A Delayed-Repeater Satellite Communications System of Advanced Design." *IRE Trans. on Military Electronics*, April-July: 195–207.

Naderi, M. and P. Kelly. 1988. "NASA's Advanced Communications Technology Satellite (ACTS)." *AIAA 12th International Communication Satellite Systems Conference Proceedings*, 204–224.

Papoulis, A. 1965. *Probability, Random Variables, and Stochastic Processes*. McGraw-Hill, Inc., New York, NY.

Proakis, J. 1983. *Digital Communications*. McGraw-Hill, Inc., New York, NY.

Rafuse, R. P. 1981. "Effects of Sandstorms and Explosion-Generated Atmospheric Dust on Radio Propagation." *Technical Report Number DCA-16*, Massachusetts Institute of Technology Lincoln Laboratory, 10 Nov.

Schwab, L. M. 1980. "A Predictor Model for SHF and EHF MILSATCOM System Availabilities in the Presence of Rain." *Lincoln Laboratory Technical Note 1980-15*, Lincoln Laboratory, Massachusetts Institute of Tech., Lexington, Massachusetts, February.

Simon, M., et. al. 1985. *Spread Spectrum Communications Volumes I, II, and III*. Computer Science Press, Inc., Rockville, MD.

Sklar, B. 1988. *Digital Communications Fundamentals and Applications*. Prentice-Hall, Inc., Englewood Cliffs, NJ.

Spilker, J. 1977. *Digital Communications By Satellite*. Prentice-Hall, Inc., Englewood Cliffs, NJ.

Stallings, W. 1985. *Data And Computer Communications*. Macmillan Publishing Co., New York, NY.

Tamir, T. 1977. "Radio Wave Propagation Along Mixed Paths in Forest Environments." *Trans. App. Phys.*, Vol. 25, July: 471–477.

Tewari, R. K., S. Swarup, and M. N. Roy. 1990. "Radio Wave Propagation Through Rain Forests of India." *IEEE Trans. on Ant. and Prop.*, Vol. 38, No. 4, April: 433–449.

Torrieri, D. J. 1992. *Principles of Secure Communications Systems*. Artech House, Norwood, MA.

Weiner, Thomas F. and S. Karp. 1980. "The Role of Blue/Green Laser Systems in Strategic Submarine Communications." *IEEE Transactions on Communications*. Com-28(9): 1602–1607.

Yeh, K. C. and G. W. Swenson. "The Scintillation of Radio Signals from Satellite." *J. of Geophysics Research*, Vol. 64, No. 12, Dec 1959: 2281–2286.

# Chapter 12

# Ground Systems

Matthew J. Lord, *Loral*

Bound by the Earth, humans operate spacecraft from the ground. The success of any mission hangs on the ability to extract information from space and get it to the user, who may be anywhere on the globe. Equipment, facilities, and communication links—collectively called the *ground system*—enable us to control the space element.

The ground system provides all the necessary tools to conduct space mission operations. It presents spacecraft operators with the information required to maintain spacecraft health and allows them to control the vehicle's attitude and orbit. It provides knowledge of the payload to mission planners and allows them to alter mission parameters. And it collects, processes, and routes data to users.

This chapter will map the path from space mission requirements to a ground architecture able to meet them. We'll investigate and trade these requirements along the way, discuss the merits of using an existing ground network, and point out when it would be more cost-effective to build a new system. But first, we need a better understanding of the ground system.

## 12.1 Defining the Ground System

### 12.1.1 Ground-System Functions and the Single-Station Model

The ground system has three main functions: telemetry, tracking, and commanding, or *TT&C*.

*Telemetry* literally means "measurement from a distance." On the ground, we acquire and process engineering or payload telemetry from a spacecraft. *Engineering telemetry* gives operators details of the condition of the spacecraft and its subsystems. Spacecraft operators monitor the spacecraft's health and status using data such as the temperature and voltage of spacecraft batteries. *Payload telemetry* contains data pertaining to the spacecraft's mission. Payload telemetry may be processed by the ground system or sent to (sometimes recorded for) the user for processing. Engineering and payload telemetry can combine in a single data stream, requiring only one set of receiving and data-handling equipment on the ground. More often spacecraft use two or more simultaneous telemetry streams, which require the same number of equipment strings in the ground system.

*Tracking* is acquiring a spacecraft and following its motion through space. We track a spacecraft by gathering data on its location and using the data later to control ground antennas. We can use different coordinate systems and frames of reference to describe an orbit, but to swing a ground antenna toward a spacecraft, we must know three points: the elevation (measured in degrees from the horizon), the azimuth (measured in degrees from true north), and the time when the spacecraft will be at that location.

*Commanding*, or controlling a spacecraft from the ground, occupies separate tasks and equipment within the ground system. We command a spacecraft by uplinking data to it. This data may contain operations the spacecraft must do, changes to onboard software, or information that spacecraft instruments will reference. The spacecraft may act on each command immediately or store them for execution at a future time, as directed by the command.

The single-station model describes the minimum set of elements we need to do ground-system tasks. It consists of spacecraft operators, hardware and software systems, and the facilities to support them. In the model shown in Fig. 12.1, we can do all critical tasks at one site. The station supports telemetry, tracking, and commanding for one spacecraft at a time. It acquires telemetry with the antenna, receives and demodulates the rf signal, processes the raw data, and displays or stores it for the user. It generates commands, modulates and amplifies them for uplink, and then transmits them to the mission spacecraft. It collects tracking data with the antenna and processes it for orbit planning.

This model has limitations that most missions will find unacceptable. A single ground station lacks redundancy, making the spacecraft it supports vulnerable to any number of environmental or human threats that could disrupt ground operations. Tracking data collected from a single station generates less accurate ephemeris. A single station also has a limited field-of-view, providing poor coverage for spacecraft in orbits other than geosynchronous. Even geosynchronous spacecraft will need more ground-observation points during the critical launch and early-operations phases. In addition, a single station is ill-equipped to handle spacecraft anomalies, when the vehicle needs more contacts throughout its orbit for troubleshooting.

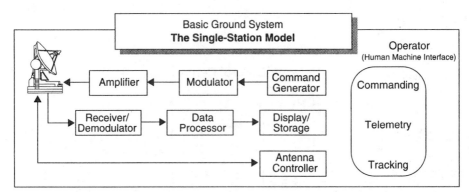

**Fig. 12.1.    The Single-Station Model.** A basic ground system does telemetry, tracking, and commanding. These functions correspond to data transport and delivery, navigation planning and analysis, and mission control, as described in Chap. 3.

Some missions can accept these limitations. Those that can't may still be able to use a single station if the spacecraft is designed for compatibility with other networks, so operators can tap these external resources during special activities. Most missions will require more complex ground systems.

### 12.1.2    Ground-System Elements and a Typical Ground-System Model

We can't fully describe a ground system's abilities with the single-station model. Figure 12.2 shows a more complete network. In this model we can support several spacecraft simultaneously by sharing network resources. Multiple ground stations provide the coverage needed for various orbits and mission architectures. To the single-station model we've added subsystems for recording and timing and more explicitly defined the hardware. This model of the typical ground system can't describe the best configuration for every mission, but it's convenient for discussing the ground system's potential functions and abilities.

We can functionally organize the typical ground system into three areas: ground stations, control centers, and communication links. A *ground station* is an installation on the Earth comprising all the equipment needed to communicate with a spacecraft [Williamson, 1990]. Antennas and TT&C hardware at the ground station transmit and receive the rf signals that operate the spacecraft. The ground station receives commands from the control center, then modulates and formats them for uplink to the spacecraft. It receives telemetry from the spacecraft and relays it to the control center (or directly to the data user). The ground station also tracks the spacecraft, providing data to the control center for orbit determination and prediction. We can double the support capacity of a ground station without doubling the cost by adding a second antenna and suite of TT&C hardware. A

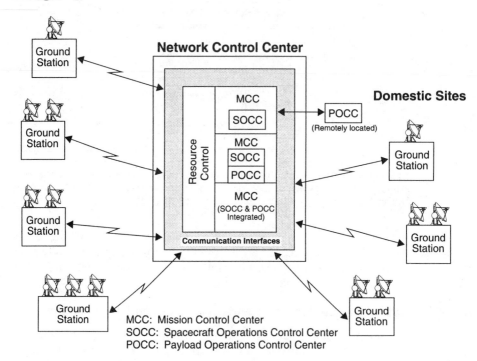

**Fig. 12.2.    The Typical Ground System.** This system supports many spacecraft simultaneously with multiple control complexes and remotely located ground stations. We can organize the control complexes in different ways, as shown, but the real network won't have this variety.

ground network may require several ground stations dispersed around the Earth to cover the globe.

The *Network Control Center* is the central node for controlling the ground stations. It maintains the communication systems that pass data to and from the spacecraft-control centers. It also provides host services for the spacecraft-control centers, including building and ground maintenance, utilities, and security. Central control of mission operations at one site minimizes costs for facility, equipment, and operations.

The *Mission Control Center* (MCC) operates the mission. Engineers and operators within the MCC create a mission plan, schedule spacecraft resources, and select ground resources to meet mission objectives. They determine the spacecraft's orbit and attitude and they send predicted values to the ground station for tracking acquisition. For simple spacecraft, the MCC carries out the mission plan.

For more complex spacecraft, a separate facility, the *Spacecraft Operations Control Center* (SOCC) controls the spacecraft's subsystems and processes its data. The SOCC generates commands and passes them to the ground station. It monitors spacecraft-maintenance telemetry received from the ground station. When the MCC and SOCC are integrated, we use the terms interchangeably.

The SOCC analyzes data only from subsystems that affect the spacecraft's health or attitude. The *Payload Operations Control Center* (POCC) analyzes the mission data. Because it's responsible for the payload, the POCC helps form the mission plan. The POCC will request command sequences to control payload instruments, but to protect the common spacecraft bus, only the SOCC is allowed to transmit these commands.

The *Resource Control Center* (RCC) schedules and monitors the use of all network resources. It accepts requests from the MCCs, and assigns blocks of time for using ground stations, communication links, and equipment strings. The RCC assigns resources to spacecraft based on priority, as described in Sec. 12.4. In smaller networks the MCC controls resources.

The MCC, SOCC, POCC, and RCC are typically co-located at the network control center, each occupying a room or several rooms. Some missions choose to place the POCC outside the network at the user's site. Smaller missions combine some or all of the centers into a single room.

### 12.1.3 Communication Links

Communication links complicate a ground system and increase its cost. We normally don't want to design communication links into a ground system unless we have to. But rarely are all ground-system elements co-located, so we nearly always need communication links to transport data between them. Figure 12.3 shows the kind of information we may need to transmit over some link depending on the location of facilities. The best type of link for a particular signal will depend on the type of information being transmitted, and how it will be used. We have several options for moving information, so we must choose between private and commercial services, satellite and terrestrial links, and switched and dedicated lines. Our solution depends on whether requirements call for voice or data signals; whether distribution is point-to-point or multi-drop; and whether speed, distance, security, or survivability are design constraints. A satellite-control network will likely employ more than one type of link to connect ground-system elements. Powers [1990] provides a good introduction to communications.

**Private Data Links.** Between buildings on the same campus, or over other short distances, we normally use private communication links. These links can be installed by the operating agency or by an outside contractor. Among the oldest media for transmitting signals, copper cable is still the most popular option for requirements under a few kilometers.

Although inexpensive and simple to install, copper cable is limited by frequency and distance. Optical cable offers increased performance at prices that are

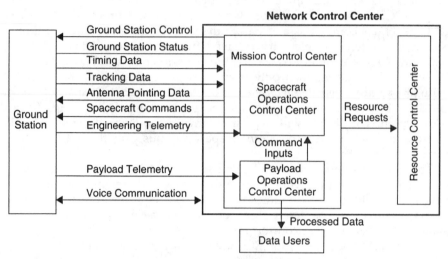

**Fig. 12.3.    Communication Links.** We must transmit various data between ground-system elements.

becoming competitive with copper. A fallout benefit of fiber is the security derived from optical isolation. Fiber cables suppress any unwanted signal riding on the transmitted data, and without a magnetic field, they are much more difficult to tap than their copper cousins. Because most data systems are electrical, signals must be converted to optical for transmission over fiber and back to electrical at the receiving end. As the advantages of fiber continue to persuade the communications industry, more vendors will offer fiber interfaces for their equipment.

The largest cost in a new cable installation is often for constructing the cable's passage (conduits, trays, or poles). Given an unobstructed, line-of-sight path between transmission points, terrestrial microwave radio is a cost-effective alternative to cable. Microwave is effective for high-bandwidth transmission, but frequency coordination is critical to avoid signal interference. The Federal Communications Commission regulates all microwave radio installations.

Table 12.1 compares microwave, fiber, and optical systems. See McClimans [1992] for more information.

**Terrestrial Commercial Communication.** When transmission distance makes a private link impractical, we use commercial services. For data rates under 20 kbps, a simple dial-up telephone line connects two modems; if we need dedicated service, we can lease the line. For higher data rates, a popular service to link two points in North America is the T-carrier. A T1 circuit passes 1.544 Mbps, and a T3 circuit passes 44.736 Mbps. These circuits can be used for one signal or shared by many signals with a multiplexer. Public carriers are working to increase the throughput of communication links. Emerging services such as the synchronous

**Table 12.1.** **Sample Comparison of Private Communications Media.** The values represent the capabilities of moderately-priced systems. We assume a small installation. As the size of the project increases, there are cross-over points at which fiber or microwave is more cost-effective.

| Specification | Copper Cable | Fiber-Optic Cable* | Microwave Radio |
|---|---|---|---|
| Data rate–mega bits per second | 10 to 16 | 150 | 90 |
| Distance (between repeaters) | 1 km | 10 to 20 km | 40 to 50 km |
| Relative hardware cost | Low | Moderate | High |

\* For multi-mode fiber. Single-mode fiber systems can transmit several gigabits per second, and unrepeated distances of nearly 100 km are achievable.

optical network (SONET), using advanced protocols like asynchronous transfer mode (ATM), promise data rates above two Gbps. More modest rates from tens to hundreds of Mbps over ATM are quickly becoming available today. The emergence of packet-based services like ATM represents a trend in ground communications—departure from dedicated links like the T1 in favor of non-dedicated, broadband services.

A potential disadvantage of commercial communications to a user like the military is the risk of not having control of all its assets. The user depends on the commercial carrier. But when dollars count, the reliability and cost-effectiveness of commercial services are convincing. It now seems to be a risk the military is willing to take.

**Satellite Communication.** Spacecraft can be the medium as well as the mission. To relay high-bandwidth data across long distances, such as between ground stations and the network control center, the spacecraft has traditionally offered the most cost-effective solution.

Today, the spacecraft competes with fiber and other media whose bandwidth exceeds that of spacecraft, without some of the problems of satellite communication. Geostationary satellites, in particular, are encumbered with a long transmission path, giving rise to a one-way delay of 0.25 seconds, which is annoying for voice communication and disruptive for computer networks. But spacecraft have unique advantages over other services. Satellite communication is mobile and flexible. Cost is essentially independent of distance and location of ground terminals. The link doesn't depend on the public-telecommunications infrastructure. And spacecraft can broadcast to an unlimited number of users.

We distinguish communication satellites by frequency, as well as the number and bandwidth of transponders. A typical transponder bandwidth for C-band is 36 MHz, which can be leased in whole or part. Ku- and Ka-band transponders offer greater capacity. The cost of a ground terminal can be moderate to expensive, depending on the bandwidth, which drives the size and complexity of the antenna and its subsystems. Chapter 11 describes satellite communication in detail.

## 12.1.4    Describing the Subsystems

We can better understand the typical ground system by examining its subsystems: the *antenna, TT&C, timing,* and *command and data processing.* Figure 12.4 describes the relationship among these groups.

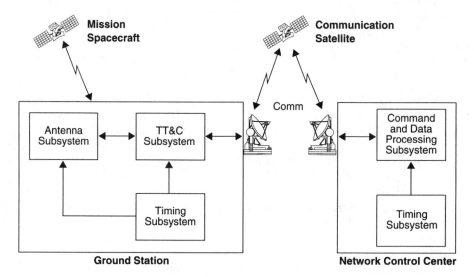

**Fig. 12.4.**    **Relationship of the Four Subsystem Groups.** Note that the communication link is not necessary in a co-located network. The next four figures show the four subsystems in more detail.

**Antenna Subsystem.** As you can see in Eq. (11.11), antenna gain is one of the few ground factors that can affect link margin. This gain is a key to whether a ground station can support a particular spacecraft in a particular orbit. Transmit gain is represented in a term called the *effective isotropic radiated power* (EIRP), which is simply antenna gain multiplied by transmit power. Receive gain is represented in the term G/T (receive antenna gain divided by system noise temperature). Together, EIRP and G/T describe an antenna's performance. For a given frequency, gain depends on antenna size and type. The parabolic-reflector antenna is the most widely used in ground systems because its relatively high gain and narrow beamwidth are attractive for satellite communication. The antenna communicates data to and from the spacecraft by converting between electronic signals and magnetic fields. With different frequencies for uplink and downlink, we can use the same antenna for transmission and reception. These frequencies are typically offset by a fixed ratio.

The uplink rf signal passes through the feed and radiates from the reflector. Downlink signals are collected by the reflector and passed through the feed to the low-noise amplifier. The antenna is anchored by the pedestal and controlled by azimuth and elevation servos, drive motors, a positioner, and a control unit. Figure 12.5 shows antenna components in the *prime-focus* configuration. An alternate configuration, called *Cassegrain*, uses a subreflector to locate the feed behind the main reflector (with the rest of the electronics), eliminating some cable loss.

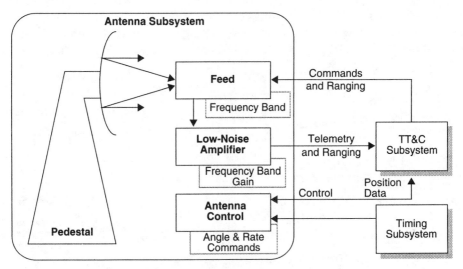

**Fig. 12.5.**     **Antenna Subsystem.** Options and considerations are boxed below each hardware component. For example, low-noise amplifiers differ by the frequency bands they pass and the amount of gain they provide.

We control the antenna through the TT&C subsystem in one of three modes: auto, slave, or manual. *Auto-tracking,* which is preferred, uses the signal strength of the downlink at the receiver to lock onto and track the spacecraft. We employ *slave tracking* to first acquire the spacecraft before we receive the downlink or as a fall-back method when the signal strength is too low. For this method, a computer programmed with predicted ephemeris drives the antenna by sending it time-dependent azimuth and elevation angles. A typical acquisition technique is to slave track the antenna into position ahead of the spacecraft track and then switch to autotrack as the spacecraft flies through the antenna beam. For geostationary spacecraft, we manually point the antenna, so tracking isn't required. The spacecraft orbit and its required accuracy determine the sophistication of tracking and ephemeris systems.

**TT&C Subsystem.** The TT&C subsystem defines the rf interface between the space and ground elements. The frequency and modulation technique supported by ground hardware must be compatible with the TT&C subsystem onboard the spacecraft. (Section 12.2 describes compatibility.) Located within the ground station, the TT&C subsystem includes a high-power amplifier to transmit the uplink signal; modulators, demodulators, synthesizers, receivers, and bit synchronizers to convert signals between rf and digital; and recorders to record and playback the spacecraft's data. Figure 12.6 shows only one of each device but most ground systems use several strings of equipment, cross-strapped through switches, for multiple, simultaneous contacts and redundancy. We configure this equipment before the spacecraft contact, after which it runs with little or no operator control. Most ground networks use computers to control these systems, either locally at the ground station, or remotely from the network control center. The same computers can automate testing of the equipment and the circuits that connect them.

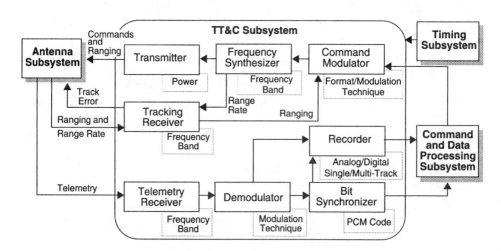

**Fig. 12.6.    TT&C Subsystem.** Frequency band and modulation technique are prime considerations for selecting TT&C hardware. Options and considerations are boxed below each component.

The TT&C subsystem includes hardware to gather orbit data. What we require depends on whether we need tracking data simply to locate the vehicle or to calibrate spacecraft sensors. In some cases, the mission needs nothing more than the course pointing angles the antenna provides. Most missions need more precise orbit data. We can determine the spacecraft's position from the ground by using optical, radar, and radio techniques. Optical tracking works only if the spacecraft is sunlit and visibility conditions are good, but radar works in any weather. Optical and older radar systems treat the spacecraft as a passive reflector; with modern radar systems, the spacecraft receives and retransmits the signal. Radar is typically used by sites independent of the ground system. Most ground stations use the spacecraft's rf link to measure the distance, or range, and radial velocity, or range rate, of the vehicle relative to the ground station. *Range, r,* is calculated by measuring the round trip delay, $t_r$, of a signal transmitted on the uplink and returned on the downlink:

$$r = ((t_r/2) - t_p) c \qquad (12.1)$$

where $c$ is the speed of light, and time, $t_p$, accounts for ground and vehicle processing time and other delays.

*Range rate* is calculated by measuring the Doppler shift of the downlink frequency. The Doppler shift, $\Delta f$, is directly proportional to the component of the spacecraft's velocity, $\dot{r}$, in the direction of the ground station

$$\dot{r} = \Delta f \cdot \lambda \qquad (12.2)$$

where $\lambda$ is the wavelength of the downlink frequency at the spacecraft transmitter. If our spacecraft's downlink frequency is $f_d$, and we measure a Doppler-shifted frequency on the ground of $f_m$, we can define the Doppler shift as

$$\Delta f \equiv f_d - f_m \qquad (12.3)$$

and we can solve again for range rate (recall that $c = \lambda f$):

$$\dot{r} = (1 - (f_m/f_d)) c \qquad (12.4)$$

In practice, the ground system does more to calculate range rate than Eq. 12.4 implies, mostly because we can't rely solely on the downlink frequency. When range-rate measurements aren't being taken, the spacecraft can operate in *noncoherent mode*, in which an onboard oscillator will generate the downlink carrier frequency. This onboard oscillator isn't stable enough to provide accurate Doppler measurements on the ground. When gathering range-rate data, the spacecraft must be configured for *coherent mode*, in which the spacecraft transmitter is phase-locked to the received uplink carrier. In coherent mode, the spacecraft will shift the received uplink frequency by a fixed ratio and modulate the downlink signal onto this carrier. Table 12.2 lists the downlink-to-uplink frequency ratios and the types of signals used for range measurement at three existing networks.

**Timing Subsystem.** Many ground-system functions are time critical. Tracking data is of little use unless we know when the data were collected, and to track a spacecraft by computer we must tell the antenna when to move. We must time-tag telemetry at the recorder and telemetry processor (unless the spacecraft has embedded a time code in the data itself). A mission operator needs accurate time displays to coordinate events during the contact. The timing subsystem does these tasks.

**Table 12.2.    Range and Range-Rate Data for Three Existing Networks.** Networks operate in the coherent mode to provide accurate Doppler measurements.

| Network | Range Measurement Signal | Downlink-to-Uplink Carrier Frequency Ratio, $r^*$ |
|---|---|---|
| Air Force Satellite Control Network (AFSCN) | 1 Mbps pseudorandom noise (PN) code | 256/205 |
| NASA Tracking and Data Relay Satellite System (TDRSS) | 3 Mbps PN code | 240/221 (S-Band) 1600/1469 (K-Band) |
| NASA Deep Space Network (DSN) | 1 Mbps PN code plus eight ranging tones, 8 Hz to 500 kHz | 240/221 (S-Band) 749/880 (X-Band) |

\* That is, $f_d$ (downlink) = $r \cdot f_u$ (uplink)

Subsystems on the ground have been synchronized with calibrated atomic clocks and with land-based transmission systems such as LORAN C. The Global Positioning System (GPS) has reduced the cost and increased the accuracy and simplicity of the timing function by an order of magnitude. Racks of timing gear have become just a GPS receiver and time-code generator, connected as shown in Fig. 12.7. (An optional frequency standard allows the system to maintain accurate time if no GPS signal is present.) GPS satellites provide an accuracy within a few hundred nanoseconds of a world time standard, such as Coordinated Universal Time (UTC). When ground stations are remote from the network control center, each will have a separate timing subsystem to account for the transmission delay between them.

**Command and Data Processing Subsystem.** This subsystem defines the data-format interface between the space and ground elements. Commutation, coding, and encryption schemes employed on the ground must match those on the spacecraft. Located within the MCC, computers and supporting systems generate commands, compute orbit dynamics, and process baseband telemetry from the spacecraft and payload into user-required formats. The type and complexity of systems varies greatly but often includes the tasks illustrated in Fig. 12.8 and discussed below.

The broadcast nature of satellite communication is especially vulnerable to unwanted data monitoring, or worse, illegal commanding. Whenever we need security, we encrypt signals before transmission (usually in the MCC), and decrypt after reception (in the MCC, POCC, or at the user's location). The spacecraft decrypts commands and encrypts telemetry. The military routinely uses encryption; NASA and ESA don't.

Before we can process telemetry, we must format and time-tag it. Formatting includes frame synchronizing and decommutating, which converts serial bit streams into discrete logical measurands.

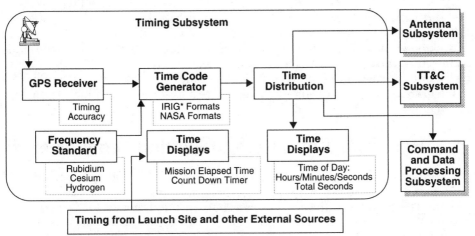

**Fig. 12.7.    Timing Subsystem.** Various time codes and display options allow distribution to the many systems that require accurate time. Today, we can buy the GPS receiver, frequency standard, and time-code generator as a single unit. Options are shown in boxes below each component.

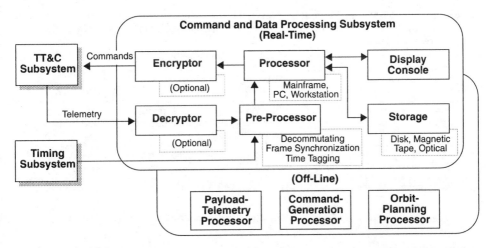

**Fig. 12.8.    Command and Data Processing Subsystem.** This subsystem consists of real-time and off-line equipment. We may process payload telemetry in real time or after the spacecraft contact. Most modern systems connect the workstations and their peripheral devices through a local-area network. Options are shown in boxes below each component.

The volume and complexity of the telemetry will drive the ground processors' cost and complexity. Large mainframe computers have long helped us do the core tasks of mission operations. Today, the flexibility, computing power, and low cost of desktop workstations commonly drive us away from a centralized (mainframe) architecture. A distributed architecture on a local-area network (LAN) allows us to install and maintain ground processors in modules. Standard interfaces allow us to integrate commercial off-the-shelf hardware and software products from various vendors.

## 12.2 Hardware, Software, and Staffing Requirements

With an understanding of the ground system's tasks and components, and with a mission in mind, we may begin to bring mission requirements to the ground. This highly iterative process must balance requirements for the spacecraft and mission objectives with cost and risk. It must weigh factors such as desired orbit, data rates, and system complexity against each other. The far-thinking mission operations manager will trade requirements between ground-system components and also between the space and ground elements. Table 12.3 lists the steps that lead to a ground system which can support mission objectives. The overlap in trading system requirements suggests we'll run through the table several times before reaching the most cost-effective solution. Table 12.4 summarizes the key influences on our the assessment of the ground system.

Table 12.3.    **Specifying a Ground System Based on Mission Requirements.** This process defines a ground architecture (existing or new) to support the mission objectives.

| Step | Description | Reference |
|---|---|---|
| 1. Characterize the mission | Class and type of mission, type of supports | Chap. 2 Sec. 12.2 |
| 2. Define key requirements | Type and frequency of supports, geography, availability, timeliness, bandwidth, compatibility | Chap. 11 Sec. 12.2 |
| 3. Identify what the network and associated ground elements must do | Number of ground stations, type of comm links, user connectivity | Chaps. 3 and 11 Sec. 12.2 |
| 4. Do key trades | See Tables 12.4 and 12.5 | Chap. 2 Sec. 12.2 |
| 5. Choose existing or new network | AFSCN, TDRSS, NAVSOC, DSN, or other | Sec. 12.3 |
| 6. Evaluate complexity and cost | Calculate the initial cost and model the operational cost | Chap. 5 |
| 7. Document and iterate | | |

**Table 12.4.   Mission Requirements.** Thoroughly characterizing the mission and analyzing key system requirements lay the foundation for developing the ground-system architecture. In some cases a trade for one requirement becomes a system requirement.

|  | Parameter | Options | Key Trades |
|---|---|---|---|
| **Mission Definition** | Mission class | • Military<br>• Commercial<br>• Scientific | • System complexity<br>• Security/survivability |
|  | Mission type | • Data collection<br>• Communication | • User connectivity |
|  | Communication and orbit architecture | • Constellation: single or multi-spacecraft<br>• Orbit: LEO, GEO, Molniya | • System complexity<br>• Frequency of supports |
|  | Mission complexity* | • Simple<br>• Complex<br>• Autonomous | • Frequency of supports<br>• Ground system complexity |
| **System Requirements Analysis** | Type of supports | • Launch and early orbit operations<br>• Routine on-orbit operations<br>• Anomaly operation | • Frequency of supports<br>• Staffing and resources |
|  | Frequency of supports | • Once per orbit<br>• Once per day<br>• Variable | • Spacecraft design<br>• Geography<br>• Orbit<br>• Staffing and resources |
|  | Geography | • Domestic/foreign<br>• Ground stations<br>• User locations | • Orbit<br>• Frequency of supports |
|  | Availability of resources | • Shared network<br>• Dedicated network | • Network augmentation |
|  | Timeliness of data processing | • Real-time<br>• Store and forward<br>• Record at ground station | • System complexity<br>• Communication bandwidth<br>• Geography |
|  | Communication bandwidth | • Variable data rates and modulation techniques | • Power and antenna size<br>• Equipment complexity<br>• Communication links |

\* See Chap. 5 for a list of complexity metrics for the ground system.

If we choose an existing system, we establish compatibility between the space and ground systems and determine if the existing system must be modified to meet the needs of the mission. If we decide to build a new system, we begin the detailed design. We focus on developing the architecture—designing a ground system isn't within the scope of this chapter.

### 12.2.1    Defining the Mission

Ultimately, the mission drives all ground-system requirements and its size, cost, and complexity. Missions differ by sponsor, type, orbit, complexity, and other factors.

**Mission Sponsor.** Missions are either military, commercial, or scientific. Each type has its own philosophy of operations. Military missions have historically favored mission objectives over cost. As defense budgets decrease, this philosophy will shift, but military missions will continue to require ground systems that are impervious to natural disasters, terrorist attack, power outages, and equipment failure. The ground system reflects this security and redundancy, having multiple ground stations, back-up control centers geographically separated, and redundant communication links and equipment strings. On the other hand, profit drives commercial missions just as it does other commercial ventures. Commercial missions operate their spacecraft with far fewer people than military missions, using a philosophy that is good advice for any mission: don't contact the spacecraft any more than you have to. Scientific missions aim to maximize information return but often under strict budgets. Indeed, the limited funding for some scientific missions severely restricts ground-system operations. These missions may emphasize end systems that can process and archive large amounts of data.

**Mission Type.** The type of mission affects how ground systems look, including user connectivity. Chapter 2 defines five mission types: communications, navigation, remote sensing, scientific, and interplanetary exploration. In a data-collection mission, the user may be connected directly to the network to share data conditioning and processing hardware. The users of a communications or navigation mission, on the other hand, need real-time connectivity, so they will operate transmit-and-receive or receive-only stations that are independent of the ground system which controls the spacecraft.

**Mission Orbit.** This includes both the number of spacecraft in the constellation and the orbits. Examples of mission architectures for communications satellites include store-and-forward, three-satellite geostationary, and Molniya-orbit, (see Fig. 11.11). The number of spacecraft needed to do the space mission strongly affects the quantity and complexity of ground resources. Equally significant is the orbit. The orbit determines how much time the spacecraft is in view of a ground station and how long it will be before the next pass.

**Mission Complexity.** The complexity of operations and of spacecraft subsystems influences the ground system's complexity and its number of operators. For example, the level of support goes up when the mission requires frequent orbit maneuvers but goes down when the spacecraft uses autonomous navigation. Mission designers trade complexity in the ground system for complexity on the spacecraft. Spacecraft subsystems that process data onboard greatly reduce or eliminate requirements on the ground system.

### 12.2.2   Analyzing System Requirements

By analyzing system requirements, we can describe the ground-system architecture at a high level. Below are the key system requirements that drive a ground system.

**Type of Supports.** The ground system must support the type of routine and special activities required during the life of the mission. Special activities demand resources beyond those for routine operations. Assets are required for launch, for example, that may not be necessary during the rest of the mission. Controllers and systems must be prepared to support planned special activities, as well as anomalous conditions.

**Frequency of Supports.** The ground system must support the number of spacecraft contacts required to operate without significantly restricting mission objectives. The quantity and perishability of data to be downlinked, spacecraft design, and orbit contribute to the frequency of routine operations. Special activities such as orbit and attitude maneuvers or resolving anomalies increase the frequency of supports, whereas spacecraft designed for autonomy depend less on ground commands to do mission tasks. How often we must contact the vehicle drives the number of ground stations and staffing levels.

When downlinking of mission data drives contact frequency, we may trade the spacecraft's data rates and storage capacity with orbit and the number of ground stations to reach the desired level of support. As altitude decreases, station visibility decreases, leaving less time per contact. With a large amount of data to downlink, a spacecraft in low-Earth orbit requires more contacts than at higher altitudes. When increasing the altitude is too expensive, or when we can't design for a different orbit without significantly altering mission objectives, we may trade contact frequency with the data rate. We can define the number of contacts, $C$, required to downlink a given amount of data, $Q$, in terms of the average duration of a spacecraft's visibility, $V$, and its data rate, $R$, as shown below:

$$C = Q/(RV) \tag{12.5}$$

For example, let's consider a spacecraft in LEO with an average visibility to a ground station of ten minutes. Our spacecraft can store one gigabyte (or eight gigabits) which is filled every three orbits. The data rate is five Mbps. The number of contacts required before spacecraft memory overflows is therefore 2.67, or nearly one ccntact every orbit. If the ground system doesn't have the availability for this contact frequency but can support higher data rates, we should design the spacecraft for a higher telemetry rate. Note that the available time to downlink data is normally less than the total duration of visibility. Acquiring the spacecraft, sending commands, and checking its health can reduce the available time by several minutes. At higher altitudes, this reduction is less significant.

**Geography.** Ground stations must provide enough coverage to satisfy mission requirements. As discussed above, the frequency of ground supports dictates the number of ground stations, whereas the spacecraft's altitude and inclination largely determine their locations, within geographical and political constraints. A single station can fully cover only spacecraft in geosynchronous orbit. All other orbits will likely require several dispersed stations for full coverage. We like to construct ground stations in remote locations to avoid frequency interference with terrestrial microwave transmissions. But we want control centers in populated areas to take advantage of cheaper public telecommunications and support services. We discuss coverage more fully in Chap. 10.

**Availability of Resources.** The ground system must have enough resources to make the required contacts at or near the time desired. The ground system's availability depends on the quantity of resources, other spacecraft supported by the network, the ability to conduct multiple contacts and process multiple data streams, and the entire system's reliability. Reliability depends on the likelihood of components or systems failing and the time required to repair them. When a network formally defines its availability as, say, 0.9999, it is referring strictly to its reliability. An availability of 0.9999 means the network experiences 53 minutes of cumulative downtime per year, but not that each user has access to all network resources the rest of the time. We can calculate availability from figures on mean time between failure (MTBF) and mean time to repair (MTTR) for components and the system. Most ground systems increase availability with designed redundancy in all elements, including the power subsystem. An uninterruptable power system (UPS) uses batteries and often a diesel generator to provide from 30 minutes to continuous backup power to critical equipment during an outage. We must carefully weigh availability requirements because they strongly affect system costs.

A decision that greatly affects resource availability is whether to use a shared or dedicated network. Spacecraft supported by shared resources must be scheduled based on equipment requirements, station availability, and mission priority, as described in Sec. 12.4. When requirements exceed the ground stations' abilities, we can install more equipment, communication links, or even ground stations.

**Timeliness of Data Processing.** The ground system must meet the users' requirements for timely data routing and processing. Missions with perishable data must be supported by system elements able to communicate, process, and display data in sufficient time, and at a sufficient rate, so ground buffers and processors don't overflow and data isn't delayed. Weather data, for example, is almost useless if more than an hour old. This requirement can severely tax ground-system coverage, or more likely, result in an architecture in which users receive data directly from the spacecraft at stations independent of TT&C on the ground.

Data that isn't time critical allows more flexible, and thus less expensive, ground control. We can record this type of data onboard the spacecraft and downlink it less often (a mission architecture called *store and forward*). Or we can record it at the ground station and play it back after the contact or mail tapes to the users.

If a communication link doesn't exist, or the bandwidth isn't available, real-time transmission can be expensive. Whenever possible, mission planners should investigate alternatives to real-time data communication. A user requirement for relaying telemetry in real time may actually mean they need one piece of data that an existing voice network can relay in near real time.[*]

Regardless of the mission, TT&C operations are normally in real time. To maintain the spacecraft's health, the control center must be able to monitor spacecraft subsystems and change parameters while in contact with the vehicle.

**Communication Bandwidth.** The ground system must have enough bandwidth to transport TT&C signals to and from the spacecraft, as well as to support the user's data routing and processing needs. A spacecraft's data rates for TT&C can't be outside the ground system's maximum and minimum rates for generating and transmitting commands or for receiving and processing telemetry. They must also match the communication bandwidth between segments for data relay. As data rates increase, so do antenna size, transmitter power, and costs for equipment and communication links. We can lower data rates with more frequent contacts, but this approach places more stress on system resources and increases staffing levels. Techniques for data compression or onboard processing lower data rates but add complexity and cost to both the space and ground elements.

The required bandwidth of the space-to-ground link mainly depends on the telemetry data rates. The number of spacecraft instruments to monitor and the amount of payload data to collect affect the downlink bandwidth, which is normally much greater than the uplink bandwidth.

The required bandwidth of the ground-to-space link, essentially the command rate, depends on the spacecraft's complexity. Most spacecraft require a very low command rate—a few hundred to a few thousand bits per second.

The required bandwidth of the communication link between the ground station and the control center (or user) is the sum of telemetry streams, tracking data, voice, and ground-system status. When telemetry rates exceed this bandwidth, some alternatives may be cheaper than installing or upgrading a communication link. A research spacecraft to be supported by the AFSCN, for example, has a five Mbps telemetry stream that can be received by the ground station but not relayed in real time over the 1 Mbps satellite link to the control center in California. The solution is to record the telemetry at the ground station and mail the tapes to the control center. Another option is to play the data back to the control center at a slower rate (in this example, one-fifth the rate at which the data was recorded). This option keeps processing speeds down but ties up resources longer.

---

[*] *Real-time* data is less than a few seconds old, or the time it takes for transmission, processing, and display. *Near real-time* data has been further delayed, such as telemetry data processed at the control center and then reformatted and transmitted to a remote user.

### 12.2.3    Functional Requirements and Compatibility

The high-level requirements discussed thus far help determine the ground system's basic structure, such as number and location of ground stations or control centers and the size of the communication links between them. The next step in developing the ground architecture is to derive and trade the low-level requirements. By analyzing what the ground system must do, we can determine what we need on the ground. Table 12.5 lists some of the more important functions and related design issues. Of course, we must always consider and trade cost. Refer to Chap. 3 for a detailed description of each function.

**Table 12.5.    Ground-System Functions.** The functions required for a mission drive the ground-system design.

| Function | Considerations and Constraints | Key Trades |
|---|---|---|
| Data Transport and Delivery | • Quantity and rates of data<br>• Location of ground system elements<br>• Compatibility between space and ground elements | • Process telemetry at ground station vs. control center<br>• Choose type of communication links<br>• Design spacecraft for compatibility vs. modify ground system |
| Mission Control | • Complexity of mission<br>• Operations and maintenance philosophies | • Shared vs. dedicated resources<br>• Redundancy vs. allowable system downtime |
| Spacecraft Planning and Analysis | • Complexity of spacecraft bus<br>• Orbit | • Level of ground automation<br>• Sophistication of software |
| Payload Planning and Analysis | • Type of payload<br>• Orbit | • Level of onboard autonomy<br>• Level of ground automation |
| Data Processing | • Location of users (co-located or external)<br>• Quantity of payload data | • Process data in Mission Control Center vs. dedicated Payload Operations Control Center<br>• Process data in real time vs. post-pass |
| Navigation Planning and Analysis | • Orbit<br>• Required knowledge of orbit | • Internal vs. external orbit determination (e.g., NORAD)<br>• Ground vs. onboard processing (e.g., GPS)<br>• Antenna angle data only vs. ranging and Doppler systems |
| Archiving | • Quantity of data<br>• Compatibility with existing recorders (e.g., at the user's facility)<br>• Duration of storage | • Store raw vs. processed data<br>• Type of storage media<br>• Type of distribution and location of storage (transportability) |

We begin the design process by analyzing functions. For example, if we must process payload data at a remotely located POCC in real time, we can begin looking at suitable data-processing systems, as well as communication systems to connect the POCC to the network. And if the mission requires us to archive this data in the POCC, we can begin looking at options for recorders. By analyzing all critical functions in this way, we begin to shape the ground system. From here, we can eliminate existing networks that don't meet our needs, begin investigating ones that do, and compare their cost with the cost of building our own.

**Compatibility.** Before we know whether the ground architecture can support the mission spacecraft, we must establish compatibility between the two. It may help to understand compatibility if we look at what happens to commands and telemetry as they pass between the spacecraft and the ground system. The goal of any transmission is to receive what was sent and be able to understand it. To achieve this goal, satellite transmission requires us to encapsulate information in successive layers. We first code raw information and format the data, possibly encrypt and encode it, and then modulate the data onto a carrier frequency. Table 12.6 describes this process in seven layers. Note that the spacecraft and ground system must be compatible at every layer. A single incompatibility means we can't contact the spacecraft.

**Table 12.6. Seven-Layer Model for Satellite Transmission.** Each layer becomes an element of compatibility. We define each layer in the text.

| Layer | Element | Options | Where Discussed |
|:---:|---|---|---|
| 7 | Data rate | Different rates for engineering and payload telemetry, commanding | 11.3 and 12.2 |
| 6 | Data code | PCM codes: NRZ (L, S, M), Biphase (L, S, M), Miller | 11.3 and 12.2 |
| 5 | Data format | Binary, framed, packetized | 12.2 and 13.3 |
| 4 | Encryption | No encryption or various military and commercial algorithms | 11.1 and 12.2 |
| 3 | Encoding | No encoding or convolutional (with Viterbi decoding), Reed-Solomon | 11.2 and 12.2 |
| 2 | Modulation technique | Analog: AM, FM, PM<br>Digital: FSK, PSK, BFSK, BPSK, QPSK | 11.3 and 12.2 |
| 1 | Frequency band | UHF, L-Band, S-Band, X-Band, Ku-Band, Ka-Band, SHF/EHF, V-Band | 11.2, 12.2, and 12.3 |

*Layer 7: Data Rate.* The data rates supported by a ground system are rigid. If the command and telemetry rates for a mission are far outside those supported by an existing ground system, changing the ground system probably won't be practical. The maximum and minimum data rates supported by a ground system are an integral part of too many devices to allow for an easy upgrade.

*Layer 6: Data Code.* Telemetry data collected from subsystems on the spacecraft bus, or from the payload, are digitally coded into one of several formats. Pulse-code modulation (PCM) offers a standard set of codes, the simplest of which is "non return to zero-level" (NRZ-L). The spacecraft may use codes other than NRZ-L to increase bit density (a measure of the number of transitions in a binary stream) and thus improve the link integrity. Because ground processors expect to see data in the NRZ-L format, all other codes must be converted to NRZ-L at the bit synchronizer or elsewhere in the ground system. Code conversion is a relatively simple operation, so most ground systems will support many PCM codes.

*Layer 5: Data Format.* After individual telemetry points have been PCM coded, they combine in one of several formats, each with its own protocol. Two generic telemetry formats are framed and packetized. *Framed telemetry* allocates specific locations within a repeating frame to data collected on the spacecraft. Each frame will look like the last, with changes only in the value of the data points. *Packetized telemetry* more efficiently uses the downlink by sending only those data points usable on the ground. A packet may vary in length, but if it doesn't, ground hardware can be simpler. Packetized telemetry may follow such protocols as synchronous data-link control (SDLC) or be unique to the spacecraft. Emerging standards, such as the packet-based telemetry format of the Consultative Committee for Space Data Systems (CCSDS), may simplify design choices. But for now, we must pay careful attention to format compatibility between space and ground systems. Because of the complexity of the hardware and software involved, a change in format is harder to respond to than, say, a change in PCM code or modulation technique.

*Layer 4: Encryption.* Both the uplink and downlink are sometimes encrypted for security. Many devices and algorithms are available, supporting various data rates and operational configurations. We must ensure compatibility not only between devices but also between the keys* loaded into those devices. Every spacecraft link has its own unique key, so we can test the transmit and receive keys for a particular link only against each other.

*Layer 3: Encoding.* A communication link, and the ground system as a whole, will normally specify a bit-error-rate below $10^{-5}$ (1 of every 100,000 bits in error) for voice and between $10^{-6}$ and $10^{-8}$ for data. To improve link performance (at the expense of bandwidth) signals are sometimes encoded for transmission using convolutional, Reed-Solomon, or other error-correction methods. If a spacecraft encodes its downlink or expects to decode its uplink, we have to install compatible hardware on the ground.

*Layer 2: Modulation Technique.* To transmit a signal over the high-frequency spacecraft link, we must superimpose information on the carrier frequency using modulation. We use amplitude modulation (AM), frequency modulation (FM),

---

* Keys are codes necessary to successfully encrypt and decrypt data on a particular link. [See Chap. 11]

and phase modulation (PM) for analog signals, but PM is most common. Modulating digital signals requires frequency-shift keying (FSK) or phase-shift keying (PSK). To improve efficiency, we may use the more complex binary phase-shift keying (BPSK) or quadrature phase-shift keying (QPSK). We trade modulation complexity with its use of bandwidth and its bit-error-rate performance. We also have to consider the presence of subcarriers and the modulation index. The modulation index, measured in degrees, is the amount of phase shift used in modulating data onto the subcarrier. Most ground systems allow various modulation techniques. A single receiver can handle several modulation types by employing plug-in cards.

*Layer 1: Frequency Band.* The most important layer of compatibility is in the frequency band used to carry signals between the spacecraft and the ground station. Antenna systems and TT&C hardware have a narrow tuning range, so ground systems support only a select number of frequencies. Spacecraft that will be fully supported by the AFSCN, for example, must communicate within SGLS frequencies*. In most cases, it's not practical to change an existing ground system to support new frequencies.

**Compatibility Testing.** It's difficult or impossible to recover from an incompatibility after launch. And with so much invested, it's a tremendous disappointment when the only thing collected from the mission is the insurance. We must discover any problems with the space/ground interface before launch, and preferably, before the spacecraft has been delivered to the launch site. No interface-design document or computer simulation will reveal as much about a ground system's ability to support a mission as an actual test. The compatibility test proves that commands, telemetry, and ranging signals can be communicated through rf channels and correctly interpreted by the end systems.

We commonly do compatibility testing twice: first at the spacecraft factory, then again at the launch site. Finding incompatibilities is much cheaper at the factory, where the components, tools, and designers are in place to make changes. Launch-site testing ensures systems are still working and enables last adjustments to the hardware.

A simple test is to transfer telemetry and commands by recording the unmodulated data and playing it back directly into the baseband equipment on the spacecraft and at the control center. Most missions would want more elaborate tests that involve the rf equipment end to end. A good test will validate compatibility for

- Command and telemetry formats and telemetry modes
- Control and display equipment
- Cryptographic equipment

---

* SGLS, the Space-Ground Link Subsystem, specifies 20 uplink frequency channels between 1.75 and 1.85 GHz, and 20 downlink frequency channels between 2.2 and 2.3 GHz.

- Radiofrequency interface
- Ranging
- Recording

We can test with deployable ground systems brought to the spacecraft or we can bring the spacecraft to a fixed ground station.

### 12.2.4    Staffing

In evaluating staffing requirements for a ground system, we must consider both the level of support and staff skills. Large, existing networks come fully staffed with operations and maintenance (O&M) people 24 hours per day. Small or dedicated networks may require less than 24-hour support. A smaller network may save money by using contractors for tasks such as maintenance, rather than supporting a full-time staff on site.

The frequency of spacecraft supports and the complexity of space and ground systems determine the required size and skill of the operations staff. Complex spacecraft that lack autonomy require more people, although more complex ground systems can reduce this number. People are expensive. If we can reduce the number of operators by automating acquisition and tracking, data recording, routing, and processing, we can save a lot of operating money.

We should also seek ways to reduce the skill levels required of operators. One way to do this is by using computer-based expert systems to make many of the decisions. Expert systems help spacecraft operators and those who control and monitor the ground network. Systems programmed with detailed knowledge of the spacecraft or ground system can now interpret data instead of requiring operators to do so. Simple text-based displays are giving way to graphical displays that correlate data in strip charts, bar charts, or cross plots. Expert systems require more up-front software development, but we recapture the cost in a lower staffing budget and fewer operational errors.

Besides requiring O&M people, a ground system needs engineers, technicians, and support staff to modify and test systems, upgrade facilities, and train operators.

## 12.3  New Systems and Existing Networks

Early in a space mission the mission operations manager must decide whether to use existing ground resources or build a new system. This decision affects all facets of the mission, from spacecraft design to the operations philosophy. In the early stages of the mission it's easy, but unwise, to ignore the ground system for what appear to be more pressing concerns of the spacecraft and its launch vehicle. Constructing a new ground system can be as long and complicated as constructing the spacecraft. And, with an existing system, it's better to know the interface requirements before we design the spacecraft. Without the ground system in mind, we can make an arbitrary decision about the spacecraft design that later causes a costly change to the ground system.

Some missions are born to a ground system. At the first mission-concept meeting for NASA's Galileo probe, for example, no one had to ask if the vehicle would be supported by the Deep Space Network. But other missions should begin investigating support alternatives by asking

1. Is an existing network able and eligible to support the mission?
2. What are the costs of operating in that network?
3. What constraints does that network place on the mission?

For most missions, the constraints of compatibility with an existing network don't harm the mission, provided we define the interface early enough. Other missions can't match any existing network without severely changing the mission's intent and function. These missions must build a dedicated ground system, designed specifically around the space element, for more specialized support.

### 12.3.1    New Systems

A good argument for constructing a new ground system is that the mission requires little operational support. A spacecraft that needs only infrequent monitoring from a single ground station may be able to afford independence. If we accept the risks and limitations of a single ground station (discussed in Section 12.1.1), building such a system can be more cost-effective than using an existing network. But only if it takes advantage of modern technology.

Missions that can build their own ground systems avoid the burden of history: the legacy of old systems. Once, we had no alternative but to custom design the pieces of a ground system. The hardware was unique, and the software wasn't portable. Operations were built around a central mainframe or mini-computer, and everything was proprietary, including data storage and printer interfaces. It was expensive and inflexible. Since this time, a few things have changed. Microprocessor-based systems have undergone an explosive increase in computing power, with equally dramatic reductions in size and cost. At the same time, open architectures with standard interfaces have been promoted. We're no longer forced into ground-up design with special-purpose components, and we're no longer locked to a single vendor. Today, we can build an entire ground system out of commercial, off-the-shelf (COTS) products.

COTS products keep development and operational costs in check. One costly driver of ground-system design is software development. The arduous task of designing, debugging, and documenting thousands of lines of code doesn't mate well to a success-oriented schedule. We can minimize the amount of software to develop if we buy commercially available software for command and control, mission planning and analysis, and orbit prediction; graphical window managers; and databases.

Another way to reduce development costs is to adapt existing software from previous missions. A better idea still is to restrict vehicle-specific information to the databases and keep the code itself generic. This way the same software can support different spacecraft at the same time.

Turn-key systems take reliance on COTS products a step further. They're complete, ready-to-use systems for spacecraft command and control. They're generic and reusable, and they leave little for the user to develop. Many vendors will even offer to install their systems or help the user integrate them with existing ground hardware. Although most turn-key systems don't include front-end equipment, such as receivers and bit synchronizers, we can buy an entire transportable ground station. Such a system will include a foldable antenna and TT&C systems that can be easily shipped and deployed. A good turn-key system will

- Adapt from pre-launch testing to launch and on-orbit operations
- Tailor to different spacecraft and existing ground systems
- Run on a distributed network to allow easy expanding and sharing of peripherals
- Support open-architecture standards to allow third-party peripherals and software
- Provide security features such as access control, with special restrictions for operations like commanding

Users of turn-key systems depend more on their vendors than users who design and build their own systems. For this reason, a good turn-key vendor provides comprehensive testing and training at installation, as well as strong technical support throughout the mission.

### 12.3.2    Existing Networks

In most existing networks resources are shared among many users. This is both a blessing and a curse. When we share resources we also share the cost of operations. What we lose in such a network is full control of scheduling and changes without arbitration. Without a powerful reason for independent control, the cost-effectiveness of shared resources convinces most missions to use an existing network.

Existing networks have other advantages. There is less risk to the spacecraft's development schedule, and ultimately the launch date, by using a ground network with known capabilities and interface requirements. In addition, training time and costs are lower. Choosing an existing network to support a space mission means contracting for a service that provides nearly everything. With the facilities and ground hardware comes the staff who will operate the spacecraft and help plan the mission. The host network will often require users to contribute equipment or other elements to the ground system if their configuration is unique, but the host will provide system engineers to integrate the mission into the network and test the spacecraft for compatibility.

The cost for the service from existing networks can be little to nothing. The AFSCN, for example, doesn't charge qualified missions, although the government is discussing doing so. Other networks have a defined cost. TDRSS charges non-NASA users a per-minute rate from under $10 to over $400, depending on the type of service [NASA, 1994].

Determining which existing network to use depends somewhat on network abilities and loading but mostly on eligibility. Exclusions have always kept commercial spacecraft from contracting with a military network for operational support, but this view is changing. When there were fewer networks the division was clear: NASA and related scientific missions were supported by the Spaceflight Tracking and Data Network (since replaced by TDRSS), or for planetary missions, the Deep Space Network. Department of Defense missions flew mostly out of the AFSCN. While this tradition still holds, missions have crossed these lines. Today the Navy and corporations like Hughes operate capable TT&C networks, and the government is opening up to commercial use of its assets. Many of these networks are connected to each other, allowing for hybrid use of their resources.

Hybrid ground architectures combine different existing networks and sometimes dedicated assets to exploit the advantages of each. A spacecraft launched in 1995 is a good example of the kind of mission that benefits from a hybrid ground system. Its payload has a data rate of 25 Mbps, which can be downlinked once every orbit. But maintaining the spacecraft's health requires contacting it throughout its orbit. The AFSCN does the TT&C to maintain the spacecraft, while a dedicated facility in Southern California receives and processes the payload data.

### 12.3.3   Changing Existing Systems

After deciding to use an existing ground system, the mission operations manager has another decision: use only the resources this system provides or augment it with other resources. There are benefits to designing a spacecraft for 100% compatibility with all existing ground-system elements. Eliminated are the costs, risks, and scheduling that come from system modifications. Added, though, are potential compromises to mission objectives. When these compromises become too great, we must change the system.

An existing system may not meet the needs of a particular mission for many reasons, such as incompatible telemetry formats, data rates, encryption, or complexity of command or telemetry systems. Because the core ground system may have been installed long before we design a spacecraft, the spacecraft commonly employs newer technology than the ground system. Large ground systems undergo major upgrades every 10 to 20 years, in a slow evolution that provides predictable, stable services to its users. Upgrading the entire network with every change in technology is too expensive. When the operator requires the most modern hardware and software tools on the ground to use advanced spacecraft technology, it's cheaper to augment the ground system as needed with mission-unique systems. Besides installing mission-unique systems, a common network upgrade is installing a communication link to support spacecraft testing or to relay data to an external user in real time. The user bears most of the cost for these changes, but the host network often provides technical services and integration support.

Regardless of the modification, we must allow enough time for planning. Because shared ground systems support many missions at the same time, we must evaluate changes to the network for their effect on other users. Networks strive to reduce cost by combining requirements from different users into one comprehensive solution.

### 12.3.4    The Air Force's Satellite Control Network (AFSCN)

The AFSCN is a global network of remote ground stations and control centers to command and control US spacecraft. Under the direction of US Air Force Space Command, the AFSCN operates crewed and uncrewed, DoD and non-DoD, space missions.

Operations are directed from two network control centers located at Onizuka Air Station in Sunnyvale, California, and Falcon Air Force Base in Colorado Springs, Colorado. Most missions use only one network control center but some missions use one center as primary and the other as backup. An integrated mission-control center supports TT&C for the bus and the payload. Less often, payload data goes from the control center to an external payload-operations control center for processing.

The control centers connect through DOMSAT links and land-lines to sixteen ground stations at nine geographic locations. Table 12.7 shows the antenna characteristics for these ground stations, and Table 12.8 lists some of the network's characteristics. In the early 1990s the Air Force upgraded most of the ground stations to Automated Remote Tracking Stations, thereby increasing their capacity and availability and reducing costs for operation and maintenance. The AFSCN, like many networks, is migrating toward full automation.

The AFSCN supports SGLS and non-SGLS spacecraft, which have one uplink and up to four simultaneous downlinks. Ground stations can record telemetry rates up to 5 Mbps but can relay telemetry to the control center in real time at no more than 1.024 Mbps.

### 12.3.5    The Naval Satellite Control Network (NSCN)

The NSCN operates spacecraft for the Naval Space Command. Control of the network is centered at the Naval Satellite Operations Center Headquarters (NAVSOC HQ) at Point Mugu, California. The NAVSOC operates three ground stations at Prospect Harbor, Maine; Laguna Peak, California; and Finegayan, Guam. The NSCN can do three simultaneous, independent contacts—one at each ground station. The network provides hardware and software for standard data formats. Users with unique data formats must provide their own ground equipment that will either integrate into the NSCN or have its own communication link if located elsewhere. Tables 12.9 and 12.10 show the NSCN's characteristics.

**Table 12.7. Characteristics of AFSCN Antennas.** Operations are directed from Onizuka Air Force Station, California, or Falcon Air Force Base, Colorado.

| Site | Location and Altitude | Size (m) | G/T (dB/°K) | EIRP (dBW) |
|---|---|---|---|---|
| New Hampshire (Manchester, NH) | 42°56.9' N, 71°37.6' W  201 m<br>42°56.7' N, 71°37.8' W  191 m | 18<br>14 | 26.4<br>24.6 | 85.3<br>77.3 |
| Vandenberg (Lompoc, CA) | 34°49.4' N, 120°30.1' W  269 m<br>34°49.6' N, 120°30.3' W  266 m | 18<br>14 | 27.1<br>25.2 | 83.7<br>77.2 |
| Hawaii (Kaena Point, Oahu) | 21°33.8' N, 158°14.5' W  428 m<br>21°34.1' N, 158°15.7' W  318 m | 18<br>14 | 26.1<br>24.6 | 79.1<br>77.6 |
| Guam | 13°36.9' N, 144°52.0' E  218 m<br>13°36.9' N, 144°51.3' E  209 m | 18<br>14 | 26.6<br>23.1 | 83.8<br>76.5 |
| Indian Ocean (Mahe, Seychelles) | 4°40.3' S, 55°28.7' E  561 m | 18 | 26.4 | 84.1 |
| Thule (Greenland) | 76°31.0' N, 68°36.0' W  132 m<br>76°30.9' N, 68°36.0' W  132 m<br>76°30.9' N, 68°35.0' W  132 m | 7<br>14<br>10 | 20.4<br>24.4<br>21.6 | 73.5<br>75.4<br>76.6 |
| Oakhanger (England) | 51°06.8' N, 00°52.7' W  82 m<br>51°06.3' N, 00°52.9' W  91 m | 18<br>10 | 24.8<br>21.5 | 78.8<br>75.2 |
| Colorado (Colorado Springs, CO) | 38°50.1' N, 104°49.2' W  1959 m | 10 | 21.6 | 74.8 |
| Diego Garcia | 07°16.0' S, 72°22.4' E  5 m | 10 | 21.6 | 74.1 |

**Table 12.8. Characteristics of the AFSCN Network.** The network supports both space-ground link subsystem (SGLS) and non-SGLS spacecraft.

| Parameter | Uplink | Downlink |
|---|---|---|
| Frequency | 1750–1850 MHz | 2200–2300 MHz |
| Modulation | SGLS: AM/FSK/PM<br>Non-SGLS: BPSK or BPSK/PM | SGLS: BPSK, BPSK/PM<br>Non-SGLS: BPSK, QPSK, FM, FM/FM, PM |
| Data rate | SGLS: 1,2, or 10 kbps<br>Non-SGLS: 100 bps to 100 kbps (BPSK)<br>100 bps to 256 kbps (BPSK/PM) | SGLS: 1.024 Mbps<br>Non-SGLS: 5 Mbps (to ground station only) |

Like the AFSCN, the NSCN uses SGLS to communicate with spacecraft. But unlike the AFSCN, the NSCN does most data processing and commanding at the ground station, not the control center. The Integrated Satellite Control System, a network based on microcomputers, evaluates the telemetry. It strips from the state-of-health telemetry the redundancy that results from slowly changing spacecraft measurements such as battery temperatures. Only changed telemetry points go to the NAVSOC HQ for analysis and archival. In addition, NAVSOC ground

**Table 12.9.   Characteristics of NSCN Antennas.** Operations are directed from the Naval Operations Center Headquarters at Port Mugu, California.

| Site | Location and Altitude | Size (m) | G/T (dB/°K) | EIRP (dBW) |
|------|----------------------|----------|-------------|------------|
| Laguna Peak, CA | 34°06' N, 119°04' W   450 m | 18 | 25 | 66 |
| Detachment A (Prospect Harbor, ME) | 44°24' N, 68°01' W   7 m | 5 | 15 | 62 |
| Detachment C (Finegayan, Guam) | 13°34' N, 144°50'E   151 m | 5 | 12 | N/A |

**Table 12.10.   Characteristics of the NSCN Network.** The NSCN is highly compatible with the AFSCN.

| Parameter | Uplink[*] | Downlink |
|-----------|-----------|----------|
| Frequency | 1750–1850 MHz | 2200–2300 MHz |
| Modulation | PM | BPSK, PM |
| Data rate | 1 kpbs or 2 kbps | 2 Mbps maximum |

[*] Uplink capability currently not available at Detachment C.

stations are fully automated, thus eliminating the need for voice communication between the ground station and NAVSOC HQ. As a result, the NSCN can reduce its communication-link requirements between the ground stations and the control center to a full-duplex, leased line operating at 56 kbps.

## 12.3.6   NASA's Tracking and Data Relay Satellite System (TDRSS)

A shortfall of any moderately sized ground system is that it covers only a small percentage of the orbit for low-altitude spacecraft. NASA overcame this deficiency with the TDRSS Network. Although it rejects the label "ground system", the TDRSS Network satisfies our definition. TDRSS monitors and controls spacecraft, and provides mission data to users by employing two geostationary spacecraft rather than multiple ground stations. Located at 41° west longitude (TDRS East) and 174° west longitude (TDRS West), the TDRSS satellites provide nearly 100% coverage[*] for subsynchronous spacecraft. A TDRSS satellite processes no data onboard because it's simply a bent-pipe repeater.

---

[*] The locations of the TDRSS satellites result from a cost-saving design that requires only two satellites and one ground station. However, these locations create a small zone of exclusion, in which the Earth blocks communication with user spacecraft orbiting below 1200 km in a region over the Indian Ocean. Three more satellites, positioned between the other two, are spares that can be repositioned if an operational satellite fails.

A single ground station at White Sands, New Mexico, supports the TDRSS satellites. The White Sands Ground Terminal communicates user-formatted data to and from the TDRS relay and operates the TDRSS satellites themselves. Through the forward link, White Sands transmits commands and a pseudorandom noise (PN) ranging code to the TDRS and down to user spacecraft. On the return link, telemetry and the PN turn-around code are received by the TDRS from user spacecraft and relayed to White Sands. The ground terminal at White Sands links with the TDRSS Network Control Center at NASA's Goddard Space Flight Center in Greenbelt, Maryland. Goddard manages and operates the TDRSS Network, including scheduling and configuring network resources, monitoring various telemetry streams, and computing tracking and orbit data. The Center connects to users through the NASA Communications Network. The user can locate its POCC at the Center or, with the appropriate communication links, at the user's site.

The TDRSS Network offers two services, summarized in Table 12.11. The Single-Access Service relays data at a high rate to a single spacecraft. For this service each TDRS has two steerable, 4.9-meter, parabolic antennas which operate in S-band and K-band. The Multiple-Access Service relays data at a low rate to many spacecraft at once, using a fixed, phased-array antenna which operates in S-band.

**Table 12.11. TDRSS Services.** The single-access service provides a high data rate to a single user while the multiple-access service provides a low data rate to many users.

| Service | Frequency Band | Number of TDRS Links | Forward Data Rate (White Sands to User Spacecraft) | Return Data Rate (User Spacecraft to WSGT) |
|---------|----------------|----------------------|----------------------------------------------------|--------------------------------------------|
| Single-Access Service | S | 2 forward 2 return | 100 bps to 300 kbps | Up to 6 Mbps |
| | K | 2 forward 2 return | 1 kbps to 25 Mbps | Up to 300 Mbps |
| Multiple-Access Service | S | 1 forward | 100 bps to 10 kbps | -- |
| | | 20 return | -- | 100 bps to 50 kbps |

### 12.3.7    NASA's Deep Space Network (DSN)

Spacecraft circumnavigating the solar system require more powerful ground antennas than most satellite control networks can provide. NASA's DSN operates such antennas. The DSN supports tracking and communication for all of NASA's interplanetary spacecraft.

The DSN consists of three multi-station complexes spread across the globe. Each of these sites—Goldstone (California), Canberra (Australia), and Madrid (Spain)—is equipped with one 70 m, two 34 m, and one 26 m antenna, described in Table 12.12. The 70 m and 34 m antennas mainly help us communicate with spacecraft at distances greater than two million km from Earth. The 26 m and other smaller antennas typically support Earth-orbiting satellites. At each site, the 70 m antenna can array with either or both of the 34 m antennas for improved telemetry performance. Table 12.13 lists some of the DSN's characteristics.

**Table 12.12. Characteristics of DSN Antennas.** Operations are directed from the Jet Propulsion Laboratory, California.

| Site | Antenna No. | Location and Altitude | Size (m) | G/T (dB/°K) S-band | G/T (dB/°K) X-band | EIRP (dBW)[*] S-band | EIRP (dBW)[*] X-band |
|------|-------------|----------------------|----------|--------|--------|--------|--------|
| Goldstone | 12 | 35°18.0' N, 116°48.3' W  1001 m | 34 | 41.7 | 52.2 | 98.2 | N/A |
| (California) | 14 | 35°25.5' N, 116°53.3' W  993 m | 70 | 50.1 | 61.1 | 105.7 | N/A |
| | 15 | 35°25.3' N, 116°53.2' W  984 m | 34 | 40.2 | 53.5 | N/A | 110.1 |
| | 16 | 35°20.5' N, 116°52.4' W  973 m | 26 | 31.5 | N/A | 91.5 | N/A |
| Canberra | 42 | 35°24.1' S, 148°58.8' E  656 m | 34 | 41.7 | 52.1 | 98.2 | N/A |
| (Australia) | 43 | 35°24.2' S, 148°58.8' E  670 m | 70 | 50.1 | 60.9 | 105.7 | N/A |
| | 45 | 35°24.0' S, 148°58.6' E  655 m | 34 | 40.2 | 53.4 | N/A | 110.1 |
| | 46 | 35°24.4' S, 148°58.9' E  658 m | 26 | 31.5 | N/A | 91.5 | N/A |
| Madrid | 61 | 40°25.8' N, 4°14.9' W  796 m | 34 | 41.7 | 52.1 | 98.2 | N/A |
| (Spain) | 63 | 40°25.9' N, 4°14.8' W  812 m | 70 | 50.1 | 61.1 | 105.7 | N/A |
| | 65 | 40°25.7' N, 4°15.0' W  781 m | 34 | 39.6 | 53.4 | N/A | 110.1 |
| | 66 | 40°25.8' N, 4°15.0' W  797 m | 26 | 31.5 | N/A | 91.5 | N/A |

* Calculated for a 20 kW power amplifier at the 70 m and 34 m antennas, and a 10 kW power amplifier at the 26 m antenna. A 400 kW power amplifier is also available for the 70 m antenna.

**Table 12.13. Characteristics of the DSN System.** Available data rates are low for the DSN because of the large distance from the spacecraft to the ground station.

| Parameter | Uplink | Downlink |
|-----------|--------|----------|
| Frequency | 2025 to 2120 MHz 7145 to 7190 MHz[*] | 2200 to 2300 MHz 8400 to 8500 MHz |
| Modulation | Carrier: PM/PSK, FSK, AM, or Sum | PM, FM, or AM |
| Data rate | 1.0 bps to 2000 bps | 8.0 bps to 6.6 Mbps |

* X-band transmit capability only available on the 34 m antenna.

Command and control of deep space missions are centered at the Jet Propulsion Laboratory in Pasadena, California. The Lab's people are normally present 24-hours per day, every day, to monitor spacecraft and manage ground systems.

Saying that DSN antennas are heavily scheduled is an understatement. Years before launch, mission designers will investigate the predicted loading of DSN assets for the period of their mission, hoping to minimize contention for antenna time. The best launch period for an interplanetary mission depends mostly on solar-system geometry, but planners will adjust the launch date to align the spacecraft in a part of the sky that has less competition with other spacecraft.

## 12.3.8    The European Space Agency's Network

The European Space Agency (ESA) is an international body of 13 member nations with a common goal: to give Europe independent access to space. The ground system operated by ESA provides TT&C, as well as Doppler and ranging services, for Earth-orbiting satellites.

The network consists of eight ground stations and three control centers. Two of these control centers, at Redu (Belgium) and Villafranca (Spain), are co-located with ground stations. The primary control center is at the European Space Operations Center (ESOC) in Darmstadt, Germany. Within ESOC are two mission control centers: the first dedicated to Meteosat and the second shared by all other missions.

**Table 12.14. Characteristics of the ESA's Antennas.** Control centers are at Redu, Belgium; Villafranca, Spain; and Darmstadt, Germany.

| Site | Location and Altitude | Size (m) | G/T (dB/°K) | EIRP (dBW) |
|---|---|---|---|---|
| Kiruna (Sweden) | 67° 51.4' N, 20° 57.8' E  402 m | 15 | S-band: 28.9<br>X-band: 34.0 | 71.0 |
| Kourou (French Guiana) | 5° 15.1' N, 52° 48.3' W  15 m | 15 | S-band: 27.4 | 74.3 |
| Malindi (Kenya) | 2° 59.8' S, 40° 11.6' E  12 m | 10 | S-band: 21.3 | 68.9 |
| Maspalomas (Grand Canary) | 27° 45.8' N 15° 38.0' W  205 m | 15 | S-band: 26.0<br>X-band: 36.4 | 79.7 |
| Odenwald (Rehbach, Germany) | 49° 42.8' N, 8° 58.5' E  347 m | 15 | S-band: 25.0 | 76.9 |
| Perth (Australia) | 31° 48.2' S, 115° 53.1' E  22 m | 15 | S-band: 28.0<br>X-band: 38.0 | 69.5 |
| Redu (Belgium) | 50° 0.1' N, 5° 8.7' E  385 m | 13.5 | Ku-band: 39.3 | 93.8 |
| Villafranca (Spain) | 40° 26.8' N, 3° 57.1' W  664 m | 15 | S-band: 27.5 | 79.0 |

## 12.3.9    International Tracking Stations

Tables 12.15 and 12.16 describe additional international tracking stations.

**Table 12.15.  International Tracking Stations for Earth-Orbiting Satellites.** This table lists additional stations for Earth-orbiting spacecraft. All sites provide telemetry and commanding. (See Chap. 18)

| Site/Agency | Antenna Diameter (m) | Locations | | Frequency Band |
|---|---|---|---|---|
| | | Latitude | E. Longitude | |
| RSA (Russia) | | | | |
| Evpatoria*, Ukraine | 25 (2) | 45  11 | 33  11 | $P^\dagger$/C |
| Ussuriisk, Russia | 25 (2) | 44  00 | 131  45 | $P^\dagger$/C |
| Tshelokovo, Russia | 25 (2), 12 | 56  01 | 37  52 | $P^\dagger$/C, $P^\dagger$ |
| St.Petersburg,Russia | 12 | 68  02 | 33  09 | $P^\dagger$ |
| Jusaly, Kazahstan | 12 | 45  19 | 64  03 | $P^\dagger$ |
| Kolpashevo, Russia | 12 | 58  12 | 82  35 | $P^\dagger$ |
| Ulan-Ude, Russia | 25, 12 | 51  33 | 107  24 | $P^\dagger$ |
| Petropavlovsk, Russia | 25, 12 | 53  18 | 158  26 | $P^\dagger$ |
| CNES (France) | | | | |
| Aussaguel, France | 11 | 43  26 | 01  30 | S |
| Kourou, Fr. Guyana | 11 | 05  06 | 307  22 | S |
| Hartebeestoek, S. Africa | 12 | −25  53 | 27  42 | S |
| ESA | | | | |
| Maspalomas, Spain | 15 | 28  00 | 344  42 | S |
| Kiruna, Sweden | 15 | 67  52 | 20  57 | S |
| Perth, Australia | 15 | −31  48 | 115  53 | S |
| Villafranca, Spain | 15 | 40  27 | 356  03 | S |
| NASDA (Japan) | | | | |
| Masuda, Japan | 18, 13 | 30  33 | 131  01 | S |
| Katsuura, Japan | 18, 13 | 35  12 | 140  18 | S |
| Okinawa, Japan | 18, 18 | 47  53 | 11  06 | S |

\* Operated jointly by Russian Space Agency and Ukrainian Space Agency.
† 157/184 and 745/930 MHz (uplink/downlink).

**Table 12.16. Additional Complexes for Deep-Space Communication.** The listed complexes support spacecraft in deep space (distances greater than 100,000 km from Earth). All sites provide telemetry, commanding, radiometry, and very-long-baseline interferometry (VLBI). (See Chap. 18)

| Site/Agency | Antenna diameter (m) | Locations | | | Frequency Band |
| --- | --- | --- | --- | --- | --- |
| | | Latitude | E. Longitude | Altitude | |
| Evpatoria, Ukraine/UkrSA & | 70 | 45 11 22.0 | 33 11 19.0 | 5 m | $P^*/C/X^†/(L^†)$ |
| RusSA | 32 | 45 11 22.0 | 33 11 19.0 | | $P^*/C$ |
| Ussuriisk, Russia/RusSA | 70 | 44 00 57.0 | 131 45 22.0 | 75 m | $P^*/C/X^†/(L^†)$ |
| | 32 | 44 00 57.0 | 131 45 22.0 | | $P^*/C$ |
| Bear Lakes (Moscow), Russia/ RusSA | 64 | 55 51 57.0 | 37 57 17.0 | 152 m | $P^*/C/X^†/(L^†)$ |
| Usuda, Japan/ISAS | 64 | 36 07 56.9 | 138 21 45.7 | | $S/X^†$ |
| Weilheim, Germany/DLR | 30 | 47 52 52.3 | 11 06 01.2 | | $S/X^†$ |

\* 157/184 and 745/930 MHz (uplink/downlink).
† Receive only; $(L^†)$ - frequency is not allocated for deep space communications, receive only.

# 12.4 Operational Concerns

After launch, a new set of issues becomes important. We'll look at three ground-system concerns that become critical during the operational phase of the mission: ground anomalies, maintenance, and resource scheduling.

## 12.4.1   Ground Anomalies

Anomalies are the bane of the mission operations team. An anomaly, by definition, is a problem or an event with an unknown cause. The detection of a problem jerks the operations team from routine operations and propels it into a search for the cause. If the anomaly is severe enough, the team won't be able to proceed with the mission until the problem is resolved.

When an anomaly occurs, we sometimes don't know right away whether the problem is in space, on the ground, or somewhere in between. When the satellite downlink drops out, do we suspect the satellite transmitter? Do we blame the ground receivers and demodulators? Or do we look at the rf compatibility between the two? The answer, of course, is that we look everywhere at once. Solving the mystery of an anomaly is much like solving any mystery: begin with the most probable causes and successively eliminate the innocent. Chapter 16 treats spacecraft anomalies in considerable detail. We'll look briefly at ground anomalies here.

Ground anomalies are fundamentally different from spacecraft anomalies— given enough time and money, we can always resolve a ground anomaly. There is more time, more information, and less panic in the investigation of a ground anomaly. The ground system has no safe-hold mode. Usually, it's easier to find the root cause of a ground anomaly because maintainers can access all ground equipment for testing. They can visit even a remote, unstaffed system during the investigation. Common types of ground anomalies are

- Hardware failures
- Configuration errors (human errors)
- Compatibility problems between space and ground
- Flaws in the software design or database errors

A hardware failure can happen at any time. So can configuration errors, but these, like compatibility and software problems, are more likely to occur during launch and early orbit, when procedures are new and, for the first time, systems undergo operations.

We attempt to cover future anomalous situations with contingency plans. These tell the operator what steps to take when data isn't at a terminal, or whom to call if a communication link seems to have failed. It's wise to be prepared, but these plans do little to calm the mission manager during the support. Even the simplest of problems can cause the loss of a spacecraft support when the pass is only ten minutes long. Instead, we should concentrate on prevention. Table 12.17 summarizes how to prevent and resolve common ground anomalies.

Table 12.17. Ground Anomalies. Prevention is the goal.

| Ground Anomaly | Prevention Technique | Resolution Action | Relative Cost (Typical) |
|---|---|---|---|
| Hardware failure | • Redundancy | • Replace or repair | Medium |
| Configuration error | • Training<br>• Automation | • More training | Low |
| Compatibility problems between space and ground | • Testing<br>• Use proven systems | • Replace ground equipment<br>• Change operation procedures | High |
| Flaws in software design or database error | • Testing<br>• Use commercial products | • Code or data base modification | Low – medium |

Redundancy in the design of all ground elements helps to prevent anomalous conditions. With the ability to patch around a suspect piece of equipment, or route data through a backup communication link, we can quickly bring a system back on line. The operations team can then resume their tasks while the maintenance staff investigates the problem.

It's impossible to prevent a human operator from ever making a mistake. We can reduce errors by designing systems with simple, logical human interfaces and training people well for less simple systems. If we can automate these tasks, we avoid the problem altogether.

Perhaps the most difficult anomalies to resolve are compatibility problems. On orbit, they are difficult to find and probably more difficult to fix. Through pre-launch testing we can discover and prevent incompatibilities, as described in Sec. 12.2. We can also minimize their occurrence by favoring proven technology in the TT&C subsystems of the space and ground segments.

Software won't fail spontaneously, but its design may be faulty. More likely than that, the database will contain incorrect parameters, or parameters in the wrong locations. Thorough pre-launch testing should unearth such errors before they cause an anomaly. We can reduce the probability of flaws in the software by moving away from custom coding and relying more on commercial, off-the-shelf products.

The ground system itself can help a lot, or not at all, in detecting and resolving anomalies. Older and simpler ground systems don't help: humans must discover all problems. Discovery typically comes through a secondary characteristic, so people have to trace the root cause of the problem. Modern systems will detect malfunctions and report them to their human attendants. A control-and-monitor system that remotely monitors equipment through independent signal lines is one example. More sophisticated systems will detect and resolve equipment failures and then notify humans of the action. This kind of automated redundancy is common at the equipment level. For example, many devices will have redundant power supplies within a chassis because they have a high-failure rate. Failure of one power supply requires no human intervention to keep the system on line. Automated redundancy is less common at the system level, but as the cost of such systems falls, replacing human maintainers will be cheaper.

## 12.4.2   Maintaining and Sparing

Often overlooked in planning and designing a ground system is long-term maintenance. The cost of maintenance can be high, but the cost to the mission of not maintaining a system can be even higher. The level of acceptable risk determines the level of required maintenance. We define this risk by the amount of time a system can be down before the impact to the mission becomes unacceptable, and we weigh this downtime against cost. Different users will have different requirements for system availability. Some military users can't afford more than 0.01 percent downtime, whereas some university-sponsored missions can't afford to provide the 100% sparing such a low downtime percentage would require. For missions in which real-time data processing isn't critical, we can relax maintenance requirements for data relay and processing, provided the ground station records the data.

Steps in developing a maintenance plan include determining acceptable downtime, recognizing existing redundancy that we can exploit, and calculating the cost of each option. The costs include people, training, equipment, and documentation. We should develop the maintenance plan before installing the system because it often affects the design.

We can define three levels of maintenance, as shown in Table 12.18 and illustrated in Fig. 12.9. In Level 1, downtime is unacceptable. An equipment failure must not block data on its way to the user. We accomplish this level of maintenance with redundant strings of equipment and hot spares. We install hot spares in the same way as the equipment they are meant to back up, but we don't use them unless the primary device fails. They can be switched into the circuit with minimal delay, sometimes automatically.

**Table 12.18. The Three Levels of Maintenance.** We must trade initial cost with acceptable risk, using mission requirements.

| Level | Allowable Downtime | Characteristics | Initial Cost | Risk |
|-------|--------------------|-----------------|--------------|------|
| 1 | None | Redundancy, hot spares | High | Low |
| 2 | Short (0 – 24 hours) | 100% on-the-shelf spares | Medium | Medium |
| 3 | Long (days to weeks) | Repair as required | Low | High |

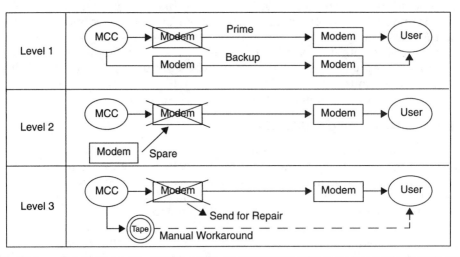

**Fig. 12.9.**     **The Three Levels of Maintenance Illustrated with a Modem Link.** In this example, users must determine how critical the data it receives from the Mission Control Center is before they can decide which level they prefer.

Level 2 quickly turns around an equipment failure, usually within 24 hours, by using spare hardware on the shelf and an on-site maintenance staff. For many missions, this compromise between cost and risk is the best maintenance solution.

Level 3 provides no sparing but repairs or replaces components as required. We can use this plan for non-critical elements having operational workarounds. In Fig. 12.9, for example, users can wait while a tape with recorded data is shipped to their site.

For all levels of maintenance a well documented system is the key to successfully troubleshooting and resolving component failure. Unfortunately, documentation can also be the key to a blown budget. It shocks taxpayers to read about the military's $500 hammer, but those familiar with strict documentation requirements have accepted this news. They know it's not unreasonable to pay more to document a component than to buy the component itself. Still, we must temper the tendency to document too much by paring down redundant documents and eliminating unnecessary ones. Today, even the government understands the need to reduce documentation. To keep in step with a shrinking budget, the Department of Defense now requires that government agencies and their contractors use commercial specifications and documentation whenever possible, rather than creating their own [Perry, 1994].

### 12.4.3    Scheduling Resources

In a shared network, spacecraft must compete for resources. We invariably lack antennas, communication links, and data-handling hardware to support every desired spacecraft contact without some compromise. Therefore, we must schedule all ground resources, usually days in advance, to use them best. Resource scheduling ensures that, although users may not get every contact they want, they will get every contact they need.

We assign ground resources to spacecraft based on availability and mission priority. The user submits a request for resources within some window, and in most cases, we can simply adjust the schedule to meet it. When two or more missions vie for the same resources in the same period, mission priority determines which gets them—in a process called deconflicting. No absolute priority scheme guides planners to the most efficient schedule. A rough mission hierarchy gets planners started: (1) crewed space flight; (2) launch and early orbit operations; (3) normal operations, in the order of low orbit, medium orbit, then high orbit; (4) preflight testing or training. Physics gives spacecraft in lower orbits a derived priority because they have fewer chances to contact a ground station. In addition to these categories, some missions are regarded as having more importance to the network than others; for example, a spacecraft supporting national defense will influence the schedule more than a spacecraft researching the atmosphere. But none of these guidelines will cover every case. In the end, only the event's priority—the relative importance of that particular spacecraft contact—matters. In the civilized arguments of deconflicting, users (with an arbitrator's help) determine whose event is really more important.

Two ground-system specifications influence the availability of a network. *Turnaround time* is the time required between successive contacts to reconfigure computer databases and equipment settings and to change recorder tapes. We must consider this figure, typically several minutes, when scheduling back-to-back contacts. *Maintenance downtime* is the scheduled removal of a system from operational status for calibration, cleaning, or repair. The time required varies greatly from system to system but can reduce the availability of a network by several hours per month.

We use hardware and software tools to decrease the demands on the operators tasked to schedule spacecraft contacts. Commercially available software that graphically displays orbits and ground-station visibilities decreases the required skill level of this operator, and more sophisticated systems can eliminate the operator. Like so many other ground system functions, scheduling is a candidate for automation. In an automated system, users submit requests for resources and let the computer schedule their contacts. The need to resolve conflicts between two missions however, may force some networks to keep a person in the loop.

## References

Klements, H. D. 1992. *Air Force Satellite Control Facility Space/Ground Interface*. TOR-0059(6110-01)-3. El Segundo, CA: The Aerospace Corporation.

McClimans, Fred J. 1992. *Communications Wiring and Interconnection*.NewYork, NY: McGraw-Hill.

National Aeronautics and Space Administration. 1994. *Mission Requirements and Data Systems Support Forecast*. 501-803. Goddard Space Flight Center, Greenbelt, MD: National Aeronautics and Space Administration.

National Aeronautics and Space Administration. 1988. TDRSS Users' Guide. STDN No. 101.2. Goddard Space Flight Center, Greenbelt, MD: National Aeronautics and Space Administration.

National Aeronautics and Space Administration. 1989. Deep Space Network/Flight Project Interface Design Handbook. Vol. I and II. JPL-DSN 810-5, Revision D. Jet Propulsion Laboratory, Pasadena, CA: National Aeronautics and Space Administration.

Naval Satellite Operations Center. 1993. *Naval Satellite Control Network to Satellite System Interface Document*. Point Mugu, CA: Naval Satellite Operations Center.

Perry, William J. 1994. "Specifications & Standards--A New Way of Doing Business." Memorandum issued June 29, 1994. Washington, DC: The Secretary of Defense.

Powers, John T., and Stair, Henry H. 1990. *Megabit Data Communications*. Englewood Cliffs, NJ: Prentice Hall.

Pratt, Timothy and Charles W. Bostian. 1986. *Satellite Communications*. New York, NY: John Wiley & Sons.

Williamson, Mark. 1990. *Dictionary of Space Technology*. New York, NY: Adam Hilger, IOP Publishing.

# Chapter 13

# Processing Data and Generating Science-Data Products

William Emery, *University of Colorado*
James Green, *Goddard Space Flight Center*

This chapter discusses the spacecraft's entire data system, including onboard processing, data relay (downlinking), ground reception, and data processing. Our discussion concentrates on relaying and processing sensor data, but it also includes the housekeeping data needed to work with the sensor data. We can't quantitatively analyze data from the spacecraft sensors without knowing the spacecraft's systems, their on-orbit functions and behavior, and their effects on the sensor data of interest. All of the data systems on the spacecraft link together, so knowing one usually means knowing something about the other. Once you understand data from science sensors and spacecraft housekeeping, you'll appreciate their importance in planning a space mission.

## 13.1 End-to-End Data Flow

Before addressing data systems and data-processing requirements, we need to review the overall or *end-to-end data flow* (EEDF). In this section, end-to-end means spacecraft to user. Here the user is either a member of the mission science team or a general user who is obtaining mission data. The *end-to-end data system* (EEDS) refers to the collection of hardware, software, and communication links that support the flow of data from the spacecraft to the user. We don't consider commanding in this chapter.

Too often space missions create an EEDS which doesn't generate mission data that is useful to the user. Development of the EEDS concept is best when it includes potential users of the mission data. An effective tool to ensure the system meets the user's needs is the EEDF diagram. While even an EEDF diagram won't guarantee the user will get the mission data in the form required, it's an essential starting point in determining the cost of mission data.

In NASA, the EEDS is documented in the Project Data Management Plan or PDMP [Green and King, 1988, Futron, 1992]. A draft PDMP should be completed during concept exploration and signed off by the Project Team and NASA Headquarters before launch. A PDMP covers the following areas:

- Mission objectives
- End-to-end data flow
- Policies
  - Proprietary rights
  - Data discarding
- Other data activities (e.g.: guest investigations)

- Instrument overview
- Description of data products
- Archiving
  - Information archive
  - Data archive

## 13.1.1 ⁀ Data Flows

EEDF diagrams are used to understand the data flow from the spacecraft to the data users. These diagrams contain several distinct systems we must evaluate for cost. Figure 13.1 shows these systems, which include the spacecraft, data reception, product generation, data archival, and distribution to the science user. Once the sensor data is available on the spacecraft, the systems in Fig. 13.1 manage, transform, and transport it. *Data management* includes how data is stored, the data volume, its accessibility, and locating individual data attributes (data finding). *Data transformation* occurs when data goes into a system in one form and comes out in another form. Finally, *data transport* involves the timely delivery of data from one point to another. How and where these functions occur in the overall EEDS is an essential part of our cost/benefit trades in the EEDF plan.

The EEDS transforms data at various times, so we have to make sure we don't do transformations at one location that are undone at another location or at a later time. This approach should be common sense, but these problems occur often in space missions, usually because funding and responsibilities overlap for teams working on the data system. Developing a data-flow diagram, as shown in Table 13.1, is an important way to identify these types of potential problems.

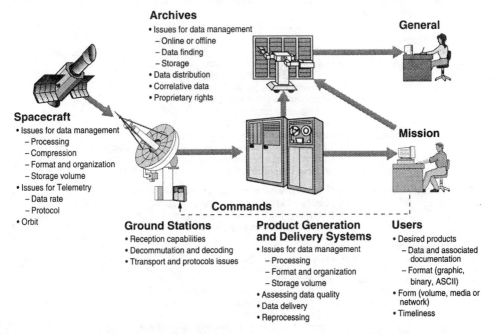

**Fig. 13.1.**    **Diagram of End-to-End Data Flow.** At each node we must address the key issues and how they relate to managing, transforming, and transporting the data.

## 13.1.2    Systems for Generating Products and Delivering Data

These key systems provide needed data products to the mission users. They can be extremely expensive and require a lot of development. Over the last 20 years hardware costs have plummeted while costs to develop and maintain software remain high or have increased because people want more complex data analysis [Boehm, 1981].

Consider a space agency's astrophysics spacecraft in a low-Earth orbit (about a 90-minute period), which makes stellar observations that can be 4 to 5 hours long. These observations require extensive mission planning with detailed time lines because they're spread over multiple orbits. The ground software that processes this data to generate products must align the observation time and combine individual segments of the observations to obtain one single file of combined data before it does the calibration. This software consists of more than 350,000 lines of code developed over seven years at a total investment of 109 person years. Using higher orbits to allow long-exposure stellar observations with much less complicated time lines would have been an important trade and may have significantly reduced overall cost.

**Table 13.1. How to Handle the End-to-End Data Flow.** Developing and evaluating diagrams of end-to-end data flow.

| Steps | Comments |
|---|---|
| Define data requirements | • Define the users of the data<br>  – Flight operators (housekeeping)<br>  – Investigators (science data)<br>  – Engineers (housekeeping)<br>  – Scientists (archived data)<br>  – Public (science data)<br>• Match data products with users<br>• Determine time scale for delivery (real-time, delayed)<br>• Determine quality of data and associated documentation<br>• Address other issues (security, proprietary rights) |
| Characterize the data flow to each user | • Determine what processing is needed and where it's done<br>• Identify what associated data is required and its format<br>• Determine transport or delivery protocol<br>• Determine derived requirements |
| Evaluate approach | • Determine cost and schedule<br>• Identify driving requirements |
| Develop alternative scenarios and refine through iteration | • Improve systems<br>  – Standards (interface, transport, form/format)<br>  – Commercial off-the-shelf software<br>  – Software reuse<br>  – Hardware technologies migration strategies<br>• Improve process<br>  – Co-locate functions (mission control center)<br>  – Distribute functions (national laboratories, universities, industry)<br>• Reduce requirements |

Systems for generating products vary tremendously in complexity and size—from 10,000 lines of code for the Scanning Multi-channel Microwave Radiometer (SMMR) instrument on the Nimbus 7 spacecraft to well over 300,000 lines of code (without comments) for the Position Sensitive Propositional Counters instrument on the Roentgen spacecraft. In a recent analysis of five product-generation systems, Price, et al. [1995] found that 10%–15% of the code is for science algorithms, 75% is for managing data (of which input/output itself constitutes 50%), and the remaining 10%–15% goes to navigation and other functions.

Goddard Space Flight Center has developed a Data Systems Dynamic Simulator (DSDS) which evaluates the operational performance of candidate data systems to support requirements definition, preliminary design, trade studies, and cost-benefit analysis. The DSDS allows us to develop the best system and to know performance characteristics before developing it. The DSDS has even simulated the processing and distributing of data for the Earth Observation System (EOS) [Bedet, et al. 1993]. This type of tool is important for estimating the cost of ground data processing.

### 13.1.3    Principles for Data Management

*Principle #1:  All data aren't created equal.*

To generate higher-level data, we may also develop various intermediate products. These intermediate products may have value for a time. A good example is Level 0 data. Usually Level 0 data is for a specific instrument and time period; it's not calibrated, and it doesn't include orbit or attitude information. A more useful product to an investigative team is Level 1 data with all necessary ancillary data such as orbit or attitude. Researchers will use this data to begin any detailed analysis or reanalysis. NASA typically archives the Level 0 data, storing it for the life of a mission. Only rarely, during the life of the mission, is this Level 0 data needed past when the next level of processed data has been generated. The project must therefore decide how long to keep any of its data products because it has to pay for their storage, physical management, and access.

Some of an instrument's data products may be requested more than others. For instance, people rarely request fully calibrated data from the Nimbus-7, Total Ozone Mapping (TOMS) instrument at the highest time resolution (which covers only a swath in longitude). But they often ask for the daily combined TOMS data product over the entire Antarctica, where the ozone hole has been discovered. This data on the ozone hole from the TOMS is available over the internet and on CD-ROM.

*Principle #2:  Mission-instrument data will need reprocessing.*

Users of mission data will understand a spaceborne instrument's responses and characteristics better over time, so they'll need to reprocess much if not all of the mission data. Thus, we need to plan product-generation systems so they'll allow this reprocessing. Also, reprocessing should take advantage of advances in standardizing scientific data. The total volume of data from an instrument, including all levels of processing, is typically between two and three times the volume of Level 0 data.

*Principle #3:  A data set not used by its creator is notoriously unreliable.*

In some cases, mission investigators will create data sets for archives rather than their own use. This practice increases mission cost. Instead, we should move data to the long-term archiving center at the earliest possible date—when the data is still in active use—before undergoing recalibration or reprocessing.

### 13.1.4    Data Archiving

Many missions provide data to users other than those directly involved in the mission. These general users get the data from either active or long-term archives (see Sec. 3.9). Archives that follow our first principle will provide different levels of archive service for different data sets. Table 13.2 compares the four levels as archive users would see them.

**Table 13.2. The Levels of Archive Service from the User's Perspective.** These levels provide users with a large variety of data access and services. The cost of archiving data increases with the level.

| | Level 1 | Level 2 | Level 3 | Level 4 |
|---|---|---|---|---|
| **Storage** | • Off-line, off-site storage | • Off-line, on-site, online limited | • On-site, off-line, some near-line, limited online | • Near-line or online |
| **Response** | • 2–4 week response | • 1–7 days | • Immediate to hours | • Immediate to minutes |
| **Products** | • Simple copies (or loan) | • Copies or simple subsets | • Flexible subsets, merged data sets, same mission | • Merged data sets from multiple missions and sources |
| **Delivery** | • Off-line | • Mostly off-line, limited online | • Mostly online | • Mostly online<br>• Archive support to accept user-derived products |
| **Catalog** | • Simple online inventory | • Simple online inventory | • Online granule inventory<br>• Online detailed catalog | • Search across multi-mission granule inventories<br>• Online detailed catalog |
| **Supporting Information** | • Paper documents as received from data producers | • Paper documents as received from data producers | • Paper documents as received from data producers<br>• System users' guides | • Product and system users' guide, formats, algorithm descriptions<br>• Accept user-produced information, data |
| **User Support** | • Minimal | • Staff | • User support office | • Research support office |
| **Typical Use** | • Bulk data delivery - low priority, mostly older data, mostly lower level | • Data delivery - high priority, low-level products with limited user selectivity<br>• Middle priority high-level products | • Data delivery - high priority, high-level products<br>• Some active browse<br>• Basic interactive data search and selectivity for users | • High priority, high-level products, individually low volume<br>• Active research, tools for research data planning and analysis, manipulation<br>• User-instructed automated data delivery |

An archive cost model [Klenk et al., 1990] has been developed which predicts the costs of managing data as a function of levels of archiving. Some groups in NASA, NOAA, and CNES use this software model to estimate archive costs. Active and long-term archives can use it. The major cost drivers for any archive are the levels of service, the data volume, complexity of the data (homogenous in form and format), and the amount of information (such as catalog, documentation, and descriptions) that must accompany the data to distribute it in a useful form and to manage it.

Information about archived data must be generated and archived along with the data. It describes specific formats, catalogs, instruments, and data anomalies at a very high level. All new data coming into the NASA data environment must have a Master Directory entry. It contains brief high-level information about data sets, allowing users to determine where they can get further details. It's an international directory with identical directories in Canada, Japan, Europe, and the United States, and is readily accessible on the internet.

## 13.2 Data Systems

Most spacecraft couple the spacecraft-bus or housekeeping data with the payload data, particularly as they are transmitted to Earth. A separate circuit handles uplinking commands to the spacecraft. We'll discuss all three components of the spacecraft's data systems. Later we'll turn to ground facilities for data handling and discuss processing of spacecraft sensor and housekeeping information along with preparing science-data products from the sensor data. Often, computing these science-data products requires information from the sensor's and the spacecraft's data streams.

### 13.2.1    Onboard Systems

Onboard systems consist of sensors, the computers used to collect data, recorders, downlink or telemetry systems, and the spacecraft-bus sensors and data streams. Before describing a generic science sensor, we'll address the nature and handling of the spacecraft data. All of this data is important to the spacecraft's daily operation, and some helps us process the spacecrafts sensor's data. The spacecraft-management data comes from the systems for power, thermal, electronics, communications, and attitude determination and control. Of these, we typically use only the attitude-determination data to process and analyze the science data from sensors. We'll briefly address these topics to show you how they're handled. Let's first look at them generically and then discuss an example from the NCAA's polar-orbiting spacecraft called ATIROS-N (TIROS = Television InfraRed Observing Satellite; A = Advanced and N = New). This series of spacecraft began operating in the fall of 1978 simply as TIROS-N.

**Subsystem for Attitude Determination and Control.** This component of spacecraft-housekeeping information is often critical to interpreting and applying sensor data. For most imagers, it's critical to know just how the sensor is pointing so we can properly map the image to its target. This amounts to understanding the relationship between the sensor optics or antenna pointing and the orientation in space. Thus, both the attitude and orbit information are important. Even if pointing control is limited, pointing knowledge is important. Most systems, however, are designed to control the spacecraft's attitude within some pre-specified limits rather than calling for control from the ground. Ground operators can narrow or broaden these requirements on the attitude-control system. See Chap. 15 and Larson and Wertz [1992] for descriptions of attitude-control systems.

Of all the housekeeping data downlinked, information on the spacecraft's attitude is of greatest value when we process the science-sensor data. We must know the spacecraft's attitude precisely to point to the instrument exactly. If the spacecraft's onboard system for determining attitude can specify this information, it's much easier to correct the sensor-image data for errors in ground location. Without this external source of attitude information, we must use the image itself to correct for variations in the spacecraft's attitude. We do so by locating ground reference points on the target image and then, using an orbital model for this spacecraft, computing the attitude necessary to yield the observed displacements of the reference locations. We need two ground reference points to specify completely the spacecraft's roll, pitch, and yaw [Rosborough et al., 1994]. The problem with this method occurs when the target area has no reference points. A prime example is the open ocean, where there may be many targets of interest but none that can serve as a ground reference point. In this case, we must compute the attitude for the orbit section that was over land and then propagate the attitude information forward to the time when the spacecraft is over the open ocean [Baldwin and Emery, 1994].

Problems arise when the telemetry data for the attitude information doesn't represent the spacecraft's actual orientation. For example, on TIROS-N the attitude information in the downlinked data stream, appears to show that the attitude-control system is working to minimize some error limits [Baldwin and Emery, 1994]. These limits differ greatly from any of the attitude information derived from images, at least for 1990 [Baldwin and Emery, 1994]. In some cases, such as LANDSAT, the attitude data is automatically incorporated into the georeferencing processing at the ground site. In general, the attitude information is not available for processing the image or other sensor data.

In locating the spacecraft sensor's data on the Earth, *geolocation* (also called image navigation) is critical to making the data useful for geophysical interpretation. But we must know the spacecraft ephemeris information to compute the geolocation. Usually, we calculate this data on the ground from spacecraft tracking information, then send it back up to the spacecraft over the command and control link and store it onboard the spacecraft. Modern spacecraft have also taken advantage of the Global Positioning System's (GPS) space-qualified receivers, which can use differential location to compute the spacecraft's position within tens of meters. These systems make it possible to compute onboard the ephemeris information needed to geolocate the sensor data. At present, even this GPS data usually goes to the ground for processing and computing the ephemeris elements. Onboard processing units aren't powerful enough yet to process all data needed to accurately locate the sensor data on the Earth. CPUs on future spacecraft may be able to use GPS sensing units to do geolocation on orbit.

**Onboard Processing Functions for Sensor Data.** Besides collecting, storing, and transmitting routine and housekeeping data, we need to consider other onboard processing needs. For example, we should be able to calibrate science-sensor data more efficiently onboard, especially because the calibration data may come from the sensor itself or from an external source. The advantage of calibrating onboard is that all users can access the calibrated data. Unfortunately, they can no longer change or improve that calibration as new information becomes available. Thus, we definitely need to be able to change the calibration algorithm as we get new information. In other words, we have to uplink the calibration routine from the ground to incorporate new knowledge in the calibration procedure. We may also need to communicate new information from data sources external to the spacecraft if they're necessary to improve or modify the calibration.

Sometimes the volume of data generated by the sensors exceeds the downlink capacity, and only onboard data compression can handle it. The compressed data can then be downlinked and decompressed on the ground. In this case, we need to allocate onboard processing capability and processor time to the compression computation. Because compression techniques vary with data and application, we may need to alter the specific compression algorithm, so we must be able to control this change from the ground. We also have to know the compression algorithm on the ground to decompress the data; it's best if the compression technique is identified in the data stream itself.

## 13.2.2 Ground Systems

**Antenna Systems.** The ground system has two types of antennas: (1) uplink and control antennas and (2) receive-only, direct-readout antennas. Both of these systems communicate with the onboard data systems through the communications system. We'll concentrate here on the receive-only ground reception and data-processing system because other chapters deal explicitly with the communications, command, and control aspects. Refer to these chapters to answer your specific questions regarding the data system's radio frequency, command, and control.

Each space mission is designed with specific objectives in mind. Although the space component (the spacecraft) is likely the most costly and complex element of the mission, we can't neglect the importance of the ground system and the need to process the sensor data. The receive-only ground system begins with the antennas themselves and extends into data collection and processing. The antenna's basic nature depends on the type of orbit the spacecraft is in. For spacecraft in geostationary orbit, the antennas can be fixed in position and thus are far less complex than antennas that must move with the spacecraft. Spacecraft in highly inclined polar orbits require antennas that must move with the spacecraft as it crosses from horizon to horizon. Many years ago, people thought the only way to meet this requirement was to build *auto-tracking* antennas that would detect and follow the spacecraft. They equipped the antenna with a multi-horn feed unit that first detected the movement of the spacecraft's rf signal across the feed and then calculated the movement of the antenna necessary to maintain its focus within the spacecraft's beamwidth. Complex and very expensive, these auto-track feeds are being replaced by *computer tracking*, which uses ephemeris data and an appropriate orbital model to track the spacecraft's movement relative to the antenna location. To do this, we must input the actual latitude and longitude location of the antenna and compute the look angles (usually azimuth and elevation) of the antenna to follow the spacecraft. Modern orbital models and readily available ephemeris data make it possible to track almost all spacecraft. For specialized missions, some agency that regularly tracks spacecraft will need to generate the ephemeris data we must have to predict the antenna's position and move it.

**Overall Ground Data Systems.** Figure 13.2 shows the main components of the basic ground data system. The tracking or geostationary antenna makes up the basic physical infrastructure of the front-end of the system. The antenna is usually a parabolic dish with a simple focal point to collect the signal off the dish. Some units may be built as Cassegrain antennas. They use a reflector at the focal point rather than a feed and then locate the feed at the base of the dish. In either case the feed unit must be equipped with a downconverter to shift the rf frequency down to a level that can be handled by the data system. The front-end also usually has a low-noise amplifier that boosts the signal for transmission to the rest of the data system's components. The dramatic improvements in performance of these amplifiers over the past 15 years have revolutionized this aspect of antenna technology. Once relegated to large electronic units, they're now compact and produce excellent figures of merit. They are usually next to the feed and downconverter units.

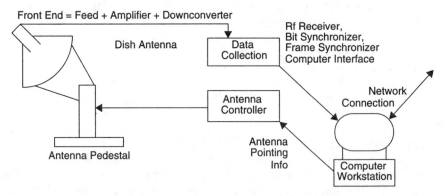

**Fig. 13.2.**    **Modern Receiving Station.** Shown are the essential components of a tracking station that transmits high-resolution pictures.

The signal feeds (Fig. 13.2) to the receiver which contains the demodulator and also compensates for the local Doppler effect. The resultant signal then passes to the bit synchronizer, which does the error checking and finds the synchronization pulse. The data strings then go to the frame synchronizer, which acts as a data buffer and accumulates image data for transfer as a file to the computer hub of the data system. Older systems used a separate computer to generate the look angles for computed antenna tracking and a dedicated workstation computer to take in and process data. The speed of modern workstations today makes it possible to track the antenna and to receive and process data from the same machine. The serial port on the workstation carries the look-angle data to the antenna-control unit while the parallel port handles data collection.

Another dramatic change has been in chip technology. Most processes can now be done in a chip rather than in a dedicated electronics unit. Today bit and frame syncs consist of preprogrammed chips in an electronics box or may even be a board installed in the system computer. All of these improvements have made it much easier to automate the collection and processing of spacecraft data.

**Levels of Data Processing.** The first step after collecting the spacecraft data is to determine where each data type should be stored or how it should be initially processed. Many steps are involved in routine processing of sensor data, so let's first talk about data processing levels and what they mean. Note that, although we commonly define data levels, everyone doesn't agree with these definitions. You need to carefully evaluate the meaning of the data and data level they're working with to precisely define what this level means and how it affects future computations and operations.

The sensor data comes down from the spacecraft in what is often called *engineering units*. These are usually voltages measured by the individual sensor. At some point these voltages are converted into a physical data product using the information on pre-launch calibration for that particular sensor. This means the sensor manufacturer has calibrated the sensor by comparing it with known (or measured) values and developing a relationship between the voltages recorded by the instrument and the physical quantity the sensor was designed to measure. One of the problems with many sensors is that they tend to drift over time, thus altering the relationship developed in the pre-launch calibrations. Only if we continue to calibrate it onboard (comparing it with a known onboard reference) can we compensate for this drift. Some people have tried to use known target values to calibrate sensors in space, but there are many questions of representativeness and contamination from the intervening atmosphere.

A good example is one of the infrared channels on the Advanced Very High Resolution Radiometer (AVHRR). The AVHRR itself has been lab or bench calibrated by the manufacturer to yield coefficients for converting the voltages read out from the infrared detector. Also, an onboard reference value measured by a Precision Resistance Thermometer (PRT) adds to the data stream. For each channel, there are two PRT measurements and a view of deep space for temperature reference values. These PRTs are also calibrated in the lab, which gives us coefficients to compute temperature from the voltage counts corrected for the PRTs viewed by the scanner. All of these calibrations are usually linear fits or at most quadratic fits to the calibration information.

Once we've applied calibration information, we have physical quantities—often expressed as a radiative equivalent of the actual quantity of interest. For example infrared sensors mainly map temperatures (cloud top or sea surface), but the initial conversion is from voltages to radiances. We need further calibration to convert the radiances to thermal equivalences known as *brightness temperatures*. Later this brightness temperature can be processed with various algorithms into geophysical quantities. These algorithms may use multiple channels of a single instrument or may involve data from other instruments or even ground measurements.

We usually refer to the first level of raw instrument voltages as *level 0 data* because it involves no processing. Once the data has been processed into radiances or brightness temperatures (depending on the nature of sensor wavelength), we refer to it as *level 1 data*. Level 1 data varies in format details. Some groups choose to geolocate the level 1 data and thus introduce new information into the data stream. We *geolocate* data by associating the data with a point (latitude and longitude) on the Earth. Because computing geolocated data typically involves both an orbital model and a reference to data on orbital parameters, adding geolocation requires external input by the navigation function data in the form of the ephemeris data (orbital-element). To distinguish level 1 data that doesn't involve other data from level 1 data that does, we've broken this category into two types. Level 1a data isn't processed to add geolocation; level 1b data includes geolocation. In some cases, the spacecraft's equipment may compute the level 1 data onboard. In this case, it usually doesn't include geolocation. Figure 13.3 summarizes this process.

Most of the above definitions apply to images and other data. If a sensor doesn't generate images, level 0 data still comes directly from the spacecraft, and level 1 data has been processed into some type of initial value that is not necessarily in geophysical units. Higher levels then depend on processing algorithms to create geophysical products. Table 13.2 lists the levels and their character.

Whether images or other data, all level 1 data from the spacecraft can be processed into higher-level geophysical products. This involves processing it in space or time (averaging, filtering, etc.); combining it with other data (spacecraft or in situ); interpreting; and mapping. An example would be to derive a snow-water equivalent from the satellite's visible and infrared images. First, we must discriminate snow cover from clouds in multi-spectral spacecraft images by combining the visible, near-infrared, and thermal-infrared channels. Although clouds and snow have similar reflective properties in the visible range, their temperatures vary greatly because most clouds are higher and colder than the low-lying snow. Thus, we can use the visible and thermal-infrared channels together to distinguish snow from clouds. Compositing over a few sequential images makes it possible to fill in for areas where the clouds blocked the snow reflection. Thus, the final snow cover is defined by a group of images over a short time.

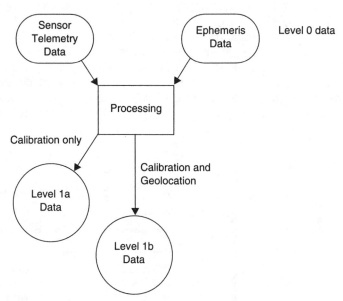

**Fig. 13.3.    Processing from Level 0 to Levels 1a and 1b.** Level 0 data comes directly from the spacecraft's sensor and is measured in voltages. Level 1 processing converts the data to other physical units.

**Table 13.3.    Data Levels.** The level and location of data processing must meet users' needs while keeping costs low.

| Level | Definition |
| --- | --- |
| Level 0 | Raw sensor voltages as downlinked from the spacecraft |
| Level 1a | Calibrated data in terms of channel radiances or brightness temperatures; pre-launch or onboard calibration information applied |
| Level 1b | Calibrated data combined with navigation data provide values for latitude and longitude, but the data is still in spacecraft perspective |
| Level 2 | Calibrated data that has been resampled to a selected map projection using inverse-image navigation; this resampled data can have a map overlaid or be combined with similarly navigated images |
| Level 3 | Geophysical parameters derived from the calibrated data that have been further geolocated; data has been converted to geophysical values and assigned a grid location |
| Level 4 | Geophysical parameters that have been computed using combinations of data sources (both spacecraft and in situ) and mapped to a grid |
| Level 5 | Definitions vary, but higher levels indicate additional processing by merging with other data, averaging over a larger grid spacing, etc. |

Finally, we must apply a relationship between snow cover and snow-water equivalent as defined by historical data relating the two, combined with information on ground slopes and topography. In some cases a hydrologic model might be run to define this relationship. After applying this relationship, we now have the snow-water equivalent and can classify it by region. Thus, the resultant product is much compressed from the original sequence of images that went into making the product. This compression is typical of most higher-level data products. Remember, though, many assumptions limit the use of this product. Figure 13.4 simply summarizes this process for image data.

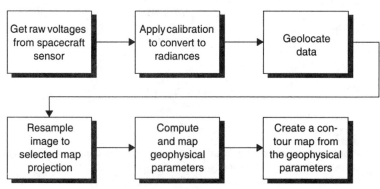

**Fig. 13.4.    Steps in Processing a Spacecraft Image.** The steps in this process correspond to higher levels of data processing and increasing cost.

Definitions of higher-level products attract debate with no general agreement. Usually, *level 2 data* has been processed into a derived geophysical product and resampled into a selected map projection. It differs from level 1 data because the latter remains in the spacecraft projection even when latitude and longitude locations have been added. Processing to a level 2 product may require more spacecraft or in-situ data, or it may imply some processing in space and time. Level 3 products are usually gridded values derived from the spacecraft, and possibly, in-situ data. In gridding we must use some spatial-smoothing algorithm such as optimum interpolation. Products with the designation *level 4* (and above) are all gridded products that have experienced further processing. Carefully examine the definitions of all higher-level products and evaluate the assumptions that went into their creation.

As we go from level 0 to levels 3, 4, and 5, we add cost in processing time and effort. Although processing is cheaper than flying a mission or collecting data, many missions have lost results because they didn't budget properly for it.

# 13.3  Data-Processing Requirements

## 13.3.1   Types of Data

**Sensor Data.** *Sensor data* refers to the raw voltages output by the detector electronics in the sensor. These voltages are strictly functions of the sensing technology in the system and don't refer to any physical value before applying some calibration procedure. Our most important concern here is to maintain the fidelity of the data and keep as many bits as were characteristic of the sensor. In some past cases, 8-bit data has been stored in only 6 bits and thus eliminated some potentially important information in the lost 2 bits.

The processing stream must also ensure the fidelity of the data transfer down from the spacecraft and into the data-processing system. Thus the receiving system must be equipped to check the data stream for accuracy and ensure the proper data goes into the processing system. We usually do so by using the synchronization pulse, which tells the receiving system when it is properly decoding the data stream. Designing this part of the spacecraft involves a trade between the desired resolution of the sensor data and the spacecraft's data rate. This trade involves considering the spacecraft's rf hardware, the ground station's receiving and processing hardware, and the processing capability.

**Onboard-Calibration Data.** This information is critical to keeping the sensor data accurate even over relatively short periods. *Onboard calibrations* are measurements we take with certain sensors—whose expected readings are known. All electronic equipment drifts over time, and a spacecraft sensor especially will change under very heavy requirements. Regular and reliable sensor calibration is the only way to account for this drift, and onboard calibration is the only way to make this task reliable.

Onboard calibration varies with the type of sensor. For infrared radiation, we need a thermal reference. We can supply a separate bolometer or use something as simple as the temperature of the instrument's backplane and monitor it to provide a temperature reference. For monitoring, we use a precision resistance thermometer (PRT), or a series of PRTs, which we can turn off or on as needed. Because the sensor views the monitored parts of the back plane (Fig. 13.5) during the internal part of the scan, the PRT temperatures become part of the scan-cycle data stream and are downlinked with the sensor data itself. A scanner view of deep space provides another reference temperature, which we assume to be a constant temperature near 0 K. Together, the PRT and data from the cold-space target allow us to compute a thermal-calibration wedge in order to adjust the voltages to temperature.

For the reflective channels, onboard calibration requires some type of light reference that we may view during the scan cycle. *Reflective channels* are the shorter wavelength channels that sense reflected sunlight from the Earth's surface, including the visible and near infrared. This calibration can get a light reference by using a mirror that deflects the radiative path to view the reflective reference rather than

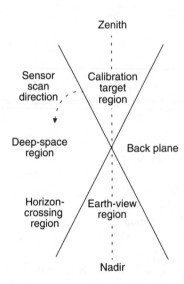

**Fig. 13.5.    Signal Position as a Function of Scan Angle.** The back plane and deep-space regions of the scan provide temperature calibrations as part of the data stream.

the ground-surface target. Of the operational environmental spacecraft, only LANDSAT allows reflective reference or calibration (Fig. 13.5). Plans for many future sensors are to have onboard calibration for the reflective as well as the thermally sensed channels.

These reference elements must themselves be calibrated in a lab before launch. Also, the data stream should include the information about which reference system we're using and what its calibration is. Temperature-reference targets using thermal-infrared calibration will likely suffer more from age-induced drift than the visible or near-infrared calibration.

**Spacecraft Housekeeping Data.** As mentioned earlier, it's important to keep good track of the data on spacecraft housekeeping to give people the chance to see how it affects the spacecraft's data records and their interpretation. This data usually resides somewhere in the onboard computer, and it's best transferred in the same downlink as the other sensor and calibration data. One of the typical problems with this information is that its format isn't widely known, and science users sometimes have trouble decoding the information into a useful form. Because data is important for many different processing applications, we must have the format well documented and available.

**In-Situ Calibration Data.** Often, we want to check on the performance of onboard calibration. We can do so by using our knowledge of some uniform ground targets. This approach is particularly useful for sensors that employ new and poorly understood technologies. An example is synthetic aperture radar (SAR), which was deployed in space for the first time in 1978 as part of the payload on SEASAT. To evaluate the performance of this new system, mission staff deployed large, corner-reflector targets in the Mojave Desert. The desert signal was fairly uniform in the SAR, whereas the corner reflectors provided a strong specular signal in the SAR data and in well-known locations on the Earth's surface. Using these strong signals over a weak desert background, operators could calibrate the SEASAT SAR data.

Many mission designers believe we must calibrate sea-surface temperatures (SST) computed from AVHRR data. They think that only SST measurements from drifting buoys are appropriate to calibrate infrared SST estimates for the spacecraft. But this assumption doesn't recognize that the spacecraft can't view below the mm-thick skin layer of the upper ocean. At the same time, the drifting buoy can't measure the temperature of the skin (contact destroys the skin layer). Thus, the calibration temperatures aren't physically consistent with the radiative temperatures measured by the spacecraft. The fundamental differences between skin and bulk SST [Schluessel et al., 1986; Schluessel et al., 1990; Wick et al., 1992] make it impossible to calibrate IR measurements of the ocean's surface-skin temperature with traditional measurements from the drifting buoys.

We must match our knowledge of the best surface targets with the data on these same targets when sampled from space. Land-surface targets may be large, homogeneous regions with fairly simple reflective properties. Deserts and salt-flats are prime examples of strong reflective targets (high reflectivity) that are very homogeneous (in character and reflectivity) over large regions of the Earth. We can expect spacecraft sampling of these targets to be constant and consistent, thus allowing us to calibrate the spacecraft data from our knowledge of the target's reflective properties.

A calibration example for a spacecraft altimeter is the need to know the sea's surface height. Usually, we use a site like an offshore oil rig to install a tide gauge and measure the variations in sea-surface elevation relative to the platform height. In this case, we must select the site so the spacecraft's orbit will pass almost directly over it. Because the spot size of the altimeter footprint is about 1–2 km, the site simply has to be within that footprint. For imaging sensors that cover large areas, our ground-truth targets can be in any part of the image.

## 13.3.2    Steps for Data Processing

We list the steps for processing data collected by the spacecraft in Table 13.4 and describe the steps below.

Table 13.4.  **Steps for Processing Data.** The order of these steps depends on mission requirements.

| Steps | Considerations | Where Discussed |
|---|---|---|
| 1. Downlink Data | • Requirements for data transport | Sec. 3.4, 11.3, and 13.3 |
| 2. Collect Data | • Resolution<br>• Format | -- |
| 3. Generate Level 0 Data | • Archiving<br>• Data compression | Sec. 3.9 and 13.3 |
| 4. Define Algorithms | • Defined by sensor manufacturer or developed by science group | Sec. 3.8 and 13.3 |
| 5. Implement Algorithms | • Combine data from different sources<br>• Levels of data processing required | Sec. 3.8 and 13.3 |
| 6. Process and Distribute Higher-Level Data | • User requirements<br>• Availability of data | Sec. 3.4, 3.8, 11.3, and 13.3 |
| 7. Validate and Update Algorithms | • Algorithms evolve over life of mission | Sec. 3.8 and 13.3 |

**Downlink Data.** Once we understand the spacecraft's sensors, we must know the downlink system, including the antenna and the rf equipment needed to read out the data on the ground. We must also know how we're controlling the antenna, the accompanying rf hardware, and the collection system. Each step influences the data, so we must be able to assess the steps in order to properly process, store, and manipulate the data.

**Collect Data.** This stage starts where the data enters the computer through the interface from the rf stream. Depending on the data's format, we usually load it into one or two bytes (each byte consists of 8 bits of data) and process it to the required level. If the raw values fill only eight bits of digital data, we can easily keep them as a one-byte array of values. But most modern sensors have 10 or 12 bits of resolution, so we must load their data into two 8-bit bytes for future processing and manipulation. Because resolution for sensors depends on the number of bits, an instrument having more bits per pixel is better able to resolve changes in the sensed parameter. Thus, modern sensors are going to greater sensitivity and, hence, more bits.

Sometimes, we separate the data into segments corresponding to data types, whereas at other times we save the entire data set together. For example, in most imagers the image data interlaces with the housekeeping and other sensor data. Users must strip out the non-imager data to view the images. All data can either be archived as a raw bit stream or take separate forms. The collection system controls both the processing and archiving of spacecraft data.

**Generate Level 0 Data.** As stated earlier, level 0 means converting voltages to engineering units—the most basic data form. We commonly archive all level 0 data at its source—usually the receiving system. This approach ensures that, no matter what happens later, the level 0 data will be available for reprocessing. It requires a lot of processing and is very expensive, but we must know the data will always be available. Experience has shown that, as science knowledge advances, many of our initial assumptions are flawed enough to require reprocessing all of the original data. In this reprocessing, access to the original data is often difficult. Thus, if our data system ensures access to the level 0 data, reprocessing will be easier.

These concerns don't apply for missions that don't create sensor data useful for retrospective analysis. Examples are communication spacecraft and military-reconnaissance spacecraft. Their information is short term, so we don't have to archive it beyond its immediate use. In this case, reprocessing clearly doesn't apply, so we can let the data go when we're finished analyzing it. This handling works for data compression as well. As we'll discuss later, we should compress science-sensor data just to the point where we don't lose any of it. Exceptions may be data from communications and military-reconnaissance spacecraft because these types of operations value immediacy of data over 100% accuracy.

**Define Algorithms.** For each product generated from spacecraft data, we must define a particular algorithm. The algorithm may be as simple as applying some constant coefficients to the level 1 data to produce a data product directly related to the sensor data. Other more complex algorithms may involve space or time gridding, combination with other spacecraft, or in-situ data. The original input data changes during this gridding, so the resultant higher-level geophysical product no longer directly represents the input spacecraft data. Because many of these algorithms are basically non-linear, we can't retrieve the original data from the higher-level product. That's why we have to permanently store level 0 and level 1 data if they are to have lasting value. Later, we may have to reprocess the level 0 data in order to interpret the old measurements in the light of new knowledge.

Algorithms may be defined by the sensor manufacturer or may be developed by an external science group and contributed to the data system. Often an instrument will have a science team who must define the algorithm intended to generate a specific geophysical product. This algorithm may apply at first, but later research may discover problems with it. In this case, new investigations can contribute new and improved algorithms for processing sensor information to higher-level products. The Special Sensor Microwave Imager (SSM/I) is a good example. The SSM/I is a microwave imager flying on the Defense Meteorological Satellite Program's spacecraft, which fly in sun-synchronous orbits much as the NOAA's TIROS-N spacecraft do. The sensor contractor, Hughes Aircraft Corp., had to create geophysical algorithms for various land, ocean, and atmospheric products computed from data collected by this new instrument. While Hughes consulted somewhat with the science community, they did most of this algorithm themselves. Later validation studies by science groups revealed some inadequacies in the Hughes algorithms, and several investigators contributed new and better algorithms. Published results of these new algorithms gave the science community a chance to judge which ones were the best. As in most things, the science community votes by how much they use a data product.

**Implement Algorithms.** Implementing an algorithm includes assigning the correct data streams to the proper processing channels. We can simply select the correct sensor channel's data and apply a few constants to generate new pixel values that represent a geophysical quantity. Or, we may combine channels while still using constant coefficients. But we may have to resample the data and compute values on a space or time grid or combine the sensor channel's data with other spacecraft or in-situ data to compute a higher-level product.

The snow-cover example we used earlier shows how this approach works. We can combine individual channel values to separate cloud cover from snow on the ground. This combination takes advantage of the fact that, whereas the reflected visible and near-infrared signatures of snow cover and cloud cover are similar, their relative infrared temperatures are quite different. Thus by combining the thermal-infrared channels with the visible and near-infrared channels, we can discriminate snow cover from cloud cover. This is true of the entire data file, whether we leave the data in the spacecraft's projection or use inverse navigation to resample the image to a chosen map projection.

If we choose to navigate (geolocate and resample) all of the images to the same map projection, we can combine images in our assessment of snow cover. This combination is necessary to map snow cover in all ground pixels in the presence of cloud features. Because clouds move around much faster than the snow cover, we can expect an image composited over a few days to pretty well cover the entire site of interest. Warm temperatures may melt light snow after it falls, but we can disregard this transitory snow to focus on the long-term snow pack, which allows us to estimate annual runoff. Temporary snow cover doesn't affect this runoff. Thus, we can composite a sequence of images over some short period to emphasize only the information we believe affects snow cover on the ground.

Moving up a level of processing, we now would like to convert snow cover to snow water equivalent (SWE). This conversion requires some knowledge of the relationship between the snow cover, the local topography, and the snow ablation from the site of interest. The simplest method is to relate the elevation of our sampling location to the historical pattern of snow cover in the area of interest. Fortunately, we already have more than 25 years of historical data on snow cover, which we can use to help define this relationship. This type of relationship ignores for the moment any type of snow ablation prompted by persistent weather patterns (for example high winds or storms).

Applying this type of elevation to SWE relationship again requires excellent image navigation for all data involved. We would want to aggregate data for particular physical regions, such as river drainage basins, which we can easily identify from a topographic map of the target area. In some cases, we'll have to discover how water may move from one river basin to another during the runoff season. If this movement is happening, it will ruin any statistical comparisons using the spacecraft's maps of snow cover and measurements of runoff.

Once we have a relationship between elevation and SWE, we can apply it to the spacecraft-derived estimate of snow cover. By merging the spacecraft image with the mountain topography, we can compute the elevation of the snow-covered regions defined by the spacecraft images. Using the SWE-to-elevation relationships for each basin, we can now convert our spacecraft's maps of snow cover to estimates of SWE for each basin. By averaging these estimates together, we can compare them with those from measured river outflow.

**Process and Distribute Higher-Level Data.** Many users aren't experienced with processing and analyzing the spacecraft's data, so they want a higher-level product which has been calibrated, converted to geophysical parameters, and mapped to a chosen projection or grid. But because each processing step can introduce errors, these users should at least be familiar with the processing algorithms and procedures used to create this higher-level product. An example of this problem is the Normalized Difference Vegetation Index (NDVI), which we can compute from the first two channels of the AVHRR sensor. A historical series of this index had been computed using level 1b data for various parts of the world. But new NDVI values using exactly the same conversion to geophysical data produced different results for the same locations. It was later found that navigation for the level 1b data was seriously in error, and all level 1a data had to be reprocessed to correct the problem.

A second main group of users are most familiar with the data stream and its processing. They're most interested in level 1 data calibrated by applying information from laboratory calibrations before launch. Level 1 data applies for most of the large-scale mapping efforts using spacecraft data. It must be available to many different users and may be distributed over the network or on media. This is usually the most basic data level used by the science community. Most research groups can't take on heavy reprocessing of level 0 data, unless a majority believes it's necessary because lab calibration data was invalid or otherwise flawed.

For some missions, data processing consists of detecting an event (such as monitoring nuclear explosions on the Earth's surface), and the sensor system is designed merely to extract that information. Still, this sensing alone isn't adequate because the location and time of the event are also critical. Thus, we need some form of onboard or ground processing to correctly locate the event in space and time. Communication satellites also don't need higher-level processing; the spacecraft mainly relays signals, so it doesn't have to process them at a higher level.

In processing spacecraft data to higher levels, we must always decide where to process it. In the past, we did all processing at a central site. But new data systems have led to new architectures for which the processing is distributed rather than centralized. This distribution avoids potential system failures at the centralized location. It also makes possible multiple and redundant processing within the system. Although we generally accept that level 1 data must be archived just as the level 0 data, it's not always clear the archiving must be permanent. Often, computations are more efficient using the original input data. This approach saves data-storage space and best uses computational abilities, which are getting better and cheaper. We must do many trade-offs when deciding whether to compute and store a product or to compute it in real-time on request.

Most spacecraft data need conversion to some form of geophysical product to interest a lot of users. As outlined above, some users can work with at least the level 1 data to generate new products, but most will depend on geophysical products a few other users create. Besides calibration, navigation, and final gridding, many products require combining either different data sets or data and related in-situ measurements. All combinations require substantial processing—particularly geographic mapping and calibration. Data sets can combine only if they cover the same space, cover similar time spans, and have the same geophysical units.

**Validate and Update Algorithms.** Algorithms for spacecraft-data products aren't static. They evolve as data applications mature. We may need to change some aspects or even all of the algorithm. Also, new information may affect the sensor data and require us to expand the algorithm to include these effects. We may update the algorithm regularly or only as new developments take place.

Algorithm developments are still part of basic science research, so this activity is difficult to control or predict. We can easily use an algorithm to develop products, but new methods require new discoveries or at least the incorporation of new data into the older algorithm. Usually, this means recalculating the entire data product. Most likely, we can recalculate using level 1 data without going all the way back to level 0. But sometimes we discover fundamental processing errors, so we have to revisit the stored level 0 data to compute a consistent time series of geophysical products.

As an example, let's again consider computing sea-surface temperature (SST) from a spacecraft's infrared imagery. Traditionally, the "split-window" technique has been used to correct for the attenuation of the infrared-temperature signal emitted from the ocean's surface because of atmospheric water vapor. Assuming that two long-wavelength, thermal-infrared channels absorb thermal energy differently as a function of the amount of water vapor, we incorporate the difference between these two channels into the SST algorithm to compensate for atmospheric moisture.

Recent studies [Emery et al., 1994] have demonstrated that this difference between infrared channels isn't enough to correct for the atmospheric water vapor, so errors in SST can be as large as $2°$ C. To better correct for this effect, we must add new information on the total moisture content of the atmospheric column, which we can estimate from nearly coincident measurements from microwave imagers. This microwave estimate for water vapor becomes part of the algorithm for SST and merges with the infrared channel's data for computing SST. Comparisons with simultaneous in-situ measurements of SST have demonstrated the significance of these corrections made possible by combining infrared and microwave data. Thus, our new algorithm will require access to nearly coincident microwave data to estimate total moisture content of the atmospheric column, which we'll add to information from the infrared channel.

With this new approach, we could use the level 1 and level 1b data from the infrared sensor. But if we have problems accurately geolocating the level 1b data (as was found for the AVHRR sensor), we can't use this data. In this case, we must return to the level 1a or even the level 0 data and reprocess it to correct for improper geolocation results. After doing so, we can compute the new algorithm for SST using merged microwave data.

### 13.3.3    Issues for Processing Data

**Requirements for a Data-Product System.** The data system must be able to handle the spacecraft sensor's data and generate related data products. Thus, we have to account for any need for ancillary data from other spacecraft or in-situ measurements and incorporate it in the system. We also must include data distribution in the design. We should avoid using any computers with operational requirements (such as the collection and tracking computer for the antenna system) to compute data products. These tasks are completely separate. Even with today's powerful workstations, we should keep them that way to improve the computing efficiency and ensure we have enough computer cycles to get each job done without interference.

**Editing and Errors.** Our data system must be able to check errors in the raw input data and all products. This error checking must be routine and must not create a processing burden for the data system's hardware. Error checking may be as simple as setting a value threshold and eliminating any values above this threshold. In this case, the difficulty is selecting the right error thresholds. Pre-launch calibration data, which defines the sensor's operational limits, may set them. Or, we may use our knowledge of the geophysical target values to set the limits on measurements. In all cases, we must set the threshold limits using some external information and not the sensor data itself. We must also recognize that our limits may be in error because of our incorrect prior knowledge. A good example is the total ozone-mapping spectrometer. In its early operation, the extremely low ozone values measured over the South Pole were rejected as bad data, effectively eliminating the ozone hole over the Antarctic from the data.

For higher-level products and their algorithms, setting error thresholds is more complex. Now, error sources in the spacecraft and related in-situ data may interact. How these errors propagate through the system to result in the final overall error for the data product is not a trivial question. Typically, we employ Gaussian error propagation, assuming normal statistics for the property in question.

These checks are intended to remove only the largest errors; they don't ensure error-free data coming from the system. As with the algorithm evolution discussed above, often new research uncovers sensor errors previously neglected.

**User Input.** Users' requirements change, so we must build an adaptable data system. In fact, the entire user community may change during the life of a particular sensor mission. New people will start using the data. They may have new requests that the original mission plan didn't consider. Therefore, the system's administrative structure must incorporate changes without disrupting everything.

Remember, any data is useful only if users employ it to answer scientific questions. Thus, we must check to see how our data system is working for the science community. If data isn't usable, we have to change the system.

**Online vs. Offline Distribution.** The large volumes of data generated by spacecraft imagers usually means that data must be distributed on some form of magnetic media. Data volumes depend on the spatial and spectral resolution of the sensor and the number of bits kept per pixel from the image. The trend is to increase image fidelity by keeping more pixels (now most systems keep 10 or 12 bits but in the future they'll be going to 14–16 bits per pixel). Images with higher spatial and spectral resolution generate large volumes of data for a sample of a relatively small geographic area. Table 13.5 gives some examples of data volumes from operating spacecraft. Individual spacecraft images that range between 70 and 250 megabytes (even at the 8- or 10-bit level) need to be stored on some form of magnetic tape. Just a few years ago the storage medium of choice would have been (BPI) 9-track computer tapes handling 6250 bits per inch. These tapes hold just over 200 megabytes, which would cover all but the largest spacecraft images. The inability of these tapes to store the spacecraft data resulted in receiving systems on early spacecraft being equipped with High Density Digital Tape (HDDT) Recorders. Previously, data was recorded on analog tape recorders and later transferred to digital tape. Both the analog and HDDT recorders used large tape reels that could record many gigabytes of data. These recorders were expensive to buy, maintain, and operate. Thus, only institutions working directly with spacecraft data used them.

**Table 13.5. Data Volume for NOAA.** Geostationary Operational Environmental Satellite (GOES) and Television InfraRed Observing Satellite (TIROS-M) are NOAA weather satellites. TIROS-M is in a Sun-synchronous orbit.

| Data Category | Data Type | Volume (per day) |
|---|---|---|
| GOES | AAA | 3.7 GB |
| TIROS-M | Global Area Coverage | 1.1 GB |
| | High Resolution Picture Transmission | 1.7 GB |
| | Local Area Coverage | 2.4 GB |
| Shared Processing | Special Sensor Microwave Temperature | 56 MB |
| | Special Sensor Microwave Imager | 210 MB |

Remember, during these early days online disk storage was also costly and not very efficient. Systems with multiple megabytes of online storage were considered large. As the online storage systems became smaller but more robust, the cost also dropped dramatically. The emergence of the Small Computer Standard Interface (SCSI) as a practical standard continued this drop in cost. Physical disk sizes have changed from large 36-inch platters to 3.5-inch enclosed disks. At the same time, new tape formats have been introduced that take advantage of helical scanning to store multiple gigabytes on each tape. These tapes are either 8 mm video tape or the new 4 mm Digital Audio Tape. They store from 2 to 20 gigabytes of digital data, making it possible to greatly expand the tape capacity for storing data. These tapes work with various readers. Lower tape costs have decreased overall storage costs. We no longer need analog tape recorders or HDDT recorders to store the data coming down from the spacecraft.

Another digital-storage development that plays an important role here is optical media. Initially only *write-once read-many (WORM) optical systems* were available; they stored about 3–6 gigabytes per 12-inch optical platter. The real advantage of the optical systems was that the data storage was presumed to have a much longer shelf-life than magnetic tapes. Tapes are known to last about 5–10 years, whereas the optical platters were rated for use up to 50 years. Thus, for lengthy storage, WORM disks were ideal. Soon *magneto-optical* (MO) systems became available; they were read/write optical systems, with 5.25-inch cartridges that hold about 600 megabytes per side. But there are problems with a lack of uniformity in the MO cartridge format and differences between MO hardware. A standard doesn't exist, and media can't be exchanged from one unit to another. Until these problems are resolved, MO won't be the medium of choice.

These systems using high-density tapes and optical disks have also led to low-cost, mass-storage systems using robot arm "jukeboxes." They combine robot-arm retrieval of tape cassettes with a number of tape drives to produce a computer-controlled system that can quickly retrieve data from many 8 mm tapes. Combined with low-cost, SCSI hard disks, these jukeboxes make it possible to build automated systems that store multiple terabytes of spacecraft data online.

**Network Requirements.** This large data-distribution capacity is important because we're living in an age when network connectivity has gone from 56 kbps to 45 and 155 Mbps. Present plans call for 622 Mbps connectivity to most sites in the US by the year 1997. In less than ten years we will have seen an increase of seven orders of magnitude in network speed and connectivity. Although there's some question how we'll pay for these services, clearly we'll be able to use the network to deliver information at unprecedented rates. Even the large volumes associated with spacecraft data can realistically be shipped over these high-speed networks. Now the problem is storing data locally, but the same technology that puts the spacecraft data online can help scientists store all data of interest at their own sites.

Just two decades ago, all data systems archived spacecraft data on magnetic tapes and distributed the tapes along with hard copy. A spacecraft archive consisted of files of images that were difficult to analyze and retrieve. The advent of digital spacecraft data shifted the emphasis to computer tapes, but these tapes were still bulky and required considerable storage space. Also, to retrieve the digital data, users had to mount the tape and search through its files. All levels required human involvement, and the final product was often hard copy.

The dramatic changes in digital storage, the speed of low-cost workstations, and our ability now to view spacecraft images on virtually any type of workstation or desktop computer have led to a new paradigm in storing, retrieving, and distributing data. The internet has also become reliable for distributing modestly sized files of all types of data including spacecraft images. Thus, we can now distribute moderate volumes of spacecraft data directly over the network, rather than copying tapes and shipping them to user sites. This new method overcomes two main barriers: (1) tapes getting lost, becoming unreadable, or having a format that other people can't read, and (2) the cost associated with distributing the data.

Reasonable size for data packets depends on network connectivity (speed) and congestion. Thus, many large files transfer at night when the daily network traffic decreases. In addition, network connectivity is changing very rapidly as the general public discovers the network. Research and education no longer exclusively drive the internet; instead, the network is responding to new pressures—from public education to online shopping services. As video-on-demand services become more widely available, they will dominate a lot of the network-performance issues. All of these paradigm shifts will require even greater bandwidth (speed or volume capacity). These changes will significantly affect how science users of networks work, when they work, and how they plan to analyze data.

The real beauty of online data distribution is that it has no overhead. Nobody has to write the tapes for shipping, read them once they arrive, or wrestle with various formats. Instead, once the data has arrived, network-connected customers themselves take care of selecting, retrieving, transferring, and locally archiving the data. The usual problem of being able to read the tape format is reduced to being able to interpret the data structure as delivered over the network. Also, this method dramatically reduces the need for tape drives at many of the user sites.

Of course, reducing overhead reduces cost for data and its delivery. Present US federal policies call for users and customers to pay the *marginal cost of reproduction* of the data—usually interpreted as the media cost of writing a tape and delivering it to the user. Delivering data over a network eliminates the need for tape copy and makes it impossible to compute the marginal cost of reproduction, so data will essentially be free over the network. Already, many federal data systems have taken advantage of this fact by distributing data over the network at no cost to users.

Another great benefit of the online data system is our ability to view or browse an image. A *browse image* is usually a highly subsampled (hence smaller) version of the original image. These smaller images are easily transferred over the network and displayed on our computers. For example, investigators interested in studying clouds would make sure the image contains the clouds they want to view. Scientists not interested in clouds could browse image to find images or parts of images without them, and thus get clear views of a planet's surface. All of these activities make users and the data system more efficient. Users find needed images faster, and data systems handle only selected images, thereby reducing traffic and bottlenecks.

Thus, whether data goes to users over the network or on 8 or 4 mm tapes depends mainly on cost and effort. If the data is already online (stored on SCSI disks or on a jukebox that can load SCSI disks), and if the user needs only to connect and have the data shipped over the wire, it may be considerably less expensive to do so. Copying tapes or writing tapes from a disk farm requires people, and people cost money. At the same time, for some data orders (say the entire globe for some property), writing the data to tape and shipping it by the US Postal Service or a courier is more efficient. The degree of effort needed to create these exchange files is also part of this decision. This boundary between online and tape delivery will likely change dramatically as new storage systems and network capacities continue to increase and improve.

**Full Resolution or Subsampled Data.** To reduce the large volumes of space-craft data, we could send only the spatial resolution the user needs rather than delivering the full-resolution imagery. The spatial resolution of the data distributed depends mostly on the application. Browsing needs only a highly subsampled image, but detailed analyses require full resolution. Some studies are interested in broad mapping and thus can sacrifice spatial resolution for overall coverage. Other projects need full (or even better) resolution to see the phenomena of interest. We simply can't generalize about which resolution is "better"—that depends entirely on needs.

**Compression or Not?** The advantages of data compression are obvious, but we must use the appropriate algorithm. For most science applications, we can use only *lossless compression*—keeping all image information. Lossless compression techniques are limited to about a 3:1 or 4:1 compression ratio. *Lossy* (loses information when reconstructed) *compression* compromises the content of the data transferred and may eliminate some features that will later turn out to be important. Even in the lossless domain, methods for compressing image data differ. Some apply to small blocks or packets whereas others are best for large files. Lossy compression may be most appropriate for some applications—to hold down the volume while still providing the essence of the images. The problem today is that most researchers working on compression use simple raster targets, such as a woman in a hat. They show how, when they're finished with their 50:1 and even 100:1 compression, the reconstituted woman in the hat looks basically the same. But scientists aren't even sure what they're looking for in the data. Thus any compression may have compromised their ability to recognize features in the spacecraft images.

Other mission applications may not be as sensitive to the compression algorithm selected. For voice communications, lossy compression is often very
effective because we're unable to detect much of the information lost by the system. Some military-reconnaissance missions can also easily tolerate the loss of
information with lossy compression. The great ratios possible with lossy compression should speed the delivery of data, which is likely to be the main concern of the
analysts using the spacecraft data for reconnaissance. Those few bits of information may not be critical for this first analysis. Monitoring of nuclear blasts or other
event detection may also tolerate some losses. Of course, if future studies want to
look more at the details of the image, lossy compression will have compromised
this effort.

The common response today for scientific applications is to use a routine compression algorithm, such as UNIX compress, to initially compress the data. Users
receiving this data can use either UNIX decompress, or some equivalent software,
to reconstruct the data. Saving volume without affecting the data puts less strain
on the network and on the people who operate the data center.

Even with lossless compression, we have to decide whether to do it in software
or hardware. Software compression is easier but slower. Hardware compression is
very fast but difficult to change. Designs must therefore trade speed against
flexibility.

### 13.3.4    Science Scenarios and Examples

One of the problems in discussing operational examples of spacecraft data systems is that most of the present systems were designed in the late 1970s. They
therefore don't take full advantage of advances in computer and onboard-storage
technology. Still, they're excellent examples of the basic components we must consider in any data system.

During the past decade, data systems have shifted from direct broadcast or
onboard recorded data to relaying the spacecraft sensor's data through orbiting
geostationary spacecraft equipped to handle large data streams. The Tracking and
Data Relay Satellite System (TDRSS) operated by NASA is one good example of
this type of system. By sending data signals from the spacecraft to an on-orbit relay
spacecraft, they don't need large amounts of onboard storage and get greater data
rates. Multiple TDRSS spacecraft allow almost continuous communication with
the operational spacecraft and downloading of all the sensor data through the
TDRSS receiving system, which is set up to communicate with the geostationary
orbiting TDRSS satellites. We'll discuss this data-relay application as it applies to
the LANDSAT series of spacecraft.

**NOAA's Polar Orbiters—TIROS-N.** Designed in the early 1970s, the Television InfraRed Observing Satellite (TIROS-N) series of spacecraft continue as the
workhorses of international weather observation. It was first launched in 1978 and
underwent slight modifications to sensors and the spacecraft in the early 1980s.
The same spacecraft bus has also been used by the Defense Meteorological Satellite

Program (DMSP) for their series of polar orbiters. The data system has remained largely unchanged since 1978 on both series of spacecraft. This system's main limitation is the Central Processing Units (CPUs) that were space qualified in the late 1970s. Since then, computing power has increased dramatically.

Figure 13.6 shows the onboard data system for the TIROS-N spacecraft. The data moves through two parallel streams. One is for the high data rate of the Advanced Very High Resolution Radiometer (AVHRR). The second stream is called the TIROS Information Processor or TIP stream; it contains all other sensor data and onboard spacecraft data. For the downlink (Fig. 13.6), the data stream again breaks into different parts. An analog broadcast of a subsampled part of the AVHRR data is known as the Applied Picture Transmission (APT). This analog stream goes down over a VHF link, allowing users to receive it with omni-directional antennas and thus eliminating the need for expensive and complex receivers. The main application of APT data is for analyzing pictures on ships at sea and remote weather stations.

A simple parabolic dish that can move either in azimuth/elevation or in x and y directions is needed to follow the spacecraft motion. For polar-orbiter tracking an azimuth/elevation system must move very quickly for overhead passes, when the azimuth changes quickly at the top of the spacecraft's pass. The x/y system can compensate more quickly for this type of spacecraft motion.

We also need a single horn feed that is sensitive to the S-band signal coming down from the TIROS-N spacecraft. Usually the downconverter and preamp are co-located with the feed at the top of the parabolic dish. This has been only possible in the past two decades, when the electronic technology became advanced enough to provide small and efficient preamps and downconverters that could be co-located with the antenna feed. Following the preamp comes the receiver, bit synchronizer, and frame synchronizer. Today, these are usually chips in the same unit, so we no longer see separate electronics boxes for these functions, and the system cost has decreased dramatically.

Data fed out from the frame synchronizer can be entered into any computer or workstation for further processing and analysis. Usually, the same computer will run the orbital-model software to generate the look-angles for the antenna and will drive the antenna from the serial port while the data from the antenna comes in through the parallel port. Further processing at the system workstation consists of separating the AVHRR and TIP data, navigating the AVHRR image data, calibrating the AVHRR's thermal-infrared channels, and preparing an archive file of all received data. The workstation will also transfer data to other computers or peripheral storage devices.

Another form of AVHRR data is known as the Global Area Coverage (GAC) data, which is a subsampled form of the AVHRR's full-resolution imagery. A sample is collected every third line and fifth pixel and stored in the appropriate array location. The result is an image with approximately a 4 km spatial resolution. This data is recorded separately on one of the recorders shown in Fig. 13.6 and is played

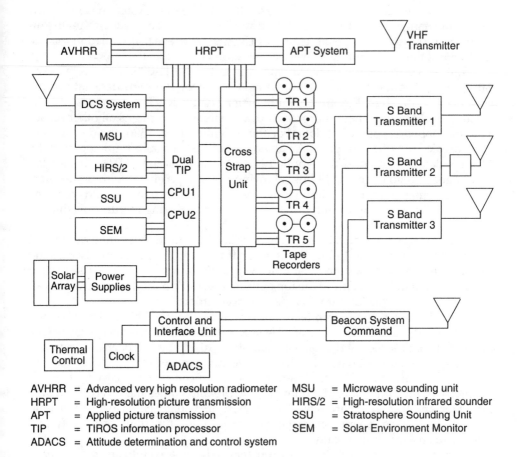

AVHRR  = Advanced very high resolution radiometer     MSU      = Microwave sounding unit
HRPT   = High-resolution picture transmission          HIRS/2   = High-resolution infrared sounder
APT    = Applied picture transmission                  SSU      = Stratosphere Sounding Unit
TIP    = TIROS information processor                    SEM      = Solar Environment Monitor
ADACS  = Attitude determination and control system

**Fig. 13.6.   Block Diagram of the Electronic System on the TIROS.** The downlink contains two parts—a VHF link that contains the APT analog data and an S-bank link that contains the HRPT digital data.

back on command at one of the NOAA-controlled HRPT stations. The recorders have enough space to record one GAC-level set of data for each orbit. As a result, a spacecraft operating in the GAC mode will create an image of the entire Earth once each day.

Recorders are also used to collect Local Area Coverage (LAC) data from the AVHRR. Like HRPT data, the LAC coverage has the full 1 km resolution of the directly broadcast HRPT data, but it's collected out of sight of a NOAA-controlled ground receiving station. Thus, the full-resolution data is recorded onboard and played back on command at a NOAA direct-readout station. The only limitation of GAC data is recorder space. As shown in Fig. 13.6, only five recorders are onboard a TIROS-N spacecraft. Two are used for the GAC data processing, and two of the remaining recorders are used for LAC data collection. One is held in reserve. The recorders can collect only 4,000 scan lines of the full-resolution LAC data, which is slightly less than the 6,000 scan lines typical of the HRPT's high-angle overpasses. In all forms of AVHRR data (GAC, LAC, and HRPT), TIP data is embedded in the data stream. Thus, even in the GAC subsampled file, a full set of TIP data provides global coverage of the other sensors on TIROS-N.

**GOES.** The primary function of the Geostationary Operational Environmental Satellite (GOES) system is to provide regular, near-real-time imagery for operational weather forecasts from a geostationary orbit. In addition, the modern derivative of the Visible Infrared Spin Scan Radiometer (VISSR), known as the VISSR Atmospheric Sounder (VAS), offers regular atmospheric temperature profiles. Because of this operational need, the main data distribution uses direct-readout ground stations. In the case of receiving the full digital data, the ground station will look very similar to that shown for the NOAA's polar orbiters in Fig. 13.2. The main exceptions will be that the antenna doesn't move to follow the spacecraft and that data will be regularly collected every half an hour. The advantage for this system is that the antenna can be simple and stationary, much like modern systems for television reception. The disadvantage is that the GOES system's very high data rates require considerable processing to create manageable data levels.

Many of the users of GOES data aren't interested in processing the full data stream from either the VAS or the VISSR itself. Instead these forecasters are interested in a processed version of the GOES image data. This service provided by NOAA is called WEFAX (for weather facsimile). It's created by ground processing of the digital GOES data stream to produce a geometrically corrected and map registered (navigated) image with map overlays and latitude/longitude markings. This data is again broadcast up to the same GOES spacecraft and then rebroadcast down to Earth as a communications signal. Thus, operational centers need only receive the WEFAX signal to acquire the information needed for most forecast operations. This is the most widely used mode of GOES data. The next most common is direct reception of digital GOES imagery or VAS profiler data.

**LANDSAT and TDRSS.** Initially designed to use direct-readout stations with high data rates, the LANDSAT series of spacecraft shifted over to using the TDRSS relay spacecraft with the advent of the LANDSAT 4 and 5 spacecraft in the early 1980s. These spacecraft were equipped with high-gain antennas to communicate directly with the TDRSS spacecraft for transferring high-rate data from the high-resolution LANDSAT sensors. This information was then relayed to the NASA/NOAA ground station at White Sands for processing and archiving. NASA's plan is to have TDRSS spacecraft continuously in geostationary orbits so one spacecraft is just west of the United States' west coast and another is just east of the United States' east coast. This separation places the TDRSS spacecraft so one or the other is usually in contact with the polar-orbiting LANDSAT spacecraft. At the same time both TDRSS spacecraft can communicate with the ground station at White Sands, New Mexico, to relay the LANDSAT imagery to the ground.

This system was developed to manage the very high rates of spacecraft data we'll soon be receiving. While TDRSS is already overloaded, its use has become a focal point of the United States' space policy for relaying and downlinking relatively high-rate data. Its existence doesn't eliminate the need for direct-readout downlinking of spacecraft data; rather, it offers another independent means of relaying the data from the spacecraft to the ground.

## References

Baldwin, D. and W.J. Emery. 1994. Spacecraft Attitude Variations of NOAA-11 Inferred from One-year of AVHRR Imagery, Intl. J. Rem. Sens. (in press).

Bedet, J.-J., L. Bodden, A. Dwyer, P. C. Hariharan, J. Berbert, B. Kobler, P. Pease. 1993. Simulation of a Data Archival and Distribution System at GSFC. Proceedings of the Third Goddard Conference on Mass Storage Systems and Technologies, NASA Conference Publication 3262, pp. 257–277.

Boehm, B.W. 1982. *Software Practices, Cost, and Engineering.* pp. 415–440. Englewood Cliffs, NJ: Prentice-Hall.

Emery, W.J., Y. Yu, G. Wick, P. Schluessel, and R.W. Reynolds. 1994. *Improving Satellite Infrared Sea Surface Temperature Estimates by Including Independent Water Vapor Observations.* J. Geophys. Res., 99, 5,219-5,236 Rosborough, G.R.,

Fortner, B. Scientific Data Formats - An introduction, Scitech Journal, 20-23, November 1993.

Futron Corporation, Guidelines for the Development of a Project Data Management Plan: NASA Office of Space Science and Applications, Contract NASW-4493, 1992.

Green, J. L. and J. H. King. 1988. Guidelines for the Development of a Project Data Management Plan, NSSDC Technical Document #88-16.

Klenk, K. F., J. L. Green, and L. A. Treinish. 1990. A Cost Model for NASA Data Archiving: Version 2.0. NSSDC Technical Document #90-08.

Larson, Wiley J. and Jim R. Wertz. 1992. *Space Mission Analysis and Design.* 2nd Edition. Kluwer Publishing. Netherlands.

Price, R., S. Vemury, and D. Love. 1995. *Estimation of Lines of Code for Large Software Systems, Static Analysis.* To be submitted to Journal of Software Practices and Experience.

Rosborough, G.R., D. Baldwin and W.J. Emery. 1994. *Precise AVHRR Image Navigation.* IEEE Geosci. Rem. Sens, 32, pp. 644–657.

Schluessel, P., W.J. Emery, H. Grassl and T. Mammen. 1990. *On the Skin-Bulk Temperature Difference and its Impact on Satellite Remote Sensing of Sea Surface Temperature.* J. Geophys. Res.,95, 13,341-13,13,356.

Wick, G.A., W.J. Emery and P. Schluessel. 1992. *A Comprehensive Comparison between Skin and Multi-Channel Sea Surface Temperatures.* J. Geophys. Res.,97, pp. 5569-5596.

# Chapter 14

# Assessing Payload Operations

Paul Ondrus, *Goddard Space Flight Center, NASA*

In this chapter we'll assume you're a new mission operations manager (MOM) who needs to know about important payload activities in order to help control costs for mission operations. Many payload issues are discussed during the mission-design phase without understanding or identifying the implications for operations. The checklists in this chapter give you a way of identifying drivers of cost for payload operations, so you can work trades with the project managers and designers during mission design.

As a MOM, one of your key responsibilities is to ensure that operational issues of a mission are identified and resolved as early as possible during a mission's development. That's because it's cheaper to solve a problem in design than in test. To fulfill this responsibility, you must emphasize payload operations and its role in end-to-end mission activities. This emphasis must start early in mission definition and continue throughout development to get the appropriate system trades from a life-cycle perspective. All too often project managers tend to trade the present-development budget for the future-operations budget and ignore life-cycle costs. Your job is to keep the project manager aware of life-cycle issues and implications.

In the past, particularly on science missions, spacecraft operations and payload operations have been distinct. The trend is to blur this distinction for smaller, low-cost missions, which make early identification of payload operations and identification of payload and spacecraft cost drivers even more important. Also, the trend to increase autonomy by automating functions and moving them onboard the spacecraft emphasizes the need for early operations involvement.

The foundation of this chapter is a mission-engineering process, based on past experience, that will help you plan a payload's operation and identify its key operational issues. These activities lead to an activity plan for operations that details key activities and how people will do them. This plan provides the framework for defining payload operations and for surfacing issues that warrant the project manager's attention. Developing this plan is closely tied to the operations concept tools discussed in earlier chapters. These tools help you understand the mission's objectives and clarify its program environment, so you can help develop appropriate trades.

Spacecraft builders tend to see a mission in terms of functions. They fragment the payload and mission operations activities into sets of thermal, power, mechanical, and control requirements and resource budgets for spacecraft subsystems. You and your support team need to work with systems engineers to translate the functions and requirements back into a timeline, scenarios, and process flows that show how to operate the spacecraft on orbit. Our mission-engineering process also gives you a structure for characterizing the mission's customer interface and seeing how the payload and spacecraft bus will carry out the mission's objectives from beginning to end.

As a mission operations manager entering a new project, you also face many ongoing, parallel activities. Our checklists guide you to key issues you must watch for and understand. They help you to avoid the tight constraints on flight resources that complicate normal payload operation—characterizing the information needs for mission planning, factoring in lessons learned from previous missions, and identifying new technology trends that could significantly change mission operations. You may not be able to change a design, but you should address effects on operations costs to encourage proper life-cycle decisions.

The mission-engineering process guides us in identifying and resolving potential issues with payload operations (see Fig. 14.1) as we move from the mission concept to actual operations. Here, we'll consider operations planning, which interacts with support for the space-element design and ground-element definition. Mission engineering emphasizes how to carry out requirements for processing the payload data, to develop payload operations concepts and scenarios, and to operationally critique the payload design. It helps us support system trades and identify drivers of operational costs. The main product of operations planning is the activities plan for flight operations. Results also influence design requirements for the ground system and design reviews for the spacecraft and payload.

Mission engineering is iterative. It uses analysis tools that help the operations staff visualize on-orbit operations. The operations concept is one such tool that we

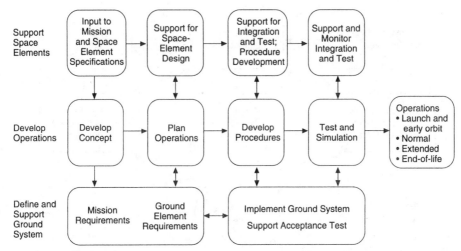

**Fig. 14.1.    Mission-Engineering Process.** We must develop the space, ground, and mission operations elements concurrently to reduce life-cycle costs.

can use to help translate requirements into timelines and operational flow. As a MOM, you must emphasize time and information flow as part of defining the mission. Also, this process is effective for both small and large missions. The level of complexity of the mission determines the level of effort required. The steps remain the same.

As you can see in Fig. 14.2, we start the concept development by understanding the payload's objective. This means understanding the payload's tasks, success criteria, constraints, and products its customers require. A MOM must keep an end-to-end systems perspective, recognize how elements interact, and understand the detailed workings of each element. This means working closely with developers and customers.

As part of this effort we need to identify and understand major program interfaces. A program's interfaces determine how much flexibility you'll have in developing the operations plan for a mission. NASA's Small Explorer (SMEX) missions, for example, tend to be high-risk scientific missions using a single payload. These missions have very few defined interfaces, so we can try innovative operations concepts on individual missions. A mission that is part of a large structured program has many fixed interfaces, so new concepts are limited. The Earth Observing System (EOS)—because of its large customer base and correlative research requirements—is an example of a program that limits flexibility.

Our assessment of payload operations divides into three main activities that lead to the operations plan: developing the payload operations concepts, assessing requirements for payload-data systems, and assessing requirements for payload operations.

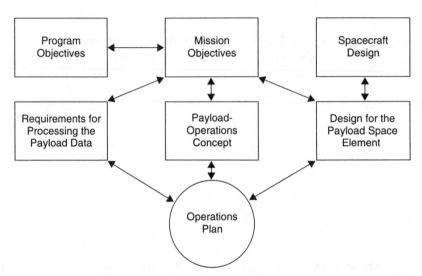

**Fig. 14.2.    How Mission Objectives Relate.** We must understand the payload's tasks, success criteria, constraints, and required products to develop a cost-effective payload operations plan.

# 14.1  Developing the Payload Operations Concept

As discussed earlier, mission engineering continually refines and adds detail to an operations concept. In most missions the concept for payload operations extends from the concept for mission operations. But to make sure the concept is complete, we should follow four steps: (1) determine the class of operations; (2) identify drivers of operations complexity; (3) develop contingency operations; and (4) analyze launch and early-orbit scenarios.

### 14.1.1  Determine the Class of Operations

We classify payloads as survey, event-driven, or adaptive. The class of payload determines specific operational requirements. *Survey payloads* tend to have simpler real-time operations and planning activities. An example is the SAMPEX payload, which is a set of space-physics instruments that continually operate and measure the same physical events throughout each orbit.

*Event-driven payloads* must operate at a specific time. They tend to be more complex and require more sophisticated planning than survey missions. The more accurate the time requirement, the more costly a payload's operation. Examples of event-driven payloads are any of the instruments on the Hubble Space Telescope. These payloads operate at specific times to observe planned astronomical targets.

The need for pointing accuracy to the milli-arc second significantly complicates payload planning.

*Adaptive payloads* transition to different states depending on unplanned events. An Earth-observing payload that automatically adapts to cloud cover is one example. If the payload operates autonomously, this type of operation will be relatively inexpensive because it has little effect on ground support and mission operations. But if the ground element needs to interact periodically with the payload concerning future operations, this type of payload can create very complex operations. This complexity arises because we must reconstruct payload operations and understand the payload's status to plan future operations.

### 14.1.2 Determine Complexity Drivers for Normal Operations

We must understand some key characteristics of payload operations to assess a payload's operational complexity. If we understand them early enough in the mission's life-cycle, we may lower life-cycle costs by changing the payload or ground-system design. These characteristics are payload planning, payload reconfigurations and calibrations, communications needs, data-product verification, and payload orientation (see Table 14.1).

**Table 14.1.**    **What Makes Operations Complex.** We include these cost and complexity drivers in our complexity metrics in Chap. 5.

| Cost Driver | Comments |
|---|---|
| Planning Process | Number of constraints and interfaces |
| Reconfigurations | Number and timing of changes |
| Communications Needs | Onboard storage and volume of payload data |
| Data-Product Verification | Data structures and timelines |
| Payload Orientation | Maneuvers, accuracy |

**Payload Planning.** This characteristic depends on the number of conditions we must take into account, interactions with other payloads on a platform, and how we optimize the payload's activity. Examples of environmental conditions that can complicate planning are the South Atlantic Anomaly, bright objects, occultations, and target visibility. Examples of margin constraints are requirements for thermal, power, and communication resources. The more environmental considerations we must take into account, the more complex and costly planning will be. We want to encourage designs that lessen or avoid these types of constraints. Also, planning activities that cross organizational boundaries, such as conflicts between science and spacecraft operations, tend to make planning more complex.

One subtle challenge to planning is the desire to overschedule the payloads on a platform, which results in marginal utility. The more we try to schedule all of the payload's abilities, the more complicated and costly the planning process. This is characteristic of deep-space missions, such as Voyager, which have had long cruise periods and short opportunities for full science during planetary encounters. Planning and replanning every millisecond of the encounter involves many subsystem and payload people, so it drives up complexity and cost.

**Number of Reconfigurations.** We're referring to the number of changes in a payload's or spacecraft's state required to operate the payload. This number has two dimensions: a payload's possible modes and timing of the reconfigurations. Because nearly all payloads need reconfiguring, we try to isolate its effect to a small part of the ground system. An example of such changes is fine pointing the Hubble Space Telescope's (HST) instruments while interacting with a science user.

The nature of the payload's job drives the number of operational modes. Examples of operational modes are to change filters, use different apertures, measure various frequency bands, or preprocess data in several ways. To save money, a MOM must make sure these different modes affect the smallest possible part of the ground system. But the increased number of modes will greatly complicate validating payload operations. One of the instruments on X-ray Timing Explorer, for example, has 256 modes which complicate science operations and analyzing instrument performance. [NASA, 1995]

Timing of the reconfigurations is our other main concern. If, for example, we need to switch a payload instrument from a survey mode to a special mode for a target of opportunity, quick action can drive up cost. A rapid turnaround time can require payload operations staff to command the mode change, force platform operators to break the existing timeline and re-plan slews or platform reconfigurations, and demand rescheduling of communication services. For many scientific spacecraft, having to respond quickly to targets of opportunity greatly increases staffing. As a MOM, you'll need to help challenge the timing for observing these targets, so you can avoid unnecessary operational costs or encourage onboard automation of functions.

**Communications Needs.** Communications are at the core of how payloads are remotely operated, directly influence payload interactions, and strongly drive a mission's operational costs. Communications requirements depend on the payload's need for uplinks and downlinks. The amount of data a payload collects and the payload's commanding needs directly drive the number of contacts. The HST's payloads, for example, require at least three daily contacts, plus back-ups to load the daily payload commands, because the onboard computers have limited memory. The payload's communications needs are one of the first system trades for a new mission. If the payload requires several manual reconfigurations or checks to verify operations, communications contacts become a major cost driver. A survey mission that requires a low bandwidth for data transmission and has appropriate onboard storage can operate with one contact per day. A number of the small explorer series (simple, single, survey payloads) can operate this way.

**Data-Product Verification.** Verification costs money because mission experts must analyze payload products and make timely, interactive corrections to the payload's operation. An orbit that allows for long contacts will help manage these costs. For example, the Solar Heliospheric Observatory (SOHO) operates at the L1 libration point, so it has continuous communications contact for up to 16 hours per day. This continuous contact allows for online reconfiguring of the payload, with little cost for instrument operations.

**Payload Orientation.** Does the platform have to provide unique orientations for the payload to work? Payloads that require pointing increase planning and on-orbit operation of the platform. The Cassini mission is a good example. To lower development costs, designers fixed the instruments to the spacecraft structure. The spacecraft will now have to do complex maneuvers in order to meet the payload's pointing needs. Also, the more stringent the pointing requirement, the more significant the cost drivers.

### 14.1.3   Develop Contingency Operations

When payload operations are normal, associated activities become routine. The challenge for a new MOM is to provide an operations scenario and staff that cover normal operations best but can handle contingencies. This means you must plan for contingencies while developing the operations scenario—planning and developing safemodes, establishing fault-isolation trees, and developing recovery operations.

Needing to respond quickly will strongly drive up cost for payload operations because it implies almost continuous ground monitoring. If we don't have safing abilities onboard, life-cycle costs for ground operations can skyrocket. All payloads must have two safing mechanisms. The first is a safemode that the payload itself enters when it detects an anomaly; the second is a safemode that the spacecraft platform can enter whenever it has a problem. *Safemode* should be a state in which the payload can't damage itself through unwanted mechanism movements and can hold indefinitely without creating thermal or power problems. Given the present state of technology, keeping a large staff waiting for anomalies is too expensive. Instead, the spacecraft or ground operations should safe the payload to a given state, which will allow ground support to take its time to find the cause and start recovery. For example, the EUVE's science-operations team has automated their ground monitoring of the EUVE's payload to reduce staffing [Biroscak, 1995].

Now we bring all raw telemetry to the ground, where we process it and isolate faults. But as onboard processing improves, we'll soon have onboard systems that can help us isolate faults and recover from them. These systems will save certain states of information and provide more functional information to ground system engineers, which implies some onboard trend analysis and inference processing. This application of technology can fundamentally shift how we operate—applying networking concepts to the forward and return links and upgrading onboard computing to move fault isolation away from ground computers.

The first step in developing contingency operations is creating fault trees for the payload. When we do so early enough, operators can walk through failure scenarios and determine if they have enough information to recognize them. The payload-failure trees should complement the failure analysis for the spacecraft platform. We should also plan to test them, whenever possible, during integration and test. Figure 14.3 shows a fault tree for an instrument with an unexpected power failure on the WIND spacecraft [NASA, 1994]. Much of the failure analysis provides a basis for simulation exercises that force operators to react to failures. The extent of this effort should match the size of the mission. If the payload is a significant instrument that will operate for many years, more effort on the failure exercises will eventually save money. If the mission is a simple, short-term instrument, we should keep the failure analysis simple.

The final consideration for contingencies is planning recovery activities, including communications needs for carrying them out on orbit. Often, the recovery procedures exercised during payload integration and test are based on continuous contact. But we can't do them during short-pass contacts, such as those for a low-Earth orbiter, so we must ensure the procedure can be done on orbit. Also, we need to control these procedures for consistency, so staff changes don't create problems caused by unfamiliarity with procedures.

To succeed in contingency operations, we must keep them simple and have the time to react properly to an anomaly. The worst danger is responding quickly and improperly to an anomaly, which usually creates even bigger problems, especially if we haven't addressed an anomaly in the failure analysis. Also, we'll encounter problems failure analysis didn't discover, particularly for payloads that use a lot of onboard software to operate. See Chap. 16 for more on handling anomalies.

### 14.1.4 Analyze Launch and Early-Orbit Scenarios

Launch and early-orbit activities are critical to all missions. Activities such as turning on the payload, outgassing, deploying appendages, and initializing processors are significant because they usually occur only once and can damage the payload if they are not done. Also, this is the time to see how the platform systems function in a dynamic environment after experiencing launch stresses that are difficult to simulate on Earth.

Payload cost drivers in this activity are special, time-driven orbital constraints for working with the payload; special platform dependencies; appendage deployments; and calibration activities. We must understand and space all these activities so breakpoints in the timeline allow us to address anomalies without heavily replanning follow-on activities.

Our first concern is any special orbital constraints or restrictions on the payload. Examples of constraints are requirements to turn on the payload during darkness, to avoid operations in the South Atlantic Anomaly, or to avoid bright objects. As a MOM, you must work with payload developers to ferret out unique requirements and to make sure the plan for launch and early orbits can support them comfortably. If we identify this activity early enough, we can help design the

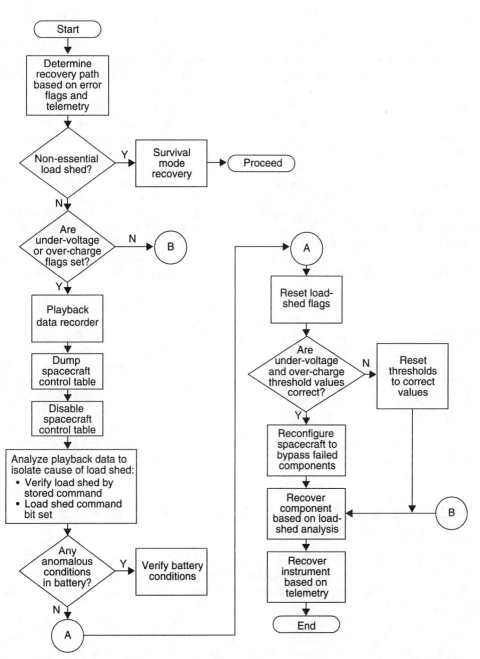

**Fig. 14.3.    Fault Tree for WIND Spacecraft.** Follow these steps to identify the cause of, and to recover from, an unexpected power failure on the spacecraft. [NASA, 1994]

payload to avoid concerns about an instrument having unique thermal or visibility problems.

Our second consideration is the platform checkout and corresponding limits on attitude, power, communications, and thermal characteristics that are present as the platform transitions to operations. We usually can't turn on the payload until the platform has reached a stable state. Payload operations must be ready to respond to any problems encountered during platform checkout, particularly if checkout is placed on hold to work a platform anomaly. This is where adequate operational margins for power and temperature can preclude complex activities to work around platform problems.

The third activity is planning for any special deployments of payload appendages or covers. A simple survey instrument probably won't have any, except disabling the latch mechanisms. But special deployments of antennas are key for communications missions. For payloads on platform appendages or having instrument covers, this effort takes on more significance. We should exercise these deployments during integration and test, using the same procedures we'll use on orbit. Operators must understand these deployments and practice contingency operations. It's important to schedule free time during which operations can resolve anomalies.

The final launch and early-orbit activity is turning on and calibrating the payload. Calibration should be straightforward. Requiring operators to use special external sources to calibrate a payload increases operational costs. The frequency and extent of calibration are also cost drivers. The accuracy requirements of the HST's instruments and their relationship to the platform created a process for calibrating and validating instruments that took months. The Gamma Ray Observatory, on the other hand, was operating within weeks. Initial accuracy requirements for calibration translate directly into operations costs.

Chapter 9 provides more detail on planning launch and early-orbit activities. Here, we've discussed allowing enough time, planning the mission to have enough thermal and power margins, and providing appropriate self-calibration activities to avoid complex calibrations.

## 14.2 Assessing Requirements for Payload-Data Systems

To do this second and probably most significant step properly, you'll need to address seven considerations: payload health and safety needs, duty cycle, data structure, data volume and timeliness, calibration and validation, product development and data completeness, and data access (see Table 14.2).

These considerations are important because they help define operations costs and fundamental mission trades. You need to visualize how they affect operations and flag cost drivers for the systems engineers and payload developers. Consider these cost drivers as part of the total system. On most scientific missions, data-

**Table 14.2. Drivers for the Payload-Data Systems.** On some systems, the cost of the payload-data systems represent 75% of the total operations budget.

| Drivers | Comments |
|---|---|
| 1. Health and Safety Requirements | Issues are autonomy, data downlinks, and payload interactions. |
| 2. Duty Cycle | Issues are data volume, communications bandwidth, and pass duration. |
| 3. Data Structure | Issues are data completeness, processing requirements, data quality, and user interfaces. |
| 4. Data Volume and Timeliness | How much data must be transferred and how soon must users receive the data? |
| 5. Calibration and Validation | Issues are frequency and complexity of calibrations. |
| 6. Product Development and Completeness | How is data translated to meet users' requirements? Are standards used? How does loss of data affect the users? |
| 7. Data Access | Who needs data and how do they get it? |

operations costs represent nearly 75% of the mission's operational costs, with mission operations for the spacecraft covering the other 25%. Controlling costs of data operations is key to controlling mission costs.

### 14.2.1 Health and Safety Requirements

Designs for the mission and spacecraft drive requirements for health-and-safety data. Basic drivers are the spacecraft's autonomy, data-downlink transfer, and payload interactions. A low-cost mission would have a payload and spacecraft with high autonomy—meeting their needs with few data dumps and having minimal payload uplinks. The Alexis mission is an example of such a spacecraft. It requires one uplink and one downlink per day. A mission such as the HST defines the opposite end of the spectrum because it requires continuous monitoring, has frequent downlink transfers, and requires regular user interaction for fine pointing and instrument verification. We can locate most other missions somewhere in this spectrum between Alexis and HST.

To lower costs, try to develop an operations concept for which health and safety operations are contained in a single shift or shared with another mission, the payload has the autonomy to do routine operations, and the payload can recognize a problem and safe itself. Your concept should also have enough onboard storage and link capacity to allow file-oriented data transfers, for which the transfer protocol verifies the quality of transfers. Because of what the payload must do, most missions can't meet these simple goals. But you still need to clearly show users and mission developers how these issues affect costs.

### 14.2.2 Duty Cycle

Duty cycle provides a way of measuring the time margin that is available to handle operational contingencies before payload data is lost. In Fig. 14.4, mission A has a low duty cycle, requiring one data transfer per day. Mission B has a high duty cycle, demanding four transfers per day. The duty cycle is a system trade that involves data storage, communications bandwidth, and ground operations.

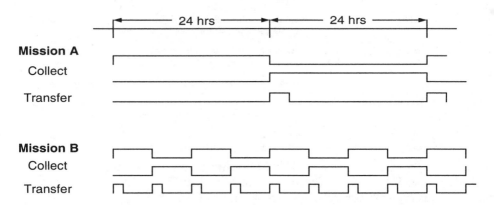

**Fig. 14.4.    Effect of the Payload's Duty Cycle on Data Systems.** Mission A has a low duty cycle and requires only one data transfer per day. Mission B has a high duty cycle and requires four data transfers per day. Variations in collect cycles represent new cycles for the next transfer of data.

Because all spacecraft payloads operate remotely, the mission designers must understand how the payload creates data volume and how the spacecraft and communications systems deliver this volume to the ground. The payload's duty cycle influences interactions with the spacecraft bus, onboard storage requirements, schedules for tracking and acquiring data, and mission-planning activities.

The major system trade on this requirement is the amount of space assets versus abilities and support in ground operations. The X-ray Timing Explorer spacecraft, for example, has a high duty cycle. This means data from the payload transmits almost continually through the TDRSS Multi-Access system. This cycle creates continual scheduling activities and ground monitoring—24 hours per day, 7 days per week. A mission with a low duty cycle, such as Alexis, collects data and dumps it once per day at a ground-station site. The onboard storage can hold one complete day's volume and requires only one tracking pass per day for data transfer. Increased mission cost up front helps lower costs for mission and operations. Automating ground activities helps to minimize the affects of a high duty cycle.

### 14.2.3  Data Structure

A major challenge for mission designers is to minimize the cost of the operational link between a payload and its users, so we must make sure the payload's data structure can handle this link at an acceptable cost. As a MOM, you must explore the operability of the end-to-end system in terms of data completeness, data-processing flow, the end-user interface, and cost.

In considering *data completeness*, you need to decide whether the payload's data structure provides for accountability and a means of measuring data quality. First, pack the payload data into logical segments that match the measurement or services the payload provides. The object is to structure the data so we can contain periodic data losses from the space-to-ground link or ground processing. Doing so prevents small data losses from contaminating a much larger segment of the data. Failure to address these needs increases the cost and complexity of data processing and staffing on the ground. Examples of logical data structures are a LandSat image or a scan of ozone data from the Total Ozone Measuring System (TOMS) instrument.

Second, analyze processing flow to determine whether you need special processing and ancillary information. The processing flow can be broken into a series of levels leading to stand alone data products for the end user (see Sec. 13.3). Critique these interactions to get the fewest retransmissions and reprocessing between levels and, therefore, the lowest costs for future operations. The trend is toward international packet standards that logically encapsulate data and allow for standard interfaces on the spacecraft and the ground. Using nonstandard data structures tends to increase the amount of unique processing software and corresponding life-cycle maintenance. On the Cosmic Background Explorer (COBE) mission, for example, processing of Level 2 and 3 data strongly drove cost because of the nature of the measurements. These processing costs were four times all other mission operations costs combined.

We must understand how our data structure will help or hinder data management during operations. Spacecraft in the past used mechanical recorders that required data reversal on replays and extensive processing to remove data redundancy, which demanded significant Level 0 processing. Solid-state memories and a logical structure for the payload packet are allowing data to be logically requested for playback and are greatly simplifying Level 0 processing. Combining file-transfer protocols and solid-state memory shows how technology enables data structures to lower operational costs. Chapter 13 discusses data-processing levels in detail.

Finally, make sure your data structure allows users access. Is the payload part of a correlative program or is it operated by a single user? The correlative program tends to impose stricter data-quality standards and costs more by increasing the number of interfaces operators must deal with. It also increases pressure to have the best-quality algorithms for payload processing available early. On low-cost missions, much of this responsibility falls on the principal investigator.

### 14.2.4 Data Volume and Timeliness

These two issues directly influence the number of spacecraft contacts and determine whether multiple shifts are required for spacecraft operations. Data volume tends to be the major driver of complex systems for transferring data from the payload to ground. Timeliness drives the system trades between the spacecraft's storage capability, downlink bandwidth, and the availability of ground resources to accept the data.

For example, even though Fast Auroral Snapshot Explorer (FAST) is a small, single-instrument mission, it creates more data during its six-month winter campaign than the Hubble Space Telescope does during the same period [NASA, 1993]. In spacecraft design, small doesn't necessarily mean simple. The volume of FAST data drives the mission to have 12 ground contacts per day, whereas a simpler survey mission may require only one contact per day. FAST's volume forces ground operations around the clock, whereas SAMPEX needs only one shift. The data volume also directly affects ground-processing operations by requiring the complete sequence of science-processing steps to be operated around the clock.

We also must consider how data volume affects storage and staging. The flow of information from the payload to the user ends up being staged on the spacecraft, at the acquisition site, and at the user's facility in order to work around various line and link failures. Staging of data at each of these facilities costs more money. The more straightforward the transfer of data to the user, the lower the fundamental operations costs. This concept is beginning to encourage direct downlinking of data to the end user's facility whenever possible because it avoids creating infrastructures that add marginal value to the overall mission. The direct downlink is possible only when the payload's orbit allows it to pass over the user's facility. Otherwise, we'd have to create networks (inefficient infrastructure) to work around visibility problems.

Also, each of these staging points is a decision point for operators. Because most spacecraft can store or record only so much data, operators must decide whether to replay data after a downlink problem or allow the data to be overwritten. This is a fundamental cost trade-off. The more timely the data, the more online processing, and thus, the higher the cost. The direct-downlink option helps to mitigate some of this cost, but forces around-the-clock operations at the user's building unless fully automated.

The marginal utility of obtaining high numbers for access to payload data also directly drives operational costs. If, for example, the payload requires 99.9% of the payload data for mission success, operations will be much more complex. This complexity is related to data structure and duty cycle, but it always requires quick decisions to avoid overwriting data at any of the staging points. Data completeness (see Sec. 14.2.6) also affects its quality. The data-packaging scheme can strongly reduce the effect of data loss on a payload by considering the error characteristics of the data link between the spacecraft and the ground.

### 14.2.5 Calibration and Validation

Payload calibration involves two actions: calibrating the payload itself and applying algorithms to the data received to convert it into usable products. This section addresses concerns about operating the instrument. The section on data products addresses ground processing.

The frequency and complexity of calibrations drive operational costs. The first cost driver is how often we need to calibrate a payload. Depending on the accuracy requirements and the payload's construction, this activity could be per orbit, daily, or weekly.

The second cost consideration is the complexity of the calibration. A calibration that requires periodic observation of an external source will complicate operations and planning. A payload that has access to an internal calibration source that has been validated externally can calibrate automatically. If we don't design calibration into the payload's operation, on-orbit operations and data processing will be more expensive.

One final operational consideration is tracking configuration changes to the payload in response to calibration problems discovered on orbit. We have to be able to match changes in configuration with the appropriate payload to avoid significant reprocessing of data sets.

### 14.2.6 Product Development and Completeness

Another challenge facing mission operators is translating the payload's data into meaningful products that are easily available to customers who can use it. Proper format and completeness are the two main issues.

The various science communities are developing standards for their types of payload data so they have common tools. The astrophysics community, for example, has settled on the High Energy Astrophysics Archive (HEASARC) format, whereas the Earth-observing community has decided on the Hierarchical Data Format (HDF). As a MOM, you must be able to work with users and understand product formats, so you can help design a mission that will develop these products correctly. Particularly important are ancillary information, such as definitive attitude information, and calibration data needed to build a meaningful data product. By working these characteristics with users early in the mission, you can avoid unnecessary data-processing steps.

Data completeness is as important as product development. Various types of payloads have different fundamental data sets: a frame for an imagery payload, a measurement at a point in time for a survey mission, or a spectrum of an object for a spectrograph. Each set of the payload data has characteristics that tie it to the fundamental activity they're measuring. As a MOM, you must be sure the data structure is properly configured, so normal transmission errors don't propagate through the data and cause unexpectedly high loss of data.

You'll need to walk through the complete data-processing system to ensure none of these problems exists. You want the actual information content to approach the level of the data collected. If data compression is used and not properly channel coded, data losses translate into a larger information loss. Although we can develop special post-processing algorithms to work around these types of data issues, it's expensive and worth avoiding. The development of packet standards has helped to improve this situation, but we still have much to learn about matching these packets to the nature of the science data.

### 14.2.7  Data Access

Missions in the past defined payload products and created significant infrastructures to deliver these products to users. In today's information-processing world, the less expensive approach is to make the data available for users to get as they need it. This fundamental shift to a file-server approach reduces the infrastructure and allows users to get only what they need, when they need it. It avoids creating mini-archives throughout the user community.

Of course, mission operators must develop directories that allow users to determine if the data they seek exists. Thus, products must be identified and available to an archive the community can access. The Earth Observing System (EOS) is trying to meet this need by creating distributed data centers with standardized interfaces for directories and common data formats.

The requirements for monitoring data-quality and the post-processing functions above level 1 fall to the data centers. This approach also allows adding data centers which can do further processing to make a higher level of data products available to a less sophisticated, but larger, data-using community.

## 14.3  Assessing Drivers for Payload and Platform Interfaces

With the operations concept and data systems in hand, you can begin the third phase of assessing payload operations. The steps for this phase are listed in Table 14.3 and described below.

This section continues the mission-engineering process outlined in the beginning of this chapter. Because this process is iterative, many of the topics seem redundant, but we're taking them to a new level of detail. These iterations allow the mission team, under your direction as MOM, to continually refine mission operations for lower cost. The final output of these iterations is the activity plan for payload operations.

The key to cost-effective payload operations is not creating large documents or lengthy analysis but identifying the key issues. The details in a payload operations plan should be consistent with the import of the mission being developed. For a simple payload that just remains on in a monitor mode, the activity plan might be only two to three pages.

**Table 14.3. Assessing Drivers for Payload and Platform Interface.** This is the third phase of assessing payload operations.

| Step | Description | Comments |
|------|-------------|----------|
| 1 | Define operational scenarios | Orbital constraints, products, timelines, and mission interactions |
| 2 | Characterize requirements for onboard command and data handling | Payload/platform interactions—level of autonomy |
| 3 | Define maintenance constraints on the platform | Power, maneuvers, and multiple payload interactions |
| 4 | Determine the need for ancillary information | Time and position accuracy |
| 5 | Define the need for special events | Safing, software updates, and fault isolation aids |

### 14.3.1  Define Operational Scenarios

Developing an operational scenario consists of (1) characterizing orbital constraints; (2) identifying all required operational products and actions; (3) determining when these products or actions are required; and (4) developing an integrated timeline that depicts these activities in relation to mission activities. The timeline should reflect the fundamental time unit of operations which, for many missions, is 24 hours. But for missions with highly elliptical orbits, it might be one orbit, which could be several days.

An operational scenario needs to address normal operations for this fundamental period and for any other key events, such as launch and early-orbit deployments and orbital maneuvers. It's important to identify the time relationships of any key activities and product development in this scenario. Operational products required quickly tend to drive up operational staff and cost.

Figure 14.5 gives a day-in-the-life timeline for X-ray Timing Explorer (XTE). The XTE is a scientific pointing mission that uses TDRSS as its main communication link. A key to properly developing a scenario is characterizing and documenting operational constraints. Operational constraints come from two sources. The first source is system-engineering trades to work around environmental events. For example, XTE added a second TDRSS antenna as a system trade to ensure contact during slewing maneuvers. Most of these events are defined by the orbit selected for the payload capability of the launch vehicle, and the capability of the communications architecture. Examples of such phenomena for low-Earth orbiting payloads are day and night periods, South Atlantic Anomaly, and guide-star availability. Spacecraft and payload limitations or problems encountered during mission integration and test are the second source of operational constraints. These constraints lead to a list of operational workarounds. It's important to maintain a complete list of these workarounds, so we can place them into the procedures, which is the next level of detail under the scenarios. One of the benefits of a core team on a small mission is that they take with them to operations any lessons they've learned from integration and test.

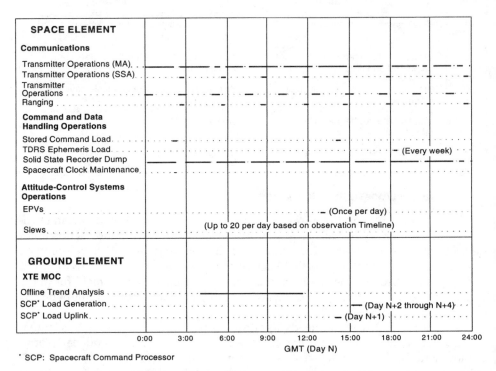

**Fig. 14.5.    Timeline for the X-ray Timing Explorer (XTE).** XTE uses TDRS multiple access (MA) and single-subscriber access (SSA) communications links. [NASA, 1995]

Another key step in developing a scenario is identifying all required products and support actions, which involves identifying schedules for communications support, science plans, and data products. This step helps define operational interfaces and sets performance needs for payload and spacecraft interactions. Examples of actions are onboard memory management of data buffers, table loads for pointing accuracy, and calibration of attitude sensors.

The final step in developing a scenario is producing an integrated timeline, which documents the effects of constraints and actions to ensure the payload can accomplish its mission. As the MOM, you must hold the timeline to ensure the mission is operable for a reasonable cost.

In summary, developing mission timelines for operational scenarios helps you and your mission operations team visualize the payload's normal operations, as well as identify external interfaces and required operational products and activities. Visualizing activities helps you lay out trades for onboard autonomy versus manual operations by showing which activities parallel one another. It provides the first meaningful analysis of mission operations staffing.

### 14.3.2  Characterize Requirements for Onboard Command and Data Handling (C&DH)

While developing scenarios, you should check them from the perspective of command and data handling on the spacecraft bus and payload. Doing so tests the latest design.

Your main concern is the bus and payload interface for command flow to, and data transfer from, the payload. This activity helps clarify the duty-cycle driver identified in Sec. 14.2.2 for this particular mission. Do this trade as early as possible so you can see operational effects and work them with the C&DH designers. An early trade also gives you the chance to walk through data-transfer activities and see if you can simplify it based on end-system needs.

Now you can identify interactions required between the payload and the spacecraft bus and operationally verify them through test scenarios. At this time, you must also establish guidelines for recovery and physical constraints. An example is characterizing payload temperature during a safing or load-shedding activity and understanding how long it could stay at that temperature without being damaged.

Also characterize the level of autonomy provided by the mission and the corresponding staffing activities by iterating the day-in-the-life scenario. The key is to be able to make end-to-end system trades that accurately reflect life-cycle operations costs. As a rule of thumb, it's best to automate routine activities the spacecraft can react to rather than developing predictive planning tools in the ground system, which is usually much more costly.

### 14.3.3  Define Maintenance Constraints on the Platform

For normal operations, we have an interface-control document that defines the interface between the payload and the bus, but we need to add platform activities that affect payload operations. Three major bus constraints are power, attitude, and multiple payload interactions. We need to assess payload operations which result from activities required to maintain the spacecraft bus.

First, consider power. Most low-Earth missions are designed to operate through various shadow events with little effect on payload activities. However, for many small spacecraft, the power margins can constrain payload operations by limiting the length of data transmissions because the transmitter typically consumes a lot of power. The power system and its limitations are key inputs to mission and payload operators, so we must put into the timeline scenario any unique limitations or power-system maintenance activities, such as battery management. We also need to understand how the platform starts shedding power loads. Presently, the EUVE spacecraft requires extensive power management because of the way its battery system is designed.

Second, consider maneuvers, including those for payload pointing and for orbit adjustment and maintenance, such as station keeping for a communications payload. Most of these missions have a maneuver mode that precludes payload operations. For many Earth-observing systems, yaw maneuvers are required twice a year to correct the payload's orientation to the Sun.

Third, consider interaction with other payloads. This interaction can be conflicts in field of view, noise interference, or contention for data-transfer resources. These fundamental issues need to be understood and worked during the concept phase. As MOM, you must advocate solving these problems early to avoid many constraints after launch.

The payload's availability to do its job is one of the measures of success for a mission. A key way to achieve success is to integrate the payload to a spacecraft bus that is almost transparent to the payload. Avoid spacecraft buses that require continuous monitoring and reconfiguration. Instead, identify bus-maintenance constraints and work with the designers to minimize them because they tend to increase mission operations costs. Move toward more autonomy by decoupling routine spacecraft-bus operations and payload operations.

### 14.3.4  Determine the Need for Ancillary Data

While considering end-to-end data processing, identify the need for ancillary information, such as time and position information required to start processing the data. With the Global Positioning System's technology, we can now do many of these functions onboard and provide the ephemeris information with the data itself. This approach eliminates costly post-processing of tracking data. It's always more costly to recreate an activity than to have it measured instantaneously.

You'll also need a data-quality statement that can be appended to the data. In the past this statement was recreated on the ground by reprocessing the raw data received at tracking stations. Now, with packetized systems, the data-transfer protocol gives us a way to develop quality statistics based on the data format. Using solid-state memory and a file-transfer protocol with retransmissions would significantly change the level 0 phase of data processing. Configuration and calibration are other types of ancillary data you must consider. It's more cost-effective to address needs for ancillary data during system definition and development, rather than trying to recreate ancillary data during the operational phase of the mission.

### 14.3.5  Define the Need for Special Events

Once we understand normal operations, we must prepare for special operations, such as safing and recovery, reprogramming the payload's processor, and isolating and trouble shooting faults.

Spell out the safing activities and their corresponding recovery in a scenario as a basis for detailed procedures. The scenario gives operators the chance to visualize how the safing should take place, how long it should take, and how it would be detected on the ground. The recovery scenario should identify the sequence of steps or stages required to bring the payload to full operations in a controlled manner, particularly for payloads that use high-voltage power in their measurements.

Almost all payloads employ processors in their operations. As payloads operate, the processor's software needs maintenance. Operators need to develop a scenario that defines how they can update the program onboard the payload to correct any problems or react to equipment anomalies. This scenario should determine the time required to complete an upload given the forward-link bandwidth and the throughput for C&DH. Not designing an efficient way to update the payload could cause complex, costly workarounds.

Finally, include fault-isolation and trouble-shooting exercises to help determine if information is available to identify the problem. Doing a few fault-isolation exercises helps to validate and refine the failure trees.

Developing meaningful operations scenarios early in a program helps you and your team appreciate the operational implications of various design decisions. By seeing these implications as early as possible in a mission, you can avoid unexpected staffing and cost just before launch.

## 14.4 Characteristics of a Good Payload Operations Plan

The payload operations plan is the end product of this phase. It's basically a more detailed iteration of the operations concept, with a complete timeline of events. The next level of mission engineering would be the detailed operational procedures required to carry out the plan.

You should match your effort on areas of this plan with their effect on the mission. Key events that mean mission success or failure warrant corresponding detail in the operations plan. The plan proposed here is organized around the mission's major timelines. The main sections are launch and early-orbit timelines, special operations, and normal operations. The payload operations plan can be part of a spacecraft operations plan, or at least complement one, depending on the complexity of the payload and the mission.

The plan's size depends on the payload's size and complexity. It could range from 5 to 100 pages. The object is not to create paper but to think through the operations and maintain a baseline of how the payload is to be operated. For many small payloads and missions, it captures the baseline of the payload's operation as the payload evolves through the mission phases.

### 14.4.1 Launch and Early-Orbit Timeline

The key events during launch and early orbits tend to focus on the spacecraft bus and turning on and checking out its subsystems. The main payload activities are usually turning it on and activating it. We have to concern ourselves with deploying appendages, such as booms attached to payloads, releasing latches, or removing covers that protect payloads during launch. Define these activities in detail and identify rest points for contingencies. Don't create a payload-activation sequence that doesn't allow time to handle anomalies.

Activities that require your special attention are turning on high voltages, initial calibration, instrument processors, and handling special orbital-activation constraints such as lighting or a calibration target. See Chap. 9 for launch and early-orbit activities.

### 14.4.2  Special Operations

Special operations occur because of anomalies or unique mission activities. In many cases, the special operations used to turn on the payload are similar to those needed to recover from safe-mode due to an anomaly. By recognizing this relationship early enough, you can avoid developing unnecessary procedures.

Special operations for anomalies need to define safemode entry and recovery. They must specify the events that will cause the payload or spacecraft platform to place the payload into a safe state. Safemode should allow ground operators the time to analyze the problem and recommend how to proceed. This is particularly true during the early phases of a mission, when there is no history of the payload problems that will occur over the life of the mission. With the increasing complexity of a payload's processing ability, we need to build the payload processor so it can gather data that will enable us quickly to determine the reason for the anomaly. Soon, the payload could actually start isolating faults and recommending actions to the ground. (Of course, we shouldn't apply this advanced, expensive technology to simple instruments.) Once we understand a problem, we need to develop a detailed sequence for recovering the instrument. Also, we must identify payload calibrations or special operations to return the payload to full operation. Examples of such special operations are decontaminating the wide-field planetary camera instrument on HST or the high-voltage turn-on sequence for any number of instruments.

The second aspect of special operations is unique mission activities. Many scientific payloads, for example, have encounters or campaigns centered on special scientific events. Examples of such events are the Cassini encounter with the Jovian system and the FAST mission's campaign for studying the northern lights during the winter. FAST depends on the winter occurrence of the northern lights for its science observations.

Targets of opportunity demand another special operation for survey missions. For these events, we must spell out in the operations plan required replanning and interruption of the payload's normal timeline (and, many times, of the spacecraft's operation). The plan should articulate timing requirements for operations and unique communications and payload-processing support. It should also detail how we'll start and carry out these interactions with integral support areas and spacecraft operations.

### 14.4.3 Normal Operations

Most missions will spend more than 95% of their time in normal operations. As the MOM, you must present normal operations for a payload to reflect a day in the life of a mission. Your key cost consideration is the level of daily support needed for payload operations. To save money, you must size the operations team for normal operations and allow time through safing sequences to handle anomalies. The faster the required response time, the costlier the mission operations. For example, operating from the Space Transportation System (STS) is a challenge because the short mission and public visibility require quick handling of anomalies. These pressures force all necessary support experts to be on hand, just waiting for an anomaly to occur.

To assess workload for normal operations, you need to develop a detailed mission profile for a 24-hour period. You may use a different period if it more aptly reflects the payload's cycle of operation. The activity plan should identify online and off-line activities and estimate the amount of labor required to do all tasks. Properly laying out the work helps identify peak activities and shows the amount of effort needed.

Your activity plan must address health and safety monitoring, mission planning, command generation, command execution, and data-product processing. For a small mission, one or two people can do these activities; larger missions may require dedicated teams for each activity. Health and safety monitoring depends on payload design. If the payload is intelligent and has a number of safe modes, we don't need to do much monitoring. Instead, we could review parameters that affect the quality of data products, thus eliminating expensive real-time operations in favor of smaller, off-line operations.

In your scenario for mission planning and scheduling, make sure you identify mission drivers that bound the operation. Examples of such drivers are orbital constraints, communications abilities, onboard data-storage ability, and the handling of anomalies in transporting mission data. Orbital constraints due to spacecraft or payload limitations are expensive to plan from the ground. Examples of such constraints are limits on pointing toward bright objects, operations in the South Atlantic Anomaly, or guide-star availability. It's a mission trade to obtain an orbit that minimizes these types of constraints in order to lower operations costs.

The commanding requirements of a mission depend on the spacecraft and payload designs. If you can preplan payload operations or execute them onboard with script direction from the ground, you can lower the level of command interaction. If the payload is dynamic and requires continuous interaction with the ground, the commanding activities will be extensive and the operations cost will be relatively high. In the first case, we can automate many of the spacecraft operations and simply have ground controllers manage tables the spacecraft processor was to operate. This activity is far less costly than having to uplink large numbers of single commands to a dumb spacecraft. The volume of data created by the payload and the spacecraft's ability to transfer it to the ground drive pass contacts.

Present network technology makes us manually isolate pass-contact faults—a costly job. The growing presence of solid-state memory makes storing large volumes of data and sending them to the ground more cost effective because it lowers the number of pass contacts.

As a MOM, you need to review the constraints and data requirements and create a 24-hour mission profile that shows their relationship. Figure 14.5 gave an example of a mission profile for the XTE spacecraft and its instruments. The top few lines list the orbital constraints, the next lines describe payload operations, and the final few lines describe network operations with the TDRS system.

The final activity is to control and monitor the transfer of the payload data to its destination for processing. For a simple survey payload, this monitoring can be very straightforward and easy to automate. Solid-state memory and file-transfer protocols are allowing this automation. Balance between the forward-link bandwidth and the downlink bandwidth is one of the limitations for this type of protocol. The key is to build the accounting capability into the payload and spacecraft, so we won't have unique processing requirements to reconstruct the payload data on the ground. Also, properly planning the data can take care of such items as time tagging and ancillary position data.

The keys to low-cost mission operations are early involvement of operators to identify development/operations trades and the discipline to enforce these trades. Users of the data must be key participants in the systems trades because they are the only ones who can assess the trades' effect on the mission objective. We must make these trades as a team and encourage alternatives to meet the mission objectives for the lowest cost. If we consider only the mission objectives, trades will be made with only development in mind and won't consider the mission's life-cycle costs.

The mission-engineering process structures iterations on the operations concept that develop increasing levels of detail to prepare for operations. The checklists in this chapter start a new MOM toward identifying cost drivers for payload operations.

## References

Biroscak, Losik, and Malina. 1995. *Re-Engineering EUVE Telemetry and Monitoring Operations: A Management Perspective and Lessons Learned from a Successful Real-World Implementation.* Publication number 669. Berkley, CA: Center for EUV Astrophysics Technology Innovation Series.

National Aeronautics and Space Administration. 1995. *Flight Operations Plan for the X-Ray Timing Explorer (XTE).* GSFC-410-XTE-031. Goddard Space Flight Center, MD.

National Aeronautics and Space Administration. 1994. *GGS Mission Operations Procedures.* Volume IV: Contingency Operations Data Base. CDRL 405, NAS5-30503. Goddard Space Flight Center, MD.

National Aeronautics and Space Administration. 1993. *Fast Auroral Snapshot Explorer (FAST) Flight Operations Plan.* FAST-OPS-006. Goddard Space Flight Center, MD.

# Chapter 15

# Spacecraft Performance and Analysis

Mac Morrison, *TRW*

15.1 How Typical Spacecraft Subsystems Work
15.2 How Typical Spacecraft Operations are Done

One of the mission operations manager's (MOM) most important tasks is to ensure that spacecraft-bus operations will maintain the spacecraft's health and safety while meeting mission objectives. Before launch, the MOM (or the spacecraft planning and analysis function) monitors the bus design and develops procedures for operating the spacecraft bus. Following launch, the MOM (or the spacecraft planning and analysis function) analyzes spacecraft-bus performance, generates commands, maintains flight software, and plans required calibrations. This chapter describes how we do the steps in spacecraft planning and analysis, as listed in Chap. 3.

Successful space mission operations require a well designed spacecraft bus for the payload or experiments. The bus typically provides power, pointing control, maneuvering ability, communications, mounting structure, and thermal control for mission operations. Failure of the spacecraft to provide these functions through design flaws, random failures, or operational errors can compromise mission success. Reviews should discover and eliminate faulty designs, but designs that work in breadboards, engineering models, and even during spacecraft assembly and test may be difficult to operate in the space environment. For instance, does the design include the ability to use ground-commanded overrides? Does it adequately cross strap subsystem elements? Does it include the right sensors to allow us to determine the spacecraft's on-orbit state of health? Good engineering practices that include review and testing throughout the design and development phases can minimize design flaws; careful parts selection and screening may lower the failure rate of piece parts; and thoughtful design of the spacecraft's onboard software and ground software, coupled with rigorous operations training, can minimize operational errors. But cost-effective mission operations means coupling the right spacecraft bus with the mission requirements and recognizing the operational effects of these designs early in a program. The earlier the better!

The MOM is a key player in determining and analyzing operations that can affect spacecraft design and reduce overall program costs. By making sure adequate plans and procedures are developed before launch, the MOM can help reduce operational errors resulting from inadequate planning, human errors in commanding the spacecraft, misinterpretation of telemetry data, and inadequate time to complete operations.

We must carefully evaluate the requirements for operating a spacecraft to determine the concept for mission operations. These requirements include proper constellation size, the right ground-control environment, and protecting mission data.

Multiple spacecraft constellations may evolve an operations concept that allows loss of individual spacecraft as a trade between spacecraft complexity and replacement costs. Concepts for programs that feature tens to hundreds of satellites in low-Earth orbit, such as Motorola's IRIDIUM and some of the now canceled SDI programs, faced this trade.

Dedicated ground-control facilities may drive trades between onboard abilities and ground-commanded abilities. If a station is available to a mission 24 hours a day, we may control many spacecraft operations in real time by uplinked commands, thus potentially simplifying onboard subsystems. Operations using institutional ground control, such as the Air Force's Satellite Control Network, must share resources. Spacecraft operations in this environment depend a lot on scheduling. The ground-control network may not provide the required coverage for uplink and downlink, thus driving the spacecraft design to more autonomy. Questions arise such as: Must we regularly interrogate the spacecraft? How immune is the spacecraft to anomalies that occur when a ground station isn't "up" on the vehicle? Does the spacecraft have built-in safe modes and can we use stored, onboard commands? However, we must consider a classic trade in spacecraft operations—ground control versus onboard control. We have to trade the one-time and recurring costs of highly autonomous spacecraft against the recurring costs of operations manpower for spacecraft that need more ground control.

Constellation design and the amount of ground control affect data loss because we may lose a spacecraft if we don't have a ground station available for downlinking data. How important is the mission data the payload will obtain —is it a one-time event, so the spacecraft must not fail during a critical period? If so, the spacecraft design and operations planning must account for this requirement. But if we can get the data regularly and can accept outages in spacecraft power or communication, comprehensive operations plans and procedures may allow for simpler spacecraft designs.

In some cases wanting to provide mission flexibility runs up against economic realities—the money just is not available to do everything we want. Spacecraft complexity (and weight) drives technical decisions on the amount of onboard redundancy to include in the spacecraft design. They also limit our choice of booster to place the spacecraft into orbit. An example of being driven to more complex designs is a mission launched and serviced from the Shuttle. The requirement of "fail operational—fail safe" drives spacecraft to triply redundant subsystems in some cases. All of these complexities mean we must know the operations requirements early in the program.

# 15.1 How Typical Spacecraft Subsystems Work

Normal operations for spacecraft subsystems are real-time, ground-controlled actions or stored, onboard actions to provide and maintain power, pointing control, maneuvering ability, communications, and some thermal control for the mission operations. Design dictates how we operate—simplified, easily accomplished spacecraft operations lead to cost-effective missions.

Although we build many spacecraft, the types of spacecraft subsystems are relatively few. Power subsystems typically provide electrical power through solar arrays, batteries, or fuel cells. For attitude determination and control, we

- Spin all or part of the spacecraft
- Use the Earth's gravitational field to point a spacecraft axis toward the Earth's center
- Use the interaction of the Earth's magnetic field and onboard torquers to align a spacecraft axis toward the Earth
- Use a combination of onboard gyroscopes or reaction wheels to provide three-axis attitude control

Spacecraft maneuver using an onboard propulsion subsystem coupled to Sun, Earth, or star sensors or to data received from global positioning system (GPS) satellites. Communications to a ground-control station or other spacecraft depend on receivers and transmitters coupled to either fixed or maneuverable antennas. Thermal control can be passive—using insulation, second-surface mirrors, and paint—or it can be active—using thermostats, heaters, or heat pipes. Table 15.1 lists spacecraft subsystems, options, and key considerations.

Normally, these subsystems provide the mission with continuous power throughout eclipse periods and maneuvers, keep a spacecraft's axis pointing in the desired direction, maintain communications between the spacecraft and with the control station, and provide a benign thermal environment for all spacecraft equipment. Chapters 10 and 11 of *Space Mission Analysis and Design* [Larson and Wertz,

**Table 15.1.    Operating Spacecraft Subsystems.** These lists of options and considerations are incomplete but represent the issues we must address for this function.

| Subsystem | Options | Key Operational Considerations | Where Discussed |
|---|---|---|---|
| Attitude control | Gravity gradient<br>Spin stabilized<br>3-axis stable<br>Sun, Earth, and star sensors | • Deploying boom<br>• Initially acquiring Earth<br>• Determining nadir angle<br>• Initial spin-up<br>• Nutation control<br>• Controlling spin speed<br>• Managing momentum<br>• Acquiring attitude<br>• Disturbance torques<br>• Losing Earth<br>• Number of independent fields of view<br>• Pointing constraints | Chap. 15; Chap. 11 [Larson and Wertz [1992] |
| Electrical power | Deployed arrays<br>Body-mounted arrays<br>Batteries | • Deploying arrays<br>• Eclipses and shadowing<br>• Solar tracking<br>• Degradation<br>• Re-conditioning batteries<br>• Charge/discharge cycles | Chap. 15; Chap. 11 [Larson and Wertz, 1992] |
| Propulsion | Cold gas<br>Mono-propellant<br>Bi-propellant<br>Electric<br>Blow down or regulated<br>Thruster size | • Location of thrusters<br>• Thrust duration and level<br>• Maneuver complexity<br>• Amount of redundancy | Chap. 15; Chap. 17 [Larson and Wertz, 1992] |
| Thermal | Passive<br>Active | • Number and location of heaters<br>• Thermal bathing<br>• Temperature gradients | Chap. 11 [Larson and Wertz, 1992] |
| Communications | Fixed antenna<br>Steerable antenna<br>Satellite cross-links<br>Operational frequency | • Encryption<br>• Pointing requirements<br>• Number of data modes<br>• Real-time or playback<br>• Data volume, data rate, and pass duration | Chap. 11 and 15; Chap. 13 [Larson and Wertz, 1992] |
| Command, control, and data handling | Onboard processing<br>Types of recorders<br>Memory available<br>Stored command files<br>Real-time data only | • Memory duty-cycle<br>• Data completeness<br>• Autonomy<br>• Pass duration | Chap. 13 and 15; Chap. 11 [Larson and Wertz, 1992] |
| Structures and mechanisms | Deployable booms<br>Articulated devices | • Coupling of articulation and attitude control<br>• Deploying boom<br>• Solar pressure effects | Chap. 11 [Larson and Wertz, 1992] |

1992] discuss these subsystems, showing how to design them and how requirements drive their design. Here, we discuss the way they operate.

Attitude determination and control for many Earth-orbiting spacecraft depends on establishing the relationship of the spacecraft body's axis to the Sun, to the center of the Earth (or the Earth limb), or to a distant stellar object. In the first case, onboard Sun sensors attached to the spacecraft body measure the angle between incident visible light and the fixed sensor, usually through a slit in the sensor housing, as shown in Fig. 15.1. Timing the occurrence of Sun pulses can establish the rotational rate. Incident light on the slit in the sensor housing provides angle data.

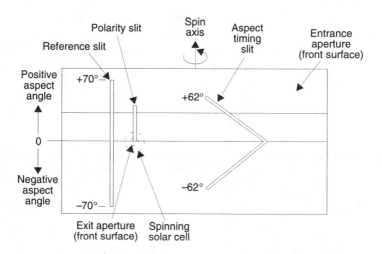

**Fig. 15.1.    Geometry of a Sun Sensor.** The geometry of the slits in the sensor housing allows us to determine the spacecraft's spin speed and the Sun aspect angle.

Earth sensors typically measure the temperature difference between the Earth's atmosphere and deep space. This difference defines the Earth's horizon and provides information to determine the geometry for spacecraft pointing. Earth sensors with oscillating mirrors view the Earth limb to limb and provide nadir pointing data to the spacecraft's attitude-control system. See Fig. 15.2.

Star sensors used on three-axis stabilized spacecraft are usually either trackers or mappers. After the tracker locates a predetermined star and tracks it, the vehicle's motion will result in an apparent movement of the star. We use this error to control the spacecraft's attitude. Star mappers use similar logic but track all stars in the sensor's field of view above a certain brightness. Typically, we use the data from two or more stars to establish the spacecraft's inertial attitude. We can couple magnetometers that measure the geomagnetic field with magnetic torquers to con-

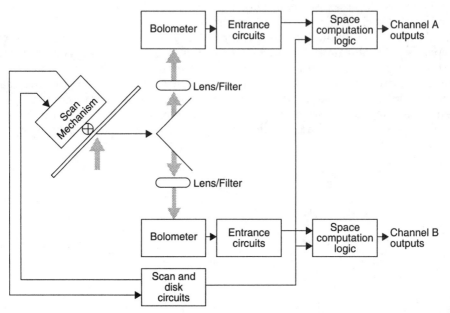

**Fig. 15.2.   Block Diagram of a Typical Earth Sensor.** This Earth-sensor design has a scanning mirror/bolometer to find temperature differences between the Earth and deep space. Redundant channels avoid the Sun's or Moon's intrusion into the sensor's field-of-view and on-orbit failures.

trol the spacecraft's attitude. However, the attitude accuracy possible with this system is less than that achievable with the other described systems. [Larson and Wertz, 1992, Chap. 11]

Propulsion subsystems enable us to maneuver the spacecraft in the orbit ($\Delta V$ maneuvers) and to control spacecraft pointing. Most propulsion subsystems for normal on-orbit operations are hot-gas, catalytic-thruster systems using hydrazine as the fuel. The system operates by passing the fuel from a pressurized source over a heated catalyst bed containing alumina particles. The hydrazine decomposes into nitrogen, hydrogen, and ammonia gases, which exit through an exhaust nozzle to supply thrust. Bi-propellant hot-gas systems using $N_2O_4/N_2H_4$ provide higher specific impulse than hydrazine alone, but the systems are typically more complex.

If it's necessary to avoid the contaminating products from a hot-gas system— as for a payload with exposed optical surfaces—we might use a cold-gas system with pressurized helium. Table 15.2 compares the specific impulse for these systems.

If the fuel source is in a tank that holds fuel and pressurant separated by a bladder, the thrusters will have decreased performance as the fuel is used and the

**Table 15.2.**   **Specific Impulses (Isp) for Spacecraft-Propulsion Systems.** Performance versus clean operation is a typical propulsion-system trade.

| Type of fuel | Isp Range | Risk of Contamination |
|---|---|---|
| Bi-propellants | 305 – 310 | High |
| Hydrazine | 220 – 240 | Medium |
| Cold gas | 30 – 70 | Low |

pressurant expands. This is called a blow-down system and requires adjustment of thruster firings (either duration or number of firings) to maintain performance. Regulated systems have a separate pressurant tank at very high pressure; the pressurant releases into the propellant tank to provide a constant pressure over the life of the mission. Figure 15.3 illustrates these two options; Figures 15.4 and 15.5 show their relative complexity.

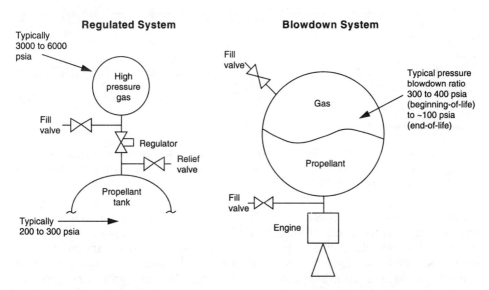

**Fig. 15.3.**   **Ways of Pressurizing Two Pressurization Subsystems.** The regulated system, though more complex, maintains a steady pressure in the propellant tank and, therefore, constant performance.

The thrust range used for on-orbit operations varies from a fraction of a newton to a few newtons for attitude control and tens of newtons for maneuvering. This means that operators typically can select the thrusters. Thrusters on the spacecraft structure provide thrust in the pitch, roll, and yaw axes for three-axis attitude

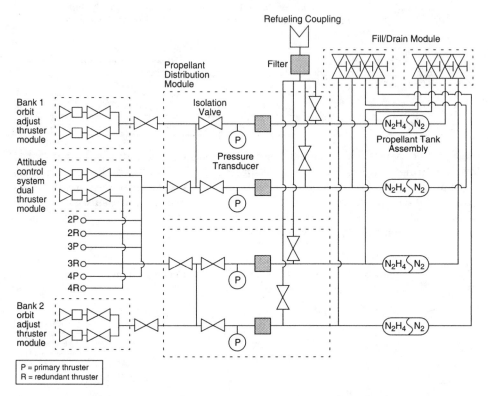

**Fig. 15.4.    Schematic of Propulsion System for the Gamma Ray Observatory.** Monopropellant hydrazine fuel with a blowdown fuel system going from a tank pressure of 400 to 100 psia. [Larson and Wertz, 1992]

control and along the spacecraft's velocity vector (+ and –) for delta velocity maneuvers. Maneuvers out of the orbital plane use thrusters mounted at an angle to the orbital velocity vector or yawing of the spacecraft. Many spacecraft have thrusters canted away from the spacecraft's x, y, and z axes, so we can use components of the thrust to control in more than one axis. Also, contamination products and heating from the thrusters may require us to move the thrusters off the space-craft's orthogonal axes. Although opening or closing valves in the system's fuel lines allow us to use redundant tanks and thrusters, we don't have to alter the sub-system's configuration unless failure occurs.

Communications subsystems—sometimes referred to as telemetry, tracking, and command (TT&C) subsystems—are the links between the spacecraft and ground-control stations (or, in some cases, from spacecraft to spacecraft). A con-cise, high-level summary of communications subsystems is in Chap. 11 of Larson

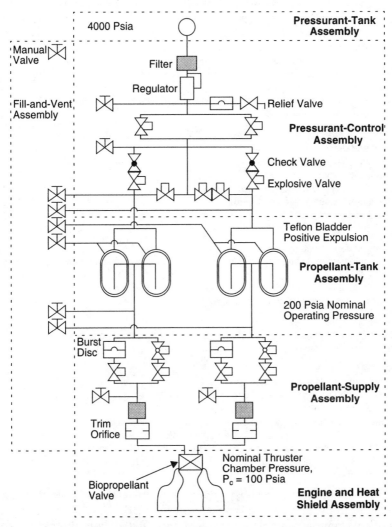

**Fig. 15.5.** **Pressure-Fed Propulsion System Using Earth-Storable Bipropellant (N₂O₄/MMH).**
A bi-propellant fuel system maintains 200 psia operating pressure through regulated re-pressurizing of propellant tanks. [Larson and Wertz, 1992]

and Wertz [1992], which we've used to give you the following information. The communications subsystem allows the spacecraft to receive and track carrier signals, receive commands, and transmit telemetry data. Table 15.3 summarizes these functions. Figure 15.6 shows a typical communications subsystem.

**Table 15.3.    What a Communications Subsystem Does.** The communications subsystem provides the ability to receive and track carrier signals, receive commands, and transmit telemetry data.

| What a Communications Subsystem Does |
| --- |
| • Carrier tracking<br> • 2-way coherent communication (downlink frequency is a ratio of the uplink frequency)<br> • 2-way noncoherent communication<br> • 1-way communication |
| • Receiving and detecting commands<br> • Acquire and track uplink carrier<br> • Demodulate carrier and subcarrier<br> • Derive bit timing and detect data bits<br> • Resolve data-phase ambiguity if it exists<br> • Forward command data, clock, and in-lock indicator to the subsystem for command and data handling |
| • Modulating and transmitting telemetry<br> • Receive telemetry data streams from the subsystem for command and data handling or for data storage<br> • Modulate downlink subcarrier and carrier with mission or science telemetry<br> • Transmit composite signal to Earth or relay satellite |
| • Ranging<br> • Detect and retransmit ranging pseudorandom code or ranging tone signals<br> • Retransmit either phase coherently or noncoherently |
| • Operating subsystems<br> • Receive commands from the subsystem for command and data handling (C&DH)<br> • Provide health and status telemetry to the C&DH subsystem<br> • Point any antenna requiring beam steering<br> • Operate mission activities through command files stored in software<br> • Autonomously select omni antenna when spacecraft attitude is lost<br> • Autonomously detect faults and recover communications using command files stored in software |

For the communications subsystem in Fig. 15.6, the uplink rf signals are received through the antennas and pass through the diplexer to the receivers. For the case of commands, the data stream is demodulated from the carrier and sub carrier and then routed to the command detector, which validates the data and forwards it to the subsystem for command and data handling. For ranging signals, the tones or pseudo random noise (PRN) codes are demodulated in the receiver and then routed to the transmitter for conditioning and modulating onto the downlink carrier. Spacecraft telemetry data that contains configuration and state-of-health information is conditioned in the telemetry-control unit(s), modulated onto the downlink subcarriers, and routed to the transmitter for modulation onto the carrier and transmission. The transfer rf switch allows us to select a redundant transmitter and antenna. The diplexer allows a transmitter and receiver to share

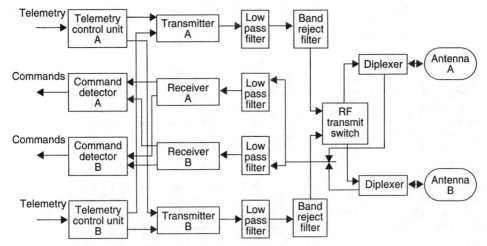

**Fig. 15.6.    Diagram of a Communications Subsystem.** To complete a communications link, both receivers must be ON at all times. The path for the uplink signal depends on which receiver locks to it first. We may select a receiver by testing for receiver sensitivity to the uplink-signal strength and lowering the uplink transmitter's power to select the more sensitive of the two units. But because we want effective mission operations, we normally set the uplink power high enough to allow either receiver to lock up on the signal.

the same antenna. It also isolates the transmitter from the receiver port at the receiver's center frequency to keep from damaging the receiver.

Table 15.4 summarizes the characteristics of three S-band and two other communications subsystems, which are standard in NASA and many military programs. The systems for spacecraft-to-spacecraft communications require more operational control than S-band systems for Earth coverage because the narrow-beam antenna requires precise pointing for link lockup.

Equipment for commanding and data handling decodes command information and routes the data to the proper subsystem for execution. Each subsystem is interrogated for equipment status, and the data is arranged and stored in accessible formats, so the communications equipment can transmit it to the control station. Command data includes information on the action to be taken (e.g. turn off a unit, select a temperature range) and routing information. Figure 15.7 compares the command formats for two different spacecraft. Note the differences between real-time commands and commands to be loaded into onboard processors.

The telemetry data formatted for downlink transmission includes equipment status (on/off) and engineering parameters (voltage, current, temperatures, pointing information). An important operations consideration in the design is the frequency of sampling individual parameters for transmission. Telemetry-data formats provide for spacecraft data to be sampled at the main frame rate or sub-

**Table 15.4.    Attributes of Some Common Communications Subsystems.** Each system can support various modulation schemes (see Chap. 11). We use Earth-coverage antennas for normal operations and omnidirectional antennas for launch and contingency operations.

| Appli-cation | Frequency | | Modulation | | Antenna Charac-teristics | Remarks |
|---|---|---|---|---|---|---|
| | U/L | D/L | U/L | D/L | | |
| Space ground link subsystem (SGLS) | S-band 1.75 to 1.85 GHz | S-band 2.20 to 2.30 GHz | FSK AM PM | PCM PM FM | Earth cover-age; omnidi-rectional coverage | SGLS standard |
| Goddard Spaceflight Tracking and Data Network (GSTDN) | S-band 2.02 to 2.12 GHz | S-band 2.20 to 2.30 GHz | PCM PSK FSK | PCM PSK PM | Earth cover-age; omnidi-rectional coverage | GSTDN is slowly phasing out. The Deep Space Net-work is absorbing some of its assets. |
| Cross-link within con-stellation | W-band 60 GHz | W-band 60 GHz | Any | Any | Narrow beam 0.1 deg typical | Modulation, coding, and en-cryption can be customized |
| Cross-link to TDRSS | S-band K-band | S-band K-band | QPSK Spread spectrum | QPSK Spread spectrum | Narrow beam | TDRSS User Standard (See TDRSS Users' Guide) |

Legend:
U/L = Uplink                                          PCM = Pulse-code modulation
D/L = Downlink                                        FM = Frequency modulation
FSK = Frequency-shift keying              PSK = Phase-shift keying
AM = Amplitude modulation                 QPSK = Quadrature phase-shift keying
PM = Phase modulation

commutated for less frequent sampling. One bit allows us to determine data that represents equipment status (on/off), so a telemetry word can contain data on the status of several items. State-of-health data may require one or more telemetry words. Operations requirements on the type and sample frequency of data contribute heavily to the complexity and cost of spacecraft and ground-telemetry processing, so we must carefully analyze them early in a program to make good design trades.

Electrical-power subsystems usually combine solar arrays that convert sunlight into electrical power through the photovoltaic process with a source of stored energy to provide power when the spacecraft is in the Earth's or Moon's shadow. Rechargeable batteries or fuel cells typically give us this stored energy. The equipment in this subsystem also controls and distributes the electrical power throughout the vehicle. The solar arrays must face the Sun to provide power. On

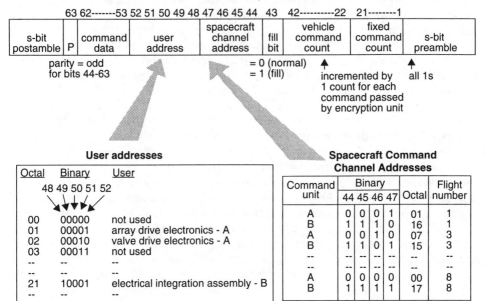

**FltSatCom Normal Command (all real-time commands)**

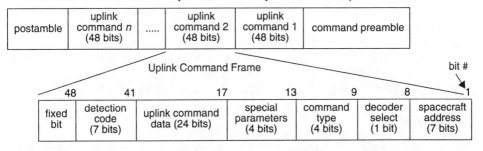

**Fig. 15.7.** **Command Data Formats for FltSatCom and the Total Ozone-Mapping Spectrometer—Earth Probe (TOMS-EP).** We first verify the content of command-format data during factory testing of the spacecraft. Operationally, a mnemonic or a command number usually identifies each command. [FLTSATCOM and TOMS, 1994]

spin-stabilized spacecraft this usually means covering most of the spacecraft's body with solar panels to meet power requirements. Compared to spinners, arrays on articulated booms allow for a smaller total array area to meet requirements because we can point the arrays at the Sun. We operate electrical-power subsystems by pointing arrays and switching to stored power when needed.

While the basic elements of spacecraft in low-Earth and geosynchronous orbits are similar, their operations are different. For geosynchronous spacecraft, the batteries are required during the 45-day eclipse seasons around the times of the vernal and autumnal equinoxes. Thus, a major seasonal operations task is preparing the batteries for these periods. For spacecraft operating in low-Earth orbits, solar arrays go into shadow on every orbit, so operators must recondition batteries almost continually.

The FltSatCom spacecraft has two solar arrays that are clocked to point toward the Sun. They supply primary power through regulators to the main spacecraft bus and three nickel-cadmium batteries rated for 34 ampere-hours. The Total Ozone-Mapping Spectrometer—Earth Probe's (TOMS—EP) power subsystem uses two solar-array wings to supply power to the spacecraft and to recharge the battery so the spacecraft can operate during eclipse. By mounting solar cells on both faces of the wings, we can generate power throughout the orbit without moving the arrays. A "super" nickel-cadmium battery with 22 cells, rated for nine ampere-hours, supplies stored energy. Solar-array regulators allow us to transfer power from the arrays to the main power bus in the power-control unit. For more on power systems and power regulation and control, see Larson and Wertz [1992].

We control the spacecraft's thermal environment passively, actively, or by combining both techniques. Passive techniques include using multilayer insulation (MLI) on the structural members, shades and baffles to protect equipment from direct sunlight, mirrored surfaces to reflect incident energy, and energy-absorbing or reflecting paint and coatings. Heat pipes use a liquid-vapor cycle to transmit heat to a radiator. For these techniques, design analysis determines the thermal control, so we don't need on-orbit control. Active thermal control uses commandable electric heaters, thermostatically controlled heaters, and in some cases, duty-cycled equipment.

The spacecraft structure is the load-carrying framework for attaching subsystem equipment. Structural mechanisms, such as booms, may deploy and fix in place throughout the mission. If we can retrieve the spacecraft for repair, refurbishment, or upgrading, we may make booms retractable and redeployable (e.g., the Gamma Ray Observatory). Other mechanisms include motors to rotate solar arrays, gimbals to move antennas, and Sun shades.

Describing how individual subsystems work doesn't give a complete picture of spacecraft operations. We must understand how maintaining power, attitude, and communication links affects all subsystems. Powering equipment on or off changes the temperature environment within the spacecraft. Operating heaters to supply heat lost when other units are off is part of many operations procedures.

Attitude-sensor data that goes into the attitude processor affects how we operate subsystems for propulsion, electrical power, communications, heating, and even structures. Suppose we're operating a three-axis-stabilized, geosynchronous spacecraft (orbits with an inclination near zero) within a small deadband area defined by predetermined limits. In this case, disturbances caused by solar torques and geomagnetic field irregularities make the spacecraft drift in longitude and latitude. When it reaches the deadband limit, we may need to activate the propulsion system to fire thrusters (in most coordinate systems, pitch or roll thrusters) that will counteract the spacecraft's motion. Usually, we may select a smaller, tighter deadband, which requires more thruster firings but gives us more accurate pointing. Interaction of Sun sensors and propulsion thrusters permits control of the spin speed and spacecraft nutation resulting from external torques. When we move spacecraft-body axes, communications coverage and link margin may be affected, and antennas mounted on articulated appendages will require repositioning. For communication subsystems with narrow beam coverage on downlink antennas, this control is essential to mission success. Table 15.5 describes some typical interactions between the communications subsystem and other subsystems.

**Table 15.5.    Interactions of Spacecraft Subsystems with a Communications Subsystem.** This table shows interactions and their affect on normal operations.

| Subsystem | Interaction | Operations Impact |
|---|---|---|
| Attitude Determination and Control | Spacecraft pointing and attitude knowledge for fixed antennas | Link losses may require tighter attitude control by reducing deadband |
| Thermal | Frequency shifts may occur on non-oven controlled equipment | Use heater control to stabilize temperatures |
| Electrical Power | Power required to operate communications equipment | Manage power needed by pointing arrays; manage battery conditioning; possibly duty-cycle equipment |
| Structures | Clear field of view for antennas | Plan for obstructions to field of view for gimbaled antennas during all contact periods |
| Command and Data Handling | Onboard command routing affected by link errors | Verify commands by telemetry response—load critical commands into onboard processors for later execution |

# 15.2 How Typical Spacecraft Operations are Done

Spacecraft operations has been described as long periods of boredom punctuated by moments of panic. Successful operations extend the long periods and eliminate the panics. This means all the equipment is operating normally and careful planning is in place to anticipate approaching events. In fact, spacecraft operations can be thought of as a series of tasks that prepare the vehicle to respond to planned external requirements. These requirements could be caused by orbital conditions (spacecraft eclipsed by the Earth's shadow); changes in mission that require us to alter the spacecraft's position or attitude (a delta velocity maneuver to change in-plane orbital location); or a need to manage equipment because the spacecraft is aging (changing thruster-firing durations over the life of a blow-down propulsion system).

These tasks are common to all spacecraft. The challenge is to do them within the constraints of the mission requirements and the ground-station environment. Doing routine operations tasks requires careful, detailed planning to make sure the spacecraft acts on the correct command data. Commands must certainly be valid, but if operators have to monitor and verify telemetry, they also have to plan carefully in order to know when they can halt an action or abort a procedure without harming the spacecraft. Table 15.6 lists some typical spacecraft activities and their effects on mission operations.

An example of a medium-complexity task is battery reconditioning, which places the spacecraft's stored-energy system in readiness to supply power during periods of shadowing by the Earth. For spacecraft operating at geosynchronous altitude, eclipses occur around the spring and fall equinoxes. At these times the Earth's equator lies along the ecliptic, so spacecraft operating with very low inclinations are in the Earth's shadow for part of the orbit. We determine the beginning of the eclipse period by analyzing the spacecraft's orbit. The eclipse seasons start with partially shadowed orbits (penumbral eclipses), move to fully shadowed orbits (full eclipses), and end with more penumbral eclipses.

For geosynchronous spacecraft, two eclipse seasons last approximately 45 days each—centered on the equinox. Therefore, the battery system should be fully reconditioned 22–23 days before equinox. First, we determine the period of shadowing by analyzing the ephemeris data. Then we do detailed scheduling of spacecraft contacts required to command and monitor the spacecraft and finish by analyzing the battery's state before the eclipse season begins. Figure 15.8 shows typical battery discharge/recharge performance. Table 15.7 shows a procedure for reconditioning one of the batteries on a geosynchronous communications satellite.

Battery reconditioning is just one example of the tasks people in spacecraft planning and analysis must do. Several other typical tasks are listed in Table 15.6, but this list is certainly not complete. The MOM must develop a complete list of spacecraft-bus operations, create a procedure for doing each task, identify the information required to complete the task, and establish a way to verify procedures. The MOM must also decide if each task will be automated on the spacecraft, automated on the groundstation, or done by operators.

**Table 15.6.     Spacecraft Operations.** The operating tasks listed in the table are the same ones discussed in Sec. 15.1—maintain communications, power, pointing, and thermal control. The "actions to take" column includes both routine responses to the tasks and first-order responses to anomalies. We've rated the complexity of the tasks from low to high based on normal operations and the need to respond in case of abnormal operations.

| Operation Task | Operator Action | Data Collection and Trending | What to Look For | Action to Take | Com- plexity |
|---|---|---|---|---|---|
| Transmit commands to spacecraft | • Monitor data receipt through telemetry response | • Correct number of commands transmitted | • Correct telemetry response, if applicable, for transmitted commands | • Retransmit command <br> • Switch to redundant equipment | Low <br><br> Low |
| Receive telemetered data from spacecraft | • Monitor data and compare to expected values <br> • Monitor against alarm limits | • Plot data for diurnal variations <br> • Plot data for seasonal conditions <br> • Compare spacecraft configuration status to expected | • Out of tolerance conditions <br> • Onboard equipment failures <br> • Trends in data that could lead to anomalous conditions | • Archive data <br> • Modify telemetry limit checking for aging/ seasonal conditions <br> • Determine if onboard anomaly has occurred | Low <br> Low <br><br><br><br> Low to high |
| Maintain power | • Monitor output of solar array <br> • Monitor all bus voltages, currents | • Beginning-of- life output vs. predicted end-of-life <br> • Spacecraft configuration status | • Aging due to UV exposure <br> • Difference in configuration <br> • Equipment failures | • Possibly duty-cycle equipment <br> • Command redundant equipment on-line | Medium to high <br><br> Medium to high |
| Prepare batteries for eclipse season | • Recondition batteries | • Battery volt- age, current, temperatures | • Battery discharge— recharge performance | • Command battery to discharge— recharge | Medium |
| Maintain power during eclipses | • Monitor battery performance | • Beginning-of- life perfor- mance vs. life- time | • Cell failure <br> • Cell memory | • Actively control battery | High |

**Table 15.6.    Spacecraft Operations. (Continued)** The operating tasks listed in the table are the same ones discussed in Sec. 15.1—maintain communications, power, pointing, and thermal control. The "actions to take" column includes both routine responses to the tasks and first-order responses to anomalies. We've rated the complexity of the tasks from low to high based on normal operations and the need to respond in case of abnormal operations.

| Operation Task | Operator Action | Data Collection and Trending | What to Look For | Action to Take | Com-plexity |
|---|---|---|---|---|---|
| Maintain operating temperature | • Monitor temperature sensors on equipment or compartments | • Diurnal variations<br>• Seasonal variations<br>• Equipment on-off status | • Degraded passive thermal control<br>• Failure of thermostat<br>• Failure of equipment unit<br>• Unpredicted change in spacecraft pointing relative to Sun angle | • Turn equipment on/off | Low to medium |
| Maintain pointing | • Monitor Sun, Earth, star sensor data<br>• Monitor wheel speed | • Pointing x, y, z<br>• Wheel speed variations<br>• Sun's or Moon's intrusion into the Earth sensor's field-of-view | • Alarms<br>• Excessive thruster firings | • Switch to redundant sensors<br>• Switch to redundant wheels<br>• Switch to scanning by redundant Earth sensor | Medium<br><br>Medium<br><br>Medium |
| Spacecraft repositioning | • Command thrusters to start and end delta-velocity maneuvers ($\Delta V$) | • Thruster firing duration<br>• $\Delta V$ start and stop times<br>• Thruster performance<br>• Fuel use | • Correct spacecraft repositioning<br>• Proper thruster performance | • Determine position from ranging data or GPS data<br>• Archive performance data | Low to medium<br><br><br>Low |
| Load onboard computer | • Prepare command file<br>• Transmit load commands<br>• Verify correct load | • Old and new contents of memory | • Errors in onboard load | • Verify load commands<br>• Retransmit required commands | Low<br><br>Low to medium |
| Move appendages | • Reset solar arrays to Sun line<br>• Point antennas | • Output of solar array<br>• Strength of receiver signal<br>• Bit error rate | • Increased output<br>• Increased strength<br>• Reduced rate | • Calculate Sun angle on array-reposition<br>• Switch to redundant array drive<br>• Recalculate and repoint | Medium<br><br><br>Medium<br><br>Medium |

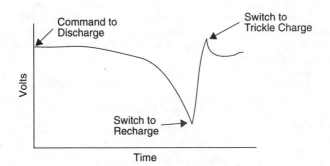

**Fig. 15.8.** **Battery Reconditioning Performance.** Voltage, current, and temperature are monitored as a normal state-of-health function. We then command the battery to DISCHARGE by switching the reconditioning circuit into the system. The battery voltage falls to a predetermined level and is switched to RECHARGE.

**Table 15.7.** **Reconditioning Procedure for the FltSatCom Battery.** The time required to recondition one battery with this procedure is over 90 hours, and this spacecraft has three batteries. This means reconditioning operations must start more than 11 days before the first eclipse (no later than mid-February for the vernal-eclipse season). [FLTSATCOM, 1986]

| Command Transmitted | Operations Action |
|---|---|
|  | Verify battery is on charge channel |
|  | Configure for battery discharge |
| Select channel automode | Verify charge channel to automode |
|  | Verify charge channel in trickle charge (initially charger may be in full charge for a short period before switching to trickle) |
| Disconnect channel battery charger | Verify channel battery charger is disconnected |
| Connect redundant AC source to battery |  |
| Disconnect redundant AC source |  |
| Start reconditioning discharge with auto stop | Begin reconditioning |
|  | Verify the channel battery current shows discharge |
|  | Monitor battery voltage, current, and temperature until discharge ends automatically (about 75 hours) |
|  | Verify recondition discharge status; wait 30 minutes. (If reconditioning discharge hasn't stopped and the battery voltage is less than 16 volts, transmit STOP RECONDITION DISCHARGE and verify the telemetry parameters.) |

**Table 15.7.    Reconditioning Procedure for the FltSatCom Battery. (Continued)** The time required to recondition one battery with this procedure is over 90 hours, and this spacecraft has three batteries. This means reconditioning operations must start more than 11 days before the first eclipse (no later than mid-February for the vernal-eclipse season). [FLTSATCOM, 1986]

| Command Transmitted | Operations Action |
| --- | --- |
| Switch channel to trickle charge | Verify that the channel is on trickle charge |
| | Verify the battery-current monitor shows approximately 0.24 A charge; wait ten minutes |
| Select automode for channel | Verify the channel is in automode |
| | Verify the channel is on full charge |
| | Verify the battery-current monitor shows full charge—about 2.1 A |
| | Monitor battery voltage and charge current to make sure it has automatically switched to trickle charge—about 18 hours |

# References

FLTSATCOM Orbital Operations Handbook. Vols. 1 &2.

Orbital Requirements Document Fleet Satellite Communications Program (FLTSATCOM). Aerospace Report No. TOR-0076 (6724-01)-2. Reissue B, 1*, 15 October 1986.

Larson, W. J. and J. R. Wertz. 1992. *Space Mission Analysis and Design.* Netherlands: Kluwer Academic Publishers.

Total Ozone Mapping Spectrometer (TOMS) Earth Probe (EP) Orbital Operations Handbook. 1994.

# Chapter 16

# Spacecraft Anomalies

Emery Reeves, *United States Air Force Academy*

16.1 Defining Anomalies
16.2 Resolving Anomalies
16.3 Planning for Anomalies
16.4 Case Studies

This chapter discusses what the ground operators and systems do when things go wrong with the spacecraft. As discussed elsewhere, the ground element exists to place the space element into orbit and conduct routine operations. It also exists to correct failures or aberrant behavior of the spacecraft. Although spacecraft don't fail often, they do fail occasionally, and a prime reason the ground element exists is to correct the failure and restore operation. Some people assert that impending failure may be anticipated by close attention to data trends, but most failures are unanticipated and catch the ground crew by surprise. The ground element's ability to react to the unexpected is an important measure of the mission operations manager's competence and indeed the adequacy of the whole operations element.

## 16.1 Defining Anomalies

Anomalies are extraordinary spacecraft events or occurrences. These include out-of-tolerance measurements, off-nominal telemetry points, and in a larger sense, any unexpected or abnormal behavior. Table 16.1 gives some examples.

An *anomaly* is anything that is wrong, or seems wrong, or is not quite right. In classifying an occurrence as an anomaly, suspicion should be the rule. It's better to be too worried than to overlook a seemingly minor occurrence that later kills a mission. There is no such thing as a glitch. An out-of-tolerance data point is an anomaly and has a reason. The only question is how much effort you can afford to put into understanding it.

**Table 16.1.    Examples of Anomalies.** Any data or performance that is not nominal or is unexpected is an anomaly.

| |
|---|
| A temperature or a group of temperatures too high or too low |
| An attitude-error signal larger than allowable or expected |
| No activity on an attitude-error signal |
| Received signal strength too high or too low |
| No indication of appendage deployment |
| Unexpected configuration or uncommanded configuration change |

This chapter deals with Earth-orbiting spacecraft that are either under ground control or can be accessed rapidly by ground control. It's also based on the fact that anomalies don't occur very often even under the very general definition provided above. A well designed spacecraft may hiccup once every few months, but if it does so more often, the design is seriously flawed. We also need to distinguish between launch anomalies and anomalies that occur after the spacecraft has been placed on orbit and checked out. Launch is a particularly traumatic event. Environmental stress on the equipment and emotional stress on the launch and operations crew are very high. Many anomalies occur during launch. Fortunately, the launch crew is conditioned to react to launch anomalies, and the operations crew is usually augmented with engineering support to handle them effectively. Even though anomalies occur with greater frequency during launch, the techniques used to investigate and resolve them are no different from those used later in operational life.

This chapter also assumes the operations crew has engineering support. When an anomaly occurs, the initial actions always fall on the operations crew. However, the detailed investigation of the anomaly is best conducted by people who aren't the minute-to-minute operators of the spacecraft. Engineering support—and in some cases, outside people such as the spacecraft designers—should do the detailed analysis of the anomaly.

## 16.1.1    The Fundamental Rule

*The fundamental rule in anomaly resolution is that an anomaly, no matter how complex, has one and only one cause.*

Many times we can synthesize failure scenarios involving multiple events that explain a set of anomalous data. Unless the scenario reduces to a single event, it is wrong. Multiple failures, no matter how attractive, don't occur unless they cascade from a single root cause.

## 16.1.2    Categories of Anomalies

The most important way to categorize anomalies is by criticality. Table 16.2 gives standard criticality definitions. The most severe anomalies can destroy the spacecraft or cause loss of mission. Less critical anomalies cause out-of-tolerance performance, loss of functional redundancy, momentary or partial loss of function, or simple annoyance. Time is also an important consideration. By the criticality definition, an anomalous condition that takes a week to kill the spacecraft has the same criticality as one that can destroy the machine in five seconds. However the danger associated with the shorter response time (not to mention the panic level) is clearly much higher. Table 16.3 considers possible reaction times and ability of a trained ground crew to respond. Even the best of ground crews typically can't react in a few seconds. Such reaction is possible only when we've loaded and pre-approved a canned reaction plan and the operator's finger is on the button. Even then, in most cases, the reaction is delayed. On the other hand, a trained crew should be able to render a spacecraft safe within an hour, and several days would be "fat city."

**Table 16.2.   Criticality Definitions.** Criticality is a key measure of an anomaly's importance and a guide to the effort we ought to expend in preparing for an event.

| Level | Category Description |
| --- | --- |
| 4 | Loss of complete mission. Single-point failure. Loss of life. |
| 3 | Degraded payload performance outside specified limits. Total loss of operational mode(s). Loss of channel(s). Major injury. |
| 2 | Loss of the payload's or spacecraft's functional redundancy within specified limits. No loss of modes or channels. Minor injury. |
| 1 | No mission impact. No-effect failure conditions. No loss of functional redundancy. |

Table 16.4 lists dangerous types of anomalies. We can use this list as a framework for evaluating the seriousness of an anomaly and in designing techniques (either onboard or on the ground) to protect the spacecraft. Most high-reliability spacecraft are designed to detect these dangerous anomalies (or some of them) and change to a safe operating mode—a *safe haven*. Table 16.5 presents typical safe havens.

The simplest of anomalies involve only a single subsystem. Compiling data and finding appropriate people to analyze the data is direct and readily done. Of single-subsystem anomalies, propulsion or attitude control are the most difficult. Propulsion anomalies are difficult because they tend to result in excess momentum or explosion; attitude-control anomalies are tough because of the complexity of closed-loop operation. Anomalies which have symptoms that cover multiple subsystems or diverse technical areas are harder to work. A system engineer or system manager—someone with a broad technical background—must interpret symp-

**Table 16.3. Reaction Time.** Examining required reaction time for anomalies helps define where we can best apply planning and practice.

| Category | Definition | Examples | Comments |
|---|---|---|---|
| **Very rapid** | Response required in a few seconds or a few minutes | • Thruster stuck in on position<br>• Runaway chemical reaction<br>• Runaway propellant temperature or pressure | Nearly impossible for ground crews to correct. Automatic safing of the spacecraft is the preferred approach. Propulsion and attitude control are main sources. |
| **Rapid** | One hour to a few hours | • Temperature near upper limit and climbing<br>• Battery-cell voltage approaching lower limit<br>• Excessive power drain | Well-trained crew with good technical data and crisp decision process should be able to safe the spacecraft and prevent mission loss |
| **Quick** | One day | • Spacecraft thermally unbalanced with slow rise or fall in temperature<br>• Power imbalance with failure to fully recharge battery<br>• Spacecraft in a damaging orbit (belt flying for instance) | A day should be long enough for a trained crew to safe the spacecraft and prevent mission loss. It's not long enough to scramble a crew from scratch or to get technical data that aren't prepared. |
| **Leisurely** | One week | Similar to those for quick reaction time | A week is enough time to marshal people and find data |
| **Steady state** | Not time critical | A spacecraft that is in safe haven and awaits further actions | No immediate action needed |

**Table 16.4. Dangerous Anomaly Types.** This table lists types of anomalies that can cause loss of mission.

| Type | Description |
|---|---|
| **Momentum** | Anomalies that accelerate the spacecraft or spin it up fast enough to produce destructive loads |
| **Temperature** | Anomalies that cause components to overheat or get too cold |
| **Power** | Anomalies that cause excessive power or energy drain |
| **Command** | Anomalies that block out the command system or prevent acceptance of corrective commands. Telemetry blockouts might also be included. |

toms and allocate actions (detailed investigation or analysis) to subsystem people or technical specialists.

Sometimes an anomaly results from properly operating equipment; it's anomalous only because ground operators don't expect or understand it. A useful term for this type of anomaly is *pseudo anomaly*. Early in the life cycle of a new system, many (perhaps most) anomalies occur because of improper expectations and are

**Table 16.5. Safe Havens.** Operating modes or states that are inherently benign are called safe havens.

| Type | How detected | Actions |
|------|-------------|---------|
| **Momentum** | Attitude-error signal out of limits Thruster temperature too high | Close propulsion isolation valves |
| **Temperature** | Temperature measurement too high or too low | Turn off equipment Turn on heaters |
| **Power** | Voltage too low | Turn off loads |
| **Command** | Command not received in timed interval | Turn on command receivers Couple receivers to omni antennas |

thus pseudo anomalies. But anomalies resulting from incorrect commanding or procedures are true anomalies, even though the anomalous data may come from good equipment operating properly. They may be as dangerous or damaging as an equipment failure.

# 16.2  Resolving Anomalies

Table 16.6 lists the steps for resolving anomalies. First, evaluate the danger to the spacecraft. If the operating mode or the symptoms indicate danger to the spacecraft's health, command it into a safe haven. Most of the time, spacecraft are operating in relatively benign modes, so you can work many anomalies without changing operating modes. If in doubt, however, abort the current operations and seek safe haven.

**Table 16.6. Steps for Resolving Anomalies.** This table suggests a common-sense approach based on experience.

| Step | Action |
|------|--------|
| 1 | Safe the spacecraft. Get it into a safe operating mode so you have time to think. |
| 2 | Get all the data, even data from unrelated subsystems and sources. |
| 3 | Establish accurate timing. |
| 4 | List possible causes. Canvas all sources for candidates. |
| 5 | Analyze the data, examine the possibilities, eliminate possibilities until you find the culprit. Validate your answer, by analysis if necessary but preferably by experiment. |
| 6 | Figure out how to fix the problem, check out your fix, and do it. |

Having determined the spacecraft is safe, collect all data concerning the anomaly and establish an accurate timeline. Organize an anomaly team and start searching for explanations of the data. The search for anomaly causes is a form of organized invention (also known as a "group grope"). The anomaly captain or team leader convenes knowledgeable people and then canvasses them for possible scenarios. Analyze and investigate these scenarios in detail until you identify the root cause of the anomaly. For complex anomalies this is a lengthy process. An important part of solving anomalies is verifying the answers. This verification isn't always possible, but even if you can't do it completely, investigate and verify your answers as much as possible.

### 16.2.1    Safe the Spacecraft

Clearly immediate action is necessary when failure to act will cause loss of mission or endanger people. Any team that operates a space system should have prepared plans *(contingency plans)* for anomalies that require immediate action. The table listing categories of dangerous anomalies (Table 16.4) is a good starting point for preparing contingency plans. Given infinite resources, you'd like to list everything that can go wrong and what to do. But given finite resources, you must identify the top contenders and prepare contingency plans for these. A Failure Modes and Effects Analysis (FMEA) is also a way to prioritize contingency planning, but you need to augment the FMEA by identifying corrective action.

Anomalies occur about once for each seven months of spacecraft operation, as suggested by a recent survey of several operating space systems (See Tables 16.7 and 16.8). The data covers two years of operation of four separate systems (13 separate spacecraft). It includes two launches, one of which was unsuccessful, and one mission ending in failure.

**Table 16.7. Distribution of Spacecraft Anomalies by Criticality.** Criticality levels are defined in Table 16.2. This table includes 13 spacecraft and two years of operation.

| Type of Anomaly | 1 - No Mission Impact | 2 - Loss of Redundancy | 3 - Degraded Performance | 4 - Loss of Mission |
|---|---|---|---|---|
| Number of Anomalies | 4 | 11 | 4 | 2 |

**Table 16.8. Spacecraft Anomalies by Subsystem or Element.** The same anomalies shown in Table 16.7 are distributed by subsystem. Note lack of structural anomalies.

| Location of Anomaly | Launch Vehicle | Electric Power | Data | Payload | Attitude Control | Thermal | Ground |
|---|---|---|---|---|---|---|---|
| Number of Anomalies | 1 | 2 | 9 | 5 | 2 | 1 | 1 |

## 16.2.2    Get All the Data

Spacecraft-status telemetry is by far the most important source of anomaly information. Table 16.9 summarizes types of telemetry.

**Table 16.9. Types of Telemetry.** This table reflects the classical separation of data into discrete (bi-level) and analog types.

| Type | Description | Examples |
|------|-------------|----------|
| Measurement | Value of a signal typically 8 bits. Basic signal may be analog or digital but is digitized for telemetry transmission. | Temperatures, pressures, signal strength, voltages, currents, and attitude-sensor output |
| Bi-level | Condition or state of a bi-level signal such as a switch | Relay state (energized/de-energized) Switch state (on/off) (closed/open) Microswitch |

From an anomaly standpoint, bi-level measurements tell us the state of the spacecraft, and other measurements tell us the value of variables within the system. Although most anomaly investigations center on the behavior of the variable signals (such as voltages, temperatures, and error signal values), we must also know the system's configuration (state). Ask these sorts of questions: "Which side of redundant equipment are we using? What mode are we in? What is our source of voltage? Which command channel are we using?" The answers to these questions affect how you'll interpret the data. Sometimes, telemetry measurements depend on operating mode; that is, the designers assign different signals to a specific telemetry word depending on operating mode. If the operating mode on the spacecraft differs from the mode used in the ground station, serious confusion can result.

Several measurements tell us much about the system even though they may not be directly concerned with a particular anomaly. These include supply voltages, temperatures, and rf signal strength. Nearly all systems telemeter voltage and current for the primary power supply—sometimes in expanded scale. From these measurements, you can often determine when transient events occur, such as when a switch closes or a new load is turned on. You can also verify power drain and compare it with the equipment you believe is activated. Bus voltage and current may also indicate spacecraft attitude. Secondary voltages usually are also telemetered but secondary currents are only rarely telemetered. Secondary voltages can show equipment operating status, or if out of tolerance, can indicate where a failure is.

You can use temperatures to infer operating status of electronic equipment. Deviations from expected temperature can also indicate mechanical damage (particularly to insulation blankets). Propulsion-component temperatures can reflect leaks by abnormal cooldown rate or low overall value. On well-designed spacecraft, actual temperature profiles should be within about 5° C of projected levels. Regard deviations beyond this level as abnormal.

Rf signal strength is usually measured on the ground and in the spacecraft. The spacecraft measurement is usually telemetered. These measurements provide information about the spacecraft attitude and can reflect the antenna's state of health. You can usually measure spin rate by observing the periodic variations in signal strength. Additionally, you can infer antenna blockage by observing the variation of signal strength within a spin cycle. You can sometimes establish the exact timing of mechanical damage by fluctuation in signal strength.

Although not commonly used, direct observations of the spacecraft can provide anomaly information. Such observations are possible using ground cameras and ground radars. Optical resolution depends on range and is therefore most useful for spacecraft at low altitudes. Routine film (or video) of launches can provide information about launch anomalies.

### 16.2.3    Establish Accurate Timing

A complex anomaly can have various symptoms. Out-of-tolerance measurements can appear almost simultaneously in a number of subsystems or components, and the plethora of data can be overwhelming. Establishing the exact timing of occurrences will be central to your sorting out cause and effect.

Telemetry measurements are usually sampled and formatted in a telemetry frame for transmission. The position of a given word in the frame tells when the measurement was sampled, and such timing can be important. Subcomutated measurements are sampled at multiples of the frame period; their position in the subcom frame also tells the sample time. The current trend toward packet telemetry tends to obscure event timing. Packets can be multiplexed with other data and transmitted with an intervening delay. Be sure to insert a precise time standard in the raw telemetry stream if you're placing the data in packets or multiplexing it with other data streams. The Space Shuttle's telemetry system is particularly bad about telemetry timing. By the time Houston gets through handling the data, it may be several seconds old and very difficult to interpret from a timing standpoint.

One of your first steps for investigating an anomaly is to establish a time base. Make a single person responsible for this time base who will correlate the various standards. All telemetry systems used today have time standards. Range telemetry at the Eastern Test Range provides an Inter-Range Instrumentation Group (IRIG) channel that establishes timing to Greenwich Mean Time (GMT) with microsecond accuracy. The Goddard Spaceflight Tracking and Data Network (GSTDN) and the Satellite Control Facility (SCF) have similar standards. However, if the telemetry isn't received in real time, these accurate standards may be lost. Normally the spacecraft telemetry stream gets its timing from an internal oscillator. One of your challenges in establishing event timing is correlating this internal oscillator to GMT or some other universal standard.

Some types of telemetry record events that occur between sampling intervals. Thruster firings are a prime example. Firings of thrusters for attitude control are often quite short (10 to 50 msec typical). Telemetry sampled at 0.5 sec will miss most thruster-firing pulses. For this reason, thruster-firing indications usually remain in memory from the time of occurrence until telemetry reads it out. Sometimes, the thruster-pulse length is quantized, so telemetry provides pulse length as well as occurrence during the sampling interval.

### 16.2.4    List Possible Causes

This key step is essentially an inventive process. You can poll the people working the anomaly for possible causes and then tabulate and discuss them. If you do the polling in an interactive meeting, you'll normally raise more questions than answers. In this process, your role as anomaly team leader is to keep the discussion focused on finding the important possible causes without discarding any likely candidates. As in any group interaction, some people want to adopt favored explanations and others tend to go off on tangents. Organizational conflict or an attitude of "who's the guilty party" may lead to recriminations or other destructive interpersonal behavior that gets in the way of orderly discussion. Structuring the discussion to keep the emphasis on compiling possibilities is sometimes a daunting management challenge. You'll need to take detailed discussion off line and, as the investigation progresses, winnow down the list of possibilities until only the most likely are left.

### 16.2.5    Analyze Data and Find the Cause

The general approach to anomaly analysis is to gather the data, establish event timing, categorize observations by subsystem or equipment group, postulate failure modes or scenarios, look for confirming or contradictory evidence, discard scenarios until a base set is reached, and devise tests or experiments to further prune the scenario set if necessary. While gathering data and categorizing observations, people usually ask more questions than they answer. Typical questions are: "What was the exact mode or setting of signal-flow switches? How is the spacecraft wired? What exactly does a particular component look like? How are the wires (insulation, plumbing) routed?" Sometimes, you'll find answers in the telemetry data, but often you'll need information from the spacecraft-build chronicles.

Anomaly data includes all pertinent telemetry data, command records, system state diagrams, records of personal observation (including operator and specialist logs), and range data such as signal-strength readings, photographs, and radar data. Normally, you'll investigate anomalies off line with people who aren't involved in day-to-day operation of the spacecraft. This is not to say that you shouldn't use operators or draw on operator experience. But you should investigate as much as possible in parallel with continuing spacecraft operation, not within it. A separate work area is almost mandatory and you'll need some way to log and keep track of data. In a well-ordered operation these preparations are all in place as part of the setup, so you'll simply invoke them whenever you need them.

Often the recording of telemetry-data streams doesn't support anomaly investigation. Typically the senior operator has an event display—a display of telemetry measurements that change in value—and selectable page displays that show measurements by subsystem or other logical grouping. Often the page displays are selectable and sometimes are supported by alert lights or switches that signal out-of-tolerance conditions and allow page call-up by pressing the indicator. Other operators have similar displays that may include time records (strip charts or computer display monitor equivalent). These displays have measurements reduced to engineering units and are quite general and powerful. However, they aren't recorded as presented and hence aren't usually available for anomaly investigation. Only the raw telemetry stream is recorded. Often getting the telemetry off the primary record and into a usable form (delogging) is time consuming and difficult. Clearly, capturing the raw-telemetry record is an important step in accumulating anomaly data. Methods of delogging the data and providing engineering output are also important.

Aerospace hardware is manufactured according to carefully prepared and controlled engineering data. These data include drawings, specifications, and procedures. Schematics and signal-flow diagrams are sometimes used in design but aren't usually controlled or used to construct hardware. For anomaly investigation, however, you'll find schematics and signal-flow diagrams are of utmost importance, whereas fabrication drawings are much less useful. Wire lists (connection data for intercomponent wiring) are another matter. These are typically formal engineering data used to make and test wiring harnesses. They're extremely useful in anomaly investigations.

True manufacturing data sometimes can help your anomaly investigation because it records how a component was built. Manufacturing and test-data packages typically include manufacturing travelers, inspection records, and test records. Sometimes, an assembly whose manufacture is particularly sensitive is designated as a special attention item. These items include single-point-failure parts or assemblies and any areas whose performance depends on precise assembly. A special attention item typically has a control plan prepared for it which delineates extra inspections, special tests, and other precautionary measures. You can use records of these measures, normally included in the manufacturing data, as evidence of actual construction. Photographs of the component during manufacture are an even better source of as-built information. In fact, photographs are so valuable that I started the practice of including 8.5×11 inch color photographs of each assembly in the manufacturing-data package.

To find a pattern in an anomaly indication that may reveal the root cause, you might construct a *failure tree*—a systematic way to list possible causes of anomalous data. Presented with a single out-of-tolerance reading, you can develop a list of failure modes that might have produced it. Because the failure modes are interrelated, a useful way to depict them is as an organizational block diagram or tree structure. Initially, your tree may be very large, but using additional observed data may allow you to eliminate or prune many of the possible causes from the failure tree. Systematically list everything that might have gone wrong; then, eliminate suspects until only one remains.

It would be nice if an anomaly investigation could come up with only one plausible cause of the observed data and everyone would agree that's what happened. Most of the time you can. Several keys to reaching a conclusion are: (1) Identifying a single root cause. Scenarios that require multiple events are incorrect. (2) Survival under challenge. The postulated root cause should hold up under examination or challenge by everyone involved. (3) Support consensus. The anomaly team should be able to agree on the culprit. In a lot of cases, you can pin down the exact semiconductor that has failed. In nearly all cases, you can narrow the cause to the switchable assembly causing the difficulty. In a few cases, you won't get a unanimous opinion, so you'll issue minority reports.

If you can't pin down the exact root cause from the data available, can you devise an experiment to verify or refute any of the candidates? Experiments can sometimes be done on the flight vehicle, but the preferred method is to use a test article such as a qualification spacecraft or component, or a subsequent unit in the production cycle.

### 16.2.6    Determine the Corrective Action and Carry It Out

After you've figured out what's wrong, how do you fix it? The simplest thing to do is to replace the broken element. High-reliability spacecraft are almost completely redundant, so you can replace almost any failed component. Whenever a direct replacement isn't available, you can sometimes use a different operating mode to accomplish the mission. In either event, a series of commands to the spacecraft and, in some cases, reconfiguring the ground or user elements are required. This means preparing a command plan or a reconfiguration plan and processing it through the operations segment of Ground Control.

In devising a scheme to correct an anomaly, you may need to experiment. If so, try to experiment with test hardware or a simulator rather than the flight spacecraft. If you need to experiment with the spacecraft, you'd better be careful not to foul it up by sending destructive commands. Most anomalies involve extraordinary operating modes or equipment sets, so you may not understand exactly what a command sequence will do. Be cautious.

In extracting your system from an anomaly, you're often going from a safe haven to a more dangerous operating mode. If you can reconfigure using successive safe modes, do so. Take your time. Verify each step before taking the next step. If you must pass through a dangerous configuration, consider having an operator prepared to abort the sequence and safe the spacecraft. (This is called a "dead man switch" after the practice on the Manhattan Project of having a person suspend a weight over the nuclear pile. If the pile went critical, the dying person would drop the weight which would disrupt the pile and prevent more serious damage.)

## 16.3  Planning for Anomalies

To work an anomaly you need a set of competent people and a good leader. A complete anomaly team would have engineers responsible for each subsystem; system engineers who understand orbit mechanics, rf-link characteristics, controls analysis, thermal design, structural design, dynamics, and computer hardware and software; and design-integration engineers who know how the spacecraft is built. Table 16.10 summarizes crew composition.

**Table 16.10. Crew Composition.** The composition and organization of an anomaly team affect the team's efficiency and ability to reach conclusions quickly.

| Members of Team | Description of Team Members |
|---|---|
| Team Leader or Anomaly Captain | Senior person having enough overall knowledge of the system and management ability to partition the work, assign actions, evaluate progress, and control the investigation |
| System Engineer(s) | Person(s) who know orbit mechanics, rf-link characteristics, controls analysis, mechanical design and dynamics, and computer resources |
| Design-integration Engineers | People who know how the spacecraft is put together. Mechanical-design integration involves location, mounting, and routing. Electrical-design integration involves wiring and signal characteristics. Thermal-design integration involves temperature ranges, heat dissipation and flow paths, and thermal-control methods. |
| Subsystem Engineers | Specialists in the spacecraft's individual subsystems |

Part of developing a space system is preparing an operations manual which is the main reference for the operations crew and their engineering support. Materials for training the operations crew may supplement the operations manual. The engineering-support team will have prepared specialist manuals to help analyze flight data. They should also have a data center that contains spacecraft drawings, schematics, and wire lists. They also need configuration-management indices that relate drawing numbers to specific components. Even relatively simple spacecraft may involve 10,000 drawings or more. Despite this mass of data, a good data center should be able to locate any drawing in the system within a few minutes. You'll also need telemetry and command lists (books) or a database. These lists should show how the telemetry confirms command receipt and execution. You can use telemetry-calibration plots and schematics showing the telemetry-signal conditioning to interpret the data. Table 16.11 lists documents, manuals, and engineering data useful for anomaly investigations.

**Table 16.11. Documents for Anomaly Investigations.** Complete and accurate documents must be available for an anomaly team to operate efficiently.

| Document | Description |
|---|---|
| Operations Manual | A manual describing the spacecraft and its operation. Typically several thousand pages long and containing detail down to the block-diagram level. May summarize operational procedures and contingency plans. |
| Training Manual | Materials used to train operators |
| Command and Telemetry Documents | Command and telemetry lists sorted by number, subsystem, alpha-numeric designator, and ground-system designator. May be augmented by expected telemetry response for each command, telemetry-calibration files, or schematics for the telemetry-conditioning electronics. |
| Procedures | Step-by-step procedures used by spacecraft operations |
| Specialist Manuals | Manuals or notebooks prepared by subsystem and system engineers to help them understand and analyze system performance |
| Schematics, Wire Lists, Signal Flow Diagrams | Aid in tracing signal flow through the spacecraft |
| Engineering Data | Drawings, specifications, and procedures. Indexed by indentured drawing lists and configuration guides |
| Manufacturing Data | Usually available from the spacecraft manufacturer. Data packages should include manufacturing and test data for each component and assembly, plus integration and test data for the spacecraft. May include as-built photographs. |

You must train the engineering-support team for anomaly investigation. For launch-phase support and for first launch in particular, you can draw engineering support from people in spacecraft design and test. For continuing operations, engineering support may be by contract. Prelaunch training should involve rehearsal for the entire crew and should have both nominal and off-nominal (anomalous) sequences. Sometimes a program will have the foresight to build a simulator which you can use to model spacecraft performance and rehearse realistically. If a simulator isn't available, you can use recorded or canned telemetry streams to educate the team on what to expect. A well-conducted rehearsal exercises the entire crew—all subsystems and, because multiple shifts are usually involved, all shifts.

The training of engineering-support people during continuing operations is pretty much a hit-or-miss proposition. For some systems, people are rotated through sessions with a simulator in which anomalies are programmed and the team response is evaluated. However the more normal situation is to use engineering-support people to analyze trends or do other make-work jobs while not resolving anomalies. The infrequency of anomalies tends to make even diligent support people rusty, so periodic rehearsals are a good idea.

Practically any form of operational gaming of a space system will reveal a set of anomalies or equipment failure modes that are both likely to occur and serious. Prudence requires that you prepare plans for these occurrences (contingency plans). Some ways to identify contingency candidates are:

1.  Identify what to do in case any box fails. This is relatively simplistic but it's a start. The next level below the box level is to plan for each switchable assembly.

2.  Use the Failure Modes and Effects Analysis (FMEA) to identify critical or probable failure modes.

3.  Use operational procedures and identified decision points as likely sources.

4.  Use people's experience. If there is a corporate database, review it for propensities to failure.

The standard operator displays are event screens and system or subsystem pages. The standard medium is a computer monitor, and the displays are normally high-resolution color. Telemetry measurements are shown in engineering units and changes or out-of-tolerance conditions are sometimes annotated in color or by a blinking marker. Operators can usually call up (by push button or by keyboard) various display pages so they can rapidly scan the spacecraft's status. Back-room displays are similar to the operator displays but usually lack command capability.

Anomaly investigators also use dynamic displays—strip charts or history plots. These plots are invaluable in examining control-system performance.

A well-prepared anomaly team also has general-purpose tools available for analysis. These tools include computers with standard application programs and special-purpose simulation programs for on-the-spot analysis. Table 16.12 summarizes some of the useful analysis tools.

**Table 16.12. Analysis Tools.** The ability of an anomaly team to function depends on access to usable analysis tools.

| Analysis Tool | Description |
|---|---|
| Orbit Mechanics | Any ground station worth the name has a powerful, general-purpose, orbit program. In addition, the anomaly team should have quick-and-dirty, conic-fit programs or tools to rapidly assess orbit changes. |
| Controls Analysis | A method of sanity checking the attitude-control response to known transients (appendage deployment for instance) and unknown disturbances |
| Power Profile | Either a simulation or tabulation of power vs. equipment complement |

## 16.4   Case Studies

### 16.4.1   Anomaly Case Study #1: FLTSATCOM 5

FLTSATCOM 5 was launched on August 6, 1981. The engineering-support team for the launch was deployed at the Eastern Test Range (ETR), the Satellite Control Facility (SCF), Sunnyvale, CA), and Space Park (TRW contractor plant in Redondo Beach, CA). Almost all members of the team had worked on previous FLTSATCOM launches.

The Air Force's Satellite Control Network collected telemetry data and relayed it to the SCF, where it was decommutated and displayed for the operators and the engineering-support team. Coverage was almost complete except for a small section of the parking orbit over the eastern Atlantic and Africa, when the spacecraft and the Centaur booster were out of sight of ground stations. Although telemetry from this section of the parking orbit was not relayed to the SCF, an ARIA aircraft orbiting over the Atlantic recorded the data. During this uncovered section of the orbit, the spacecraft and Centaur passed over the equator, and the Centaur executed its final burn, putting the spacecraft in a transfer orbit to geostationary orbit. The Centaur then oriented the spacecraft for its orbital-injection burn and released the spacecraft. After release, the spacecraft fired thrusters to spin-stabilize its attitude. A short time after spinup, the spacecraft came into view of a ground station, and telemetry started to arrive at the SCF [FLTSATCOM OOH].

For the previous launches, which had all been nominal, the arrival of telemetry at the SCF was signaled by a rise in received signal strength and increased activity on the monitors. This activity was garbled at first but rapidly cleared to the expected pattern as the decommutator and the telemetry stream synchronized. For FLTSATCOM 5, the readouts remained garbled, so anomaly action began. A precanned contingency plan existed for resetting the spacecraft equipment converter. Because the converter was the key component in the spacecraft's fault-protection design, its upset was the most probable cause of the observed data. The contingency plan was executed, and the telemetry stream locked up.

Initial telemetry showed that the spacecraft was spinning and that the appropriate thrusters were warm (indicating they had fired). The spin-attitude telemetry was not nominal, raising the possibility that mechanical damage had altered the moments of inertia. Telemetry from the electric-power subsystem was seriously anomalous. Battery-charge indications were fluctuating violently, as were main bus current and voltage. Plans were immediately drafted to connect the batteries directly to the main bus, thus bypassing the onboard charge-control channels in favor of direct ground control of the batteries. After these actions, the power-subsystem's telemetry stabilized, but there were still large fluctuations in main bus and battery currents. The subsystem engineers were canvassed to explain the data. They concluded the solar arrays were damaged near the solar-cell strings used to charge the batteries. It was even possible to map the exact location of damage to the arrays using wiring data the subsystem engineer had included in his specialist notebook.

With the spacecraft in a more-or-less safe and stable condition, it was possible to review the data, organize an anomaly team, and plan future actions. The top priority was planning the spacecraft apogee-kick motor's (AKM) orbital-injection burn. The spacecraft was in a highly elliptical orbit, passing through the Van Allen belts twice on each revolution. Furthermore, the orbit was decaying with each perigee passage, so injecting into final orbit was somewhat urgent. Stability during AKM firing required an increase in the spacecraft's spin rate from the initial 30 rpm to a final value of 60 rpm [FLTSATCOM OOH]. This was a normal command sequence, so operators decided to do it. The spacecraft responded normally up to about 45 rpm; there it went into nutation and commanding was suspended. But the nutation damped out, so commanding was resumed up to 60 rpm.

Injection into the proper final orbit required knowing the spin attitude. Because the spin-attitude telemetry was anomalous and mechanical damage to the spacecraft was highly probable, the adequacy of the normal methods for measuring and controlling spin attitude was questionable. To bound the accuracy of spin-attitude knowledge, people began a dynamic analysis at the spacecraft contractor's plant. This analysis, completed in a matter of hours, showed that the spacecraft's spin axis might be misaligned from the AKM axis by as much as 15°. If this were the case, the velocity increment produced by the AKM would be grossly inadequate, and 24-hour orbit wouldn't be possible. To avoid this possibility, the anomaly team devised an alternate targeting strategy. By aiming the spacecraft for injection at an inclination of 7° instead of the nominal 2°, the spacecraft's final velocity would be close enough to a 24-hour period so the liquid propellant on the spacecraft could make up the deficit. If, on the other hand, the AKM achieved full performance, the final velocity would exceed that required for 24-hour orbit but would still be within the liquid system's correction range. The team carried out this strategy by retargeting the spacecraft to 7° inclination. They achieved almost full efficiency of the AKM. After correction, the orbit's inclination was 5.4° and it had a 24-hour period.

After orbital injection, the spacecraft appendages were deployed. The solar arrays, despite damage, deployed successfully. The receive antenna and the transmit antenna failed to deploy despite execution of contingency plans aimed at spinning the spacecraft fast enough to dislodge them if they stuck. Test of other spacecraft equipment showed it all to be working correctly except for failure of all receive channels that used the UHF receive antenna and reduced gain of the UHF transmit antenna. Patterns of the omni antenna showed asymmetry, which was attributed to a bent mast on the UHF transmit antenna.

Failure analysis after the fact revealed that a high-energy event had occurred during booster-powered flight. Synthesized failure scenarios revealed that the inner face sheet of the aerodynamic shroud had explosively delaminated. The face sheet hit the solar arrays, the receive antenna, and the transmit antenna. The receive antenna, although broken at its base, remained attached until the spinup maneuver, when it broke free at 45 rpm.

### 16.4.2 Anomaly Case Study #2: FLTSATCOM 1

FLTSATCOM 1 was launched on February 9, 1978. By all indications, the launch was nominal until about 2:00 A.M. on February 10 when the TT&C subsystem engineer informed me that the telemetry was indicating a command count of zero. Because the command link was encrypted, and proper operation of the decrypter depended on knowledge of the command count, this was disturbing. We had drafted a contingency plan for enabling the backup command channel, so we used it. The backup channel responded properly, so we knew we had command capability and were in no immediate danger. But we still didn't understand the anomalous command-count telemetry. We pulled the schematics and examined the circuitry. No explanation suggested itself. We called the engineering-support office at Space Park and asked them to find the unit engineer for the command unit (the decrypter interface box). Although we had prepared a telephone list of responsible design engineers (RDEs), the command unit's RDE had recently moved and didn't have a phone. We dispatched a runner who got the RDE to a neighbor's phone at 4:00 A.M. He couldn't explain the data. At this point, I ordered the engineering-support office at Space Park to pull in a test crew and run a test on the qualification spacecraft. This involved bringing the test crew in (about a dozen people), writing a test procedure, and doing the test. It was complete by 8:00 A.M. and showed that when the uplink rf-signal strength dropped below a threshold (a condition that occurred each time we changed the ground station), the telemetry indication of command count reset to zero. But the command count in the decrypter didn't reset, and on the next successful command, the telemetry would reset. What we were seeing was proper operation of the equipment, so we properly classified this anomaly as a pseudo-anomaly.

### 16.4.3 Anomaly Case Study #3: FLTSATCOM 2

FLTSATCOM 2 was launched on May 4, 1979. On the launch of FLTSATCOM 1 during powered flight, we had observed a single telemetry measurement of high bus current. On FLTSATCOM 2 there was an extended period (approximately ten seconds) of high current drain (10–20 amps). It was frightening, but the spacecraft seemed to survive and checked out perfectly on orbit except for a telemetry measurement of solar-array temperature.

An off-line anomaly team was designated to analyze the data. It determined that an exposed steel deployment cable running behind the folded solar array could vibrate and strike exposed connections on the array. The spacecraft was designed to clear inadvertent short circuits by dumping battery current through the short and burning it out. This had apparently happened and vaporized the cable. Experiments duplicated both the magnitude and duration of the short. Furthermore, clearing the short would also burn out a ground wire in the solar array's slip-ring assembly, which provided the ground for the array-temperature measurement. Thus, the investigation was able to trace disparate data to a single root cause—demonstrating our fundamental rule.

## Reference

FLTSATCOM Orbital Operations Handbook (OOH), Vols. 1 & 2.

# Chapter 17

# Interplanetary Space Mission Operations

David W. Murrow, *Jet Propulsion Laboratory,*
*California Institute of Technology*

If you become the mission operations manager for an interplanetary mission, you should understand basic elements of the interplanetary environment and how they affect mission operations. Because this book emphasizes Earth-orbiting missions, this chapter focuses on differences between those missions and interplanetary missions. After reading this chapter, you'll understand these differences and how operations change throughout an interplanetary mission. Where appropriate, "rules-of-thumb" and tables help you see how different mission types or planets affect operations.

Many exploratory missions to other planets have been launched since Mariner 2 travelled to Venus in 1962. The United States, the former Soviet Union, and the European Space Agency have combined to visit all but one of the planets, as well as some comets and asteroids. Interplanetary exploration will continue, but with smaller, more focused spacecraft and lower mission-operations costs. Plans include the NASA-sponsored Discovery program for small missions to the planets and international missions to Mars, Saturn, and other targets.

Originally, each interplanetary mission—Pioneer, Mariner, Voyager—was the first chance to explore a new planet. But expanding costs now make missions appear to be the last chance to visit a planet. This is particularly true of ambitious missions to the outer reaches of the solar system, such as Galileo and Cassini. Future planned missions should be neither the first nor the last chance, but one chance in an ongoing program of exploration.

Both first-chance and last-chance missions strive hard for maximum performance. Scientific payloads tend to be comprehensive, adaptive planning is desirable, and the use of resources is optimized. Optimizing leads in turn to operating with small margins in most spacecraft resources, so managing these resources requires a lot of labor. In this chapter we'll explore how optimizing affects operations.

First we'll discuss the key elements of interplanetary missions and detail the key differences between Earth-orbiting missions and interplanetary missions. Then, we'll look at mission operations activities throughout the mission. You can best use this chapter by becoming familiar with interplanetary objectives and architectures in Sec. 17.1. Then, compare the information in Sec. 17.2 to the rest of the book's material on Earth-orbiting missions. Section 17.3 is a generic plan that identifies typical activities throughout an interplanetary mission, so you can use it to plan the life cycle of your own mission.

# 17.1    Elements for Interplanetary Mission Operations

Tables 17.1 and 17.2 guide you to the key differences between interplanetary and Earth missions and show where you can find them in this chapter. They match the space mission elements in Chap. 1 and mission operations functions in Chap. 3.

## 17.1.1    Orbit: Types of Interplanetary Missions

For this discussion, we'll divide missions into five types: flyby, mapping, tour, landers, and probes. Table 17.3 summarizes the different types and some of their key characteristics.

*Flyby* missions have widely different spacecraft-target ranges during science gathering times. Flyby missions were among the first missions flown to other planets and have short periods of detailed observation compared to the time for the trip to the planet. Flyby missions have been useful for discovery or initial exploration of the solar system. A flyby will likely discover something at a distance which requires plans to change for the near encounter. An interplanetary flyby results in a gravitational impulse to the spacecraft, and we can string flybys together if the geometry allows, as in the Voyager "Grand Tour" of the outer solar system. Navigation to accurate flyby conditions is important because it creates a link between the downlink part of mission operations and the planning.

*Mapping* missions—most like Earth missions in their repetitive nature—have been used for more detailed exploration of the solar system. Examples include Magellan at Venus and the planned Mars Global Surveyor. They use orbital geometry much like that for Earth missions to view the entire planet's surface at similar

**Table 17.1. Differences Between Interplanetary and Earth-Orbiting Missions.** Each mission element differs based on the differences in mission subjects.

| Space-Mission Element | Key Differences | Where Discussed |
|---|---|---|
| Subject | • Lack of knowledge of target<br>• Users represent scientific community | Sec. 17.2.3 |
| Orbit | • Long cruise times<br>• Very high activity periods<br>• Interplanetary flybys and gravity assists<br>• Different coordinate systems which change with mission phase and subject | Sec. 17.1.1 and 17.3 |
| Space Element | • Historically small and negative resource margins<br>• Environmental extremes<br>• Lack of adequate solar power at outer planets | Sec. 17.2.7 |
| Launch Element | • Highly constrained launch window<br>• Higher launch energy required | Sec. 17.2.1 and 17.2.2 |
| Ground Element | • Different activity levels for different mission phases<br>• Very few capable antennas for all missions, causing scheduling conflicts | Sec. 17.3 |
| Mission Operations | • Range and light time delay<br>• Lack of knowledge of target<br>• Uniqueness of opportunities<br>• Time-critical operations | Sec. 17.2.1, 17.2.4, and 17.3.5 |
| Command, Control, and Communications | • Range and light time delay<br>• Very low data rates | Sec. 17.2.1 |

resolution. This similar geometry also allows for the same kind of operations. Differences from Earth-mapping missions include telecom links beyond Earth ranges and constraints imposed by solar illumination or solar conjunction.

*Tour* missions contain repeated gravity assists using the planet's natural satellites. The satellite flybys are used for trajectory shaping as well as for science gathering. Tour missions, such as Galileo at Jupiter and Cassini at Saturn, stay at their target for a long time, but the geometry is less repetitive than for a mapping mission. The repeated gravity assists cause us to operate the mission like several flybys in a row, but tour missions still focus on the primary body between satellite encounters. Also, the geometry for flying by a satellite changes very rapidly in a flyby mission, which makes navigation and timing accuracy important. To plan the mission, we must compromise between the objectives of a satellite flyby and our future objectives. For example, Cassini tour flybys of Titan begin at a high altitude to guarantee operators can control the spacecraft as it encounters the atmosphere. This desire to maintain the long-term goals of our mission conflicts with the desire to capture low-altitude, high-resolution imaging and aeronomy data as early as possible.

**Table 17.2. Key Differences in Mission Operations Functions.** Chapter 3 defines the differences in each of the 13 operations functions.

| Mission Operations Function | Key Differences | Where Discussed |
|---|---|---|
| Mission Planning | • Range and light time delay<br>• Adaptability to changing environment or updated knowledge | Sec. 17.2.1, 17.2.3, and 17.3.8 |
| Activity Planning and Development | • Range and light time delay | Sec. 17.2.1 |
| Mission Control | • Critical or unique commands | Sec. 17.3.5 |
| Data Transport and Delivery | • Range and light time delay<br>• Small or negative margins | Sec. 17.1.3 and 17.2.1 |
| Navigation Planning and Analysis | • Number of coordinate systems<br>• Critical maneuvers for flybys | Sec. 17.1.1 and 17.2.1 |
| Spacecraft Planning and Analysis | • Small or negative margins<br>• Environmental extremes | Sec. 17.2.1 and 17.2.7 |
| Payload Planning and Analysis | • Small or negative margins<br>• Lack of prior knowledge of target | Sec. 17.2.7 and 17.2.2 |
| Payload Data Processing | • Variety of users<br>• Need for level 0 data [*] | Sec. 17.2.4 |
| Archiving and Maintaining the Mission Data Base | • Need to archive level 0 data | Sec. 17.2.4 |
| Systems Engineering, Integration, and Test | • Small or negative margins<br>• Environmental extremes<br>• Long cruise phases before key engineering and science events | Sec. 17.2.1, 17.2.7, and 17.3.3 |
| Computers and Communications Support | • Range and light time delay<br>• Software uploads during mission | Sec. 17.2.1 and 17.3.3 |
| Developing and Maintaining Software | • Software uploads during the mission | Sec. 17.3.3 |
| Managing Mission Operations | • Workload difference between long cruise phase and brief encounter<br>• High turnover rate in team personnel due to long cruise phase | Sec. 17.3.3 and 17.2.4 |

[*] Project scientists usually want level 0 data.

To maintain a planned satellite tour, the spacecraft must undergo propulsive maneuvers with rapid turnaround before and after flying by the natural satellite. A maneuver before the flyby ensures the flyby is as close as possible to the prearranged science sequence. Following the flyby, targeting errors which result in velocity errors on the trajectory (because the satellite's gravity is different than

**Table 17.3. Types and Characteristics of Interplanetary Missions.** Each of the five mission types has a distinct profile for science gathering and time for cruising between planets.

| Mission Type | Examples | Science-Gathering Life | Cruise Duration | Key Effects on Mission Elements |
|---|---|---|---|---|
| Flyby | • Mariner to Venus, Mercury, and Mars<br>• Pioneer to Jupiter and Saturn<br>• Voyager to Jupiter, Saturn, Uranus, and Neptune | months per encounter, split between far and near encounter | months to 2 years between encounters | • Mission Operations: Short burst of high activity<br>• Space Element: Desire for large data storage<br>• Subject: Usually little prior knowledge |
| Mapping | • Magellan at Venus | >3 years actual | 15 months | • Mission Operations, Orbit: Repetitive<br>• Command and Control: Continuous tracking required |
| Tour | • Galileo<br>• Cassini | 22 months<br>4 years | 6 years<br>7 years | • Mission Operations: Generally high activity, with burst of higher levels<br>• Subject: Changing target from day to day |
| Lander | • Mars Pathfinder<br>• Viking | 30 days<br>>1 year | 9 months<br>9 months | • Mission Operations: Interplanetary daily cycle vs. Earth day, low navigation requirement<br>• Orbit: On surface during prime mission |
| Probe | • Galileo<br>• Cassini/Huygens | 90 minutes<br>3 hours | 6 years<br>7 years | • Space Element: Multiple spacecraft |

planned) are corrected quickly to keep their cost from becoming prohibitive. The cost grows with time from the flyby for two reasons. First, a velocity error propagates into larger position errors with time. Second, the energy change that results from flyby errors is better corrected near the satellite, where the differential energy change ($\Delta E \sim V\Delta V$) is larger due to the spacecraft's higher velocity.

*Lander* missions, such as Viking or Mars Pathfinder, explore surfaces while surviving for a long time after landing on a planet's surface. Their "buttoned-up" configuration, which makes cruise activities manageable, limits visibility during cruise. Usually the only required activity is a periodic checkout of the lander systems. The entry, descent, and landing of the lander require many activities, but we must preprogram them into the onboard sequence because of visibility and timing constraints. Post-landing activities are similar to post-launch activities for orbiters, except that navigation is required only once to determine the lander's location on the surface. The communications link is usually direct, with the planet's rotation modeled in the visibility algorithm.

*Probe* missions require the space vehicle to hit the target body. Probe missions are similar to landers in that they enter the atmosphere, but their planned operations occur only during the atmospheric descent. The spacecraft takes scientific data through direct sampling of the atmosphere or remote sensing of the surface. The cruise is typically very long compared to the encounter. A relay spacecraft receives, records, and protects the data before playing it back to the Earth.

For operations, the cruise duration is a key difference between mission types. As a mission operations manager, you must keep the teams engaged enough during the cruise to perform at their peak during the high-activity encounter. All interplanetary spacecraft cruise several months before operations, and some cruise up to seven years. Additionally, the proportion of time in cruise to time during encounter is very high for flyby missions and probe missions. Tour and mapping missions have more science-gathering time but still have relatively long cruises. For example, the Galileo and Cassini tour missions cruise for six and seven years, respectively. Following cruise, the prime mission lasts 2–4 years. Many of the mission operations functions during cruise are dedicated to planning for the encounter, but advance planning competes against the desire to retarget observations when we discover new phenomena during the encounter.

### 17.1.2    Subject: The Planets

The subjects of interplanetary missions are widely varied, largely unknown, and very distant from the Earth [Beatty, 1990]. These three attributes set the tone for interplanetary mission operations. First, varied targets lead to different operations strategies and mission types, which in turn reduce inheritance from one mission's operations to the next. Common ground elements can offset this effect somewhat, but the space element and the mission timelines are often unique. Second, unknown targets require adaptable operations and close coupling between the uplink and downlink functions. Finally, the distance from the Earth has many implications for operations. In Sec. 17.2, we'll talk more about these three differences as they relate to the mission-operations elements.

### 17.1.3    Command, Control, and Communications and Ground Elements

Command, control, and communications for interplanetary missions usually requires a limited number of antennas on the Earth directly linked to each spacecraft. The number of spacecraft has so far kept any ongoing spacecraft-to-spacecraft links from being cost effective. Because interplanetary spacecraft are distant, the Earth's rotation dominates their apparent motion, so antennas can view the spacecraft from horizon to horizon without requiring large angular rates. However, because of the distance, the antenna aperture must be able to receive a weak signal. These combined effects lead to a few large antennas for commanding and controlling the interplanetary spacecraft.

NASA's Deep Space Network (DSN) collects telemetry and radiometric navigation data and radiates commands to the spacecraft. The DSN is made up of three antenna complexes spaced around the Earth—at Goldstone, California; Canberra, Australia; and Madrid, Spain. A key difference between Earth-orbiting and interplanetary missions is that the DSN provides a limited number of tracking facilities. Several 26 m and 34 m antennas, plus one 70 m antenna, are at each complex, each with different capabilities and commitments. Planning to use the DSN starts at the earliest stages of a mission's development, sometimes 20 years before the tracking is required. Because of the high demand and planning based on uncertain missions, scheduling the DSN resource is a key driver for interplanetary mission operations. (See Chap. 12 for a description of DSN capabilities.)

A DSN station's performance depends on the declination of the target, which can be thought of as the latitude of the spacecraft in an Earth-centered frame. Much as the length of day increases during the summer, the duration of a spacecraft pass increases if it is at the same latitude as the station. Thus, for a spacecraft at 23° declination, a northern-hemisphere pass will be 12–13 hours, whereas a southern-hemisphere pass will be 8-9 hours. Similarly, same-hemisphere coverage will result in higher maximum elevations and therefore higher data rates.

An aggressive operations concept will count on the best station, whereas a conservative concept will plan for some off-peak performance. If the mission is at a critical point, the DSN will be firmly committed. But if the mission is in a routine phase such as mapping, we may substitute or cancel the DSN antenna's time if another spacecraft has a higher priority. If the mission return has counted on the most capable station, substituting a shorter or a less capable pass will require some replanning.

## 17.1.4    Mission Operations Element

Interplanetary space missions typically have a large flight team devoted to uplink and downlink. The uplink portion of the flight team does everything required before the spacecraft executes an activity. The downlink portion does everything after execution. The work of the two teams interact when the spacecraft's performance, characterized by the downlink, affects planning by the uplink function. For interplanetary missions this is a strong interaction, because the uplink element will typically re-plan an activity when performance estimates or navigation predictions change.

The uplink design includes incorporating the spacecraft and instrument performance into an integrated plan. The best information is synthesized about the spacecraft, the tracking configuration, and the science desires for a certain amount of time. Deciding on the boundaries of the time blocks is another, very early function of the uplink team. When the plan is at a stage where few changes are expected, and conflicting science desires have been reconciled, the team generates and simulates a spacecraft command file. After a successful simulation, the team stores the command load until just before its execution, then transmits it to the spacecraft. Uplink also includes adding so-called "real-time commands" to the spacecraft memory while a stored sequence is executing.

For downlink, we monitor the spacecraft's health and safety and track its movement. We periodically schedule subsystem calibrations and then fold detailed analytical results into future plans. When the Jet Propulsion Lab flies interplanetary missions, they focus on the timing of subsystem performance for two main reasons. First, they must understand the spacecraft and anticipate any future changes in observed parameters. Second, tiny changes in subsystem behavior can strongly affect a plan optimized for performance.

Navigation tracking is also key to the downlink. The navigation team determines the spacecraft's trajectory, maneuvers to optimize future geometry, and updates future plans as required.

### 17.1.5    Launch Element

The launch element for an interplanetary mission consists of the launch vehicle itself and any supporting services needed to deliver the spacecraft onto an interplanetary trajectory. An interplanetary trajectory requires higher launch energy to escape the Earth's gravity and a specific alignment of the planets. The latter means we can launch the spacecraft only at certain times. Because the launch energy varies with the launch day, we can get several possible launch days by restricting the spacecraft's mass to the lowest allowed value (corresponding to the highest required launch energy) during the desired number of days. Sections 17.2.1 and 17.2.6 cover the launch element.

### 17.1.6    Space Element

The space element of an interplanetary mission is a spacecraft (or perhaps two) which is uniquely designed for the mission being flown. While there may be heritage in parts of the hardware or software, the spacecraft usually is unique enough to require a dedicated, specialized flight team. In the early days of space exploration, pairs of interplanetary spacecraft typically flew to increase mission reliability. As missions became mere complex, single spacecraft (such as Galileo) flew. The key distinguishing factor in the spacecraft is its ability to survive lengthy space exposure, do large orbit-insertion maneuvers, and communicate over large ranges. For a description of the engineering systems and subsystems, see Larson and Wertz [1992].

The science portion of the space segment can be very complex and comprehensive. Many times, multiple science objectives are satisfied by a payload which has differing observation types and spacecraft and operations drivers. These comprehensive payloads may lead to conflicts over requirements and resources. Table 17.4 describes the typical science characteristics and their operational requirements.

*Remote-sensing instruments* require a spacecraft to point at the target. They can be either passive, where the instrument receives energy directly from the target, or

**Table 17.4. Instrument Types and Operations Characteristics.** The type of science instrument affects both the operations activities and the spacecraft design. If more than one type of instrument is flown, operations activities must often be structured to provide different spacecraft or target conditions.

| Instrument Type | Characteristics |
|---|---|
| Remote Sensing<br>　Passive<br><br>　Active | • Resolution proportional to distance from target; requires target fixed pointing<br>• High power due to active transmission; requires target relative pointing |
| In-Situ<br>　Particles<br><br>　Magnetic and radio wave<br>　investigations | • Sensitive to spacecraft contamination; requires orientation to acquire samples<br>• Requires magnetically clean spacecraft; may require special calibration to remove spacecraft signature from data |
| Detection of High-Energy<br>Particles | • Low time constant for three-dimensional temporal phenomena; may require rapid scanning |

active, where energy emitted by the spacecraft is reflected by the target and received by the spacecraft. Additionally, the electromagnetic spectrum is covered by different instruments (microwave, infrared, visible, or ultraviolet).

Optical remote-sensing instruments (cameras) are passive, receiving light from the target. The resolution is proportional to distance from the target, so the mission will try to fly as close to it as possible. The target's apparent motion is directly proportional to the flyby distance. If the relative motion is too high and the exposure is long, the image may be smeared. The tradeoff between unsmeared and high-resolution images will determine the best flyby distance for an imaging instrument.

Imaging instruments have large data rates and very high public interest. For example, consider an imaging frame 1000 by 1000 pixels, with each pixel represented by 12 bits. If the framing time (time between images) of the camera is one second, the data rate is 12 Mbps. For a 20-Watt, X-band radio system from an outer planet, a typical data rate is 100 kbps. Obviously, a camera with these characteristics can oversubscribe the downlink if it takes data continuously.

Active remote-sensing instruments send a signal from the spacecraft. The most common active remote-sensing instrument is a radar, which is used to image the surface of a body obscured in the optical wavelengths. For example, Magellan at Venus and the Cassini Radar at Saturn's moon, Titan, will do this experiment [Young, 1990]. Operationally, the most noticeable feature of a radar is its use of the spacecraft's high-gain antenna as the instrument aperture, which requires data storage because the antenna can't simultaneously view the Earth and the target.

*In-situ instruments* measure the fields they're in or particles the spacecraft samples. Unfortunately, because the spacecraft contaminates its surroundings, it can't produce an environment identical to the one of scientific interest. For example, the thruster plume will affect a sensitive mass spectrometer, so we must take data when the thrusters are inhibited, which may imply allowing less control than nominally planned. A magnetic-field instrument is also susceptible to contamination by spacecraft fields from operating current loops. Because eliminating current loops is not practical when science activities are high, we require a magnetometer to calibrate sources of the loops on the spacecraft. In other words, we'll have to run a calibration sequence when activity is low, cycling subsystems on and off, so the magnetometer may characterize the fields. The characterization allows later removal of the fields from the science data set.

Because the particles being measured by radiowave sensors have a very short time constant and are three-dimensional, a scanning motion may be required during data gathering. If the spacecraft can't independently sweep the instruments through space, we may move the entire spacecraft, as in the case of the Cassini spacecraft. Operationally, we must trade this scanning motion against the target's fixed attitude required for remote-sensing measurements.

Infrared-sensing instruments operate at cool temperatures, thereby requiring less solar illumination and a thermally stable platform. Infrared spectrometers are typically the most sensitive, operating at 70–100°K. To operate at this temperature the detector connects by a cold finger to a radiator, which we then point at deep space. The plan for operating the spacecraft must ensure the radiator doesn't see a warm body such as the Sun or a planet. If it does, the viewing time is shorter. The energy input scales as the square of the solar range, so the exposure limits change during the mission.

Remote-sensing instruments typically use a lot of data. Therefore, they drive the downlink design and, if they over-subscribe the bus, the data-management system. Users of less data, such as fields, particles, and in-situ instruments, are easier to manage operationally. And radio science is easiest of all because it has no telemetry modulated on the downlink. However, if radio science demands highly stable frequencies, we must operate the spacecraft while keeping it as motionless as possible. Thruster firings which induce a translation are undesirable, and even reaction-wheel control can cause a frequency shift. This shift occurs because the antenna is offset from the spacecraft's center of mass, so a pointing offset of the spacecraft introduces a translational motion in the antenna. For an antenna offset one meter from the spacecraft's center of mass, an angular pointing correction of one microradian in one second would induce a motion of 0.000001 m/s, which corresponds to a doppler frequency shift in X-band of 0.0001 Hz. This frequency shift is comparable to the signal our radio-science experiment will be trying to measure, so we may need to do special attitude control.

The number of instruments increases operations costs because we have to do more integration and resolve more conflicts between the instruments. If staring instruments fly with scanning instruments, we must plan the spacecraft's attitude to accommodate both. The mission plan of integrated activities results from compromises we typically develop through negotiation and advance planning. If the planning is premature, the operations team may do the planning over again. Conversely, if our planning is too close to the event, we'll have trouble finding compromises.

# 17.2    Differences Between Interplanetary and Earth-Orbiting Missions

The interplanetary environment, mission duration, and mission objectives cause most of the differences from Earth-orbiting missions. The spacecraft's design for interplanetary missions is also different, and mission operations must deal with these differences.

## 17.2.1    Orbit Differences

**Range and Light Time Delay.** The distance from the Earth to the target is the most dramatic difference between interplanetary and Earth missions. Earth missions vary in distance from hundreds to thousands of kilometers, but interplanetary missions range over millions of kilometers. The basic distance unit used is the Astronomical Unit, or AU. The AU is defined as the semi-major axis of the Earth's orbit, equal to $149.6 \times 10^6$ km, or $92.96 \times 10^6$ miles. The range for the Earth to the spacecraft changes because of the spacecraft's trajectory and the Earth's motion around the Sun [Bate, 1971].

The most apparent difference that the increased range makes is that we have no real-time visibility into the spacecraft's operation. As shown in Table 17.5, even for a mission at the closest interplanetary range (Venus at conjunction), an event seen by the ground has already occurred 2.3 minutes ago. As a result, for any operation which requires ground interaction, we must include at least a round trip light time in the plan. The first one-way light time is used by the signal arriving from the spacecraft to the ground operations; the second half of the round trip is for the return command. A handy rule of thumb is that the speed of light $c \sim 3 \times 10^5$ km/s = 0.12 AU/min, or light time ~8.3 min per AU. Similarly, light time is 3.33 seconds per million kilometers. For a Saturn mission, then, we need to know only that the semi-major axis of Saturn's orbit is 9.5 AU to know that one-way light times to Earth will be in the range of 90 minutes. Therefore, each interplanetary mission has a time constant associated with its target.

The change in Earth range changes the telecom signal strength during the mission. The cruise phase of the mission will have ranges near zero close to launch and during any Earth flybys, and the science-operations phase will have a range which itself changes by the Sun-Planet distance ±1AU, as the Earth travels around the Sun. As shown in Table 17.5, the telecom signal strength varies in strength from the range alone up to 16 dB as the Earth and the target planet approach and recede. For a flyby mission, the telecom signal changes less during the observation of one target because the time is short compared to one year, but the signal strength will vary from target to target.

**Table 17.5. Interplanetary Distances and Resulting Environments.** Sun-to-spacecraft and Earth-to-spacecraft ranges vary with the planet and the duration of a mission, affecting designs for thermal management, telecommunications, and radiation tolerance.

| Planet | Sun Distance (AU*) | Earth Distance (AU) | One Way Light Time Range (min) | Telecom Space Loss Difference Min - Max Range (dB) | Solar Flux at Mean Distance (W/m$^2$) | Synodic Period[†] with Earth (years) |
|---|---|---|---|---|---|---|
| Mercury | 0.31–0.47 | 0.53–1.47 | 4.4–12.2 | 8.79 | 9066 | 0.32 |
| Venus | 0.72–0.73 | 0.27–1.73 | 2.3–14.3 | 16.07 | 2601 | 1.60 |
| Mars | 1.38–1.67 | 0.38–2.67 | 3.2–22.2 | 16.88 | 586 | 2.13 |
| Jupiter | 4.95–5.45 | 3.95–6.45 | 32.9–53.6 | 4.26 | 50 | 1.09 |
| Saturn | 9.0–10.0 | 8.0–11.0 | 66.5–91.7 | 2.78 | 15 | 1.04 |
| Uranus | 18.2–20.3 | 17.3–21.3 | 143.7–176.8 | 1.80 | 4 | 1.01 |
| Neptune | 30.0–30.3 | 29.0–31.3 | 241–260 | 0.66 | 2 | 1.01 |
| Pluto | 29.5–50.0 | 28.5–51.0 | 237–424 | 5.06 | 1 | 1.00 |
| Geostationary | 1.0 | 42,162 km | 0.14 sec | 0 dB | 1358 | N/A |
| Moon | 1.0 | 384,400 km | 1.27 sec | 0 dB | 1358 | N/A |

\* 1 AU = 149.6 × 10$^6$ km
† Synodic period is the time required for any phase angle to repeat

**Environmental Extremes.** The change in Sun range, also shown in Table 17.5, changes the thermal environment experienced by the spacecraft. This thermal change leads to more environmental variation than during Earth missions. The spacecraft must operate thorough the cruise phase in environments which are different from that at the primary target. Earth missions may have to operate with variations in solar input because the length of occultation varies during the mission, but these durations are typically very short, and the spacecraft re-enters the sunlight before it reaches a steady-state temperature. Radiation and particulate environments also vary during the cruise phase.

The thermal environment requires us to choose the spacecraft's attitude so key areas are in the shade, and some subsystems are powered down to prevent operations (and heat generation) during peak heating periods. Thermal variations during cruise dictate attitudes and hence maneuver and telecom profiles, as opposed to Earth missions which have no cruise phase. The telecom profile is much lower if the spacecraft is not allowed to point its high-gain antenna toward the Earth, as is the case for a Sun-pointed spacecraft with a fixed antenna.

The Galileo and Cassini missions, which contain Venus flybys to supplement propulsion and therefore allow the spacecraft to have more mass, must survive the solar environment from 0.7 to 5 AU and 0.67 to 10 AU, respectively. The effective input of solar thermal input then ranges by up to a factor of 100 from Venus flyby to their final destination. As a result, the operations system has to follow flight rules which depend on the mission. In this example, when near the Sun, the spacecraft must be pointed toward the Sun to hide the electronics and instruments behind a shade. This pointing complicates the maneuver design and forces our fault protection to be able to quickly reacquire the Sun, which might not be necessary during the prime mission.

We must also take into account the solar cycle. Solar activity fluctuates over 11-year cycles. The most recent maximum of the solar cycle was in 1991, so the next solar maximum will fall in 2002. If a mission lasts past the solar maximum, the spacecraft will experience less solar radiation.

**Opposition and Conjunction.** The relative geometry between the Sun, spacecraft, and Earth for an interplanetary mission changes in ways Earth missions don't see. Figure 17.1 shows the geometric relationships. When the spacecraft and the Sun are aligned as seen from the Earth, the spacecraft is said to be in *conjunction*. When the spacecraft's conjunction places it on the far side of the Sun from the Earth, it is *superior conjunction*. Superior conjunction restricts communications with the spacecraft due to interference with the radio signal as it passes close to the Sun. Mission planning will disallow or restrict activities around the time of superior conjunction. When the Sun is on the side of the sky opposite from that of the spacecraft as seen from the Earth, the spacecraft is at *opposition*.

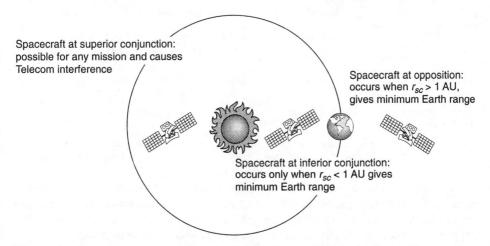

Spacecraft at superior conjunction: possible for any mission and causes Telecom interference

Spacecraft at opposition: occurs when $r_{sc}$ > 1 AU, gives minimum Earth range

Spacecraft at inferior conjunction: occurs only when $r_{sc}$ < 1 AU gives minimum Earth range

**Fig. 17.1.    Conjunction and Opposition.** At superior conjunction, when the Sun is between spacecraft and Earth, solar activity in the signal path disrupts communications. Conservative interference regions are ±10° for S-band and ±3° for X-band.

**Coordinate Systems.** Having to model coordinate systems is another way an Earth-orbiting mission differs from an interplanetary mission. Understanding the different coordinate systems helps us avoid confusion and potential mistakes during operations. For an Earth mission, an Earth-centered coordinate system is the only one necessary to describe a satellite's motion. Ground stations are modeled in a local geographic coordinate system. For an interplanetary mission, however, the central body of the coordinate system changes, and the reference plane may be a planet, a planet's orbit, or the spacecraft orbit itself.

As shown in Table 17.6 and Fig. 17.2 and 17.3, several different coordinate systems are used. An Earth-equatorial coordinate system is appropriate for launch and for most applications involving tracking of the spacecraft. For interplanetary cruise, the spacecraft is usually in a plane quite close to the ecliptic plane (the plane of the Earth's orbit), so we use an ecliptic-referenced coordinate system for the trajectory and maneuver modeling. For an interplanetary approach, and for the orbital phase of a mapping or a tour mission, the target planet's equatorial coordinate system is the logical choice because this coordinate system most easily represents the orbit dynamics. As shown in Fig. 17.2, the transformation between coordinate systems centered on a planet's equator and the ecliptic coordinate system first requires a rotation through the obliquity of the planet's rotation. The *obliquity* is the angle between a planet pole and the normal to the ecliptic plane. (The ecliptic is the plane defined by the Earth's motion about the Sun.) The obliquity of a planet's rotation causes the planet to have seasons.

**Table 17.6. Common Coordinate Systems for Interplanetary Missions.** The coordinate system we choose depends on the data we're collecting, the mission phase, and the customer.

| Coordinate System | Reference Plane | Reference Direction | Use |
|---|---|---|---|
| J2000 (or EME50) | Earth's equator | Vernal equinox on Jan. 1, 2000 (or 1/1/1950) | Earth-referenced position, DSN viewing geometry |
| EMO2000 (or 1950) | Earth's orbit (ecliptic) | Vernal equinox on Jan. 1, 2000 (or 1/1/1950) | Interplanetary trajectory |
| Interplanetary equator of date | Planet's equator | Planet's prime meridian of date | In-orbit geometry |
| B-plane | Perpendicular to incoming asymptote | Intersection of B-plane with planet's equator | Flyby and orbit Insertion targeting |

For targeting to the interplanetary approach, we use the B-plane (Fig. 17.3), referenced to the target planet and the incoming spacecraft's hyperbolic orbit. The encounter conditions are described in terms of the aimpoint or the point where the asymptote crosses the B-plane. The aimpoint can be described in polar coordinates as shown (B and Q), or as a projection on the R and T axes referred to as B · R and B · T, respectively. The third dimension of the target point is defined in terms of the arrival time, which is equivalent to a displacement along the S direction. The

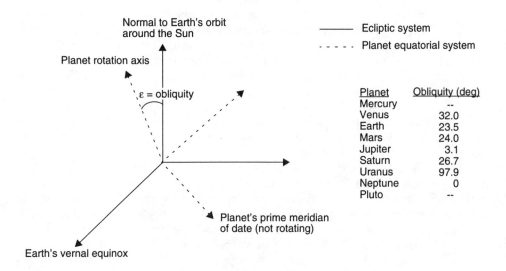

**Fig. 17.2.     Ecliptic and Equatorial Coordinate Systems.** The interplanetary trajectory and interplanetary orbits are expressed in an ecliptic coordinate system. Once in orbit, the trajectory is described in a planet-centered equatorial coordinate system. The two coordinate system differ by the planet's obliquity and the reference direction.

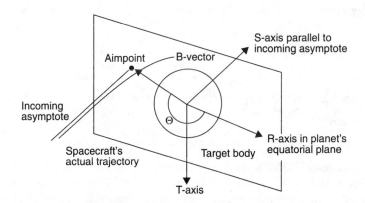

**Fig. 17.3.     The Planet Target Plane or B-Plane.** This coordinate system describes the geometry of an interplanetary encounter by specifying the "aimpoint" in a plane perpendicular to the approach asymptote and containing the target body's center.

incoming asymptote, not the trajectory, defines the B-plane. Thus, the target plane itself doesn't change if we select a different aimpoint.

**Differences in Orbit Mechanics—Interplanetary Escape.** We must place the spacecraft on a hyperbolic trajectory to escape Earth and start its interplanetary transfer. That trajectory, of course, is still in orbit around the Sun in an elliptical path, which will intersect the target planet's orbit at the arrival date. Launching to the hyperbolic escape trajectory requires the launch site to be in the plane of the hyperbolic orbit at launch. Because the launch site is rotating with the Earth, this requirement constrains the time of day for launch. A typical ascent profile will place the spacecraft into a low-Earth coasting orbit. The rocket's lower stages then fire to ensure the proper direction of the escape asymptote. Finally, an upper stage fires to provide the escape energy.

**Discussion of Gravity Assist.** An interplanetary trajectory to the outer planets may use intermediate gravity assists to increase its energy. The gravitational attraction of the flyby body changes the direction of the spacecraft's heliocentric trajectory and therefore can increase or decrease the energy of the spacecraft orbit. The Mariner 10 mission first used a gravity assist from Venus to target it to a subsequent Mercury flyby. For the Voyager mission, repeated gravity assists ensured the flyby of four outer planets. Later missions such as Galileo and Cassini rely on successive flybys of the inner planets to achieve the required energy for transfer to Jupiter and Saturn, respectively.

**Table 17.7. Trajectory Types for Different Interplanetary Missions.** The trajectory type for any mission varies with the target, the spacecraft mass, the launch vehicle's abilities, and the allowed mission duration.

| Trajectory Type | Mission Application | Characteristics |
|---|---|---|
| Ballistic/Direct | Inner planet, low energy trajectory | Few maneuvers, short flight time |
| Delta Velocity-Earth Gravity Assist | Asteroid flyby, outer planet | At least one, large, deep-space maneuver, set up for interplanetary encounter |
| Interplanetary gravity assist (e.g. Venus-Earth-Earth Gravity Assist) | High-energy trajectories to the outer planet | Repeated interplanetary encounters, maneuvers to avoid striking the planet, long flight time |

**Orbit Insertion.** Orbit insertion is necessary for any mission which plans a long residence time at the target. The orbit-insertion burn mainly removes energy from the spacecraft's orbit, so we want to place it closest to the point at which the spacecraft approaches the target body. That's because an orbit's energy change is related to the applied $\Delta V$ through an amplification by the actual orbital velocity ($\Delta E \sim V \Delta V$), a differential form of the vis-viva integral. To meet this timing requirement, we insert the spacecraft into orbit by using a sequence stored onboard—the light-time delay prevents real-time commanding. Figure 17.4 shows an orbit-plane view to the geometry of the orbit-insertion sequence. We target the incoming hyperbolic orbit to the desired inclination and periapsis altitude, and the burn results in the desired capture orbit.

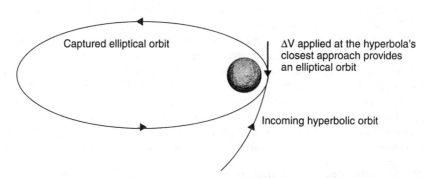

**Fig. 17.4.    Orbit-Insertion Geometry.** The spacecraft does a retro-maneuver to reduce energy at its closest approach to the planet and therefore goes into a planet-centered orbit.

## 17.2.2    Differences in Launch Element

**Launch Window.** For an Earth-orbiting mission, the launch window is usually constrained by geometric considerations such as Sun exposure and lighting at the launch or contingency landing sites (for manned vehicles). Also, because the desired geometry largely repeats from day to day, launch elements don't restrict the launch date. In contrast, the launch period for an interplanetary mission is constrained first by the interplanetary alignment and second by the orientation of the interplanetary transfer plane. Geometric considerations such as lighting then become constraints on the spacecraft and operations design. Section 17.2.1 describes the launch geometry, as do many references on orbital mechanics, such as those listed at the end of Chap. 10.

If we miss the defined launch period, we must usually wait months or longer before the interplanetary alignment allows another opportunity. The Voyager mission, for example, took advantage of a interplanetary alignment that won't be repeated for another 176 years.

The availability of a launch vehicle affects mission planning, as in the case of the Shuttle/Centaur loss to Galileo. We plan a mission assuming a certain capability for the launch vehicle, but program changes sometimes require us to change these capabilities before launch. In this case, we must replan parts of the mission before launch, often having to assume different capabilities for different phases. A longer cruise, which was required for Galileo, means that the time to wait for science gathering is either extended, if cruise science is disallowed, or shortened, if we allow science gathering early in the encounter, as Galileo did at the Venus flyby less than five months after launch.

The launch vehicle's performance affects post-launch operations mainly in the time before the first post-injection maneuver. If the time to go before encountering a target body is large, we can change the target conditions a lot with a small $\Delta V$. We require post-injection maneuvers because of errors in our injection maneuver. When an upper stage causes an inaccurate injection, we must correct the problem quickly so the error doesn't continue and require a larger $\Delta V$ later.

We can use low-energy launch vehicles or massive spacecraft by doing several gravity assists during the cruise phase. This approach complicates operations by extending the cruise duration and by increasing the thermal range of exposure. For example, the Galileo trajectory used one gravity assist from Venus and two from Earth to make up for the energy shortfall of the Shuttle/IUS combination compared to the direct trajectory achievable with the Shuttle/Centaur. Galileo's trajectory had three more years of cruise, and the spacecraft had to fly at a Sun range from 0.7 to 5 AU rather than from 1 AU to 5 AU as it had been designed to do.

### 17.2.3     Differences in Subjects

**Unknown Targets.** Interplanetary missions are exploratory, so mission operations must adapt. In contrast, Earth-science missions know a lot about the target and can make decisions before flight. The uniqueness of interplanetary encounters and wanting to get the most information from a new discovery often require us to retarget observations. We must therefore design the observation strategy with the flexibility to re-target in mind, all the while not wasting any valuable time close to the target.

The target body's position and velocity are also uncertain. This uncertainty is especially true with first encounters of a planet, such as the early flyby missions, but even a return to a planet requires some improvement in planet ephemeris. The observation strategy in this case must allow for errors in one of two ways. As Galileo demonstrated while encountering the Gaspra asteroid, we can take enough pictures, including those of dark sky, to ensure that the target will be in at least one frame. Or, if we have enough information, we can insert timing and position changes into our sequence of observations just before the encounter.

As illustrated, operations for interplanetary missions require continuing navigation, whereas most Earth missions demand only station keeping. The pre-launch plans usually include a 99% certainty that the mission will have enough propellant. Successful navigation can save propellant and, therefore, extend the mission. We can keep the operations team in place for an extended mission, and some of the team can plan the extended mission during the primary mission. The primary example of an extended mission is the Voyager II's encounters with Saturn and Neptune, which were enabled by successful encounters with Jupiter and Saturn. In fact, the aimpoint at Saturn was chosen as a compromise between Saturn science and future science goals. The Voyager's operations team successfully dealt with this late change. [Kohlhase, 1989]

**Degree of Adaptability.** Some interplanetary missions must be very adaptable because the target is unique and unexplored. Other missions, particularly for mapping, can have repetitive operations. Repetition saves workforce, so we use it as much as possible through pre-tested command files. Science planning for some interplanetary missions must adapt to unexplored targets. For some observations, we purposely build in time and other margins to allow changes to the part of the target system we're viewing. New discoveries, such as the volcanoes on Io or Titan's atmosphere, lead to major changes in mission philosophy. Many interplanetary missions have a payload that attempts to comprehensively investigate the target. The different objectives and observation types of the instruments create the need for a large committee, the Project Science Group, which defines priorities and resolves conflicts. Sometimes the mission objectives lend themselves to a more fixed observation design. A pair sent to the same target might be an example for which we use results from the first mission to adapt on the second. Missions with either one instrument or a mapping-style objective could have very repetitive observations.

### 17.2.4   Differences in Command and Control Architecture and Ground Element

**Continuous Spacecraft Visibility.** A key driver is whether the Earth can contact the spacecraft at any time. A spacecraft with a scan platform for instrument pointing allows us to do real-time downlink and science gathering at the same time. Simultaneously pointing the spacecraft body at the Earth and instruments at the target requires pointing planning to satisfy more constraints, and it may require more simulation. But data management is simpler if capabilities for downlink and data gathering are equal.

If the spacecraft can't communicate and gather science at the same time, the opposite is true. Fewer axes will require less pointing planning and constraint checking, but data management may be more complex.

**Stored Commands Versus Real-Time Operations.** Because we can't practically send commands in real time, we place most spacecraft activities in a stored file and plan them in advance. The spacecraft carries them out at the appropriate time. We have to work around the basic tension between advance planning and adaptation.

For time-critical, mission-critical events, mission planning must make sure the command file will execute with zero defects. Examples of this type event are orbit-insertion burns, for which the time window to start and the light time are just minutes. Typically the spacecraft has a stored sequence of commands, including potential faults and the spacecraft's response to those faults. The planning, simulation, and execution of an orbit-insertion burn is among the most costly parts of mission operations.

Few commands are sent in this fashion. Usually, only mission events which are important, but not time-critical, fall into this category. Examples include deploying a probe, such as the Cassini Huygens probe, for which we allow a contingency release at some cost in mission performance. Another example is a $\Delta V$ maneuver that isn't time critical, for which we can abort a maneuver at some cost in propellant if anomalous conditions result.

**Telecom Link.** The downlink is subject to down time due to the limited number of antennas. The DSN is available to many projects, but they must compete for resources. The DSN sponsors a resource-allocation process in which all users participate, including those from ESA and Russia. If the spacecraft is undergoing a critical activity, the DSN will likely assign a high priority to the mission and tracking will be provided. If the spacecraft is doing a routine activity, its pass is more likely to be sacrificed to another mission's critical activity or anomaly. Mission planning must account for the DSN's availability and allow for the loss of some planned tracking passes.

Some missions have managed tape recorders to save data in the case of a DSN outage or to return high-priority data first. Positioning the tape recorder and managing tape travel across the heads are time consuming. Future missions will likely use solid-state recording devices, which won't need tape management because they'll use software-controlled memory management.

## 17.2.5   Differences in Mission Operations

**Engineering Health and Safety.** Spacecraft monitoring relies on alarm limits set for key subsystems. It usually requires a trained flight team on station during the mission to filter out false alarms and reset limits as the subsystems become better characterized. They also characterize the spacecraft by trending, or examining the performance of critical parameters over time. Trending spots latent problems in the spacecraft's performance and keeps the team familiar with the details of the spacecraft to help resolve anomalies if necessary.

**Reconstruction and Analysis.** We must provide for access to preliminary and calibrated science data. The preliminary data, sometimes at level-0, helps us assess the health and safety of instruments, just as the flight team assesses spacecraft health. We must also provide spacecraft and mission data to properly interpret the science data. A navigation reconstruction of the trajectory provides positions for the spacecraft, target, and Sun. Also, by analyzing downlink telemetry, we can give the science community parameters for spacecraft pointing and magnetic-field effects.

**Data Release.** In most interplanetary missions, the science data undergoes a proprietary period, during which only the investigation team has access, to reward them for the time devoted to the project. During this period, the science teams often publish their first and most important conclusions, and we often have to participate in public forums such as press conferences. After the proprietary period, the science data goes to the public through the interplanetary data system, so we've completed our task of delivering the data to the taxpayers who funded the program.

### 17.2.6 Differences in the Space Element

**Small Margins.** Nearly all resources for interplanetary missions have small operating margins, mainly because we want the best return from a unique mission opportunity. Operating with small or negative margins affects mission operations in three ways. First, there is little room for error in any spacecraft operation, which forces levels of checking and simulation unnecessary in missions with margin. Second, we use the actual performance of the spacecraft and the instruments to update models for future activities, forcing significant replanning. Third, we operate the spacecraft as a set of subsystems, rather than getting the best performance from the whole system. This approach increases workforce and cost.

**Power Generation.** Solar power is the usual power source for Earth missions, and it's also practical for missions to Venus and Mars. Analyzing a power system for an orbiting mission to one of these planets is similar to that for an Earth mission. The only real difference is that we must size the array to the worst-case combination of Sun range and power usage at that phase in the mission.

The large solar ranges of the outer planets disallow use of solar power, so we typically use Radioisotope Thermoelectric Generators (RTGs). An RTG's power output degrades very slowly with time—at a rate of 3–4 Watts per year. Thus, for missions to the outer planets, we don't consider solar power or reductions in power from occultation.

If no battery is flown, as has been the case with many outer-planet missions, we must manage the power peaks very carefully with time. This management results in power cycling if the payload can over-subscribe the power bus, or in software limitations on spacecraft-controlled power users, such as reaction wheels. We use fault protection to guard against overloads but carefully manage power overall to avoid safing conditions. When we have small margins to return the most science, power management uses a lot of resources.

**Thermal Control.** The variation in Sun range over the cruise requires us to use different rules near the Sun. For instance, the spacecraft may not be able to point towards the Earth for high-gain antenna communications without exposing its thermally sensitive areas. On those areas, we have to count integrated solar input and thermal cycles as mission consumables. We often avoid damage to science instruments by using deployable covers and special fail-safe operations.

Engineering subsystems are often coupled thermally, and varying solar input requires us to maintain proper conditions. For RTG-powered spacecraft, we sometimes use waste heat from unused electrical power to heat engineering subsystems. This approach further couples the spacecraft-engineering and science operations because the waste heat is inversely proportional to the amount of power used for the payload. An example is the use of RTG waste heat for Galileo's propulsion system. In this case, the heat input varies the operating range of the engines, so modeling of propulsion performance must include the entire spacecraft payload.

**Resource Monitoring.** For interplanetary missions, the cruise (non-operational) time often dominates the spacecraft's lifetime, so we use strategies to conserve resources during the cruise phase. For example, we can use a different attitude-control deadband to reduce thruster pulses, employ thrusters rather than reaction wheels to reduce wheel revolutions, or leave systems off. We must continually evaluate resource use versus cruise performance to get the best total return from the mission.

We have to model resources in advance and account for them after carrying out an activity. If we have resource margins, we simply check high-level telemetry. However, if we have small margins to capture the most science, we must model and count more accurately. Table 17.8 describes some of the typical resources interplanetary missions must model and shows how we treat them differently from their counterparts for Earth-orbiting missions. The resources are renewable or consumable. We budget renewable resources at different times, whereas consumable resources have only one budget for the entire mission.

**Table 17.8. Typical Operational Activities to Best Use Spacecraft Resources.** A resource-constrained interplanetary mission may increase mission return by intensifying operations activities.

| Renewable Resources | | |
|---|---|---|
| **Resource** | **Comment** | **How Interplanetary Missions Differ from Earth Orbiters** |
| Command bandwidth | Assign commands execution time within spacecraft frame time | Transmission delays prevent real-time execution |
| Telemetry bandwidth | Model link performance with high fidelity to ensure data capture—remove modeling if adequate margins are available | Telecom range varies less with Earth orbiters over both pass and mission duration |
| Command memory usage | Repeatedly estimate as planning develops—simulation verifies estimates before load is transmitted | We develop some software after launch and upload it to the spacecraft during the mission |
| **Consumable Resources** | | |
| Fuel | Use worst case for preliminary planning; replan or extend mission when worst doesn't occur | Navigation use is usually much larger than attitude control |
| Thermal cycles | Plan options for both heating and cooling | Thermal environment more extreme |

# 17.3   Operations Activities Throughout the Mission Life Cycle

## 17.3.1   Pre-Launch

**Designing the Operations System.** The time leading up to launch goes to building the ground data system and team structures for mission operations. If mission operations gets staffing early enough in the project-development phase, we can engineer systems for the ground data and flight data at the same time. This style of project development may be the most cost effective in the long run, if operations costs are part of design trades for the spacecraft and ground system. We also may use this development style if the ground system has a lot of inheritance. For many interplanetary missions, however, the operations element is largely unstaffed early in the project. Thus, the ground-system design must react to the spacecraft data system and may not take full advantage of inheritance.

During the assembly and system test of the spacecraft, we generate spacecraft activities used to test the spacecraft and simulate key mission events. We also test critical-event activities, such as launch, maneuver, and orbit insertion. These tests verify that the spacecraft will behave as anticipated and that its computer can read the product of the command-generation software. The simulated activities used to test the flight team contain one or more "hiccups," which make the test more like real life. In this fashion, we develop and have pre-launch procedures. Also, the flight team, who are different from the spacecraft-development team, can learn expected spacecraft characteristics.

**Mission Planning Before Launch.** A key pre-launch activity is integrated activity planning for mission operations. An Integrated Mission Operations Plan (IMOP) will include activities for people who handle the spacecraft, DSN, science gathering, and ground systems. The IMOP allows us to determine the spacecraft and ground resources needed for the planned activities. Before launch, planning is at a fairly high level, so future changes won't affect it much. As in spacecraft design, it helps to have experienced operators participating in this planning, so our staffing and cost figures will be realistic. The mission plans after launch will then be more robust, and the mission will be less susceptible to cost overruns from grandiose plans or to reduced return from insufficient staffing.

## 17.3.2   Launch

The launch phase includes the launch event and initial characterization of the spacecraft. After lift-off, we insert the spacecraft into its parking orbit, properly align the parking orbit with the interplanetary transfer orbit, and inject the spacecraft into the transfer orbit. The first major events after injection are deployments, early pressurizing and priming of the propulsion system, calibrating attitude-control systems, and deploying of booms and antennas.

The events in the initial stages of the mission are limited to those which support the first trajectory-correction maneuver. Compared to an Earth-orbiting mission, whose main propulsive events may be complete within a few days of launch, an interplanetary mission usually runs the first of many propulsive maneuvers at 10–25 days following launch. The first correction maneuver is mainly to correct injection errors, which develop from errors in the amount of burn or pointing of the launch vehicle.

Other than initializing the spacecraft, preparations for the first maneuver include determining the initial orbit by using the DSN tracking stations. Navigation will solve for trajectories and use the final available one before the maneuver to design a correction maneuver back to the desired trajectory. Spacecraft outgassing, which produces a force like that of solar pressure on the spacecraft, complicates determination of the initial orbit. We need the spacecraft's area and reflectivity to model the effects of solar pressure, but we know them only roughly until after launch.

**Launch Approval.** Launch approval is different for US Air Force, non-US, and human-rated vehicles but in all cases we need to plan for contingencies or failure. For Shuttle launches, we must plan for interplanetary injection on more than one revolution of the parking orbit as well as for landing with the spacecraft still in the Shuttle bay.

Using RTGs to generate power significantly complicates launch approval. For an RTG-powered mission, we must take special precautions to ensure RTG safety after launch. Mainly, we have to bias the trajectory or aim the spacecraft away from the desired flyby point at the next flyby. Biasing reduces the spacecraft's probability of hitting a planet, but it requires many more larger, pre-determined maneuvers, which may be critical to the mission.

**Activities During the Launch Period.** The spacecraft's angular rates during injection require acquisition aids and a hand-off between 26 m and 34 m or 70 m coverage. Sometimes, Advanced Range Instrumentation Aircraft (ARIA) first acquire the spacecraft during the launch-vehicle upper stage's coast and injection. Spacecraft telemetry routes through the upper stage to provide visibility. Following the injection burn, we use the predicted state vectors for the spacecraft to provide pointing for the DSN's 26 m antennas. The 26 m antennas provide a higher angular rate than the larger 34 m antennas.

Control of the spacecraft passes between several agencies and tracking networks. Sometimes, different centers control the launch vehicle and upper stage, and a host center such as JPL controls the spacecraft.

We cover the spacecraft continuously during the critical first month of operations, which characterizes its outgassing, trajectory performance such as solar-pressure effects, and attitude-control performance. By continuously receiving spacecraft telemetry, we can compare actual against predicted performance to understand behaviors of the spacecraft subsystems. The spacecraft's performance is largely unknown at launch, including outgassing and uncalibrated performance that we must correct.

## 17.3.3    Cruise

Cruise is sometimes the longest phase, typically beginning after correcting the injection and lasting until primary science observations begin. For multiple flyby missions such as Voyager, the period of up to two years between flybys is really a cruise phase as well.

Typical activities include calibrating and maintaining the spacecraft and payload, plus continued navigation. We use the cruise phase to learn how the spacecraft flies, which is usually somewhat different from the design specifications. Other engineering activities may include deployments delayed for some reason from immediately after launch. An example is the Galileo antenna, which operators tried to deploy 18 months after launch, when the thermal environment was within the design tolerance.

Cruise science includes intermediate interplanetary encounters, as well as observing fields and particles. For many missions, the cruise science is relatively benign, but it offers a chance for the science-operations team to learn their instrument and its operational procedures.

**Effect of the Launch Period on Cruise Operations.** Continuous coverage must span the potential first 30 days for a 30-day launch period. We have to negotiate for the coverage from the DSN and be prepared to release coverage once we know the launch date. This may include strategies to retain coverage only around the most significant events, such as the first trajectory-change maneuver for launches that occur late in the launch period.

The coast duration from Earth orbit to interplanetary orbit varies across the launch period. We provide targeting information to the launch vehicle, typically in the form of time-varying polynomials over the launch period. For a shuttle launch, the performance varies depending on the deploy orbit, and not all orbits achieve the right conditions. The target polynomials, and the initial tracking and staffing plans, must account for the 90-minute difference (roughly the Shuttle's orbit period) between injection orbits. For some launch vehicles, the azimuth of the first-stage launch varies across the daily launch window, which may change our strategy for tracking or acquiring the launch vehicle.

If there are intermediate encounters, their dates will change with different launch dates, implying different coverage schedules and spacecraft sequences. For example, during Galileo's month-long launch period, the date of the Venus flyby varied by approximately one week. Most of the variation was handled by a low-activity stored sequence between the Earth and Venus, which was variable in length.

If science is planned for intermediate encounters, it must be somewhat flexible until after the launch date. This is especially true in the case of targeted observations, which will change not only in time over the launch period but in pointing direction and resolution as the flyby altitude and lighting conditions change.

**Navigation.** Navigation is one of the primary cruise functions; it includes calculating orbits, designing maneuvers, and propagating trajectories. Data for orbit determination contains different information for interplanetary missions than for Earth-orbiting missions. Two-way coherent Doppler data is the main type used, with two-way ranging data adding information.

The quality of the orbit determination is a function of the amount of data and the trajectory's characteristics. In particular, low-declination trajectories, or imminent interplanetary flybys, require special strategies to collect the data for orbit determination. The Doppler data has a singularity at zero declination relative to the Earth and we use additional ranging or different data types to make up for the problem. We have to plan for the additional data, and for the performance impact if critical events occur with this geometry.

The first maneuver following injection is typically large, may remove some injection bias, and is the first on-orbit use of the propulsion system. The cruise phase of a mission contains many maneuvers to shape the trajectory, correct navigation errors, and maintain biases that keep the spacecraft from hitting a planet. Compared to an Earth-orbiting mission, the propulsion system is larger and more complex, and it interacts more with future mission operations. The interaction comes about because using updated performance estimates for the mission leads to different estimates for propellant margins and then to different strategies for propulsion. A key way to save money in the operations phase is to limit changing plans based on updated performance. This would reduce a key operations task—reconstructing the maneuver performance using data for subsystem performance and navigation tracking.

Around each targeted flyby, we must do a sequence of targeting and post-flyby cleanup maneuvers. The final targeting maneuvers are small but are required to remove most of the previous errors in carrying out maneuvers and calculating orbits while allowing little time for remaining errors to corrupt the flyby. Post-encounter maneuvers mainly correct for excesses or shortfalls in energy caused by the flyby. The maneuver occurs shortly after the flyby to regain the nominal trajectory before velocity errors propagate with time into larger position errors.

**Maintaining Engineering and Science Equipment.** Both science and engineering equipment must be checked out, maintained, and calibrated during the cruise phase. We don't clearly understand the physics of many mechanisms which may cause damage during the long cruise, largely because we can't test these mechanisms for the same amount of time before launch. Therefore, we do prudent but sometimes costly maintenance, such as regular motor motion.

Earth-orbiting missions rarely wait years before using key engineering capabilities. An example is an engine used only for orbit insertion and not for regular trajectory maintenance during cruise. Magellan, for instance, used a solid rocket motor for orbit insertion after storing it while cruising for 15 months [Young, 1990]. Other than for orbit insertion, we typically use engineering equipment right away and check it out quickly.

Science instruments are mostly idle during cruise, but early checkout provides practice and often exposes problems. Routine maintenance and calibration of science instruments often cross the line into real science collection, with full demands on operators.

Maintaining and developing flight software compose another activity during cruise. We have to maintain software because we discover errors or learn more about the spacecraft during the mission. We develop software for three reasons: planned-for capability has been deferred from before to after launch, spacecraft anomalies force a rework of algorithms, or the mission has been extended into a regime we haven't planned for. All three cases require us to know all of the spacecraft's inner workings—in effect, to become development engineers. If the original development team has disbanded, operators may face a significant challenge.

### 17.3.4     Intermediate Encounters

For the operations team, intermediate gravity assists mean that the spacecraft needs accurate navigation at intermediate points during the mission. For a mission such as Galileo, navigation must operate fully for a six-year cruise and a two-year tour. Also, the availability of a gravity-assist trajectory drives cost throughout the mission. Galileo's direct trajectory required a three-year rather than a six-year cruise.

Observing intermediate flyby planets during the flyby adds value for three reasons. First, the instruments flown on interplanetary spacecraft are often more comprehensive than those on Earth-orbiting spacecraft, allowing simultaneous observation of the target in several wavelengths of energy levels. Second, intermediate encounters provide science, which adds value to the mission and the interplanetary science data. Third, intermediate encounters give the instrument and flight teams valuable information about checkout and calibration.

### 17.3.5     Arrival

Arrival at the target body is crucial for practice, early observations of an unknown target, and the last science before orbit insertion. If the mission is a flyby, we must plan for science on the inbound and outbound leg of the flyby. If it's eventually an orbital mission, we'll have to trade between ensuring a successful orbit-insertion maneuver and doing pre-encounter science with a healthy and understood spacecraft. Also, unique science opportunities may occur near orbit insertion because the target planet is so close.

**Targeting.** Early observations of the target (optical navigation) and continued radiometric tracking combine for navigation to the target. The radiometric data determines the spacecraft's Sun-centered trajectory. Optical data for navigation, which images the target planet in the same frame with a known star, positions the spacecraft with respect to the target body. The information combines to improve the target's Sun-centered position or the planet's ephemeris. With this combination, the spacecraft and the target are mutually positioned at the insertion time with respect to one another.

With additional data, we improve our ability to hit the desired aimpoint. We use several pre-encounter maneuvers to achieve this aimpoint. Early maneuvers can change the spacecraft aimpoint a lot but are therefore susceptible to large errors. Later maneuvers have less leverage but are more accurate. A typical sequence for a interplanetary encounter includes maneuvers at −60, −30, and −10 days.

**Orbit Insertion.** Orbit insertion is a unique opportunity because it requires the spacecraft to have a certain energy with respect to the target body and the Sun. The best point is where the heliocentric orbits of the spacecraft and the target planet are nearly aligned and are nearly the same size. That's where we need the least velocity change to orbit the planet. In a planet-centered frame, the most efficient point is at the closest approach point, where the velocity, and hence the effect on orbital-energy change, is greatest. We time the insertion burn by trading between the most efficient mission and one that allows early completion of the science, which may drive an earlier, less efficient, insertion strategy.

Due to light-time delay, this event happens in the blind, so we have to balance anxiety while no link is available against response time in case something goes wrong. We must therefore test the stored sequence for the event against any possibility of problems. Also, the spacecraft is usually in a special fail-safe mode for this critical event. This mode may include powering down science instruments to create extra power margin or powering up otherwise unused subsystems for engineering backup.

### 17.3.6    Encounter

Encounter lasts from hours to years for the different mission types and is the main period for science gathering. Thus, operations have historically been fairly intensive for this phase, when we take very little risk with spacecraft or science data. Reducing risk drives up the cost of operations.

**Checkout on Orbit.** We'll first use many instruments and observations strategies during the initial orbital phases. For some experiments, the pre-encounter time is enough to wring out problems with operational instruments. Instruments that require a target to operate successfully, such as an in-situ or active remote-sensing instrument, must be checked out early in the orbital-operations phase.

**Science Operations on Orbit.** Sharing all of the time in a mission between different instruments gives us the best science data but uses all margins and a lot of dollars. Allocating time instrument by instrument is inefficient but decreases integration costs.

**Strategy for Engineering Maintenance.** Continuously calibrating engineering subsystems increases performance but costs time in planning, integration, and analysis of the downlink. Also, in the past, we've used continuous engineering data to get the best returns from a mission. Doing so requires more uplink and downlink activity for characterization and analysis, plus replanning of subsequent activities to take advantage of performance changes.

### 17.3.7     Extended Mission

An extended mission is often possible, but we must take more risks to achieve it. Interplanetary missions are often extended because we've used worst-case or 99% planning throughout the mission. If the mission ends without significant problems, the spacecraft is usually proven reliable and the consumables aren't exhausted.

An extended mission means two things for mission operations. First, mission planners always keep the extended mission in their thinking. An example is the Voyager flybys of Saturn and Uranus, which were constrained because of the chance of going to another target. (The Jupiter flyby was similarly constrained because of the planned Saturn flyby [Kohlhase, 1989].) Second, costs tend to exceed pre-launch estimates. But we can reduce this effect by accepting more risk and shrinking the flight team as the Magellan mission did.

### 17.3.8     Ongoing Mission Planning

Mission planning is continually revisited on most interplanetary missions because of increasing accuracy of the predicted spacecraft trajectory. These updates can result in either large or small changes in the mission plan. If, for example, the spacecraft performance is markedly better (or worse) than predicted, targets may be included (or deleted). An example of this is Galileo's flyby of Ida's asteroid, which was enabled by a launch early in the month-long launch period, and also by favorable spacecraft performance. The propellant budget then allowed the "detour" from the optimal trajectory to fly by the asteroid.

Small updates to planned activities result from collecting and processing navigation data. The navigation aspect of the pointing problem is essential due to the pointing of most observations relative to the target, coupled with the unknown position of the spacecraft relative to the target. A late navigation solution is folded into the observation design by adjusting the start time of an entire block of observations.

Pre-encounter analysis may allow us to account for expected navigation errors by planning an observation strategy which includes all possible positions of the target. In this case, the mission-planning function will be in an iterative loop with the performance of the navigation system throughout the mission, in order to drive out inefficiencies with more accurate predictions.

Many interplanetary missions have used most of the spacecraft's resources at the expense of ground resources, which means that operations plans were very tightly coupled and therefore very sensitive to change. For example, a change in DSN station from one hemisphere to another will result in a shorter pass and a lower peak data rate. If we've planned the data flow to use all of the planned capability, the change in station will force us to replan activities.

Improved understanding of performance leads to changes in the telecom link, which may change the observation strategy if a planned scenario is optimized to link performance. If we don't need top performance, small changes in the link performance wouldn't result in observation changes.

We may improve the spacecraft-pointing performance by frequently calibrating elements such as gyros, but we must trade time for this activity against time for science observations.

We can adapt to some changes in the trajectory which improve science at an encounter without seriously affecting future objectives. In an orbital-tour mission, we can sometimes best correct navigation errors by reoptimizing the trajectory rather than by returning to the nominal trajectory.

### References

Bate, Roger D., Donald D. Mueller, and Jerry E. White. 1971. *Fundamentals of Astrodynamics.* New York, NY: Dover Publications, Inc.

Beatty, J.K., B. O'Leary, and A. Chaikin. 1990. *The New Solar System.* Cambridge MA: Cambridge University Press.

Kohlhase, C. ed. 1989. *The Voyager Neptune Travel Guide.* JPL Publication 89-24. California Institute of Technology, Pasadena, CA.

Larson, Wiley J. and James R. Wertz. 1992. *Space Mission Analysis and Design.* 2nd ed. Netherlands: Kluwer Academic Publishers.

Young, C. ed. 1990. *The Magellan Venus Explorers Guide.* JPL Publication 90-24. California Institute of Technology, Pasadena, CA.

# Chapter 18

# International Space Mission Operations

Karen R. Altunin, *AR Space Enterprises*
Valery I. Altunin, *Jet Propulsion Laboratory,*
*California Institute of Technology*
Mikhail Artukhov and Konstantin G. Sukhanov,
*Lavochkin Association (Moscow)*
Viktor D. Blagov, *Flight Control Center (Kalliningrad)*
Viktor I. Glebov, *All Union Research*
*Institute of Electronics (Moscow)*
Dennis Taylor, *Space Systems Engineering, Ltd.*

The threat of having a mission descoped or ended because it costs too much to operate has motivated managers of current projects and designers of future missions to search for solutions in foreign space projects. Many alternatives exist including jointly operating a mission with domestic and foreign people or with foreign ground support. Increasingly, Europe, the US, Russia, and Japan are investigating and formally agreeing to such alternatives.

The topics discussed in this chapter should help you, as an operations manager or mission designer, in deciding whether to seek or secure foreign partners and in structuring cooperative missions. The topics include: (1) how operations for international space projects have been and can be structured; (2) the challenges which arise when attempting to develop a mission operations system that reflects each participating country's unique culture, history, and space infrastructure; (3) recommendations for constructing an international space partnership as a money-saving venture; and (4) operations practices in the space missions of Russia and the European Space Agency.

Proposals for developing space resources and pursuing scientific advances have steadily become grander, more complex, and therefore more costly. Costs have increased at a time when people across the globe have demanded large cuts in government spending. Also, the number of nations interested in pursuing space ventures has increased because most industrialized nations recognize they must be in space to compete in the global economy. These nations want to spread the financial burden among all interested parties. Many nations, including the United States and Russia, want to symbolically bury the vestiges of the cold war by embarking on a joint scientific (and peaceful) mission with their former foes.

International cooperation in space brings significant benefits: (1) Grandiose goals need financial backing from several governments, rather than one. Thus, more complex missions, such as the Space Station, can be pursued on a grander scale. Others, such as the Mars Exploration Initiative, couldn't go on without this backing. (2) Technical advances will proceed globally at a greater pace if nations share technology and stop wasting limited resources on redundancies. (3) Scientific knowledge will increase significantly if we seek it openly, make it accessible to all, and share expensive resources (such as tracking networks). The money saved can go to more experiments. And, most importantly, (4) world peace will stand a greater chance if nations band together to pursue knowledge and wonder.

## 18.1 Overview of Cooperative Missions

### Karen R. Altunin and Valery I. Altunin

All space missions consist of a set of elements or components forming the space mission architecture.[*] They include the mission objectives, orbit(s), space element (payload and spacecraft bus), launch element, ground element, mission operations, and command, control, and communications. All of these elements are candidates for international cooperation, depending on the mission concept. Once the concept is developed, a project must assign performance requirements or expectations to each international partner based on the abilities of the international partners and the known constraints. Abilities and constraints include those driven by mission resources and by user requirements.

International cooperation can occur at three different levels:

    1.     Subsystem—partner supplies instrument to be integrated with spacecraft provided by another partner, e.g., Earth Observing System project [Smith, 1993].

---

[*] *Space Mission Analysis and Design*, 2d edition, eds. W. J. Larson, J. R. Wertz, (Microcosm, 1992) pg. 10.

2.  System—spacecraft supplied by one partner and controlled or tracked by ground element of another partner, e.g., Radioastron project [Altunin, 1990; Altunin, 1992; Altunin, 1993].

3.  Mission—each partner supplies own spacecraft and ground element but shares the same mission objectives with other nations, e.g., Halley's Comet Campaign [Armand, 1986], International Mars Exploration Initiative.

Once we've decided the level of participation, we can determine the level of involvement in mission operations. The partners may choose to share operations or they may decide to delegate this responsibility to one particular partner.

To assign the responsibilities, we need to consider how choosing a particular partner will affect certain costs: (1) The cost of equipping the nation with necessary hardware or other materials unavailable at its operating centers, (2) the availability and cost of insurance to protect the hardware and people donated by the participating countries, (3) delays in the mission because local operators lack experience and because of the time for shipping and equipping the operations centers with any necessary hardware, and, (4) the cost of training (including transporting, housing, and feeding) the foreign operators at the operations site.

### 18.1.1    Mission Operations Functions in International Missions

Mission operations functions exist at either the subsystem, system, or mission levels of international cooperation. (See Chap. 3 for description of each function.) Table 18.1 describes how most mission operations functions can be shared depending on the level of participation in the mission architecture. Functions such as spacecraft planning and analysis or computer and communication support are usually not subjects for cooperation in an international mission. But mission planning and management are candidates for cooperation at all three levels. Table 18.2 describes the relationship between international management of mission operations and planning and scheduling.

**The Ground Element and International Cooperation.** The ground element divides into three main parts: (1) spacecraft-tracking stations (networks), (2) control centers (spacecraft, payload, and navigation centers), and (3) communication links between stations (see Chap. 12, Ground Systems). Ground stations and their main activities (tracking, acquiring spacecraft data, and navigation measurements) are prime candidates for international cooperation because their geographical locations make them valuable and it's expensive to build and maintain them. A tracking station's location is important for lengthening communication sessions with the spacecraft—particularly for Earth-Observing Satellites, very-long-baseline interferometry (VLBI) with high data rates, and crewed missions—or for navigation measurements such as VLBI or three-way Doppler.

In general, the existing ground networks or satellite-communication facilities sponsored by different agencies fit four categories: deep-space communications;

**Table 18.1.** **Levels of Cooperation and Mission Operations Functions.** How we assign mission operations functions depends on the level of participation in the mission architecture.

| Functions | Level of Cooperation and Mission Operations Functions | | |
|---|---|---|---|
| | Subsystem Level | System Level | Mission Level |
| Mission Planning | X | X | X |
| Activity Planning and Development | X | | |
| Mission Control | | | X |
| Data Transport and Delivery | | X | |
| Navigation Planning and Analysis | | X | X |
| Payload Planning and Analysis | X | | X |
| Archiving and Maintaining the Mission Database | X | X | X |
| Systems Engineering, Integration, and Testing | X | | |
| Managing Mission Operations | X | X | X |

**Table 18.2.** **International Management of Mission Operations and Planning and Scheduling.** The level of international cooperation determines the appropriate management elements and mission operations functions.

| Level of Cooperation | Management Element | Responsibility/Mission Operations Functions |
|---|---|---|
| Mission | International mission's steering committee or project offices in participating agencies | 1. Coordinating mission objectives (experiment selection)<br>2. Resolving conflicts<br>3. Allocating resources |
| System | International mission's operations group, spacecraft-operations group, or operations group for the ground element | 1. Long-term planning<br>2. Supporting users |
| Subsystem | Operations subgroup (commanding, housekeeping telemetry, navigation, or payload telemetry) | 1. Scheduling<br>2. Planning activities |

Earth-orbiting uncrewed; Earth-orbiting crewed; and special systems (e.g., communication, navigation, and weather satellites). The basic characteristics of the major ground networks sponsored by the various space agencies for deep-space missions and Earth-orbiting spacecraft are in Tables 18.3 and 18.4, respectively [CCSDS, 1990; Smid, 1991].

**Table 18.3.** **Complexes for Deep-Space Communication.** The listed complexes are the only ones in the world that can do operations for spacecraft in deep space (distances greater than 100,000 km from Earth). All sites provide telemetry, commanding, radiometry, and very-long-baseline interferometry (VLBI).

| Site/Agency | Antenna diameter | Locations | | | Frequency Band |
|---|---|---|---|---|---|
| | | Latitude | E. Longitude | Altitude | |
| Goldstone, California/US NASA | 70 m | 35 25 33.3 | 243 06 40.6 | 1019 m | S/X*/(L*) |
| | 34 m | 35 25 18.8 | 243 06 49.1 | | S/X |
| Tidbinbilla, Australia/US NASA | 70 m | −35 24 14.4 | 148 58 58.1 | 818 m | S/X*/(L*) |
| | 34 m | −35 24 00.1 | 148 58 35.2 | | S/X |
| Madrid, Spain/US NASA | 70 m | 40 25 56.6 | 355 45 11.0 | 791 m | S/X*/(L*) |
| | 34 m | 40 25 42.1 | 355 44 59.7 | | S/X |
| Evpatoria, Ukraine/UkrSA & RusSA | 70 m | 45 11 22.0 | 33 11 19.0 | 5 m | P†/C/X*/(L*) |
| | 32 m | 45 11 22.0 | 33 11 19.0 | | P†/C |
| Ussuriisk, Russia/RusSA | 70 m | 44 00 57.0 | 131 45 22.0 | 75 m | P†/C/X*/(L*) |
| | 32 m | 44 00 57.0 | 131 45 22.0 | | P†/C |
| Bear Lakes (Moscow), Russia/ RusSA | 64 m | 55 51 57.0 | 37 57 17.0 | 152 m | P†/C/X*/(L*) |
| Usuda, Japan/ISAS | 64 m | 36 07 56.9 | 138 21 45.7 | -- | S/X* |
| Weilheim, Germany/DLR | 30 m | 47 52 52.3 | 11 06 01.2 | -- | S/X* |

\* Receive only; (L*) - frequency is not allocated for deep space communications, receive only.
† 157/184 MHz and 745/930 MHz (uplink/downlink).

**Table 18.4.** **Tracking Stations for Earth-Orbiting Satellites.** This table lists the main civilian operations complexes for Earth-orbiting spacecraft. See CCSDS, 1990 for a listing of all tracking stations. All sites provide telemetry, commanding, and Doppler tracking.

| Site/Agency | Antenna Diameter | Locations | | Frequency Band |
|---|---|---|---|---|
| | | Latitude | E. Longitude | |
| NASA (USA) | | | | |
| Goldstone, California | 26 m | 35 09 | 243 07 | S |
| | 11 m | 35 09 | 243 07 | X/Ku |
| Tidbinbilla, Australia | 26 m | −35 13 | 148 59 | S |
| | 11 m | −35 13 | 148 59 | X/Ku |
| Madrid, Spain | 26 m | 40 14 | 355 44 | S |
| | 11 m | 40 14 | 355 44 | X/Ku |

**Table 18.4.**  **Tracking Stations for Earth-Orbiting Satellites. (Continued)** This table lists the main civilian operations complexes for Earth-orbiting spacecraft. See CCSDS, 1990 for a listing of all tracking stations. All sites provide telemetry, commanding, and Doppler tracking.

| Site/Agency | Antenna Diameter | Locations | | Frequency Band |
|---|---|---|---|---|
| | | Latitude | E. Longitude | |
| RSA (Russia) | | | | |
| Evpatoria[*], Ukraine | 25 m × 2 | 45 11 | 33 11 | P[†]/C |
| Ussuriisk, Russia | 25 m × 2 | 44 00 | 131 45 | P[†]/C |
| Tshelokovo, Russia | 25 m × 2, 12 m | 56 01 | 37 52 | P[†]/C, P[†] |
| St.Petersburg,Russia | 12 m | 68 02 | 33 09 | P[†] |
| Jusaly, Kazahstan | 12 m | 45 19 | 64 03 | P[†] |
| Kolpashevo, Russia | 12 m | 58 12 | 82 35 | P[†] |
| Ulan-Ude, Russia | 25 m, 12 m | 51 33 | 107 24 | P[†] |
| Petropavlovsk, Russia | 25 m, 12 m | 53 18 | 158 26 | P[†] |
| CNES (France) | | | | |
| Aussaguel, France | 11 m | 43 26 | 01 30 | S |
| Kourou, Fr. Guyana | 11 m | 05 06 | 307 22 | S |
| Hartebeestoek, S. Africa | 12 m | −25 53 | 27 42 | S |
| ESA | | | | |
| Maspalomas, Spain | 15 m | 28 00 | 344 42 | S |
| Kiruna, Sweden | 15 m | 67 52 | 20 57 | S |
| Perth, Australia | 15 m | −31 48 | 115 53 | S |
| Villafranca, Spain | 15 m | 40 27 | 356 03 | S |
| NASDA (Japan) | | | | |
| Masuda, Japan | 18 m, 13 m | 30 33 | 131 01 | S |
| Katsuura, Japan | 18 m, 13 m | 35 12 | 140 18 | S |
| Okinawa, Japan | 18 m, 18 m | 47 53 | 11 06 | S |

[*] Operated jointly by Russian Space Agency and Ukrainian Space Agency.
[†] 157/184 MHz and 745/930 MHz (uplink/downlink).

Cooperation for ground systems supporting an international project can occur at the

1. Subsystem level—Tracking station of one agency becomes part of the ground-tracking network of another agency. For example, during the Voyager Neptune encounter, the US DSN antenna in Canberra

(Australia) was arrayed with the Japanese DSN antenna in Usuda and the Parkes radio telescope (Australia) to achieve greater sensitivity for receiving the spacecraft signal.

2.  System level—One agency's ground network of tracking stations supports operations of another agency's spacecraft. For example, a network of three US DSN antennas will provide data-acquisition and navigation support for Russian (Radioastron) and Japanese (VSOP) VLBI missions. Also, US DSN provided VLBI navigational support for Russian spacecraft Vega and Phobos, Japanese spacecraft Planet A, and ESA spacecraft Giotto during the Comet Halley campaign.

3.  Mission level—Ground networks of different agencies provide coordinated support for a mission. For example, a world-wide network of fifteen radio telescopes, plus US and Russian DSNs, acquired data from Venus Balloon (part of the Vega mission).

The main elements of any future Mars mission will be ground support and deep-space communications. The DSN will maintain communications with the relay satellites that support the Mars landers, maintain mission support during cruise, and directly communicate with instruments on the surface of Mars. It will also execute scientific experiments, such as radio science, celestial mechanics, and planetary research. Due to the growing demand for US DSN support from US deep space missions (Galileo, Ulysses, Voyager, Cassini, Pluto, NEAR), resources for supporting the Mars mission will be limited. But foreign complexes for deep-space communications will be able to overcome this shortage and provide these key benefits:

- Lower 70 m loading of the US DSN
- Increased coverage
- Improved navigation quality
- Increased sensitivity through arraying of deep-space antennas
- New science

## 18.1.2    Requirements and Constraints on International Missions

The requirements imposed on each international partner depend on the partnership's level of cooperation. For example, (1) at the subsystem level, whoever provides scientific instruments should provide the command sequences and telemetry analysis for the instrument; (2) at the system level, each subnetwork of an international tracking network should use the same reference frame to most accurately position the spacecraft, and (3) at the mission level, partners should share mission costs or offer services at a reduced cost in exchange for participation.

International partners must cooperate to achieve:

**Compatibility.** Incompatibility arises from technical differences in hardware and software or from differing management structures, professional language, or operating methods. Cultural differences and different levels of funding in each sponsoring nation often drive these incompatibilities.

**Noninterference of Elements.** Each partner contributing an instrument to the spacecraft must ensure it won't keep the spacecraft or other instruments from operating properly. If an international mission involves several spacecraft, the owners of each spacecraft must comply with requirements for proper orbit and frequency separation established and enforced by international bodies such as the World Administrative Radio Conference and the Space Frequency Coordination Committee.

**Security and Reliability.** The high cost of space missions requires all partners to protect the mission. Thus, the partner who approves the final commands must have simulated them so no instruction to the spacecraft or science payload will damage it. Ideally, they'll simulate hardware and software. All participants should share the cost of developing and operating these simulators because all benefit from a secure mission.

**Mission Efficiency.** We define *mission efficiency* as the amount of useful data returned from the spacecraft and its payload over a given transmission time. International partners must agree before launch what level of efficiency is both necessary to satisfy all the sponsoring agencies and possible given the mission resources. They must then develop an operations system that achieves this level. For example, several strategies have helped resolve the dilemma of inadequate coverage by tracking stations. On the space VLBI mission, Radioastron, science data moved at too high a rate (128 Mbps) to be stored onboard. Thus, to capture enough science data to meet goals for mission efficiency, planners had to establish a new tracking network and globally integrate the existing network.

**Mission Commitment.** Prestige and cost demand firm commitments from all partners—a difficult goal considering the dynamics of global economies and national priorities. But the US and Russia often use international participation to protect their investment and secure future government funding. Examples are the US's Space Station and Russia's Spectrum series of scientific spacecraft, which international participation has helped to keep alive. ESA is less vulnerable to project cancellations because they have inter-governmental agreements and long-term funding commitments.

In international missions, constraints often drive requirements:

**Language.** To overcome the language barrier, operations teams from all participating nations should train together before launch. They should also have technical dictionaries, including the definitions of acronyms used in the project. All operations computers should contain translation software, and operators should have classes in foreign languages.

**Management Structures.** Each partner needs to know the decision makers and the organization tree of all other nations. Partners should develop these trees from scratch, using titles and roles everyone understands.

**Management Styles.** Decision making and operational dynamics depend partly on culture, so contrasting management styles can often lead to misunderstandings and inefficiencies. For example, if one country's engineering teams usually dominate decision making for a space project, placing them in an organization for which science teams dominate could be counterproductive. Often, international missions require a new management style that reflects traditions, sensitivities, and attributes of each partner.

### 18.1.3    Allocating Resources and Responsibilities Among Partners

These allocations depend on how the partnership forms. The mission may have started as a national mission and then added foreign partners—an evolved partnership. For example, the International Space Station began as the US's project. Or the mission may have been cooperative from the beginning—an original partnership.

For an original partnership, allocations will depend on assessing what is the mission's most critical element and how much of the available funds its development will consume. For example, for the International VLBI Satellite (IVS), the most critical element was the science payload—the radio astronomy antenna. At first, the mission considered a 25 m antenna, but cost analysis showed that size antenna would consume 80% of the money from sponsoring agencies. So they chose a 15 m antenna. Once they met the requirements of the most critical element, they could set priorities for supporting elements (such as the mission operations system) and allocate the remaining funds to each one.

Allocations for evolved partnerships depend on when foreign partners join the project. If the original partner has already designed and developed the critical element or supporting elements, we simply need to set priorities and assign the remaining elements to particular partners.

Evolved partnerships experience several problems: (a) equitable allocations may be more difficult because the initial sponsor most likely won't get back the development money they spent before the mission became cooperative; (b) initial sponsors may have trouble distributing responsibilities, and partners may balk at constraints, such as incompatible standards, which weren't part of the project's early planning.

An operations concept for an international mission should describe the products, interfaces, and timelines which will govern the operational relationships between the partners. It should also provide a mission architecture that will underpin future detailed descriptions of mission requirements.

Mission operations concepts evolve as mission resources become firmer and as the partners better understand their respective roles. One of the main challenges early on in an international mission is to integrate the mission operations concepts from all the key participants. Rather than having each participating agency generate its own concept and then reconciling them, representatives from each agency should work together from the beginning to create one concept. Teams more easily resolve conflicts. Also, all team members should review the concept often, so they benefit from combined wisdom [Altunin, 1992]. As described in Sections 18.2, 18.3, and 18.4, developing a mission operations concept differs greatly in Russia, the US, and ESA.

## 18.2 US–Russia Joint Missions

### Karen R. Altunin and Valery I. Altunin

Since the mid-70s, Russia has joined the US in several missions:

- Apollo Soyuz (the 1975 crewed mission) [Froehlich, 1976]
- Cosmos 782 (the 1978 US biomedical experiment on a Russian spacecraft)
- Sarsat/Cospas (1980–present, US-Russia-ESA search and rescue program)
- Vega (the 1985–86 Venus balloon and pathfinder mission) [Armand, 1986]
- Phobos (the 1988–89 Phobos lander) [Sunyaev, 1988]
- Shuttle-MIR (1995)

Table 18.5 summarizes the key characteristics, architecture, and mission operations functions of some prior cooperative missions.

**Table 18.5.**   **Cooperative Space Missions Involving Russia and the US or Europe.** Future cooperative missions will likely have similar joint elements in the mission architecture and operations functions. See Table 18.6.

| Joint Mission/Experiment | Joint Elements of the Mission Architecture | Joint Mission Operations Functions |
|---|---|---|
| Apollo-Soyuz (1975); US and Russian crewed spacecraft meet in space, united crew performs joint experiments | • Orbit - two spacecraft must have the same orbit<br>• Mission Operations - must work independently but coordinate fully | • Mission planning<br>• Payload (crew work) planning and analysis<br>• Navigation<br>• Management |
| Vega-balloon (1985); Russian spacecraft carries ESA/CNES science payload VENUS-balloon. The balloon flies freely in the atmosphere of Venus and transmits signals to the Earth | • Ground Element - worldwide tracking network, including US DSN, supports data acquisition from balloon<br>• Mission Operations - ground network must coordinate with the balloon's transmitter operations | • Mission planning<br>• Data transport and delivery<br>• Data processing<br>• Archiving<br>• Navigation<br>• Management |

**Table 18.5.    Cooperative Space Missions Involving Russia and the US or Europe. (Continued)** Future cooperative missions will likely have similar joint elements in the mission architecture and operations functions. See Table 18.6.

| Joint Mission/Experiment | Joint Elements of the Mission Architecture | Joint Mission Operations Functions |
|---|---|---|
| Vega-Pathfinder (1986); Russian spacecraft was the first to encounter the Halley comet and accurately determine its position, allowing GIOTTO (ESA) to navigate as closely as possible to the comet | • Orbit - Russian spacecraft's measurements determines the orbit of ESA's spacecraft<br>• Ground Element - US DSN takes navigation measurements for Russia's and ESA's spacecraft<br>• Mission Operations - coordinated operations: Russia, ESA spacecraft, and US DSN | • Mission planning<br>• Archiving<br>• Navigation<br>• Management |
| Granat (1989–present); Russian spacecraft Granat carries telescopes, particularly the SIGMA X-Ray telescope (CNES), from different countries | • Space Element - SIGMA X-ray telescope onboard Russian spacecraft<br>• Mission Operations - spacecraft (Russian) and telescope (French) teams jointly operate the telescope | • Mission planning<br>• Activity planning and development<br>• Payload planning and analysis<br>• Systems engineering, integration, and test<br>• Management |
| Shuttle-MIR (1995); A US astronaut did experiments on the MIR Space Station, and Russian Cosmonauts returned to Earth on the Shuttle | • Orbit - Shuttle and MIR must have same orbit to dock<br>• Space Element - A new docking mechanism was developed<br>• Mission Operations - coordinated operations for docking | • Mission planning<br>• Activity planning and development<br>• Navigation<br>• Data transport and delivery<br>• Management |

The US and Russia benefit significantly from combining their experience in space exploration. Each nation has developed superior hardware and techniques for maximizing the returns from their respective missions. With political barriers removed, the two nations want to share their vast reserves of technical know-how. The US can benefit by reducing mission costs with bargain-basement Russian hardware and services; Russia can benefit by using US dollars to finance ongoing and future missions.

### 18.2.1    Requirements and Constraints on US–Russia Missions

**Compatibility.** Hardware from the US and Russia are often incompatible. For example, Russia uses C- and P-bands for spacecraft operations, whereas Western agencies use X- and S-bands. Resolving this type of incompatibility requires some adapting: (1) Installing an L-band transmitter on the Vega Balloon to make it compatible with a global network of radio telescopes. (L-band [1.67 GHz] is used for radio astronomy but not deep-space communications.) Thereafter, both the US and Russian DSNs installed L-band receivers on their respective DSN antennas to support this mission. (2) Using C- and P-bands for spacecraft operations on Mars

'96 and adding an X-band so the US DSN can determine the spacecraft's position. (US DSN antennas don't have C- or P-bands). (3) Choosing C- and P-bands for operations on Radioastron, but equipping the science payload with X- and Ku-bands so tracking stations can cover it world wide. Resolving the incompatibility of radio channels has also increased reliability and flexibility for mission operations by adding channels.

**Spacecraft Security.** Previous missions show we must carefully develop operations sequences to decrease risk. The Phobos project is a case in point:

### The Loss of Phobos I

After the Vega mission's great success in 1986, the Soviet Academy of Sciences decided to observe the Martian moon, Phobos. As with all Soviet space projects, two spacecraft, Phobos I and II, were built and later launched in 1988. The Phobos spacecraft had many foreign instruments onboard. The participating scientists were to select a code and develop the instrument commands. Because they lacked time and staff, the spacecraft-control group never fully tested the science-command loads before uplinking them. Because the commands for the science payload and spacecraft were written in two different codes, the uplink operator was supposed to instruct the spacecraft's command-processing computer that the upcoming commands were for either the science payload or spacecraft. Unfortunately, during cruise, two mistakes occurred. The operator didn't instruct the spacecraft that the upcoming commands were for the science payload, and the transmitted science commands (when translated into spacecraft code) matched instructions to the spacecraft to turn off its pressure-fed propulsion system. The terminal command load was uplinked at the end of the day's communication session. By the time the spacecraft controllers returned to work the following day, the spacecraft's batteries had already drained. As a result, Phobos I was deaf to any further communication from the ground.

**Language.** Both nations have insufficient operators who are fluent in English and Russian. While this problem also exists on missions between the US and Japan or Europe, those partnerships fortunately share technical terms. For example, the Japanese have adopted Jet Propulsion Laboratory terms to discuss their mission-planning methods. Because the Japanese and Europeans entered the space race after the US, they modeled many aspects of their programs after US programs. But the Russians have developed their own technical terms, which often are dissimilar to US concepts.

Recommendations: Although experienced translators can be worth their weight in gold, the project shouldn't depend on them. We can never foresee how many interactions will require translators, who are expensive and unable to work hour after hour without relief. Under-estimating needs and avoiding costs may cause managers to hire too few translators for a project. Even if a joint project were blessed with several talented translators, subtle nuances in the two languages may lead to misunderstandings and problems. Unless translators have lived and worked for several years abroad, they're unlikely to understand these nuances.

Because translators may be unavailable or capable of error, we need common dictionaries which team members can refer to during operations. And the terms defined must be consistent with those in *all* the documentation operators require. Of course, operations people from both countries must train together long before launch to overcome language barriers and make joint missions successful.

**Access.** As recently as 1992, communication with Russia was wrought with frustration. A shortage of telephone circuits into Russia led to seemingly endless busy signals. Fax machines were turned off at night at the space agencies because of paper shortages. Inefficiencies in the Russian postal system made mail deliveries unreliable, and transmitting an electronic message from a US space agency to a Russian space agency required prior governmental approval. Fortunately, many of these roadblocks are well on the way to disappearing, but the most reliable way to communicate with a Russian counterpart on a space project is through dedicated voice and data channels. These are essential to the success of any joint mission involving Russia or any other former members of the Soviet Union.

**Management Structure.** Russia's structure differs from the US's in several ways. Russia tends not to have mission operations managers (MOM). Instead, the director of the General Operative Control Group (GOCG, the lead operations organization for uncrewed spacecraft in Russia) plays this role, with several key differences. The GOCG director represents all current uncrewed science missions, so his attention can be divided among several missions at any given time. Also, engineering teams dominate the GOCG, which arguably leads it to favor more conservative spacecraft operations (for example scheduling fewer science activities or extending the checkout period in orbit). In the US, the MOM handles one mission and doesn't favor engineering or science.

Recommendation: A balanced approach to operating a US-Russia mission needs similar management structures. If the US and Russia share operations, the two organizations should mirror each other, so team members aren't confused about the flow of data and decisions.

**Management Process.** Management processes in Russia also differ from those in the US for several important reasons. First, Russian projects develop different sets of documents than US projects normally do because they have different requirements and operations philosophies (such as acceptable levels of risk on a given mission). Secondly, they develop concepts, procedures, and documents on a different timeline. Russia's operations teams are accountable to their own "NASA," which in turn must respond to requirements from its own "Congress." The Russian versions of NASA and Congress, of course, differ considerably from the US versions [Bogdanov, 1993].

Recommendation: Staff from the US should clearly state what documentation they need and work closely with their Russian counterparts in preparing it. They may need to assemble the data themselves and provide it to the Russians for review and approval. Each cooperative mission should also have a joint working group that meets regularly and often to exchange and review information.

**Function Versus Form.** The US's space teams tend to look for large-scale technical advances with each new project, whereas Russia's teams are satisfied with making smaller, steadier progress. Scientific space hardware evolves much faster in the US than in Russia. For example, while the US space program spent a large amount of money developing a pen that would operate in a zero-gravity environment, the Russian space program decided to equip its cosmonauts with pencils. These contrasting approaches also are apparent in each country's launch vehicles. US rockets are like "Ferraris, highly tuned, complicated, and fragile" whereas Russian rockets are more like "trucks—simple, rugged, and economical."[*] These differences are deeply rooted in the contrasting histories of the two countries. Whereas planned obsolescence and product "bells and whistles" are essential to a successful consumer-based economy, Russian designers have had to emphasize function over form.

<u>Recommendation</u>: The Russian approach is definitely more economical than the US approach. For instance, Russia's Proton launch vehicle spends only a few hours on the launch pad and has a launch crew of fifty, whereas the US's Titan 4 launch vehicle spends at least three months on the pad, with a crew of one thousand.[†] But the "small is beautiful" attitude has gained momentum in the US since 1990, so the two approaches are becoming more compatible. At the same time, joint missions should encourage one US attribute—striving to invent new strategies for mission operations.

**Customer Orientation.** Russian scientists have had to place their instruments on existing spacecraft, which were designed without their specific instruments in mind because the design industries weren't accountable to the Russian Academy of Sciences or other scientific institutions. For example, the Russian astrophysical-series of spacecraft (Spectrum-X Gamma, Radioastron, Ultra-violet, Infrared) was actually based on the spacecraft bus developed for the military. When Russian scientists complained about being forced to use existing space buses, the Russian Academy of Sciences (Space Research Institute or "IKI") began to develop a new spacecraft, Regatta, for scientific use (e.g., solar plasma exploration, astrometry). (As of the publication date of this book, none of the Regatta series has launched.) In the US, scientific spacecraft tend to compliment specific instruments.

Also, in Russia the organizations that operate a spacecraft have been part of the industries that built it. For example, the GOCG is a part of the Lavochkin Association, which has built most of the uncrewed scientific spacecraft. The Flight Control Center in Kalliningrad (near Moscow) is managed by an institution closely related to the Energia Corporation, which has developed and managed crewed missions.

---

[*] Washington Post, John Mintz, January 12, 1994
[†] Idem

Thus, Russian development and operation of scientific spacecraft has been oriented towards engineering, with science being almost an afterthought—opposite to the attitude in the US that the scientist is the customer.

**Exchange of Funds.** With the Russian ruble devalued, Russians prefer to receive US dollars for goods or services, but transferring dollars from the US to Russia is expensive. Russian banks which accept electronic transfers of US funds do so at a premium. And, once the funds are transferred, they distribute dollars instead of rubles to the intended recipient only for an additional fee. Checks aren't accepted. This constraint should disappear as Russia moves further into the free-market system.

Recommendation: US organizations that need to exchange funds between the US and Russia should hire an intermediary to transfer the money. Two other alternatives are: (1) agree with the Russians to transfer technical hardware or consumer goods instead of dollars or (2) agree to match resource contributions (such as people, launch, or tracking services) so neither country needs compensation.

**International Science Planning.** International missions need a science strategy before launch which establishes guidelines for selecting the onboard instruments and experiments, as well as a science plan for operations. This plan requires timelines for science activities and targets for specific experiments.

An International Science Steering Committee (ISSC) selected experiments for the Vega, Phobos, Granat, and Mars '94–96 missions. This committee had representatives from the science teams of each country which helped to develop the instruments. Selected experiments during mission development were based on the mission's known technical capabilities and constraints.

For the Halley's Comet Campaign, an Interagency Coordinating Group for Space Science was created in 1982, with representatives from each space agency sponsoring the mission. The committee coordinated the entire mission. They used information from one spacecraft to navigate another spacecraft and combined measurements from each spacecraft to provide physical parameters and images of the comet from different perspectives. At this stage, the participating nations must agree to what extent mission operations will support all the experiments selected.

**Activity Planning and Development.** Activity planning and development on Russian projects starts when the payload integration with the spacecraft begins: usually, 1.5 to 2 years before launch. In the US, this effort usually begins earlier. In Russia, payload integration normally is complete about a year before launch. The spacecraft team then provides the science team with a timeline that shows when the spacecraft can support experiments. In America, the science teams usually have more latitude in scheduling their experiments because projects serve science as much as possible.

The lead organization for planning activities on Russian scientific spacecraft is the General Operative Control Group (GOCG) (See Sec. 18.3). As in the US, a science-operations team representing the participating scientists develops commands to operate the science payload and to process science data. The GOCG ensures the commands from the science group don't jeopardize the spacecraft's health.

### 18.2.2    Routine Operations

**Command and Control.** The US relies heavily on software to plan, generate, and validate command loads. Due to the shortage of high-speed computers and desktop systems, relatively inexpensive technical labor, and the government's emphasis on full employment, people did much of the operations planning and command generation in Russia. Today, software handles more of this work because computers have become more commonplace.

Final command loads in Russia are formed only three days before uplink. Because the science targets and spacecraft health are both dynamic, this approach avoids having to change a planned load. It saves money but forfeits some reliability because there's not much time to simulate the commands. Also, planned experiments or observations are more subject to being bumped from the activity plan because problems with command loads may not surface in time to be resolved before uplink.

<u>Recommendation</u>: The type of mission determines when to form the command loads. If a science mission and the experiments are unique (e.g., Voyager's flyby of Neptune), we must support all planned experiments. In such a case, we have to spend more money during operations to develop the command loads with enough lead time to fully simulate them.

While a Russian planetary or interplanetary spacecraft is in routine operations, the spacecraft-control team and science-payload representatives are together at the Satellite Control Center in Evpatoria, the Crimea (about 600 miles south of Moscow). Both groups are part of the GOCG. A housing shortage in Evpatoria keeps entire families from moving there, so the GOCG rotates its operations teams monthly. For crewed space missions, the science and engineering teams are at the Flight Control Center in Kalliningrad (about 10 miles from Moscow).

Monitoring uncrewed scientific spacecraft 24 hours per day isn't routine in Russia partly because the location of Russian tracking stations keeps them from covering the entire Earth. Technicians seated at consoles scan the day's spacecraft telemetry data to see if limits have been exceeded. Software doesn't flag limit violations as it does in the US. Instead, technicians rely on their own memory of the proper limit values.

Command and control for Russia's uncrewed scientific spacecraft are executed preferably when the Ussuriisk and Evpatoria tracking stations share the same visibility. This method provides extra reliability because a backup exists if problems arise at one of the stations during a communication session.

Russian command and control uses C- and P-bands whereas the US uses S- and X-bands. During past joint missions the countries agreed that the spacecraft's owner should do the commanding. Reasons included issues of spacecraft security, incompatible wavebands, and the high costs of building and equipping a C-band transmitter at the US DSN and S- and X-band transmitters at the Russian DSN. The Phobos lander was an exception. The US DSN placed a low-power (10 kw) transmitter on the Goldstone 70 m antenna to do commanding and Doppler measurements with the Phobos lander. Because the lander had only a few instruments and couldn't interfere with the Phobos spacecraft, Russia felt comfortable allowing a foreign agency to command it.

**Navigation.** For increased reliability and parallel processing of navigation parameters, the Russians rely on two navigation centers: one managed by the Flight Control Center (under the Russian Space Agency) in Kalliningrad and the other managed by the Institute of Applied Mathematics (under the Russian Academy of Sciences) in Moscow.

Russian navigation centers (known in Russia as *ballistic centers*) support multiple crewed and uncrewed missions, and people aren't dedicated to one project. These centers plan trajectory maneuvers and measurements, collect and process navigation data, and produce the spacecraft trajectory. The navigation subgroup, within the spacecraft-operation group, plans and carries out maneuvers to correct the spacecraft's attitude.

Navigation support is one of the most developed areas of international cooperation. Several examples of cross support between the US and Russian Deep Space Networks (DSNs) exist. The former Soviet Union equipped its deep-space, 70 m dishes with an L-band receiver, and the US equipped some of their stations with a C-band transmitter and L-band receiver. As a result, the US DSN could measure navigation parameters for the Vega and Phobos spacecraft. Based on this experience, a US-Russia navigation team developed procedures for planning resources and a compatible format for exchanging navigational data.

For the future mission, Mars '96, Russia placed a special X-band transponder onboard the spacecraft so the US DSN can measure its position. VLBI measurements by the US DSN enable Russia to reduce the fuel requirement for trajectory maneuvers and substitute more science instruments for the fuel's mass.

Recommendation: For added reliability, two navigation centers (one each in the US and Russia) should support joint scientific missions because every spacecraft maneuver or reorientation changes the navigation parameters. The Radioastron mission, for example, will have two navigation groups—one at the Jet Propulsion Laboratory and one at a Russian center previously mentioned.

### 18.2.3    Data Management

**Data Transfer and Archives.** We must create a master database to help transfer data among all the operational elements. The database should be read-only, but each group of participants should have access to files within the database that contain data required from the other participants to do its job. The files should have reference (e.g., tracking-station coordinates) and dynamic (e.g., tracking-station availability) data.

Creating an archive of spacecraft-performance data (e.g., attitude and orbit) also is essential because scientists will need it during post-processing in a scientific mission. The NAIF system [Acton, 1993], developed by the Jet Propulsion Laboratory, is an excellent example of how international projects can benefit from a common database of navigational data.

**Data Exchange.** The Vega and Giotto missions show how data exchanges between international missions have returned the most scientific information. Both Vega and Giotto did a flyby of Halley's comet. Because Halley was so optically faint, atmospheric turbulence kept observers from seeing it on Earth. The Giotto mission needed to know Halley's position to within 10 km so the spacecraft could do a close flyby. Thus, the Giotto mission used images Vega took earlier (showing the comet against a background of stars) to do navigation and attitude maneuvers which achieved the close flyby.

### 18.2.4    Operations Management

**Primary Differences.** Every spacecraft-operations group in Russia, unlike those in the US, has military and civilian components to improve the reliability of operations. Also, in Russia, one group manages operations of the space and ground elements, whereas separate groups manage those elements in the US. This is because the US DSN has had to distribute its resources to a wide range of customers whereas most Russian missions have had dedicated ground tracking facilities. The Russian scheme best distributes operations functions between the space and ground elements, but the US's approach best loads the tracking stations.

In further contrast to the US, Russian space missions haven't had project managers. Instead, the spacecraft and payload managers share overall management of Russian projects. For example, the spacecraft manager directs the GOCG; the science-payload manager heads the operations division at IKI. The manager from IKI oversees day-to-day operations of the science payload. His or her subordinates are members of the GOCG.

**Key Operations Organizations.** The following organizations have coordinated cooperative science missions involving Russia.

*Interagency Space Coordinating Group for Space Science:* This group was created for the first time for the Halley's comet campaign (1982). Later, they began coordinating astrophysical and solar physics missions. Intercosmos, a division of the Russian Academy of Sciences, represented Russian space science. Recently, an International Mars Exploration Working Group, equivalent to the Coordinating Group, was formed to coordinate international exploration of Mars. This Working Group has representatives from seven national space agencies including the Russian Space Agency.

*International Science Steering Committee:* This committee forms for each project to determine how the payload will operate. The chief scientist of the project is chairman of the committee. This organization is mainly Russian but has international representatives. They usually meet twice per year to review the project's scientific status and to plan science observations for the following six months.

*Interministry Operations Committee:* This committee oversees all spacecraft operations beginning with launch preparation. The committee has representatives from all organizations that help build and operate the spacecraft and science payload. Thus, it's mainly concerned with the mission's technical status and meets a few times per year. They decide whether the spacecraft is spaceworthy, how it should operate, how to allocate operations resources, and the expected lifetime of the mission.

*General Operative Control Group:* See Sec. 18.3.

### 18.2.5    Contingency Planning and Response

Russia usually has manufactured and launched redundant spacecraft so that, if a total failure occurs, another spacecraft will immediately be able to assume the primary mission. (The US approach has been to launch one spacecraft with many fail-safe systems.) Partly for this reason, the Russians don't simulate commands or do contingency planning as much as US teams do. Their software also doesn't use alarm limits which would help rapidly detect spacecraft problems. Finally, Russia relies more on ground processing and analysis instead of the onboard processing and analysis used in the US. The US prefers to install sophisticated computers on the spacecraft to handle, among other things, contingency analysis and response. The current trend in Russia is to employ more fail-safe systems and launch only one spacecraft, such as with Mars '96.

Occasionally, a US-Russia mission may involve a spacecraft owned and controlled by one of the partners and tracked by another, as is the case with the Radioastron mission. Russia's orbiting radio observatory is to download its science data routinely to US stations. This scenario created a potentially dangerous situation—if the spacecraft experienced an anomaly while being tracked by a US station, how would the US know? The US was to capture the science data but had no way to analyze the spacecraft telemetry in real time. The resolution agreed to by both partners was to provide spacecraft-housekeeping data in the science data header transmitted to the US station. The US station would occasionally transmit blocks of the housekeeping data to the Jet Propulsion Laboratory, Russia's partner in the mission, where they would analyze it for anomalies based on algorithms the Russians supplied. If a dangerous anomaly existed, they'd tell the right people in Russia.

Recommendation: Depending on the mission's complexity, each partner requires people who can be accessed rapidly (through beepers, etc.) to provide translations whenever an anomaly requires voice communication with the partner controlling the spacecraft. Also, coded messages which can be transmitted and translated electronically are extremely useful.

In Russian uncrewed missions, a spacecraft contingency means failure to receive a telemetry signal. To diagnose the source of a contingency [Altunin, 1992], operators

1.   Check for inaccurate state vectors or Doppler measurement
2.   Check radio equipment and receiver noise

3. Change tracking stations. If there's still no telemetry signal, assume the spacecraft has a problem. In that case:

4. Increase ground transmitter power or change frequency on the ground. If the problem continues, prepare a command file for transmission to the spacecraft. Before uplinking the commands, make sure the Control Center Chief has written verification from all spacecraft-systems managers that the commands won't damage their systems. If authorizations are complete,

5. Instruct the spacecraft to change its receiver, decoder, and transmitter. If the signal is still absent, assume the onboard power supply has a problem.

6. Instruct the spacecraft to turn off the entire science payload and all other non-critical elements. If the problem continues, assume it's in the attitude-control system.

7. Transmit commands to correct errors with the attitude-control system

The near future promises several joint US-Russia missions including planetary missions—Mars '96, the International Mars Exploration Initiative (1994-2005); Discovery-class missions—Phobos Sample Return and Pluto (1996-2005); astrophysics missions—Spectrum-X and Radioastron (1996-2000); and continuation of crewed missions—Shuttle-Mir/Space Station (1996-2005).

Table 18.6 summarizes the key characteristics, architecture, and mission operations functions of the proposed or planned cooperative missions between the US and Russia.

# 18.3 Russian Space Mission Operations

Karen R. Altunin, Valery I. Altunin, Mikhail Artukhov, Konstantin G. Sukhanov, Viktor D. Blagov, and Viktor I. Glebov

The following subsections describe mission operations for uncrewed and crewed Russian space projects: (Sec. 18.3.1) a scientific orbital observatory, Granat; (Sec. 18.3.2) a Russian remote-sensing system for Earth observations, Resurs, and a meteorological satellite system, Meteor; and (Sec. 18.3.3) the crewed orbiting station, Mir.

### 18.3.1    Mission Operations for the Granat Orbiting X-Ray Observatory

The Granat orbiting X-ray observatory started operating on December 2, 1989, and continues to do astrophysical research in X- and Gamma-wave bands. The Russian Academy of Sciences asked the Lavochkin Association to design the Granat spacecraft. Lavochkin is the leading Russian enterprise that builds and

**Table 18.6.    Examples of Future US-Russia Missions.** The missions listed include some of the future cooperative missions under development.

| Joint Mission/Experiment | Joint Elements of the Mission Architecture | Joint Functions for Mission Operations |
|---|---|---|
| International Space Station (2001): US, Russia, Europe, Japan, Canada to build jointly crewed space stations. | • Space element will consist of separate modules from each of the partners to be used for joint experiments and living quarters<br>• Launch Element - joint payloads launched from US and Russia<br>• Ground Element - US and Russian Flight Control Centers | • Mission planning<br>• Payload planning and analysis (including crew)<br>• Navigation planning and analysis<br>• Managing mission operations<br>• Mission control<br>• Archiving and maintaining the mission database |
| Mars Exploration Initiative (1996 – 2005): US, Russia, Europe, Japan will provide network of instruments (penetrators, rovers, balloon) on and above the Martian surface. Italian Space Agency will supply Mars Relay spacecraft. | • Orbit of Mars relay spacecraft<br>• Space Element - scientific instruments onboard spacecraft owned by other countries<br>• Launch Element - Russian Proton may launch spacecraft of another country<br>• Ground Element - integrated global DSN | • Mission planning<br>• Payload planning and analysis<br>• Navigation planning and analysis<br>• Managing mission operations<br>• Archiving and maintaining the mission database |
| Space Very-Long-Baseline Interferometry (VLBI) (1996 – 2000): Russia will provide Radioastron Space Radio Telescope (SRT): Japan will provide VLBI Space Observatory Program SRT: US will provide ground tracking network. Ground network of radio telescopes provided by possibly 17 different nations for science support. | • Space Element - two space radio telescopes owned by two different agencies<br>• Ground Element - tracking stations and global VLBI networks provided by different countries | • Mission planning<br>• Payload planning and analysis<br>• Navigation planning and analysis<br>• Data processing<br>• Managing mission operations<br>• Archiving and maintaining the mission database |

operates autonomous spacecraft for solar, planetary, and astrophysical research. They designed and operated Luna (Moon), Venera (Venus), Mars, Vega (Halley's Comet), Phobos (Martian moon), Astron (UV telescope), and Prognoz (microwave background radiation project). They built the orbital spacecraft Granat on the same bus as that developed for the interplanetary spacecraft Venera.

**Description of Granat Project.** Granat carries eight X- and gamma-ray telescopes for astrophysical research: four Russian, one Russian-French (Sigma), one Russian-Dutch (Watch), one Russian-Bulgarian (Sunflower), and one French.

The customer for the Granat spacecraft and builder of some of the onboard scientific instruments is the Space Research Institute (IKI) of the Russian Academy of Sciences. Lavochkin designed the spacecraft, created the mission plan, built models of the spacecraft, integrated the science payload, and did engineering tests of the spacecraft after payload integration was complete. After shipment of the spacecraft to the Baikonur launch site, it was tested again under Lavochkin's supervision. Once it was in its working orbit, the General Operative Control Group (GOCG) carried out mission operations. The GOCG mainly includes specialists from Lavochkin and the Russian enterprises which developed the spacecraft systems. A joint working group planned the scientific program and maintained interaction between the GOCG and the scientific customers—IKI and foreign scientists. This working group still creates the current scientific program. It includes specialists from Lavochkin, IKI, France, and the Netherlands.

The GOCG controls the spacecraft from the Spacecraft Control Center within the Deep Space Communications Complex in Evpatoria, the Crimea. The staff of the GOCG, which includes French specialists and scientists from IKI, also are in Evpatoria.

Granat's orbital parameters meet the program's science requirements, spacecraft constraints, payload characteristics, and needs of the ground-element. Its high-apogee orbit provides a proper environment for the payload's X-ray telescope, which must operate above the Van Allen radiation belts. Another important consideration for choosing the orbital parameters is how stable the orbit would remain during a few years of operation. Orbit stability was achieved by choosing the argument of perigee according to the position of the ecliptic plane. The longitude of ascending node depends on the constraints connected with the position of the attitude-control system's reference star, the position of the Earth and Moon, and the area of the sky to be observed by the science instruments. Along with the launch booster's characteristics, these factors determined these parameters: apogee 200,000 km, perigee 2000 km, inclination 5.5°, argument of perigee 285°, longitude of ascending node 20°, and orbital period 98 hours.

The Granat spacecraft, shown in Fig. 18.1, includes a hermetically-sealed instrument block containing the main spacecraft systems. A thermal regulation system helps maintain the required temperature inside the instrument block. The science payloads are on the surface of the instrument block. The masses of the spacecraft and science payload are four metric tons and 2.33 tons, respectively, and the spacecraft is 6.5 meters long. Radio communications between the ground and the spacecraft are on the C- and P-bands with uplink through the P-band. Both spacecraft and science telemetry use the C- and P-bands. The onboard radios transmit telemetry at 65 kbps (science telemetry) and serve as transponders for navigational measurements. Spacecraft and science-payload commanding occurs at speeds of up to ten kbps. Telemetry transmission is either in real time or recorded for later downlink.

The attitude-control system provides three-axis orientation by using optical star sensors sensitive enough to detect sixteen reference stars. The three main orientation modes are three-axis, single-axis Sun orientation, and orientation of the axes in a specific direction with stabilization relative to this direction. The pointing accuracy is about 1.5 angular minutes for three-axes pointing and 20 angular minutes for the other modes.

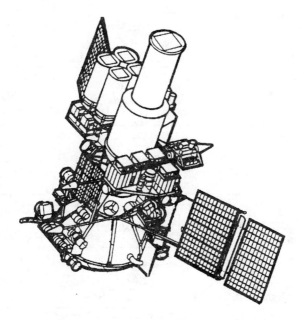

**Fig. 18.1.** **Granat Spacecraft.** The Granat spacecraft is an international, space-based, X-ray observatory which has been in operation since 1989.

The ground element, as shown in Fig. 18.2, includes the spacecraft-operations center, tracking stations, navigation centers, center for processing scientific data, and communication channels.

**Spacecraft Control Center.** The center operates all spacecraft developed by Lavochkin. The GOCG staff does long-term plans (one-month program), plans communication sessions, generates command loads, diagnoses spacecraft anomalies, and simulates commands. When a contingency situation arises, the GOCG contacts specialists from the enterprises which built the spacecraft systems and provides a model of the contingency situation. A physical model is at the Lavochkin Association. In addition to supporting mission operations, the control center supports spacecraft testing during assembly at Lavochkin's spacecraft plant and at the launch site.

**Tracking Stations.** The primary tracking station is in Evpatoria and the backup station for commanding and telemetry receiving is in Ussuriisk (the Russian Far East). The telemetry station at Bear Lakes (near Moscow) is a backup for receiving science telemetry and then transmitting it to IKI.

**Navigation Centers.** Traditionally, two navigation centers—the Institute of Applied Mathematics (Moscow) and the Center for Scientific Research (Kalliningrad)—support all deep space and Earth-orbiting missions. They plan, collect, and

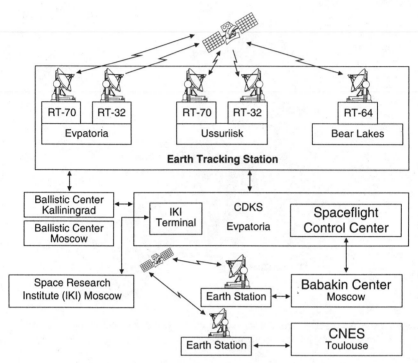

**Fig. 18.2.**    **Granat's Control Systems on Earth.** The diagram represents the typical mission operations elements for Russian space science missions.

process navigational measurements; determine the spacecraft orbit; provide pointing and Doppler information for the tracking stations; and support science-data processing. The GOCG includes a navigation team which maintains contact with the navigation centers.

    **Centers for Processing Scientific Data.** The Granat mission uses two levels of scientific-data processing: The first level is real-time processing to analyze the status of science instruments—done by the science subgroup of the GOCG (representatives from IKI and CNES). This subgroup connects with IKI in Moscow and CNES in Toulouse, France to transmit scientific data and work science-payload contingencies. The science subgroup also participates in short-term scheduling and planning, creates command files for the science instruments, and analyzes real-time telemetry to determine if the science program needs correction. The second level is post processing of transmitted science data at IKI and CNES.

    **Communications.** Data and voice lines connect all parts of the ground element through dedicated telephone channels and personal computers with modems. The data rate is 1200 bps to 9600 bps. Communications between the Spacecraft Control

Center and Toulouse go through telephone channels from Evpatoria to Moscow and from Moscow to Dubna (Moscow region). The link from Dubna to Toulouse uses an Intelsat channel.

**Organizational Structure of Flight Operations.** As mentioned above, the Lavochkin Association develops and operates spacecraft, working simultaneously on a few different projects which therefore benefit from a well-experienced staff. Lavochkin also maintains contact during all phases of the mission with the other enterprises which developed the spacecraft subsystems. One example is the Space Devices Engineering Institute in Moscow, which has developed spacecraft radio complexes for many years. This approach helps teams prepare for mission operations and develop future spacecraft. Because Lavochkin tests the spacecraft during assembly, at the launch site, and in orbit, they achieve a continuity that's impossible when different organizations do these tests.

**General Operative Control Group (GOCG).** The GOCG handles flight operations. It usually begins to work on a project three months before launch. The GOCG's people are mainly from the Lavochkin Association and other enterprises which developed the spacecraft subsystems and science instruments, and from the ground-support complexes and navigation centers. As shown in Fig. 18.3, it consists of five groups: (1) flight-control operations; (2) science; (3) navigation; (4) facility support to maintain the Control Center's computers; and, (5) ground tracking, which manages activities of the tracking stations.

The GOCG

- Plans science over the long term with Russian and foreign scientists
- Develops the monthly programs of spacecraft activities
- Develops programs for the spacecraft's communications sessions
- Operates the spacecraft during the communication session—includes managing the ground tracking station for commanding; telemetry; processing, recording and transferring data; analyzing real-time telemetry; and responding to contingencies
- Communicates with the science-operations team
- Provides detailed analysis of the spacecraft's performance after a communication session
- Changes the spacecraft's model according to information received during communication session
- Archives and distributes science and housekeeping data
- Distributes navigation data

The GOCG's staff are at the Control Center in Evpatoria. At the initial stage of flight operations, the GOCG staff includes 30 to 50 people dedicated to a particular project. After four months of early-orbit checkout for the spacecraft and science payload, the staff drops to about 15 people. This team works in Evpatoria for one month, then returns to Moscow while another team replaces it. Those returning to

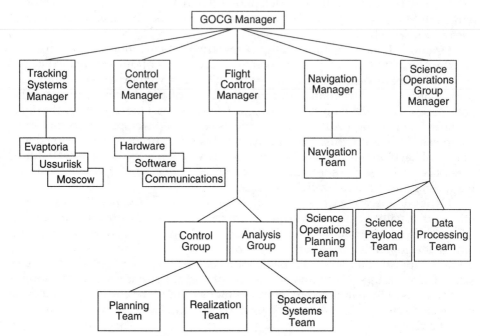

**Fig. 18.3.**   **Organizational Structure for the General Operatives Control Group (GOCG).** The GOCG controls operations for several missions. Its structure is the same for all Russian space-science missions.

Moscow work at the Lavochkin Association on other projects. This approach decreases the cost of spacecraft operations. Only seven to ten people are at the Control Center in Evpatoria now because the long Granat mission has allowed streamlining of operations. Recently, a data-processing center was completed at the Lavochkin Association in Moscow. All information from the Control Center now goes to this center, which does backup spacecraft analysis.

**Phases of Flight Operations.** Phases include launch, early-orbit checkout, routine operations, and extended operations. The launch phase includes orbit insertion and two trajectory corrections. This phase of operations ends when the spacecraft is in final orbit and has one axis oriented to the Sun. The tasks during early-orbit checkout include deploying the spacecraft elements, testing the spacecraft systems, testing the science payload, and determining the current orbital parameters. End-to-end systems tests include the ground element, Control Center, and tracking stations.

**Mode of Operations.** During the communication sessions, operators on the ground start programs stored in the onboard memory. These programs cover typical operations, such as selecting the spacecraft's orientation, and start either by command from the ground or at a set time interval. Autonomous operations were necessary because Russian tracking stations in Russia and the Crimea couldn't cover the entire Earth. They also provide spacecraft security in case of onboard contingencies because the spacecraft can operate autonomously until it receives ground communications.

**Technology of Flight Operations.** About 70 hours of Granat's 98-hour orbit are for scientific observations. This period varies with the viewing period of the ground-control station and how long the spacecraft is inside the Van Allen radiation belts. The rest of the orbit includes 20 hours inside the radiation belts, during which the science instruments must be turned off, and ten hours for data transmission and command uplink. During each orbit, at least two or three science observations take place. Communication sessions occur between these observations.

During each communication session, operators (1) Switch off science instruments and place the instrument's microprocessor in a wait state and send a command from the ground to switch on the science-data transmitter and the instrument's memory to begin transmitting data recorded during the previous observation. (2) Orient the attitude-control system to the next source for observation and turn off the instrument's microprocessor during the orientation. (3) Turn on the microprocessor, then test and load the computer's memory with commands for the next observation. (4) Transmit commands instructing the computer as to the next recording mode—first for the Sigma telescope and then for the other onboard instruments.

Additionally, during each communication session (and between observations), the teams do engineering tests on the spacecraft and select the spacecraft's next configuration. For example, they decide which section of the solar panels will operate during the next science observation. Also, they measure the range and range-rate trajectory. The communication sessions which take place before and after the spacecraft moves through the Van Allen belts have the same structure except that before entry the instruments are turned off and after exit they are turned on. The communication sessions last four to seven hours, long enough to transmit 100 to 150 commands.

**Cycle for Planning and Completing Operations.** This cycle includes these steps: (1) develop timeline for communication sessions with spacecraft, (2) develop science program based on the set of targets selected, (3) develop activities for the communication session, (4) prepare spacecraft commands, (5) simulate commands, (6) load command sets to tracking station's memory, (7) uplink, and (8) analyze telemetry after the communication session.

Two months in advance, the International Science Group develops a preliminary timeline for the communications session, including a list of sources. The GOCG uses this timeline to develop the scientific program. After this, detailed activities are developed based on the status of the onboard systems and the availability of the ground element. As a rule, the final communication session is complete three days before uplink. It includes timing of commands and commands to be uplinked, a timeline for telemetry reception, and instructions to the ground-support complex. After simulation and validation by the GOCG, the command file goes to the uplink station's memory. (And, in critical situations, to the backup station's memory.) The uplink station's computer automatically uplinks data. The GOCG uses spacecraft telemetry from the spacecraft to verify that the command file successfully loaded onto the onboard computer and to monitor the spacecraft systems and science payload in real time. The telemetry information is displayed on computers at the Spacecraft Control Center, relayed to the operator terminals, and recorded. In a contingency situation, or if a command doesn't execute as expected, the GOCG team analyzes the situation and corrects the communication-session program.

Commands from the ground begin and end the communication session and change the spacecraft's system configurations, operational modes, the telemetry transmission rate, and the configuration of the onboard radio complex. Also, during commanding, operators load information on the spacecraft to select its orientation, plus the operational modes of the onboard computer and the scientific instruments. Ground-based commands can also change the flight software.

If only minor corrections to the program are needed during the communication session, operators manually uplink individual commands to the spacecraft. In the case of a major contingency, such as malfunctions of one of the scientific instruments, operators correct or end the communication session. This decision depends on how much time a solution requires and how much time remains in the tracking station's viewing period. In the case of a major contingency, operators can prolong the communication session by two or three hours. If it's not possible to correct the failure within the current communication session, they'll cancel upcoming scientific observations to allow time to communicate with the spacecraft.

**Performance Assessment.** During the initial stage of early-orbit checkout, 14 communication sessions tested the spacecraft's housekeeping systems. Then, teams tested and calibrated the scientific instruments. During the following two months, the main goal of mission operations was starting up the scientific instruments, so they checked the control block of the scientific instruments and calibrated the star sensors. Two months after launch, they finished calibrating the telescopes, and routine operations began three days later.

Granat operations went according to plan with only a few minor changes. Occasionally, they took a month to amend the programs for the spacecraft and science payloads to take in a target of opportunity. A few times, they had to interrupt observations, mostly because of weather conditions at the tracking station or high solar activity. When this occurred, they reprogrammed the telescopes' software.

On August 1, 1990, Granat reached the end of its planned lifetime. However, because the housekeeping systems and the primary scientific instruments continued to work properly, the mission was extended. Developers of the follow-on scientific program had to use as little nitrogen as possible for attitude corrections, in order to increase the spacecraft's extended lifetime and the number of sources they could observe. For this reason, they increased the observation timeline from one to three months to best locate sources and reduce reorientations.

The first five years of Granat operations included 1072 communication sessions. Of these, 724 prepared the scientific instruments and transmitted science data. The remaining sessions were devoted to spacecraft housekeeping. Fifty planned observing sessions weren't done: 23 because of high solar activity and poor weather at the tracking station and the other 27 because of failures of the onboard or ground systems.

For spacecraft operations, 110,000 commands were uplinked, and Granat observed 132 sources—many several times. Along with the main scientific program, Granat also did radio-science experiments on the spacecraft signal's propagation through interplanetary and atmospheric turbulence. The scientific results of the Granat mission have received worldwide recognition.

## 18.3.2 Russian Remote-Sensing and Meteorological Space Systems

The Russian government developed two programs, Resurs-0 and Meteor, for Earth remote sensing and meteorological research. The Resurs-0 system includes one or two spacecraft, Resurs-01, which are continuously in orbit. When a spacecraft's operational life ends, it's immediately replaced. Resurs-01 is the second-generation spacecraft for Earth-observing research. The Meteor system includes two or three meteorological spacecraft, Meteor-3, which also are continuously in orbit. Meteor-3 is the third generation of Russian weather satellites.

**Program Objectives.** The Resurs spacecraft has remote sensors in the near and far infrared, optical, and microwave wavebands. The main goals of the Resurs program are (1) remote sensing of Earth's surface features and continental shelves to search for oil, natural gas, and minerals, (2) predicting agricultural crops, (3) predicting snow melt and planning to manage water resources, (4) range management, and (5) forest management and protection from forest fires.

The Meteor program provides global meteorological information on: (1) clouds, ice, and snow in visual and infrared wavebands, (2) the temperature and humidity of high-altitude clouds, (3) the temperature of surface water by measurements in the infrared, (4) solar radiation in the near-Earth region, and (5) the levels and vertical distribution of ozone.

**Architectures of Resurs and Meteor Space Systems.** Figure 18.4 shows these architectures. The launch facility and tracking station at the Baikonur Cosmodrome are for launch preparation, launch, and control of the booster, Zenit, which carried Resurs-01. The launch facility and tracking station at the Plesetsk Cosmodrome are for launch preparation, launch, and control of the booster, Cyclone, which carried Meteor-3. The ground tracking stations are for flight operations and flight control of the onboard systems, navigation measurements, and receiving spacecraft telemetry.

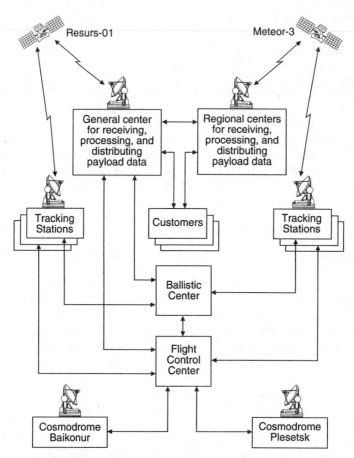

**Fig. 18.4.     Architectures of the Resurs and Meteor.** This is the typical mission architecture for Russia's remote-sensing missions.

As shown in Fig. 18.4, general and regional centers receive and process space-craft data and distribute this information to their customers. The autonomous tracking station receives information (images) from Meteor-3 in real time. The navigation center determines spacecraft positions for flight operations. All elements of the space system are connected by communication and data-transmission channels.

Table 18.7 shows the organizational structure of the Russian enterprises who develop and operate Resurs and Meteor. Figures 18.5 and 18.6 show the Resurs-01 and Meteor-3 spacecraft, respectively. Table 18.8 lists their main characteristics.

**Table 18.7. Organizational Structure for Resurs and Meteor.** The table shows how responsibilities divide between the organizations involved in Russia's remote-sensing operations. The Russian Space Agency, formed in 1991, is the customer of the organizations listed and contains other organizations not shown.

| Responsibilities of Russian Space Agencies for Remote-Sensing Missions | | | |
|---|---|---|---|
| **Space Devices Engineering Institute (Moscow)** | **All-Union Research Institute for Electromechanics (Moscow region)** | **Scientific Production Complex "Planet" (Moscow region)** | **Space Military Division of Russian Ministry of Defense (Moscow)** |
| Develop onboard remote optical and infrared sensors | Develop and operate spacecraft Resurs-01 and Meteor-3 | Receive and process data from Resurs-01 and Meteor-3 | Prepare and launch boosters with space-craft at Baikonur and Plesetsk Cosmo-dromes |
| Develop onboard systems for compressing and recording information and transmitting system with 128Mbps | Integrate remote-sensing instruments with spacecraft to observe Earth and atmosphere | Rapidly produce images of requested Earth regions | Flight control of launch booster |
| Develop ground-receiving complexes including movable, autonomous receiving stations for data receiving and prime processing | Help develop microwave remote sensors [radiometers, onboard antennas, deployment mechanisms] | Produce and distribute [sell] maps of requested Earth regions | Flight operations for spacecraft |
| -- | Develop and integrate piggy-back satellite to launch with Resurs-01 and Meteor-3 | Develop algorithms for data processing | -- |

The housekeeping systems of Resurs-01 and Meteor include the onboard control system, 3-axis gyro system for attitude control, attitude-control system for the solar panels, power-supply system, thermal-regulation system, the electro-ion system for orbit correction, onboard time-reference standard, and solar sensors.

**Orbital Systems for Resurs-0 and Meteor.** Resurs-01 was launched from the Baikonur Cosmodrome into a Sun-synchronous circular orbit with an altitude of 650 km, inclination of 98 degrees, and rotation period of 97 minutes. If two Resurs-01 spacecraft are launched, they go into the same orbit but 180 degrees from each other.

Meteor-3 satellites are launched from the Plesetsk Cosmodrome into a nearly polar orbit with an altitude of 1200 km, inclination of 80° to 83°, and rotation period of 109 minutes. The Meteor system has three spacecraft simultaneously in orbit. The angular difference between the ascending nodes of the different spacecraft is about 60°. The three spacecraft provide full global coverage twice daily.

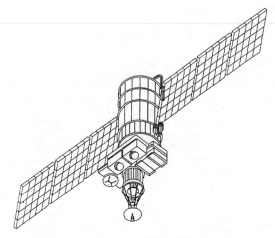

**Fig. 18.5.**    **Resurs-01 Spacecraft.** Resurs-01 is a remote-sensing spacecraft which provides images in the near and far infrared, optical, and microwave wavebands.

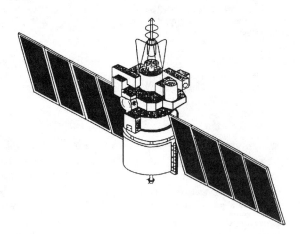

**Fig. 18.6.**    **Meteor-3 Spacecraft.** The Meteor-3 spacecraft provides global meteorological data for Russia.

**Flight Operations for Resurs-01 and Meteor-3.** The onboard-control complex

- Receives commands from ground control and distributes them to the appropriate onboard system
- Collects and transmits data from the sensors to the ground

**Table 18.8.    Characteristics of Resurs-01 and Meteor-3 Spacecraft.** The spacecraft were launched by the Russian boosters Vostok and Cyclone. The actual lifetime of these spacecraft usually exceeds the design lifetime by at least a factor of two.

| Characteristics | Resurs-01 | Meteor-3 |
|---|---|---|
| Spacecraft Mass (kg) | 1700 | 2200 |
| Payload Mass (kg) | 500 | 700 |
| 3-axis Pointing Accuracy (arcmin) | <150 | <20–30 |
| Power Supply (ave. day) (w) | 450 | 550 |
| Design Lifetime (years) | 1 | 2 |
| Possible Trajectory Correction | yes | yes |

- Verifies command reception and transmits verification to the ground
- Retransmits the signal for Doppler measurement to the ground

The ground-control complex

- Creates and transmits commands to the spacecraft
- Receives, processes, and analyzes spacecraft telemetry
- Determines the radial velocity (range rate) of spacecraft
- Verifies ground and onboard time-reference frames and corrects the onboard frame, if necessary

Operators control the onboard housekeeping and payload systems in two ways: through commands from the ground or by using the onboard microprocessor's time-program system. The first method switches onboard systems on or off and changes the onboard mode of operations. Each command set issued by the ground covers one day of operations and includes 10–15 commands. For the second method of control, the time-program system automatically issues commands stored in the spacecraft's permanent memory. The system bases these commands on time values (e.g., orbital period) ground controllers have programmed into it after launch.

These methods best distribute commanding between the onboard and ground systems and reliably control complex equipment. They improve reliability because the spacecraft needs only a few commands from the ground, so tracking stations can be simpler.

The General Operative Control Group (GOCG) handles flight operations for Resurs and Meteor at the Spacecraft Control Center in Moscow. The GOCG's structure and responsibilities for Resurs and Meteor are similar to those for the Granat mission (see Sec. 18.3.1). Managers and staff on the GOCG are representatives from the enterprises listed in Table 18.7. The chairman of the GOCG also represents the Spacecraft Control Center (SCC).

The ground element for Resurs and Meteor includes the SCC, tracking stations, the navigation center in Kalliningrad, and communication and data transmission channels. The Space Division of Russia's Ministry of Defense operates the ground element.

Spacecraft commands are generated at the SCC and transmitted to the tracking stations over telephone lines. The tracking stations work automatically according to instructions the SCC transmits. Tracking stations compress received spacecraft telemetry and send the data to the SCC. They also transmit trajectory-measurement data (range-rate) directly to the navigation center over telegraph lines. The navigation center provides to the SCC and the tracking stations all navigation data needed to support the mission.

The SCC has a computer complex for spacecraft operations and data processing which can produce 30,000,000 operations/second. Tracking stations transmit the compressed data to the SCC at 1200–1600 bps. Fully processing the data received during one communication session takes about 20–30 minutes.

Two types of tracking stations for Resurs and Meteor are near Moscow, St. Petersburg, Jusaly, Kolpashevo, Ulan Ude, Ussuriisk, and Petropavlovsk-Kamchatskii. The first type uplinks commands and does navigation measurements; the second receives telemetry from the spacecraft and payload. The tracking stations can support communications with the spacecraft from a distance of up to 6000 km, with a circular orbital height up to 1300 km and inclination of 80–98 degrees. They use the P-band (137-138 MHz and 460-470 MHz) for uplink and navigation and the X-band to receive payload telemetry. Handover time from one station to another is about one minute. The accuracy of Doppler measurements is about 0.1 m/second.

**Spacecraft Commanding.** Figure 18.7 shows the command scheme. Commanding includes developing a command set and transmitting it to the tracking station, uplinking commands according to the SCC's communication-session program, verifying through tracking stations that the spacecraft received these commands, and validating the tracking stations' verifications at the SCC.

The command sets are electronically produced. GOCG staff members prepare the computer input according to requests from the payload representatives and the status of spacecraft systems. The computer's output are the commands and instructions for the tracking stations. Commands remain in the data archive to support future data processing.

**Receiving and Processing Data.** Figure 18.8 shows how the tracking station that receives telemetry processes it. First, the station compresses telemetry received at up to 7.68 Mbps so it will fit the narrow-bandwidth telephone lines. Data compression includes capturing parameters from the data stream needed for post processing, excluding unreliable data, compressing and tagging the data with the tracking station's time reference, estimating the quality of information, and recording and transmitting data to the SCC.

The data-processing complex at the tracking station processes telemetry information from the payload sensors and the onboard computer through one of the

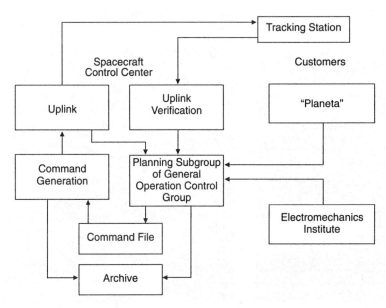

**Fig. 18.7.**     **Commanding Schemes for Resurs and Meteor.** This diagram represents the typical commanding scheme for Russia's remote-sensing satellites. The scheme includes the customers, who request activities; the spacecraft control center, which converts the requests to commands; and the tracking station, which uplinks these commands.

spacecraft's three operational modes: (1) directly transmitting data from the sensor to the ground, (2) transmitting observational data from the onboard memory to the ground, or (3) combining the two.

The SCC post processes and time tags data, archives the payload data, displays data on monitors of the GOCG and the users, and monitors status of the spacecraft systems.

**Customer-Services Complex.** The customer-services complex includes the general center for receiving, processing, and distributing payload data in Dolgaprudny (near Moscow), as well as the regional centers in Novosibirsk and Khabarovsk (Russian Far East). The general center collects user proposals, selects the science targets and observation priorities, transmits them to the SCC, and manages the activities of the customer-services complex. Figure 18.9 shows this complex's structure.

The General and Regional Centers label the data with global position and time, geometrically correct it for 3-dimensional output, and combine the various spectral filters into one picture. They then make the data available as photographs, on magnetic tape, and in numerical and tabular form.

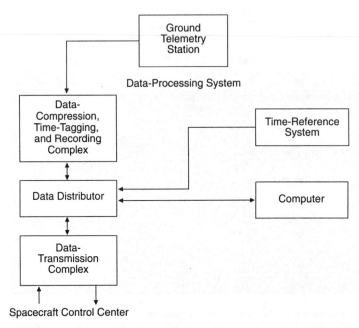

**Fig. 18.8.    Flow of Data Processing.** The diagram shows how data processing flows for instruments on remote-sensing satellites. It operates mostly autonomously. The maximum data flow from the instruments is 7.68 Mbps.

### 18.3.3    Flight Operations for a Crewed Orbital Station

Russia is the leader in long-term crewed flight, having successfully operated orbital stations for more than twenty years. The first-generation station, Salyut-1 through -5, had only one docking device, an expected lifetime of one year, and a habitation time of one to two months. The second-generation station, Salyut-6 and -7, had two docking devices and a design lifetime of 5–9 years. The current-generation station, Mir, is modular and has been operating successfully in orbit since 1986.

Flight operations have evolved with the hardware. During the first short-term flights, ground support for flight operations drew from the specialists who designed and manufactured the onboard systems. They worked only during the flight, afterwards returning to their positions. During this period, the flight operators were at one of the tracking stations in Evpatoria (in the Crimea). When the station became permanently crewed, a dedicated flight-operations group was created, the Flight Control Center (FCC) was built in Kalliningrad (near Moscow), and new flight-operations technologies were developed.

Today, flight operations for crewed missions include highly-qualified people, effective technology, a highly-developed control facility in the FCC and tracking

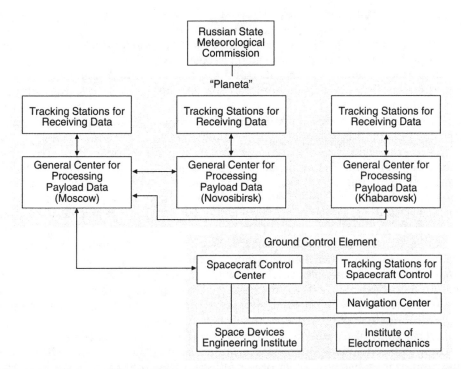

**Fig. 18.9.**    **Customer-Services Complex.** This complex has three main elements: the Russian State Meteorological Commission, which represents the customer; NPO Planeta, which is the civilian organization that receives, processes, and analyzes data; and the ground-control element, managed by the military space division, which handles spacecraft operations.

stations, and reliable data transfer using ground and satellite communication channels. The following paragraphs describe Russia's systems control for flight operations and technology for operating crewed space flights.

**Flight Operations.** Flight operations for Russia's crewed missions consists of (1) the onboard autonomous control system, (2) flight control center, (3) Baikonur launch facility, (4) ground tracking network, (5) rescue system, (6) flight-simulation facility, (7) centers for processing scientific data, and (8) the communication system, including ground and satellite (Molniya, Altair, Horizon) channels. Figure 18.10 shows the architecture for the communication network.

**Functions of the Ground-Based Elements.** The functions are (1) select objectives for each phase of flight, (2) plan flight-control operations, (3) plan support for scientific experiments, (4) keep the crew safe, (5) control and monitor onboard systems and monitor crew status, (6) control docking, (7) receive and process station data, (8) analyze situations and respond, (9) prepare commands for flight operations and scientific experiments, (10) develop recommendations for responding to contingencies, (11) manage ground-support activities, (12) resupply consumables, (13) calculate the orbit, (14) process trajectory information for docking maneuvers and crew return, and (15) distribute data to customers.

The orbital period is 90 minutes, and the communication sessions last 10–50 minutes; thus, the onboard autonomous complex must control the station for periods of 40 to 80 minutes.

**Functions of the Onboard Autonomous Complex.** The functions are (1) maintain orientation during flight, (2) control onboard systems according to the uplinked command file, (3) diagnose onboard systems, (4) warn crew when the onboard system fails, (5) switch backup systems on and malfunctioning system off during onboard failure, (6) point communication antenna to relay satellite, (7) open radio contact with relay satellite, and (8) support automatic approach and docking with cargo spacecraft, Soyuz-TM and Progress-M. Science experiments need not involve the crew in most (70%) cases.

**Crew Functions.** The main tasks of the crew are to conduct science experiments; control, monitor, maintain, and repair onboard systems; do activities outside the vehicle; and manually dock with the cargo spacecraft in case the automatic mode doesn't work. If the onboard system doesn't recognize an unexpected contingency, the crew resolves the situation, if possible, and reports their findings to the ground during the next communication session.

**Flight Control Center (FCC).** The center manages all flight-operations and tracking station activities; interacts with launch, rescue, and flight-simulation facilities; and coordinates with other organizations involved in flight operations. The FCC in Kalliningrad was created in 1970. It includes a communications center, data-processing complex, control room for spaceflight operations, and space for the FCC staff.

The FCC's data-processing complex has three tasks: planning and commanding, including long- and short-term planning, maneuver planning, simulation and exchanging data with the ground-control complex; supporting navigation—determining trajectories; and telemetry support—processing raw data, selecting and judging the accuracy of most important parameters, and analyzing the status of onboard systems. The data-processing complex uses a multi-functional computer that can do 150 million operations per second and a dedicated computer that can do 1 billion operations per second. Routinely, it processes 2 Mbps and estimates up to 20,000 parameters. It can support up to ten spacecraft simultaneously.

**Baikonur Cosmodrome.** The Baikonur Cosmodrome supports launch preparation, launch, and monitoring of onboard systems and crew status during launch and until the spacecraft separates from the booster. After separation, operations transfer to the FCC.

**Ground Tracking Stations.** Seven tracking stations support the Mir from St. Petersburg, Tshelkovo, Jusaly, Kolpashevo, Ulan Ude, Ussuriisk, and Petropavlovsk.

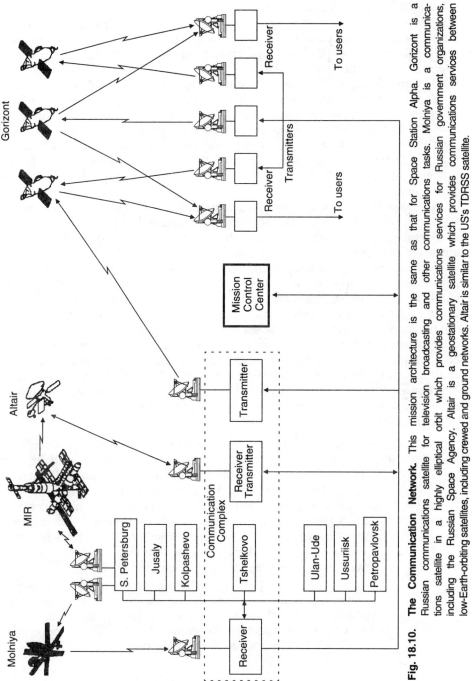

**Fig. 18.10. The Communication Network.** This mission architecture is the same as that for Space Station Alpha. Gorizont is a Russian communications satellite for television broadcasting and other communications tasks. Molniya is a communications satellite in a highly elliptical orbit which provides communications services for Russian government organizations, including the Russian Space Agency. Altair is a geostationary satellite which provides communications services between low-Earth-orbiting satellites, including crewed and ground networks. Altair is similar to the US's TDRSS satellite.

**Flight-Simulation Facility.** During launch preparation, this facility simulates Mir's behavior during routine operations and contingency situations to train and test the FCC's staff and test the FCC's hardware and software. Once Mir is in orbit, it provides predicted parameters of onboard systems to the FCC and develops alternatives for contingency situations.

**Rescue Complex.** The complex locates where the rescue and resupply spacecraft have landed upon their return to Earth, communicates with the crews of these spacecraft during landing, and evacuates crews from the spacecraft. The Rescue Complex includes helicopters, airplanes, and other ground hardware.

Customers of the Mir's science data post-process the data from the FCC. The primary customers are the Space Research Institute, Institute of Medical Biology Problems, State Center "Priroda" ("Nature"), and others (including foreign customers).

**Communications Complex.** The communications complex shown in Fig. 18.10 includes ground and satellite channels for transmitting data. Data from Mir gets to the ground station either through the data-relay satellite, Altair (one operating) or through direct transmission from Mir to the tracking stations. Information from the tracking stations moves to the FCC through ground and satellite (Molniya) channels. The Molniya system includes eight spacecraft located in highly elliptical orbits. Customer data is transmitted through the communication-satellite system, Gorizont, which includes nine geostationary spacecraft. The FCC can transmit data through the Intelsat and Eutelsat systems.

**Technology for Flight Operations.** The main objectives of flight control are executing flight programs, maintaining the crew's safety, and keeping the station operating. Figure 18.11 shows the algorithm for spaceflight control. Mir flight operations occur in one of two modes: autonomously through the onboard control system or manually by using onboard control panels. Flight control includes the following operations:

1. Pre-launch planning. Plan the flight program for each expedition. (Each expedition lasts 6–12 months). The planned program will probably change because ground or onboard systems fail, ground or flight crews make mistakes, or objectives change during a long-duration flight. For this reason, programs have a 10–15% pad. To conserve onboard resources (fuel, power, crew task time), controllers try to minimize changes. On average, 87–92% of each expedition's programs have been successfully executed. Were the programs not padded with excess activities, the success rate would increase to 93–97%.

2. Day-to-day activities. FCC computers generate programs for day-to-day activities, while also considering the dynamics of spaceflight conditions. They include instructions for the onboard autonomous complex and text files for the crew. The day-to-day program is simulated and transmitted to the orbiting station.

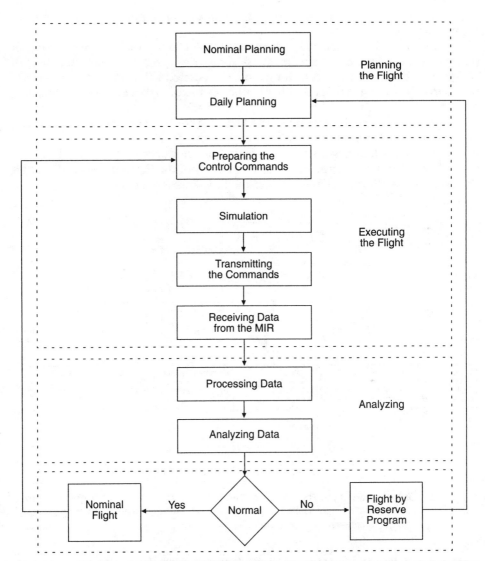

**Fig. 18.11.**   **Algorithm for Controlling MIR Spaceflight.** The figure represents the typical algorithm for controlling a Russian crewed spacecraft. The algorithm evolved over the 20 years of Russia's experience with operating crewed space stations.

On average, each day the following communications between the Earth and the space station occur: voice communications—10–12 times per day; telemetry reception—10–12 times per day; trajectory measurements—once per week; uplinking instruction for onboard autonomous complex—once per day; and television communications—once or twice per week.

Monitoring of the onboard systems from the ground includes processing of onboard data, analyzing telemetry, and comparing predicted parameters with actual ones. If the parameters differ from the predicted values, the FCC diagnoses the source of the disparity and either continues the program or corrects it. If they correct it, they'll return to the nominal program as soon as possible.

**Organization and Functions of the FCC.** Figure 18.12 shows how the FCC is organized. Flight Directors normally are cosmonauts who have flown in space a few times. Their deputies manage the main branches of operations: the FCC, the Crew Transport and Resupply Spacecraft Division, the Scientific Modules Division, the Ground Stations Division, and the FCC Personnel Division.

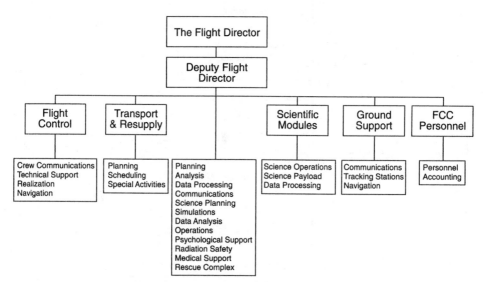

**Fig. 18.12.    Organization of the Flight Control Center (FCC).** The FCC in Kalliningrad for crewed space operations has a different structure from that of the GOCG, which operates scientific and remote-sensing spacecraft. See Fig. 18.3.

The FCC staff divide into groups for Planning, Simulation, Analysis, Experiments, Crew Communications, the Ground Control Complex, Crew Psychology Support, Medical Support, Radiation Safety, Navigation Support, Special Activity (extra-vehicular activities, fuel resupply for station jets, and docking), and Foreign Interface. Table 18.9 shows what they do.

The FCC is staffed continuously, seven days a week and 24 hours per day. People divide into five shifts, the fifth one being a backup. Each of the regular shifts lasts 25 hours and is based on the station's flight-operations sequence.

**Table 18.9.    What the Flight Control Center Does for Crewed Spaceflight.** Includes tasks for the onboard crew.

| | |
|---|---|
| Prepare Before Flight | • Acquire and analyze customers' science proposals<br>• Forecast Mir resources<br>• Forecast number of crew transport and resupply flights<br>• Train FCC staff |
| Plan Flight Operations | • Plan daily operations<br>• Prepare command files for onboard computer<br>• Assign crew work<br>• Instruct crew |
| Execute Flight Program | • Transmit instructions to crew<br>• Uplink commands<br>• Receive telemetry |
| Analyze Data | • Process data<br>• Analyze data<br>• Provide recommendations based on analysis |
| Make Decisions | • Analyze flight information<br>• Deliver instructions |
| Do Post-Experiment Activities | • Analyze results<br>• Distribute data to users<br>• Prepare report |

Attempts to decrease fatigue caused by the length of each shift have included regular breaks after each set of communication sessions, during which the staff is encouraged to exercise and relieve tension. Other techniques are a three-day break between shifts, medical surveillance, and staggering the activities within each shift so intensive work, such as command uplink, precedes less-demanding activities, such as planning for the next shift. To keep repetitive work from dulling reactions to contingencies, management occasionally creates test situations, being careful not to endanger the actual flight.

People at the FCC must respond to contingencies caused by failures of the onboard system or errors by crew and ground workers. The number of failures and human errors experienced depends on certain conditions. For example, the onboard system fails at the highest rate soon after launch, while we're testing it under different modes of operation. After this, the number of failures stabilizes at a low level. As the station ages, failures increase again; at the end of the station's design lifetime, the increase can become exponential. To prolong the station's lifetime, the crew repairs it and installs spare or new equipment.

The number of human errors increases for a new station and then stabilizes as experience increases. Human errors also tend to increase when the station's configuration changes (e.g., new module docks with station) and when new systems or new onboard software appears. At the end of the station's design lifetime, human errors increase because onboard systems fail more often. On average, flight controllers err once every three days.

When an onboard system fails or human errors occur, we lose station resources. To minimize these losses and to prevent catastrophic failures, management made sure people trained regularly—especially before important events such as extra-vehicular activities (EVA). This training helped standardize experiment procedures and simulate commands for the onboard computers before uplink.

**Flight Crew Assignments and Schedule.** The crew's daily schedule (see Table 18.10) must be routine—beginning at 8:00 AM (Moscow time) and ending at 11:00 PM (Moscow time) during the entire flight. They eat, exercise, and rest at the same times each day, except when special events occur such as extra-vehicular activities, docking, and landing. Such things as visibility and orbit determine the times for these activities. The working day lasts 8.5 hours, and sleep time is eight hours. During the first two weeks of flight, the crew adapts to the zero-gravity environment. The crew's work capacity is lower during this time, so the working day decreases to seven hours. The work time also decreases by 1.5 hours during the two weeks before returning to Earth. The crew uses this time to prepare for returning to a gravity environment.

**Table 18.10.**     **Schedule of Activities for the MIR Crew.** The schedule provides for five days of work followed by two days of rest. The crew rises at 08.00 MT (Moscow Time). Sleeping time begins at 23.00 MT. Work time decreases by 1.5 hours during the adaptation period (first 14 days of the flight) and before end of flight (last 14 days).

| Activity | Work Zone | Domestic Zone |
|---|---|---|
| Survey of station | -- | 00.10 |
| Personal hygiene | -- | 00.40 |
| Breakfast | -- | 00.40 |
| Duty communications (in the morning) | -- | 00.15 |
| Work preparation | 00.30 | -- |
| Work in accordance with the program (in the morning) | 03.30 | -- |
| Physical exercises (in the morning) | -- | 01.00 |
| Dinner | -- | 00.40 |
| Rest | -- | 00.50 |
| Work in accordance with the program (in the evening) | 03.00 | -- |
| Physical exercises (in the evening) | -- | 01.00 |
| Preparing supper | -- | 00.30 |
| Supper | -- | 00.30 |
| Preparing the report on program | 00.30 | -- |
| Review next day's program | 01.00 | -- |
| Duty communications (in the evening) | -- | 00.15 |
| Private time | -- | 01.00 |
| Sleep | -- | 08.00 |

The station design reduces crew fatigue by allowing onboard systems or ground-based people to do routine work. Also, the crew's activity schedule must include work they enjoy doing. For example, all cosmonauts enjoy activities like

EVA and visual observations. To maintain the crew's mental health, the Crew Psychology Support Group organizes video teleconferencing with family members and journalists. To maintain the crew's physical health, Mir has exercise equipment. All of these things enable the crew to remain in orbit for a year or longer.

**Interaction with Foreign Flight Control Centers.** The first joint activity with a foreign (US) control center was during the Apollo-Soyuz program in 1975. During this flight, joint communications occurred through 13 telephone and two television channels.

In 1992–93, a Russian-French crew carried out four flights. The Russian FCC in Kalliningrad and the French control center in Toulouse developed technologies for jointly operating the flight and exchanging flight data. For example, in July 1993, the two control centers communicated through one data-transmission, four telephone, and two television channels. The FCC in Kalliningrad, transmitted television, voice, navigation, and telemetry information to Toulouse. In 1995, three significant events occurred in support of the NASA-Russian Space Agency agreement to build an international orbital station. A Russian cosmonaut flew on the US Space Shuttle, a US astronaut flew on the Mir, and the Shuttle and Mir docked. Figure 18.13 shows a possible scheme for communications between the two countries' flight control centers during operation of the space station.

# 18.4 European Space Mission Operations

## Dennis Taylor

This section overviews how the European Space Agency (ESA), headquartered in Paris, organizes, prepares, and carries out space missions. It also describes the concepts ESA is using to reduce costs for designing, developing, and operating space systems. ESA applies the principle of Geographic Industrial Return to determine how much each Member State participates in each program, based on that country's contribution to each program. These international programs are more difficult to manage than a purely national program.

### 18.4.1 ESA's Approach to Developing and Operating a Space System

**Organizing.** ESA de-couples development and operation of the space and ground elements because of the traditional division of responsibilities among ESA, industry, and the ESA Directorates. The developer of the space element procures and verifies the spacecraft, then delivers and checks it out at the launch site before launch. The organization responsible for the ground element develops, verifies, and validates its own systems and prepares for in-orbit operations. Mission operations conducts actual operations, from lift-off through commissioning, in-orbit test, and the end of mission—which is usually at the end of the spacecraft's useful life.

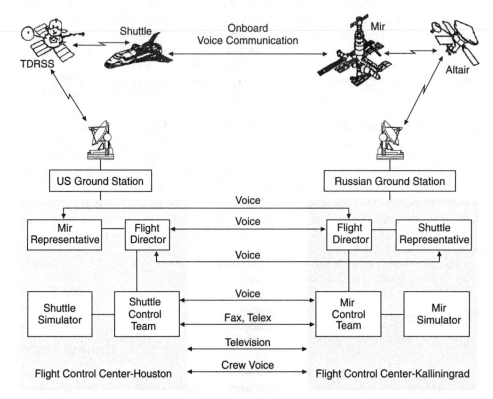

**Fig. 18.13.   Flight Control Centers in Moscow and Houston Cooperating during Mir-Shuttle Flight.** The figure shows one of the variants for communications between control groups during the future Mir-Shuttle experiments. Note that, to help operations, representatives of the Mir control team in Kalliningrad will be at the Flight Control Center in Houston, and representatives of the Shuttle control team at Houston will be at the Flight Control Center in Kalliningrad.

ESA's division of responsibilities:

- The Overall Program Directorates, the Program Departments, and Individual Project Offices are at ESA Headquarters (ESA/HQ) in Paris, France. They manage and allocate resources for whole programs, which comprise missions with common objectives. An example is the Directorate of Observation of the Earth and its Environment, which handles several Earth-observation and meteorological programs. These programs operate several spacecraft over several years.

- The Individual Project Teams are at the European Space Technology and Research Center in Noordwijk, Holland. They handle the budget and technical matters in developing the space system (space element, ground element, communications infrastructure) for a particular mission.
- The Operations Directorate of ESA is at the European Space Operations Center (ESOC) in Darmstadt, Germany. It oversees overall development of the ground element and mission operations.[*]
- The Prime Industrial Contractor—selected early in the development program—designs, develops, integrates, and verifies the space element before launch.

Figure 18.14 shows the project funding, control, and lines of communication between the above organizations for a typical program. ESA Headquarters manages the Agency's total spending profile and allocates resources to the various programs.

Program Directorates cover Scientific, Earth Observation, Space Station and Microgravity Platforms, Space Transportation, Communications, and Space Technology Programs, which in turn consist of several Program Departments and Individual Program Offices.

The Program Office controls the funding for a particular mission or series of missions, decides on the Prime Satellite Contractor at an early stage in the development cycle, and delegates budget and technical responsibility to the Project Manager at the European Space Technology and Research Center (ESTEC).

The Project Manager oversees the budget and technical quality of a particular mission and supervises the project team. Most members of the project team come from the ESTEC and remain assigned to that line. The structure of the project team varies across projects but typically includes sections responsible for project control, spacecraft bus, payloads, assembly, integration and verification, and mission operations (see Fig. 18.15).

The mission operations section interfaces with the Directorate of Operations and defines the high-level requirements on the ground element. It also monitors and controls ground-element development and manages the project-support team for pre-launch and in-orbit operations.

The Operations Directorate at ESOC is organized into departments that have the experience to design, develop, and operate the ground element. Five technical departments nominate staff to support a particular mission. The departments manage these people, and the number of staff on teams for mission preparation and operations varies throughout the project life-cycle. Table 18.11 lists the technical departments of the ESOC and their responsibilities.

---

[*] At publication, ESA is re-evaluating operations and is considering contracting out for operations on selected missions.

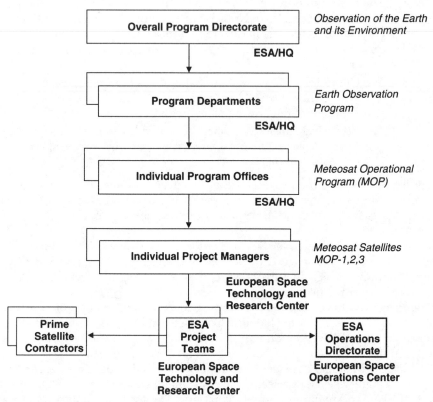

**Fig. 18.14.** **Project Funding and Management for the European Space Agency (ESA).** Meteosat is typical of ESA projects.

Besides handling all aspects (designing, developing, integrating, and validating) of the space element and delivering the spacecraft to the launch site, the Satellite Prime Contractor provides a Spacecraft User's Manual, which is the main source of information for mission preparation and mission operations. The contractor also provides support before and after launch (for Launch and Early Orbit Phase, mission critical activities, and anomaly investigation).

**Designing the Mission.** The procurement cycle begins with Phase A (Feasibility Studies), in which we review mission requirements and objectives and form a skeletal operations concept.

A competitive tender for Phase B (Detailed Definition), normally follows. Phase B produces a detailed design of the spacecraft's subsystems and payload so the Agency can accurately assess cost to completion and write a development plan and schedule. During this phase, critical reviews and trade-offs for requirements and performance specifications keep the project within the Agency's budget.

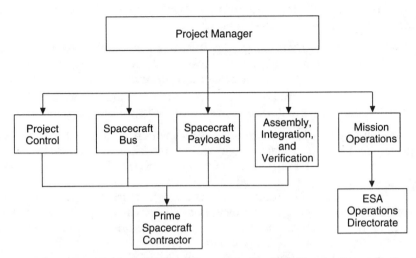

**Fig. 18.15. Structure of a Typical Project Team under the ESA.** The structure varies for different projects.

**Table 18.11. Responsibilities of the Technical Departments in ESA's Operations Directorate.** All departments report to the Directorate of Operations.

| Technical Department | Responsibilities |
|---|---|
| Mission Support Office (Technical) | • Provides long-term planning support |
| Mission Operations Department | • Provides ground-element manager<br>• Provides flight-operations team<br>• Validates ground element<br>• Conducts mission operations |
| Computer and Network Operations Department | • Processes and maintains all computer hardware and software<br>• Operates and maintains ground station and communications networks |
| Flight-Control Systems Department | • Processes spacecraft and systems data |
| Stations and Communications Engineering Department | • Develops ground station network and communications links |

Phase C/D is the build program, consisting of assembly of the subsystems and payload, integration, and final test and verification. Phase C/D concludes with delivery of an integrated spacecraft and support equipment to the launch site, followed by integration to the launch vehicle and pre-launch checkout.

Ground-element development normally lags behind development of the space element and logically splits into two parts:

- Developing hardware and software to prepare the ground element itself (for the spacecraft and mission control systems, spacecraft simulators, ground stations, and communications links)
- Preparing for mission operations, such as planning the mission, developing the database and procedures, training, and rehearsing

The development phase formally concludes with final acceptance of the whole ground element for operational use. It includes exercising operations staff and procedures. To do so, it uses a spacecraft simulator, and possibly the live spacecraft, as a data source for full-up rehearsals of the mission timeline.

ESA has a worldwide communications infrastructure, which includes the S-band tracking stations and communications network used for launch and early orbit operations, as well as the primary ground stations and networks used to transmit data to and from the control center during normal operations.

As the hub of this communications network, ESOC monitors, controls, maintains, and upgrades it. Adding new missions results in new communications requirements, which are met by building onto the existing infrastructure. ESOC also procures and upgrades ground stations, procures space and terrestrial links, and oversees the ancillary receiving and transmission equipment at the various sites. ESOC always monitors and controls the infrastructure for operational communications, even though the control center may be elsewhere.

Ideally, the Operations Concept Document would capture the plan for operations, but such a formal document is unlikely to exist for ESA programs. Instead, the roles and responsibilities of ESA and external facilities usually appear in high-level User Requirements and System-Level Design Documents for the ground element, which ESOC produces. Still, an outline operations concept for the space element must be available at the beginning of Phase B, and it should clearly state who does what.

The ground element for large projects depends on a separate Phase B from the space element and undergoes early trade-offs with the space-element design. European Space Technology and Research Center normally does these studies, with ESOC's participation.

Lately, projects having routine operations have handed the traditional role of ESOC to other ESA facilities. Examples are the ground stations at Redu (Belgium) and Villafranca (Spain). In this case, ESOC handles only launch and early operations but still develops and operates the ground element and manages operations.

**Preparing for the Mission.** ESOC defines *mission preparation* as those activities preceding launch. They include

- Developing the spacecraft and mission-control systems
- Developing a spacecraft simulator

- Developing and refurbishing ground stations
- Developing a communications infrastructure
- Producing the Flight Operations Plan and Flight and Ground Procedures
- Producing operational databases
- Recruiting and training staff

We describe below the phases that control how we carry out these tasks. For a typical medium-sized mission, preparation takes about four years. That is, requirements definition starts about four years before launch. Figure 18.16 shows how ESOC Line Departments organize to prepare for the mission.

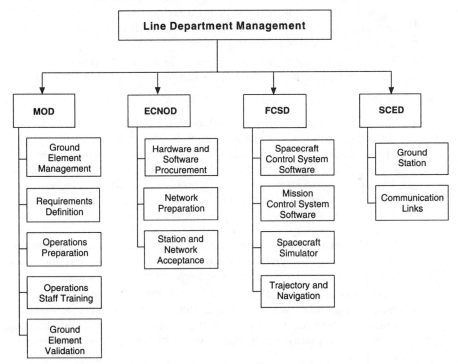

**Fig. 18.16.**   **ESOC Organization for Mission Preparation.** MOD is Mission Operations Department, ECNOD is ESA Computer and Network Operations Department, FCSD is Flight-Control Systems Department, and SCED is Stations and Communications Engineering Department.

During the concept phase, we define requirements, determine the mission operations concept, and establish the derived, top-level, system requirements on the ground element. We identify which parts of the ground-element infrastructure we'll re-use, which need refurbishment or modification, and which new elements we need to develop. We also draw up a development plan and schedule.

Operational requirements flow in both directions between the Program and Operations Directorates, and developing an operations concept is a collaborative venture. The main activities during the concept phase are to

- Define an operations concept
- Define ground-element requirements through the Mission Implementation Requirements Document. European Space Technology and Research Center (with support from ESOC) provides the highest level of operational requirements on the ground element.
- Write a Mission Implementation Plan—the formal response from the Operations Directorate to the Requirements Document which gives a high-level description of the ground facilities, the management roles and responsibilities, an overall schedule for developing the ground element, and the cost to complete it
- Issue User Requirements Documents. ESOC/MOD writes these documents, which cover in detail what all operational facilities must do.

Two other documents often appear during the concept phase for new missions:

- The Spacecraft Operations Interface Requirements Document (SOIRD), from ESOC, contains generalized requirements on the space element for all missions. These derive from current operations concepts and the intended use of the ground element's infrastructure. The SOIRD also specifies the deliverables from the space-element program to prepare for mission operations (such as access to flight hardware) and defines the support services required from the Project team and Satellite Prime Contractor for the whole mission profile. Requirements for a specific mission modify the general requirements in the SOIRD, derived from cooperation with the Project Team, and go into the Mission Implementation Requirements Document.
- The Mission Assumptions Document, from ESOC, contains the assumed operations requirements derived from project feasibility studies (Phase A). It also outlines the ground-element design based on existing and planned facilities, and it outlines the mission operations concept.

Figure 18.17 shows the sequence of activities during the concept phase, using the above documents.

The design and development phase is the build phase of the ground element, during which we

- Prepare the Spacecraft Control System (i.e flight control system, flight-dynamics system, spacecraft simulator) and Mission Control System

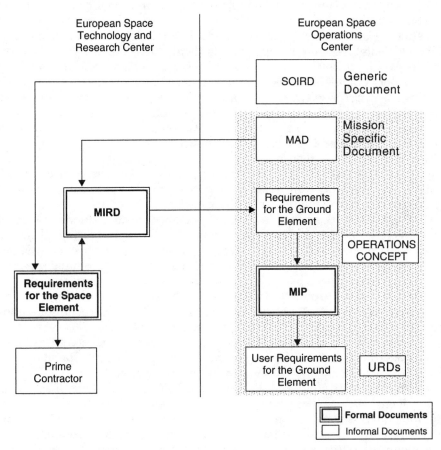

**Fig. 18.17.    Concept Phase for Developing the Ground Element.** SOIRD is Spacecraft Operations Interface Requirements Document, MAD is Mission Assumptions Document, MIRD is Mission Implementation Requirements Document, MIP is Mission Implementation Plan, and URD is User Requirements Documents.

(mission support software, such as tools for mission planning and scheduling)

- Prepare ground stations and communications links
- Prepare the communications network
- Prepare operational databases, flight and ground procedures, and the Flight Operations Plan and procedures
- Recruit and train staff

All software develops according to strict engineering standards—in this case ESA's Software Engineering Standards, called PSS-05-0. These standards are part of a set of Program Support Standards adopted by ESA to procure space elements, ground elements, and communications. Hardware items are developed and procured in parallel to the software, again according to standards. Items such as control-center computers will be long-lead items to allow software development to begin. The cycle of software development begins with the user requirements formed during the concept phase, followed by software requirements derived from these. The architectural design follows. It defines the software system's architecture and structure and details how to produce it.

The detailed design then begins, in which we design software down to the lowest module level, code it, and progressively verify it to system level. Finally, the transfer phase gradually transfers the software to the operational users, at which time the operations and maintenance phase begins.

The Flight Operations Plan is a crucial document for operations. It formally documents how the mission and space element will operate during all mission phases, from pre-launch to end of mission. Producing this plan is a major task, and a main source of information is the Spacecraft Users Manual from the space-element prime contractor. This plan contains items such as mission rules and constraints, flight and ground procedures for both nominal and contingency operations, and timelines for critical and routine phases. Spacecraft-operations engineers working under ESOC/MOD produce it, working closely with the project team at European Space Technology and Research Center. The Project Manager and the Director of Operations must sign off the Flight Operations Plan before ESOC can declare the ground element ready for operations.

**Integrating, Verifying, and Validating the Ground Systems.** Each of the major systems, such as those for spacecraft control and flight dynamics, are initially integrated separately, then integrated onto the target hardware and progressively tested to system level until the ground element is complete. Communications interfaces and protocols between the systems are tested using simulators or pre-recorded data. At the end of the integration phase, the ground element won't have been integrated as a whole, but we'll have built up the major systems at the various sites.

The verification phase really has three sub-phases: test, acceptance, and verification. The result is a formal acceptance that the design meets the specifications for each of the major systems that comprise the ground element. The verification phase is complete when we can certify that all software and hardware works within the original specifications.

Validation is a test phase during which the entire ground element (including operational teams) comes together to rehearse selected operational scenarios and the mission timeline, so ESOC can declare the ground element ready for operations. Of course, true operational validation can only occur in flight, during actual mission operations. To validate a system, we typically

1. Validate systems using simulators and pre-recorded data tapes. Include the major software systems of the spacecraft and mission-control centers, the ground stations, and the communications infrastructure.
2. Validate the spacecraft simulator
3. Validate the software, operational databases, and Flight Operations Plan using the spacecraft simulator
4. Validate the software, databases, and Flight Operations Plan using the spacecraft engineering or flight model. These tests are termed System Validation Tests, which are a major milestone in ground-element development.
5. Test data flow using simulator at the ground stations
6. Train the operations team using the spacecraft simulator and formal simulations and exercises
7. Rehearse flight readiness for the launch and early orbit phase of the mission timeline using the spacecraft simulator

A crucial document used during this phase is the System Validation and Test Plan, which ESOC produces early in the program in cooperation with the Project Team. The plan provides a detailed schedule of tests, together with required test tools, procedures, and acceptance criteria.

**Operating the Mission.** ESOC defines mission operations as activities from launch to end of mission that deliver the mission products:

- Plan the mission
- Carry out the mission
- Evaluate the mission
- Distribute mission products

To prepare in-orbit operations, we need a more project-oriented organization, which brings together the line-department staff (permanent staff and contractor staff), augmented by senior staff who work across missions, the Project Support Teams, and the Industry Support Teams. The line-department staff are key people who have been preparing for the mission. We also gradually build up the Mission Control Team during the latter part of the preparation phase. The team decreases after launch and early-orbit activities but may build up again for other critical phases of the mission. Figure 18.18 shows the structure of the Mission Control Team that operates the mission.

The spacecraft-operations manager still oversees the reduced team and mission operations but shares his or her time with mission preparation and operations responsibilities for other missions. Spacecraft-operations engineers therefore manage and control routine operations.

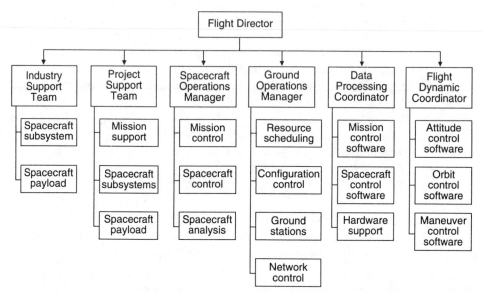

**Fig. 18.18.   ESOC Organization for Mission-Critical Operations.** The team is normally reduced for routine operations.

In the following text we briefly describe the core activities for mission operations; Fig. 18.19 shows the relationship between these activities.

Mission planning occurs throughout a mission, in addition to the once-through activities that are part of mission preparation. The planning requirements vary considerably across missions, from the relatively simple tasks for communications missions to the more complex tasks for scientific missions. Planning requirements may be so stringent that we need some onboard autonomy, whereby we prepare a detailed schedule on the ground and uplink it to the spacecraft well in advance.

Automated software tools help us do the more complex planning tasks. Typically, they accept user inputs for subsystem and payload operations, and they check the feasibility of operations against mission rules and onboard constraints. The more advanced tools produce a detailed schedule ready for uplink, which reduces the need for skilled manpower and gives us better, more reliable operations.

Activities to carry out the mission have followed the same basic concept for all ESA missions since 1977, when the Multi-Satellite Support System first started operating. Mission execution is essentially a near-real-time monitoring and control activity to support the planned timeline of events. Monitoring involves continuously observing subsystem, system, and payload telemetry in near-real-time, with automatic checking against desired state and alarm limits. The data can be dis-

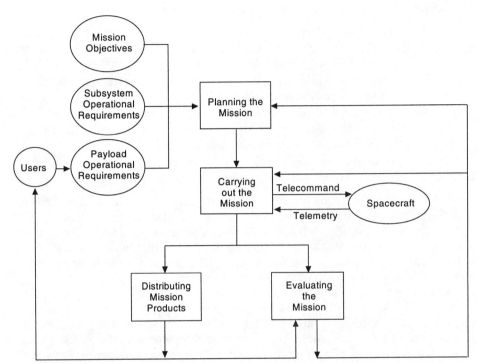

**Fig. 18.19. Relationship of Mission Operations Activities.** The four main activities—planning, carrying out, evaluating, and distributing—occur roughly in sequence, although the actual timescales vary considerably across missions.

played alphanumerically, graphically, or synoptically, and data retrieval is possible from short-term and long-term telemetry archives. Uplinked commands normally feed into the monitoring system to allow for the effect of telecommanding and subsequent onboard status changes.

Control depends on the phase of the mission. During mission-critical phases, the preferred approach is to do manual commanding—supported by the flight-dynamics, project, and industry teams. A reduced Mission Control Team does routine operations once the spacecraft is in its operational orbit and configuration. They use either a paper-based schedule, which calls up procedures from the Flight Operations Plan, or a software-based schedule, which automatically handles commands. For contingency operations, the spacecraft controllers can use recovery procedures from the Flight Operations Plan, which covers well-defined anomalies. Other anomalies require additional support, which may involve the project and industry support teams.

Support software for the control functions may include

- A command-handling system which accepts commands from various sources, checks for conflicts, encodes the commands, validates them, and finally uplinks them to the spacecraft and verifies execution
- A command-verification system, which monitors change of state according to uplinked commands
- A command scheduler, whose input can be taken automatically from the mission-planning activity or manually, and whose output is a command schedule ready for uplink
- Automatic command programs, which can build up sub-schedules based on an operational requirement or telemetry

ESOC uses a mission-specific Spacecraft Control and Operations System (SCOS-1), which can include the above monitoring and control functions. SCOS derives from common (kernel) software, which programmers tailor for each mission. SCOS-1 will eventually give way to SCOS-2, which will be a distributed system based on modern workstations, using artificial intelligence and an advanced interface between machines and operators.

Mission-evaluation activities are those associated with evaluating the performance of the space and ground elements, managing resources, investigating anomalies, and doing work-arounds. Operators can do some of them online, constrained only by the duration of the short-term telemetry archive. ESOC uses the Spacecraft Performance Evaluation System, which receives its data automatically from SCOS to help do these activities.

Distributing mission products includes sending data to the end users, such as the principal investigators in the case of science missions, or delivering continuous communications service in the case of communications missions. ESOC or other organizations who receive downlink data usually pre-process it to save the need for duplicate processing by the end users. An example is the use of the METEOSAT spacecraft as a data relay to transmit processed images to the meteorological community. The current trend is to make data available to the users electronically through a centralized database. This online access eliminates distributing data on magnetic tape. For security reasons, the computer holding the database is separate from the operational system.

## 18.4.2    Future Trends to Save Costs

We can reduce costs by reusing and enhancing concepts for space and ground elements that we've already developed for similar missions. Because mission preparation and operations are closely linked, common operations concepts will promote re-use of tools for both phases.

ESA is a relatively fixed organization, with space-element design managed by ESTEC and operations managed by ESOC. This enables close collaboration, maintains a centralized pool of development and operations expertise, and provides a means of controlling the development of common hardware and software. ESA runs projects with the smallest possible staff, especially for the operations phase, and reusing technology and expertise is common across program boundaries.

Flexibility is built into the ground element's infrastructure to support new missions, with maximum use and reuse of existing equipment, for launch and early orbit and routine operations. ESA vigorously enforces this policy for communications links from ground stations and for developing spacecraft-control facilities. For example, common software for spacecraft control has already been developed, and the S-band network is reused for launch and early orbit on every mission.

We can use common mission characteristics to develop common flight hardware and software for space platforms. This is especially true for spacecraft subsystems, on which we're adopting common standards for telemetry and data handling. In some cases, we reuse the complete platform.

These mission characteristics can also lead to common operations concepts for mission preparation and operations. As missions become more complex and require more automated tools, common concepts will help us build better software and improve mission operations.

ESA is trying to reuse software for major projects, so ESOC has developed a common Spacecraft Control and Operations System and Spacecraft Performance Evaluation System for all missions.

The next step is to develop a set of Advanced Tools for Operating Systems, which will allow us to customize complete mission and spacecraft-control systems from common software modules. ESA policy is to reuse hardware and software within, but not between, the space and ground elements. Some major programs (for example Columbus) have tried to consider all software part of a total software system and to reuse flight software on the ground. Examples of space-element software we can reuse on the ground are

- Spacecraft-checkout software, which we can reuse for spacecraft control and monitoring on the ground
- Dynamic-simulation software, which we can reuse for spacecraft simulators needed to support operations and training

ESA has worked hard over several years looking for common aspects of ground-checkout and spacecraft-control software through the Committee for Operations and EGSE Standardization. (EGSE is ESA's acronym for Electronic Ground Support Equipment, which includes all the hardware and software necessary for ground test, checkout, and in-orbit support.) Promoting reuse of space software on the ground is difficult because it further constrains space-element contractors, who aren't normally concerned with the ground element.

The concepts of commonality and reusability are often confused. For ESA projects,

- Commonality results from common functions. It may exist for technologies, expertise, organization, or actual hardware and software.
- Reusability results from commonality, where a product developed for a specific application in one environment can be completely reused in another.

Following standards and applying generalized requirements promote commonality and reusability. For example, we can reuse software developed for the space element on the ground only if the ground requirements are imposed on the contractors. ESA is stressing reuse by specifying common space-element hardware for different missions. They haven't yet followed the idea of common operations concepts, although ESOC has done some preliminary work. They can achieve truly effective trade-offs and cost efficiency in this area because costs for operations and supporting infrastructure are a substantial part of total mission costs.

### 18.4.3    Influence of Emergent Technologies

ESA is trying to transfer new technologies into spacecraft operations. For example, we've developed several prototypes of artificial-intelligence applications to aid spacecraft monitoring and control. These typically encapsulate the experience and judgment a human operator would use to make decisions. ESOC has already used knowledge-based systems offline to aid Fault Diagnosis, Isolation, and Recovery. These systems have mainly supported mission operations but they're equally relevant to mission preparation. They bear strongly on operations concepts because they influence the manpower and skill levels needed for mission preparation and operations. They also influence operations safety, reliability, and autonomy onboard or on the ground.

Operations engineers spend a long time transcribing knowledge about the spacecraft design in the Spacecraft Users Manual to the Flight Operations Plan. Normally, the spacecraft prime contractor delivers the Users Manual on paper. ESA is now requiring these contractors to present the information in a suitable electronic form, which automated tools can use to produce the Operations Plan.

We can also use knowledge-based technology for the mission database, which would be a central repository of all technical and management information throughout the project. Teams developing the space and ground elements could access this database as could others working on mission preparation and operations. This database becomes a Mission Information Management System, which would form the backbone of an open system architecture for developing future ground systems to support spacecraft operations.

ESOC is working on Advanced Tools for Operating Systems, which will employ such a management information system. It will also allow the future distributed Spacecraft Operating System (SCOS-2, see Sec. 18.4.1) to communicate with advanced tools for spacecraft operations, some of which will include artificial intelligence and knowledge-based systems. This is an important new approach because it will radically change the whole development and operations phase for the ground element (e.g. cross-mission commonalities, reusability, manpower profiles). Knowledge-based systems will eventually migrate onboard, but not before operational experience with these systems has accumulated on the ground. ESA has identified several areas in which automated tools can be useful on the ground. These tools have been prototyped and are working to a limited degree in the operational environment. The identified areas include

Mission-Preparation Activities
- Prepare operational databases
- Prepare Flight Operations Plan
- Prepare nominal and contingency procedures in the Operations Plan
- Prepare mission timeline

Mission Operations Activities
- Plan and schedule the mission
- Support onboard autonomy
- Maintain onboard software
- Control orbit and attitude
- Support fault diagnosis, isolation, and recovery
- Schedule resources for spacecraft subsystems and payloads

These tools will be subject to trade-offs because they'll reduce staffing but increase development cost, risk, and maintenance costs for operations.

### References

Acton, C. H. "Using the SPICE system to Help Plan and Interpret Space Science Observations." "Proceedings of the Second International Symposium on Ground Data Systems for Space Mission Operations." November 16-20, 1992. Pasadena, California. Pasadena: JPL Publication 993-5, 1993. Pp. 781-787.

Altunin, V. I., Anderson, B., Baars, J. W. M., et al. *IVS. An Orbiting Radio Telescope.* Report on the Assessment Study. European Space Agency SCI(91)2, January 1991.

Altunin, V. I., Robinett, K. H. "Mission Operations System for Russian Space Very-Long-Baseline-Interferometry Mission." IAF-92-0547. 43rd Congress of the International Astronautical Federation. August 28 -September 5, 1992. Washington, D.C.

Altunin, V. I., Sukhanov, K. G., Altunin, K. R. "Radioastron Flight Operations." "Proceedings of the Second International Symposium on Ground Data Systems for Space Mission Operations." November 16-20, 1992. Pasadena, California. Pasadena: JPL Publication 993-5, 1993. Pp. 185-190.

Altunin, V. I. "Technical Parameters of the Ground Segment and Data Management of the Radioastron Project." (AIAA-90-5012). The 2nd International Symposium on Space Information Systems, AIAA/NASA. Pasadena, CA. September 17-19, 1990.

Armand, N. A., Aleksandrov, Yu. N., Altunin, V. I. et al. "The Special Network for Telemetry Reception and Interferometric Measurements of the Vega Balloon Experiment." Soviet Astronomy Letters, v. 12, No. 1, 1986.

Bogdanov, Anatoly. "The Organization and the Structure of the Management of Space Affairs in Russia." Space Bulletin. Vol. 1, No. 1, 1993. Pp. 4-7

Consultative Committee for Space Data Systems. "Radio Frequency and Modulation Systems, Part 1, Earth Stations. CCSDS 411.0 G-2. June 1990.

Froehlich, Walter. *Apollo Soyuz*. Washington, D.C.: National Aeronautics and Space Administration, 1976.

Larson, Wiley J. and Wertz, James R., ed. *Space Mission Analysis and Design*, 2d edition. Torrance, California: Microcosm, Inc., 1992.

Smith, Gene. "EOS Ground Data Systems: A Description & Interface Overview." Proceedings of the Second International Symposium on Ground Data Systems for Space Mission Operations. November 16-20, 1992. Pasadena, California. Pasadena: JPL Publication 93-5, 1993. Pp. 767-778.

Robinett, K. H. "Radioastron Mission Operations Concept." Pasadena: Jet Propulsion Laboratory, 1992.

Smid, H. H. F. "Soviet Space Command and Control." Journal of the British Interplanetary Society. Vol. 44, No. 11. November 1991. Pp. 525-533.

Sunyaev, R. Á. *The Phobos Project*. New York: Harwood Academic, 1988.

# Chapter 19

# Microsatellite Mission Operations

Jeffrey W. Ward, *Surrey Satellite Technology Ltd.*
Craig I. Underwood, *University of Surrey*

Although not every mission operation manager (MOM) will work on a microsatellite mission, the space industry is moving toward smaller spacecraft built quickly for inexpensive missions. This chapter shows that such missions depend heavily on an integrated approach to mission operations planning. Thus, managers of large missions will see here a complete example of the cost-effective techniques exposed throughout this book.

Unfortunately, cost-effective doesn't necessarily mean inexpensive. Traditional space missions, even when cost-effective, are not within the financial reach of most universities and corporations or even many governments. A class of inexpensive satellites known as *microsatellites* has been developed to remove, or at least to lower, this financial barrier to participating in space missions. It's not unrealistic to plan a microsatellite mission costing only $3 to $4 million—including installation of a mission-control station and one year of mission operations.

Much of this small budget goes into getting the most from the spacecraft, and only a small fraction remains for on-orbit operations, as shown in Table 19.1. In this environment, we must take a particularly aggressive approach to cost-effective space mission operations. We've used this approach for ten microsatellite missions from 1981 to 1994. Many AMSAT missions have used similar procedures and have been so inexpensive that they are definitive examples of cost-effective operations. Our first-hand experience is mostly with UoSAT/SSTL and AMSAT missions, but guidelines in this chapter apply to other circumstances.

**Table 19.1.** **Typical Microsatellite Budget.** In a low-cost space mission less than 10% of the budget is allocated to mission operations—compared to between 20% and 50% for a typical space mission.

|  | US $ | % of Budget |
|---|---|---|
| Satellite | $2,250,000 | 62.5% |
| Launch | $750,000 | 20.8% |
| Insurance | $450,000 | 12.5% |
| Mission Control Station | $75,000 | 2.1% |
| Operations (one year) | $75,000 | 2.1% |
| Total | $3,600,000 | |

Many of these guidelines apply to concept exploration and detailed development because we use the concurrent approach to design. We have to analyze all of the mission elements of a microsatellite at once because of their small size, low cost, and short timescales. Therefore, these satellites have demanded concurrent planning for 20 years. Cost-effective mission operations result from decisions about management structure and concept development as surely as from any isolated process for mission operations design. We'll try to show how the thread of low-cost operations weaves through the fabric of the microsatellite mission.

We define *microsatellite missions* as those costing less than $5 million, using spacecraft between 10 and 100 kg mass, and moving from system definition to launch in 12 to 36 months. Microsatellites of this class have operated exclusively in Earth orbits, and usually in low-Earth orbits (LEO), because they ride as secondary payloads on commercial and military launches that often use LEO or geosynchronous transfer orbit (GTO). GTO presents a hostile radiation environment, suitable only for very brief missions or spacecraft that can maneuver in orbit. Microsatellites such as OSCAR-10 and OSCAR-13 have maneuvered from GTO to Molniya orbits, but such maneuvers are complex and expensive. The body of experience in mission operations for microsatellites derives from LEO missions, so we must be careful when attempting to translate this experience to other orbits.

Section 19.3 will show that we have to carefully select and design objectives and concepts for microsatellite missions to produce cost-effective operations. Despite constraints on mass, volume, and mechanical complexity, we can still deploy microsatellites for a wide range of useful missions. Traditionally, they've provided early flight opportunities for new technologies. But we should consider them for missions using real-time and store-and-forward communications, scientific experiments with low data rates and modest attitude-control requirements, and medium- or low-resolution remote sensing. Inexpensive yet capable microsatellites have served all of these missions—indeed, through careful spacecraft design and operations planning, a single microsatellite can serve all at once.

The ability to define a mission concept for low-cost operations that delivers high-quality data will measure the success of mission-concept development for microsatellites. In this chapter, we describe the organizational and technical elements which combine to produce inexpensive mission operations that give us a high return.

## 19.1   Microsatellite Teams

Working on short timescales within small budgets to produce a useful space mission requires teamwork. Correctly forming, training, and organizing this team are the foundations of any cost-effective space mission, but they are particularly critical for microsatellite missions. The microsatellite team must be able to cut costs radically in the space and ground-operations elements while still developing an architecture that serves the users' goals.

Clearly, to operate at low cost, saving money in the space element can't drive up costs for ground operations. To build a practical and useful spacecraft, mission operators and the end users must be part of the mission team. A well constituted team will be able to review continually how design decisions affect mission operations.

### 19.1.1   Training

Many end users, operators, and developers will enter the microsatellite arena with considerable experience in traditional space missions, so we may think they can handle the microsatellite mission with no further training. But we must train team members specifically for microsatellite missions.

Although specialists must be on the team, they must also receive general training, especially concerning mission operations. Low-cost operations are possible only if spacecraft-bus engineers, payload engineers, and operators trade ideas. The most certain way to achieve this sharing is to train each engineer in the others' disciplines.

Once we launch the microsatellite, operators take over the spacecraft bus and payloads. They have to maintain the bus and use bus services to operate the payload. The payload must return the data end users want, thus completing the mission's life cycle. The communications architecture of the bus is a critical link in this chain. Both bus and payload engineers need to design for the operator as much as possible.

A development decision made by a bus engineer intending to save money or to decrease risk may do the opposite when operators take over the spacecraft. To avoid such problems, bus engineers should know operations in general, as well as the specific mission operations concept selected for the mission. Initially, this investment in training increases cost. Ultimately, it results in a spacecraft designed to be operated, which will save effort and cost during the mission's operations and support phase. In the most extreme case, the training will help bus engineers identify operational failure modes which they would otherwise have overlooked.

What goes for bus engineers also applies to payload engineers. The payload engineers will determine how, physically, the payload is to sense, or interact with the subject. They'll do this by trading off the end user's desires against the bus's abilities and limitations, but they must also consider how their payload design affects operations. Designing microsatellite payloads is somewhat different from designing payloads for larger spacecraft. Lack of sophistication in one area doesn't mean the project generally lacks resources. Attitude determination may be better than attitude control. Store-and-forward collection of data may be more flexible or reliable than using data-relay satellites and proliferated ground stations. Constraints in mass, volume, and power may be offset by the availability of powerful onboard computers and the ability to closely integrate the payload with the bus. Operating without these insights, payload engineers may inadvertently duplicate bus services, limit the payload's operational scope, or complicate operations. Managers must give payload engineers these insights and encourage them to develop the payload accordingly.

After launch, the responsibility for a mission's success rests with the operations organization. This is particularly true for microsatellite missions, in which the operations organization may act alone over a wide range of tasks to save money. Supplied with a little high-quality technical documentation, they must define and carry out routine and emergency operations. This approach is safe and reasonable only if the operators are technically competent and have been trained to understand the characteristics of the microsatellite bus and payload. Rather than separating operations and support from other phases of the mission, the operators must participate in all mission phases from definition and development through testing and pre-launch operations. With such comprehensive training, they'll be able to operate the spacecraft safely, successfully, and cost-effectively.

The integrated mission team described above is an ideal organization for concurrent design of mission operations. Comprehensive cross-training is practical for microsatellite missions because we can keep the entire mission team relatively small. Indeed, if the mission is to be truly inexpensive, the mission team *must* be as small as possible. One way to achieve this team training is to permit team members to change roles from mission to mission. By rapidly defining and developing microsatellite missions, engineers can play several roles and see several missions to completion within only a few years. This team of richly trained engineers—each with experience spanning the spacecraft bus, the payload, and the mission operations—can produce operations that are inexpensive yet deliver high-quality data and services.

## 19.1.2    Organization and Management

Simply assembling a team of engineers with all-around training doesn't necessarily lead to success. Mission managers must organize the team effectively and manage it correctly. Although many management philosophies move into and out of fashion, the techniques described here have proven particularly effective for low-cost microsatellite missions. These successful techniques emphasize free information flow throughout the team, flexibility at all stages of the mission, and minimum bureaucracy.

**Communication.** To capitalize on the training which has given each engineer insight into all aspects of the microsatellite mission, all team members must be accessible to and communicate with the others. Regular project meetings plus modern computer tools (electronic mail, work groups, and local-area networks) create forums and mechanisms for formal communication among team members.

Formal communication—written, electronic, or in a review meeting—carries only a small part of the information flow in a successful team. Through informal daily discussions, a well-trained team can achieve the best costs, time-scales, and mission returns. Managers must create an environment in which this informal collaboration can thrive by establishing a shallow management structure and co-locating the mission team. A shallow management structure, as shown in Fig. 19.1, allows all members of the team to take responsibility and contribute to the mission. It allows the mission manager (who must retain overall responsibility) to detect these contributions and incorporate them in mission planning. It's also best to house all team members in a single building, allowing informal communication to take place efficiently and naturally. Although geographically distributed teams can succeed, they do so at increased cost and risk, and with reduced inter-disciplinary communications. Team training gives each person an understanding of the others' domains; management and organization put this understanding to use.

The need for close communication during mission definition and development is obvious. Less obvious is that this collaboration must continue during the operations and support phase. During post-launch operations, the bus and payload engineers will know the spacecraft's limitations and features. The operations organization will be learning about the practicalities of spacecraft operations (possible failures or unanticipated environmental constraints). The end users will be changing their mission goals based on initial results. Only a continuing team effort can best apply the available resources (including money) to meet the end users' goals.

**Flexibility.** Defining and developing a complete mission in as little as 12 months doesn't allow a longer mission's certainties of analysis and design. The mission team must retain flexibility for as long as possible in order to overcome late-breaking difficulties and exploit opportunities as they complete the analysis in parallel with development (see Fig. 19.2). Each area of the mission design has its own requirements for design and interface freezes which will close this period of flexibility. The interfaces between the bus and payload must be frozen relatively early, especially if different teams are developing the two. Hardware designs must

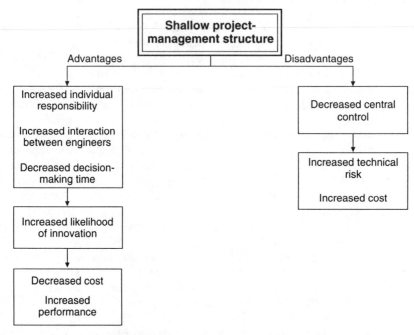

**Fig. 19.1.     Consequences of Shallow Management Structure.** Microsatellite missions depend on innovative solutions to technical problems. Innovation thrives when teams of cross-trained engineers work within a shallow management structure. To avoid losing central control, mission managers and engineers must communicate often.

also be frozen early, to allow time for manufacture, test, and integration. Software designs, on the other hand, can remain flexible throughout the mission life cycle—assuming we properly allow for uploading software after launch. Like software, the plan for space mission operations should continue to evolve throughout the mission, in response to changing goals or capabilities. Although we must have a preliminary operations plan against which to judge mission design, we must *not* freeze this plan unnecessarily. By maintaining flexibility for as long as possible, we can feed the information gained through team training and communications back into the completed space-mission product, increasing returns and decreasing costs. Feedback and flexibility should continue into the operations and support phase.

**Standards.** The flexible, team-oriented approach described above can succeed if the mission's end users and developers support it. The processes and standards used in traditional aerospace projects don't promote rapid, inexpensive development of space missions. We must apply these corporate norms and industry standards only if they lower costs.

**Fig. 19.2.** **Consequences of Late Design Freeze.** If the detailed development phase is particularly short, it's especially important to delay design freezes. Prematurely frozen designs can lead to overly conservative design and to persistent problems, both of which increase costs for mission operations.

One area in which we encounter standards is communication links. For example, NASA and the European Space Agency (ESA) have created standards for telemetry and telecommand links. Adherence to one of these standards allows us to use off-the-shelf equipment in the ground and space elements, as well as existing ground stations and data-relay satellites. But we don't know that this approach will actually reduce the cost of the ground element and post-launch operations. Accessing an existing ground element or using a data-relay satellite may cost more than a custom or semi-custom communication system directly tailored to mission requirements. Below, we'll see that we can build communications links with low and medium data rates by modifying ground communications equipment and assembling it into a data station dedicated to microsatellites. With such a dedicated system, conforming to industry standards isn't in itself an advantage, so we don't need to take it as a constraint during mission definition.

Similar arguments arise when considering NASA's or ESA's strict standards for parts procurement. Restricting designs to use only approved parts increases the cost of the space element and can also increase the cost of mission operations. This situation arises because approved parts are conservative, emphasizing flight heritage and thus restricting the use of new technologies. These new technologies, such as microprocessors, high-density semiconductor memories, and plastic-packaged components, are the very technologies which make low-cost, high-performance instruments possible (see Fig. 19.3). In particular, using special space-qualified microprocessors and low-capacity memories may increase the cost of software development. This, in turn, can drive up costs for post-launch operations.

As an alternative to a conservative parts specification, microsatellite missions have routinely employed state-of-the-art microelectronics. Relatively large solid-state data recorders (up to 384 megabits), controlled by reprogrammable onboard computers, have appeared on payloads and buses. Using these systems, mission operations managers have developed cost-effective operations plans which don't

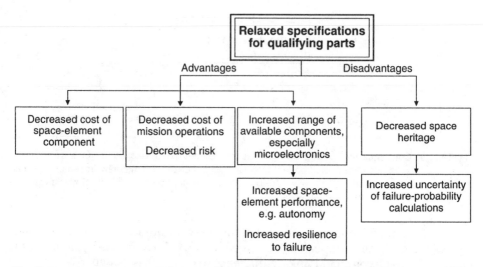

**Fig. 19.3.    Consequences of Relaxed Component Qualification Requirements.** The use of components which have not been space qualified greatly increases the number of micro-electronic components the spacecraft engineer may choose from. This increased design freedom permits rapid and inexpensive development of more complex and capable circuits.

rely on data-relay satellites or continuous ground-station contact. Thus, relaxed parts-procurement policies have saved money from operations.

Low-cost space missions must substitute the initiative and skills of a well-trained team for rigid and comprehensive standards. We can't reject standards arbitrarily but should trade off the costs and benefits of standards just as we would any other engineering decision. The end user must accept this process, which includes risks as well as benefits.

## 19.2  Exploring the Mission Concept

We may strive for cost-effective operation of any mission. We may even build small spacecraft to serve a wide range of missions. Low-cost microsatellites, however, are suitable only for certain missions. Microsatellites, and the design, construction, and operation described here, have strengths and weaknesses. The success and cost-effectiveness of a microsatellite mission will depend on how well the end user's objectives correspond to the microsatellite's strengths. Concept exploration is the time to assess this correspondence. Furthermore, if the entire mission is to be inexpensive, mission-concept development must give appropriate weight to costs of the ground element and operations. [Sweeting, 1994]

### 19.2.1    Mission Concepts for Microsatellites

In some cases, the space-mission budget will be tightly restricted so we can consider only a low-cost microsatellite. In such situations, we must identify mission concepts we can implement within constraints. At other times, the mission will be at least partially defined, and the question will arise, "can a low-cost microsatellite serve this mission?" In both instances, we'll need to evaluate physical and operational constraints.

Microsatellites are constrained in volume, mass, and mechanical complexity. If we're launching a microsatellite as a secondary payload, we must keep its volume and mass low so we have a wide range of launch opportunities. Reliable, complex mechanical subsystems are usually too expensive for a microsatellite budget, so we don't use deployable solar panels and complex stabilizing systems. These basic constraints on the system reduce available power and pointing accuracy. We can overcome mechanical simplicity and small size to some extent by creatively applying microelectronics, but the microsatellite's basic physical limitations will rule out some missions.

If initial mission evaluation shows a microsatellite bus can support the payload, we must determine whether or not a microsatellite can do the mission. This is a two-part analysis: select critical elements of the mission architecture and evaluate their effect on operations. This analysis is critical because it ensures the payload, bus, and ground element will be inexpensive and successful after launch. The main architectural elements we must select are the orbit, the data-communications system, and the attitude-control system. Once we select them, we must evaluate their demands on mission operations.

### 19.2.2    Orbit

Microsatellites have deployed most often in low-Earth orbits (LEO), but they've also worked in GTO and highly-elliptical orbits. If a mission requires an intermediate or high-Earth orbit, or even an interplanetary orbit, we shouldn't immediately assume microsatellites are out of the question. Rather, we should analyze the constraints presented by the orbit: ionizing radiation, profile of visibility to ground stations, navigation requirements, and characteristics of the communications link's path. If we can define a microsatellite mission concept which meets the users' and operators' requirements and remains affordable, the mission is feasible. Our discussion in this chapter centers on LEO because many microsatellites have operated in LEO, *not* because LEO is the only orbit suitable for microsatellites.

### 19.2.3    Communications Architecture

As discussed in Sec. 11.2, the data rate between a spacecraft and ground station depends on the spacecraft's orbit, the availability of ground stations or data-relay satellites, and the mission's data requirements. In general, a low-cost microsatellite may have one or two dedicated ground stations, with communications-link rates up to 50 or 100 kbps. Higher data rates can be supported, but with a significant increase in ground element's complexity and cost, perhaps driving the mission outside the cost-effective range. Using data-relay satellites such as TDRSS is also unlikely to be cost-effective for a low-cost mission.

A 50 kbps downlink to a single ground-control station provides a good starting point for evaluating a typical communication architecture for a microsatellite in LEO. For example, this rate gives us up to 24 megabytes of data return per day from a microsatellite in 850 km sun-synchronous orbit to a ground station at 50° latitude. This rate would permit continuous operation of a payload generating roughly 2.2 kbps (assuming that suitable data storage were available onboard). An experiment with a higher data rate could operate a correspondingly smaller fraction of the time. For example, a remote-sensing instrument taking $1000 \times 1000$ pixel, 8-bit, monochrome images could return 25 images per day.

The basis of these simple analyses is a cost-effective mission operations plan involving a single ground station operating at medium data rates. If such a preliminary analysis shows clearly that the mission will work, mission development can continue. Similarly, if the data return available is several orders of magnitude too small, development should cease. Between these limits are grey areas for which we can offer cost-effective solutions. Perhaps we can add a second ground station relatively inexpensively, or we can increase the downlink rate without unacceptably increasing the ground element's cost. We must explore these options in detail, trading the cost increases for the ground and space element against the benefit of increased data return.

### 19.2.4    Attitude Control

We also need to examine a preliminary budget for attitude determination, control, and stabilization (ADCS), beginning with the present state-of-the art for low-cost systems. A microsatellite stabilized by a magnetic-torquer-assisted system using the Earth's gravity gradient can be expected to remain nadir pointing within ±1° (in 850 km LEO), while attitude determination may be an order of magnitude better. Spin stabilization, achieved by magnetic torquing or cold-gas thrusters, is also relatively inexpensive for payloads with an inertial- or Sun-pointing requirement. Moving away from these simple systems can increase costs significantly and may again drive the mission out of the low-cost realm.

### 19.2.5    Operations

Having identified communications and ADCS architectures which fulfill the payload requirements, we must generate a mission operations concept. It's now particularly important to study how the mission architecture responds to phases of the mission other than orbital operations—especially the early-orbit phase and anomalies during orbital operations. We may have overlooked these demanding periods when we first developed the mission operations concept, which concentrates on routine orbital operations.

Microsatellites are usually released from their launch vehicles with unknown attitude dynamics and all onboard systems dormant. Even if we expect some nominal attitude, the safest commissioning plan assumes little or nothing about deployment dynamics. Thus, thermal conditions, power budgets, and communication-link margins must all remain acceptable without attitude stability. The ideal spacecraft could remain in this pre-commissioned state indefinitely to allow for difficulties in initial acquisition or software loading.

Once we've acquired the spacecraft and started nominal ADCS operations, thermal and link conditions should stabilize, permitting payload checkout and operation. The nominal scenario for orbital operations will be in effect.

It's important to analyze what will happen if hardware or software failures in the space or ground elements disturb nominal orbit operations. Chapter 16 describes anomaly investigation and recovery. In a microsatellite mission, a small team operating from a single location carries out these procedures. This arrangement may alter such matters as crew composition (Table 16.10). In a microsatellite team, the anomaly captain, systems engineer, and design engineer may all be the same person.

Given this organization, we must address several issues. Can the payload and bus return to their safe pre-commissioning state? How long can the mission remain "safed" before the mission objectives are completely lost? We need to ask similar questions about the communications links. How much data loss can the mission sustain? Are there periods during which data is absolutely critical? The answers to these questions will show whether the mission can employ a relaxed approach to fault detection, isolation, and recovery. Increasing the coverage or the speed of these procedures can completely alter the mission architecture, and drive up costs.

For example, if the mission depends critically on continuous data return, both the space and ground elements must have more redundancy. The costs associated with this redundancy might overshadow savings in other areas of the design, throwing the entire mission concept into disarray. How the development team responds to such problems will determine whether or not we can achieve a reasonable low-cost mission. The team may call for increased subsystem and mission reliability (e.g. increased redundancy and autonomous fault handling) or they may call for increased subsystem and mission flexibility (e.g. accepting some data loss). If the end user's mission requirements or bureaucracy force the system design down the path of increased reliability, a low-cost mission is probably inappropriate. But if the development team can devise a mission operations plan that compromises between reliability and flexibility, employing systems that fit a microsatellite's physical constraints, we may have an inexpensive, effective mission.

Thus, we must explore mission operations concepts and choose one that neither rules out viable missions or admits impossible ones. In some cases, we won't be able to devise a cost-effective mission operations concept which can use a low-cost microsatellite. If so, we'll need a more expensive spacecraft or ground element. This filtering must take place early in the mission to avoid severe difficulties during detailed development. In more successful cases, the operations concept will identify some mission requirements which stretch the microsatellite's abilities. During detailed development, we'll give these areas more resources, to ensure we can carry out the operations concept. Note that cost-effective mission operations turns into a design driver. Operators collaborating with developers of the ground and space elements will produce an architecture that supports cost-effective mission operations.

## 19.3  Detailed Development

The output of concept exploration will be a mission concept which uses a low-cost microsatellite and cost-effective operations. The mission now enters detailed development of the spacecraft bus, payloads, and ground element. For microsatellites, this phase is often done in 12 months.

Design decisions must be consistent with the plan for cost-effective mission operations. At all times, the team will design to cost, keeping the mission within its relatively small budget. The team must also design for operations, ensuring the space and ground elements operate cost-effectively. The following subsections discuss areas of detailed design which can strongly affect operations.

### 19.3.1    Mission Plan

We must refine the preliminary plan for mission operations to include detailed plans for routine operations and critical procedures during payload checkout (e.g. software loading). In keeping with the integrated philosophy, the entire team (ground-element engineers, space-element engineers, and mission operators) must help develop the detailed mission plan. This is another chance for the bus engineers to adapt to the needs of the payload engineers, as well as for both groups of space-element engineers to check how their design decisions affect the ground-element engineers and operators.

A detailed mission plan should be complete before freezing the space-element and ground-element designs, so hardware and software designs can form around the mission plan, rather than forcing the mission plan to meet rigid hardware and software constraints. This optimizing step will decrease operating costs and increase data return.

We must remain flexible while developing the mission plan, and the mission plan must carry that flexibility through to post-launch operations. The mission plan, developed months before launch, shouldn't be prematurely frozen, and it shouldn't dictate a static operating environment after launch.

Particularly during the checkout phase immediately after launch, we should expect the mission plan to vary daily and even from session to session. External factors, such as difficulty getting post-launch orbital elements, or problems with the spacecraft itself may slow the pace of operations. Or everything may go smoothly, accelerating the pace of commissioning. We can't slavishly follow a plan developed and frozen before the spacecraft was fully designed.

After checkout, bus maintenance will settle into a routine involving telemetry gathering and general housekeeping. Payload operations, on the other hand, may remain fluid for months or years, depending on whether the mission is experimental or operational. A microsatellite carrying a commercial communications payload can follow a relatively fixed mission plan to maximize the owner's revenue. If the owner's business plan changes, the mission plan may change. Or, as we gain experience with the spacecraft, we may improve the operations plan to reduce operations expenses or increase mission performance. A microsatellite carrying science or technology experiments is different. It will require a continuously evolving mission plan, allowing experimenters to choose the appropriate next operation after analyzing the results of each experiment run. In both types of missions, these adjustments to the mission plan get the best return on the investment in space hardware.

When developing the initial mission plan, during spacecraft design, we must include feedback paths in management, software, and hardware. These paths will allow the spacecraft's end users, operators, and sponsors to alter operations easily and inexpensively. We can build support for flexible, rapidly-changing plans into the mission architecture, as described below.

Before moving from mission planning to spacecraft design, we must plan for contingencies. In low-cost missions, a mission plan that provides detailed contingencies for all anomalies isn't cost-effective. Instead, anomaly planning concentrates on the most likely anomalies during critical periods of the mission. For example, it's valuable to have a decision tree we can follow in case of failure when we first acquire the spacecraft. Outside of these critical periods—which concentrate during checkout after launch—failures are unlikely, and the failure-resilient design of the bus should allow hours or even days of troubleshooting. If the operations group can call upon bus and payload engineers for help with an anomaly, a comprehensive contingency plan is unnecessary. In this system, we call on the engineers' expertise as needed to fix actual problems, rather than investing it in a (perhaps unnecessary) "what if" exercise aimed at producing a comprehensive contingency document.

## 19.3.2    Resilience to Failures

To reduce demands on the ground element and increase the probability of mission success, we want a spacecraft intended for a low-cost mission to be failure resilient. In other words, performance should degrade gradually in case of failures, and systems to aid troubleshooting and reconfiguration should be available. Although catastrophes will happen, we should minimize them.

In advanced microsatellites, two systems are particularly subject to temporary or permanent failure: advanced microelectronics undergoing their first space use and complex, onboard, software systems. Yet these systems can lead directly to cost-effective mission operations, so we must deploy them safely in microsatellite designs.

In this context, the main goals of an architectural design for a microsatellite should be to permit the ground element to receive spacecraft telemetry and issue spacecraft telecommands without the aid of experimental hardware or complex software onboard. Furthermore, the spacecraft and its experiments should remain safe despite temporary unavailability of such high-level functions. We can achieve these goals by adopting a spacecraft architecture divided into layers. At the lowest layer are proven telemetry and telecommand systems, mostly based in hardware. In the middle layers are onboard computers and software with some flight heritage, and at the highest layer are new onboard computers and advanced software. Reliability and dependability are greatest at the bottom, but we get the best flexibility and potential mission returns at the top. (See Fig. 19.4). Working within such an architecture, engineers can employ advanced hardware and software which can make operations less expensive and more effective. Yet, the spacecraft doesn't depend utterly on perfection of software development or operational procedures. We can handle a range of human and machine failures, fostering rapid development efforts to get the best returns.

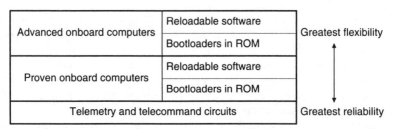

**Fig. 19.4.**    **A Layered Spacecraft Architecture.** In this architecture, proven, reliable systems control and monitor the spacecraft and therefore back up new, unproven systems that give the mission great flexibility. Graceful and rapid transitions from one layer to another are essential (downward in the case of an anomaly and upward when full functions return).

### 19.3.3    Adaptability

For a layered system architecture to be successful, the data paths within the spacecraft must adapt to different modes of operation. We should extend this adaptability to allow reconnecting of the spacecraft's subsystems in any sensible configuration—not just in the configuration for nominal operations. This adaptability implies that the spacecraft must have a wide range of optional analog and digital data paths, as well as ways of making the paths operate.

We shouldn't design the spacecraft to support a single, narrow mission plan. Instead, the bus and experiments should support several possible operating scenarios. Thus, software should play a major role in controlling experiments and gathering data, so we can change the operating scenario after launch. The spacecraft should have a general-purpose node for handling data (the onboard computer) connected to special devices for gathering data (experiments) and actuators (bus controls). The onboard computer should use a common high-level language, so many engineers and experimenters can develop their own onboard software.

If a spacecraft and its experiments support several operating modes, we must periodically switch modes using techniques ranging from simple telecommands to reloading software tasks. We'll have the most flexibility if these changes can be routine. In particular, procedures for software reloading should be practical, safe, and reliable.

Adaptable hardware and reloadable data-processing software greatly increase the spacecraft's complexity and introduce many possible spacecraft configurations which the engineering team must understand. In return for this complexity, we get wider scope for cost-effective operations. We can adapt the spacecraft to overcome failures, to serve altered mission goals, or to respond to advances in ground-element technology.

### 19.3.4    Autonomy

Microsatellites must be able to operate routinely without constant contact with the ground. Usually, onboard computers allow autonomous operation by executing closed-loop algorithms to control and monitor the spacecraft. Particularly for LEO spacecraft, which are in range of their control station infrequently, some autonomy is essential.

Autonomy can be cost effective if it covers enough of the spacecraft's monitoring and data-acquisition functions. Developing hardware and software for limited autonomy in controlling attitude, gathering data, and monitoring payload parameters is far less expensive than developing comprehensive autonomous systems, which deal in real-time with all possible failure modes and recovery procedures. The underlying failure-resilient design for the bus should eliminate comprehensive fault-detection, isolation, and recovery in real time.

We should design the software that will make a microsatellite autonomous so it will cope with the most likely scenarios. Upon detecting an out-of-bounds condition, the software may either return the spacecraft to gathering and generating data revenue or place it in a safe state. It will do the former when we expect the out-of-bounds condition and know the remedy. Safe mode is reserved for truly anomalous behavior for which no remedy is safe and sure. Figure 19.5 illustrates the associated program algorithm. For instance, the HealthSat-1 microsatellite runs a program that monitors the bus battery's charge status and the transmitter's operating parameters to use transmitter power effectively. Sometimes the bus power can't support maximum transmitter output. The software adjusts the transmitter power to match the available power. On the other hand, if the transmitter's operating parameters exceed safe values, the program places transmitters in a safe state until operators can diagnose the fault and start recovery. Thus, a combination of fully autonomous and operator-assisted operation produces the best payload operations and ensures spacecraft safety.

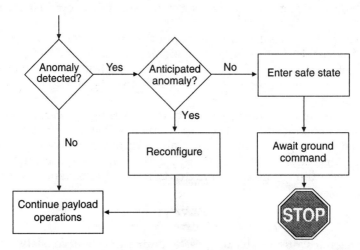

**Fig. 19.5.**     **Resolving Anomalies Onboard.** Onboard software can anticipate some anomalies and use its reconfiguration routines to keep the spacecraft operating. Upon detecting a truly unexpected anomaly, the software places the spacecraft in a safe state and awaits ground intervention.

## 19.4   Ground Element

In a space-mission budget, operations costs begin with the ground element. Although the space element may be designed with many operations-oriented features, they're not usually assigned as operations costs. The ground element for a low-cost microsatellite mission must be inexpensive in itself and must support cost-effective operations.

The key to producing a low-cost ground station is to take advantage of mass-produced consumer electronics wherever possible. Some equipment will be directly useful, but in other cases it will require slight modifications or custom-built interfaces. Even with these changes, using mass-produced electronics as the core of a microsatellite ground station saves a lot of money.

### 19.4.1 Radio Frequency Equipment

The spacecraft's communications links must be engineered to allow a low-cost ground station. Most microsatellite missions have substituted VHF (150 MHz) and UHF (400 MHz) communications links for the more standard S-Band (2.4 GHz) links. At these lower frequencies, medium-gain antennas with relatively broad beams can provide acceptable links. Broad beamwidth, in turn, permits the use of open-loop predictive tracking and low-cost antenna rotators. Selecting these lower radio frequencies eliminates the need for parabolic-dish antennas with high-speed, closed-loop tracking systems, which are ten times more expensive.

Many ground services, notably mobile and amateur radios, use the VHF and UHF bands for voice communications. Thus, there are many sources of mass-produced radios for these bands. Particularly in the Amateur Radio service, competition between radio suppliers is very fierce, resulting in low-prices and sophisticated features. The radios typically are computer-controlled so we can do Doppler corrections and automate ground stations. We can buy a computer-controlled radio with VHF and UHF chains for only one or two thousand dollars.

The main constraint on adapting these radios to data communications for a spacecraft is limited channel bandwidth. Frequency-modulated (FM) voice transmissions usually occupy 20 kHz, and readily-available equipment implements this standard. Data rates up to 10 kbps are supported relatively easily with this type of equipment using frequency-shift keying (FSK) modulation. Although this data rate may be too low for some applications, data compression, onboard data filtering and even multiple ground stations can help overcome this limitation while keeping costs very low for ground equipment.

Chapter 11 shows that communications-link design is a complex subject which can consume a lot of development effort. Standard practice for satellite communications is to use phase-shift keying (PSK) on S-Band for telemetry, telecommanding, and data recovery. Adopting FSK on VHF and UHF bands as a design baseline conflicts with this approach but results in an affordable and robust system which many have adopted for microsatellites in LEO.

### 19.4.2 Data-Processing Equipment

Consumer electronics also play a role in data processing on the ground for low-cost missions. Microprocessor-based systems, still increasing in performance and decreasing in cost, are the clear choice for all data processing and automation in a low-cost ground station. The performance and cost distinctions between workstations and personal computers are blurring, leaving system designers free to choose the platform more familiar to the programmers, operators, and experimenters who will use it. Thus, we save money not only by buying inexpensive hardware and software but also by providing ground-element developers and users with a familiar environment that doesn't require extensive retraining.

### 19.4.3    Delivering Data Products

The computers in the ground station should be connected as a local-area net-work (LAN) and equipped with communications that support access to wide-area networks (WAN). The extra cost of networking equipment is offset by more oper-ational ease and effectiveness. We can use networking to gather data, plan operations, resolve faults, and deliver data products (see Fig. 19.6).

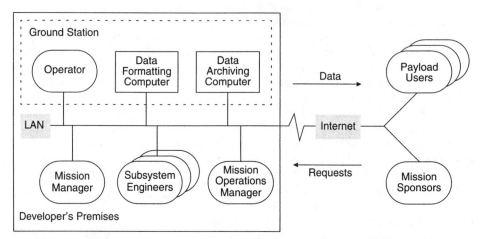

**Fig. 19.6.    Using Computer Networking in Microsatellite Operations.** The ground station for a microsatellite mission should use local and wide-area networking (LAN and WAN) to con-nect operators to other members of the mission team. Using suitable wide-area network-ing, such as the Internet, we can rapidly and inexpensively deliver operating requests from payload users and data from the payload.

If microcomputers control the ground station, we can easily give experiment-ers and engineers immediate access to all data through sharing of disk files, for which we have many well-developed commercial applications. Thus, for science or technology missions, all level 0 data should be present on the network immedi-ately after downlinking from the spacecraft. Experimenters should be able to remotely access the network (remote users on the WAN, local users on the LAN) in order to collect their data. This approach has the twin advantages of reducing workload on ground-station staff and making experimental data available as rap-idly as possible.

The infrastructure through which data is distributed can also be used by mis-sion planning to receive requests for events and payload maintenance, and to distribute mission and activity plans to interested parties. If automatic, scheduling software would validate and arbitrate requests; if manual, the operator would complete these tasks. In either case, it gives users immediate input to mission planning, which means operations will be most relevant to the users' present needs.

Finally, LAN and WAN connectivity can improve fault diagnosis and recovery by bringing all spacecraft-development engineers to bear on a problem. If the team members aren't at the ground station, they can still retrieve data to examine the problem and rapidly contribute documents or software to the ground-station staff. Effective electronic links between the development engineers and the operations group make it unnecessary for development engineers to document in detail the precise operational diagnoses and solutions for all possible spacecraft problems. These links also allow the development engineers to remain involved in operations of one spacecraft while they are concentrating on the development of another.

### 19.4.4 Automating Ground Stations

We can easily automate a ground station equipped with powerful general-purpose computers and comprehensive networked communications, thereby decreasing operating costs while increasing data return and keeping the mission safe. A typical control center for a microsatellite mission will be able to predict when the spacecraft will be visible, drive antenna tracking, establish a proper data-handling configuration, and gather data, all without human intervention. This automation makes it possible for a single operator to oversee the housekeeping and data gathering for several microsatellites. It also permits an unattended ground station to collect data.

Simple, inexpensive additions to this basic automated station can enhance mission safety. Software should confirm the receipt of telemetry when expected and check received telemetry for out-of-range conditions. When it detects an anomaly (no telemetry or out-of-range telemetry), automated paging equipment can notify an operator, who can further diagnose the problem. Thus, spacecraft in LEO may be monitored on each transit over its command station, without needing second-shift operators.

Ground-station automation often pairs with onboard autonomy. The microsatellite's onboard computer may execute a stored schedule of operations that is composed weeks or months in advance. Once a mission has settled into routine operations, it may use a single schedule indefinitely. Thus, the spacecraft will continue to carry out a varied mission to the operator's plan without the operator's constant intervention. If the ground station is autonomously capturing the generated data and monitoring the spacecraft for failure, the load on human operators drops significantly (especially if experimenters can retrieve data over the network without operator intervention). These techniques have worked routinely on LEO microsatellites which would otherwise require 24-hour shifts.

## 19.5    Pre-Launch Operations

We might assume operators and ground-element staff begin working only after we launch the spacecraft. This is not the case in a microsatellite mission. Someone must operate the spacecraft during integration, environmental testing, and pre-launch checkout. Although test engineers using special ground-support equipment could handle these operations, there are several reasons to use operators and some or all of the ground-element hardware and software we'll use after launch.

A typical team for a microsatellite mission doesn't include specialist test engineers. Instead, development engineers and spacecraft operators do subsystem-level and system-level testing. Each brings valuable information and expertise to this stage of the mission. The development engineer has detailed knowledge of the space element's hardware and software. The operators know details of the ground element and of the mission plan. This is a valuable chance for developers and operators to pass information back and forth. The presence of the specialists also increases the safety of the test period, which can be hazardous even with the most detailed documentation.

Ground-support equipment used for pre-launch operations should be a subset of the ground-station equipment for the mission, so we don't spend money on dedicated test stands and so we get the most value from pre-launch tests. Operations before launch will be the first chance to verify that the ground element and the space element are compatible. In a typical microsatellite project under rapid development, this will be an important test for new space and ground hardware and software. By conducting a single test campaign covering the space and ground elements, the team will best use all available resources.

Thermal and vacuum tests are a critical milestone in a microsatellite mission. Typically, they involve one week during which we operate the spacecraft in vacuum at various temperatures. Thermal and vacuum testing helps to identify circuit or component problems, confirms that the spacecraft meets the launch vehicle's outgassing requirements, and forces us to operate the spacecraft for a long time without physical access. Thus, it's a dress rehearsal for spacecraft operations, and we command and monitor the spacecraft only through its rf communications links. A well-constructed thermal and vacuum test should include periods of operational exercises alternating with periods of more intense checkout. Operational exercises confirm we can control the spacecraft as anticipated by the operators and the mission plan, but they don't investigate the performance of every redundant path. The checkout periods test for faults which wouldn't appear during such low-coverage tests. After doing these two types of tests for a week, we learn about the details of reconfiguring the spacecraft and ground station and confirm that the space hardware and software are operating correctly.

Just as thermal and vacuum tests are the *first* chance for full-up operational exercising of the spacecraft, the launch campaign is the *final* chance to verify that the space element, ground element, people, and procedures will work together. Even a campaign for a low-budget launch should include at least one spacecraft operator and one spacecraft-systems engineer. This team should repeat (perhaps in shorter form) the combination of operational exercises and systematic checkout developed during the thermal and vacuum tests. These tests, done at the beginning of the campaign, verify that the spacecraft has survived shipping and is ready for mating with the launcher.

We typically operate the spacecraft through an umbilical connector to avoid radiating radio signals at the launch site. But we must still get proper clearance to run a final rf checkout, confirming all systems and procedures are ready for the initial commissioning of the spacecraft after launch.

Formal life cycles of a space mission sometimes omit pre-launch operations. In a typical microsatellite mission, however, these operations are critical to the system developers and operators. If done well, they can decrease the cost and increase the effectiveness of integrated testing.

## 19.6   Post-Launch Operations

The preceding sections have discussed many parts of cost-effective microsatellite operations. Because these operations don't depend only on efficient activities after launch, we've shown how mission planning is integrated into the mission's life cycle. But we also need to understand how the mission architecture is actually used in orbit to inexpensively retrieve data or produce revenue. Thus, in this section, we overview post-launch operations and then show you a brief practical example drawn from the S80/T microsatellite mission.

### 19.6.1   Overview

Post-launch operations include commissioning, checkout, exploratory operations, and mature operations. Each phase places different demands on the mission team and the space and ground elements. During commissioning and checkout, operators will work very closely with development engineers, especially if difficulties develop. During the exploratory operations the team develops its best plan for using the end user's payload, given actual prevailing conditions in orbit. This phase involves end users, operators, and engineers. The exploratory phase gradually gives way to mature operations of the payload, which operators and end users can handle without the engineering team. The duration of each phase depends on the nature of the mission; the transition to mature operations can occur in days if a standard bus and payload are reused for a new mission.

For most missions, mature operations will occupy much of the mission's lifetime. The space and ground hardware and software will remain relatively stable, and the goal will be to use the fewest people to generate the most experimental data or commercial revenue. Typically, one operator can sustain mature operations of a microsatellite while interacting with end users. End users request operations through the electronic-communications systems discussed earlier, and the operator produces an operating schedule which is uploaded to the spacecraft. This efficient yet flexible operating mode is possible because we designed it into the mission architecture from the beginning.

Before describing how such a system works in practice, it's worth reiterating that microprocessors supporting operations automation and spacecraft autonomy are critical to cost-effective operation of low-cost microsatellites.

**Microprocessors.** More than any other single technology, the microprocessor is at the heart of the architectural design and at the forefront of post-launch operations for cost-effective microsatellites. The ground station is automated by microcomputers, linked to the spacecraft by microprocessor-controlled communications protocols, and connected to a network of experimenters, operators, and engineers through their personal computers. Onboard the spacecraft, computers in the bus and the payload execute schedules of ground commands, manage communications, maintain the spacecraft's attitude, and provide automated fault detection and recovery. These powerful, inexpensive processors and their associated software streamline every element of post-launch operations.

**Automation and Spacecraft Autonomy.** Microprocessors alone don't guarantee cost-effective systems. Two key decisions make this microprocessor-based architecture cost-effective. First, the decision to automate removes human operators from as many control and operations tasks as possible. In particular, operators never need to make real-time decisions under normal operating circumstances. Second, the decision to keep the ground station out of as many control loops as possible makes the spacecraft autonomous. This is a logical extension of the first decision—once we transfer routine operations to computers we may as well transfer them to computers onboard the spacecraft. Control loops implemented by onboard computers don't have the delays and uncertainties of loops involving the spacecraft's communications channels and ground-station equipment. While both rules have exceptions, they're critical parts of a cost-effective architecture for mission operations, as illustrated by S80/T.

### 19.6.2    S80/T—An Example of Cost-Effective Microsatellite Operations

**Mission.** The S80/T microsatellite mission was funded by the French space agency CNES to measure interference and propagation characteristics in the VHF frequency bands allocated to the Little LEO communications satellites. Matra Espace was the prime contractor, and Surrey Satellite Technology, Ltd. (SSTL) supplied the microsatellite bus, ground systems, and post-launch operations. Dassault Electronique supplied the payload—an ultra-linear transponder operating in the VHF bands of interest. The mission went from concept to launch (on the third Ariane ASAP) in just 12 months. [Allery, 1994]

**Basic Spacecraft Characteristics.** S80/T is a 50-kg microsatellite approximately 600 mm tall with a 300 mm square cross section. Four body-mounted solar panels provide power, and a 6-amp NiCd battery pack serves peak power needs during payload operations and eclipses. The spacecraft's nominal attitude is to have its communications antennas nadir pointing, with a slow spin about the Earth-pointing axis. A 3-axis magnetometer, two Sun sensors, and an Earth-horizon sensor provide basic attitude determination. A 6-meter, rigid, gravity-gradient boom topped with a 3-kg tip mass makes the spacecraft stable. The onboard computer uses 3-axis magnetic torquers in a closed-loop control system, which removes residual libration. Telemetry and telecommand links operate at 9600 bps—the downlink on UHF and the uplink on VHF. The satellite's onboard system for handling data consists of an 80C186-based primary computer for onboard control and a Z80-based secondary computer for onboard control embedded in a layered architecture. The 80C186 connects to a 16-megabyte, solid-state data recorder configured as a *RAMDISK*. The S80/T bus was based on previous SSTL microsatellites, with changes needed to achieve the primary mission.

The S80/T payload is a 5-kg, VHF/VHF, linear repeater supported by a circularly polarized measurement antenna. Because the microsatellite structure is not large enough to support two VHF antenna systems, we also use the measurement antenna for the bus telecommand uplink. In order to permit accurate interference-level measurements over a wide dynamic range, the repeater has variable gain which we can set using the spacecraft's telecommand system. Signal strength within the repeater passband can be measured through the spacecraft's telemetry system, hence closed-loop control of the repeater operating point is possible using telemetry and telecommand. Because the repeater draws 40 W from the bus (which supports only 19 W orbit average), scheduling of repeater operations and monitoring of bus performance during these operations are critical aspects of the mission operations design.

The S80/T operations concept called for two ground stations: one for mission control at SSTL and a transportable one from which CNES would make experimental measurements. SSTL provided both ground stations, using personal computers, mass-produced radio equipment, and other low-cost elements. Prelaunch plans called for the SSTL station to monitor S80/T's bus subsystems while the CNES station would conduct payload operations—keeping the ground station in the loop control while controlling the repeater through the uplink and downlink. This operations plan was based on a traditional model, using little onboard computing power. We devised a plan more appropriate to a microsatellite early in the mission, and all aspects of the mission architecture were flexible enough to support this change.

**Commissioning.** SSTL engineers commissioned S80/T from the SSTL Mission Operations Centre. This process started with downlink activation about six hours after launch. The operating sequence after downlink activation involves software uploading, bus checkout, and attitude-control maneuvers. Spacecraft-development engineers with detailed knowledge of the S80/T design controlled the spacecraft during these operations. Having engineers directly involved in commissioning operations paid off when S80/T entered an unforeseen mode of attitude stability. In this mode, the spacecraft's thermal controls were ineffective, and electronic systems were operating at the edge of their ability. The engineers on hand were able to make critical judgments in real time, devising a contingency plan, executing it swiftly, and restoring nominal operations before we lost communications with the spacecraft. The problem, the decisions, and the recovery took a matter of hours, exploiting all the available expertise and the streamlined decision processes typically found in a microsatellite mission team.

After establishing full onboard software operations and nominal attitude, operators concentrated on a 30-day checkout period, during which they exercised all primary and most redundant bus systems. Working with payload engineers from Dassault, SSTL operators also exercised the spacecraft's repeater, verifying that it had survived launch. They used onboard computer software to schedule the repeater operations and to monitor bus performance during these operations. Having the software end repeater operations at some specified time was especially important because the team hadn't verified positive uplink telecommand of the spacecraft during this operation. The software provided a safety net in case the ground element or command link failed.

**Early-Orbit Operations.** S80/T's early operations show how reloadable software and general-purpose onboard computers allow the mission team to best operate the payload after launch. The initial operation plan for the S80/T payload was traditional: the spacecraft's telemetry system would measure the payload's operating parameters and transmit these measurements to the ground station for payload operations (see Fig. 19.7). Software in the ground station would decide whether or not to adjust the payload gain or operating mode, based on the telemetry readings. If changes were necessary, it would transmit them on the spacecraft's telecommand channel. Involving the ground-based computer increased uncertainties and delays in the control loop but was the natural approach taken by mission engineers unfamiliar with the power and flexibility of a microsatellite's onboard computers.

In the final stages of pre-launch testing, we learned that interactions between the payload's rf systems and the bus's rf systems were greater than we expected. This wasn't surprising considering the nearness of the payload and bus antennas. As a result, the payload operators wanted to operate the payload without the bus's telemetry transmitter turned on. We invalidated the original mission plan, which depended on the telemetry downlink to complete the feedback loop. The solution was to use an onboard computer in the payload's gain-control loop (Fig. 19.8). We decided to use onboard control *after* the spacecraft had been shipped to the launch site.

With most members of the S80/T team occupied by the launch campaign, we hired a consultant programmer to write this critical application. This consultant couldn't access hardware models of either the payload or the S80/T's onboard computer, yet was able to develop a complete control, monitoring, and data-logging package for the payload operations. We made this development possible with

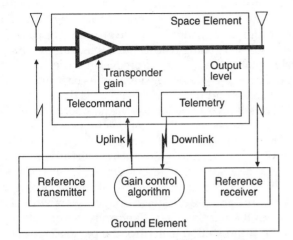

**Fig. 19.7.**    **Initial Operations Concept for the S80/T Payload.** The initial operations concept for the S80/T transponder required telemetry and telecommand links during all payload operations to provide closed-loop control of the transponder gain. Telemetry downlink transmissions during payload measurements established a high noise floor for the measurement transponder.

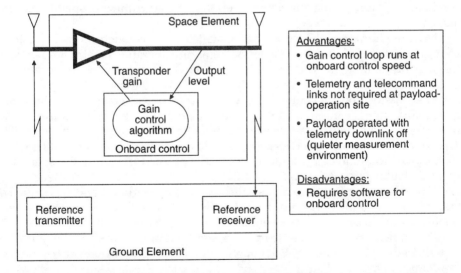

**Fig. 19.8.**    **Actual Operations Concept for the S80/T Payload.** We chose the actual operations concept for S80/T only weeks before launch. Moving the gain control loop to the onboard computer permits operation with the telemetry downlink turned off, establishing a lower noise floor for payload operation.

earlier system-development decisions: using a familiar microprocessor, structuring onboard data handling as a general-purpose system, and using standard computer languages in a familiar programming model on the ground. The consultant was thus able to write the new software using readily available tools and high-level interfaces which completely isolated the software from S80/T's specific hardware. We shipped the software to Surrey, where people verified its basic functions (apart from closed-loop control) on a ground model of the onboard data-handling system. After initial in-orbit checkout of S80/T, we loaded the new software to the spacecraft, verified it, and placed it into the normal operations sequence.

This post-launch development and verification of a large new piece of operating software illustrates how a flexible approach to operations can greatly improve mission cost-effectiveness. Without the software, S80/T would have returned fewer results, with much more difficulty. Because the spacecraft design was general enough to support onboard control, and the mission managers were willing to accept software development late in the mission, S80/T returns a lot of data and doesn't require a payload-control ground station to control the payload for every operation. This, in turn, has allowed the customer to use the payload in field trials at very low cost.

You may believe we could avoid such late changes in mission design by perfect planning during the development phase. Unfortunately, perfect planning is never possible and is especially unlikely when doing a mission with new technology on a short schedule. A successful microsatellite team will plan for change and continue to improve the mission's operating scenario during operations after launch.

**Mature Operations.** Once we had installed and validated the new payload-control software, S80/T entered a period of mature operations which has persisted for most of the mission. The operations plan extensively uses the microsatellite's abilities, so it's a good example of an effective plan for microsatellites.

The mission-control center at Surrey handles housekeeping operations and bus monitoring. The mission-control center (shared with several other microsatellites) automatically collects telemetry and other housekeeping data whenever S80/T is above its horizon. A network of personal computers collects the data, which then moves to archives on optical disks. During each transit of the spacecraft, ground-station software compares received telemetry values to the expected high and low values. If it detects extreme values, operators are automatically notified so they can examine the data and correct problems if necessary. The operators send brief monthly reports of control-center activities to CNES.

Experimenters at CNES schedule repeater operations based on the particular experiments they're running. Each operation includes special settings of the control-loop parameters, a start time, and an end time. Ten days of operations are prepared as a single schedule file for transmission to Surrey. Rather than using ground-communications links, this schedule file actually transfers between CNES and Surrey using the store-and-forward transponder onboard S80/T. Software in the CNES station automatically uploads the message, and software in the Surrey station downloads it. This software is essentially that used by other store-and-forward communications networks based on SSTL microsatellites, and is the same software used to recover the payload's data files.

When downloaded at the Surrey mission operations center, CNES's payload-operation schedules go through automatic consistency checks, and operators verify that the spacecraft has enough power for the proposed operations. The validated schedule automatically converts to the binary format needed by the payload-control task in the onboard computer and then uploads to the spacecraft, again using the store-and-forward communications transponder. The onboard computer executes the plan autonomously for one week, and payload operations can proceed with no further operator intervention. Except for the power-budget check (a bus issue), CNES themselves control payload operations, and scheduling is completely automated.

As discussed previously, each payload operation consists of a relatively brief period of repeater operation—typically one transit over a target location. During the operation, an automatic gain-control loop adjusts the payload's operating parameters by monitoring the magnitude of the repeater's input signals. The payload-control task also collects the data which must be correlated with ground-based measurements. Data stores as files in the spacecraft's RAMDISK, where the store-and-forward transponder software can access it. The CNES's experimental station is programmed to automatically download these files on the transit following each operation. Once downloaded, data is available in MS-DOS compatible files, time-tagged to within 10 ms, for experimenters to analyze. Thus, no one needs to download data, prepare it, or send it to CNES. This is an exemplary system, which can also increase the cost-effectiveness of science missions.

The final element of S80/T operations is fault detection. The payload repeater is a high-power device, with specific requirements for power-supply sequencing and maximum operating temperature. It places a large current drain on the spacecraft's batteries, which also have several critical operating limitations. All of these parameters are monitored by the payload-control task executing on the onboard computer. We set thresholds for fault detection for each scheduled operation or leave them at default values. If something is out of bounds, the payload operation ends and important telemetry values go into a log. Certain errors result in temporary suspension of the operating schedule, whereas others require operator intervention to resume the schedule. This fault-detection system is acceptable as long as the primary onboard computer is operating. If the onboard control crashes, a secondary computer serves as a watchdog. Periodically, primary and secondary computers exchange messages on the spacecraft's LAN. If either control computer fails to hear from the other, the spacecraft's telecommand system ends payload operations. The automatic fault detection, with redundancy, is the final link in the operations chain which allows long periods of autonomous spacecraft operation.

While it is certainly possible to develop and operate a microsatellite in the same manner as a larger spacecraft, this type of operation won't be cost-effective and inexpensive. In this chapter, we've discussed the specific elements of development which lead to cost-effective operations before and after launch, within the framework of a low-cost, rapid mission. We've used this system to develop and operate 11 microsatellites for various missions.

The main philosophical points we've adhered to during these missions are to keep operations in mind during all phases of development and to keep the mission plan as flexible as possible. These decisions have led to a heavy reliance on onboard microprocessors running software uploaded after launch. To make this mode safe, we develop each spacecraft to withstand temporary software failures. A flexible mission plan supported by a reconfigurable microsatellite can provide high-quality data after launch without requiring a large operations team.

In any cost-effective project, we're attempting to spend the available money for the greatest return. An inexpensive and cost-effective project is particularly difficult to manage—especially in a field such as space, where the norm is for each project to be expensive and conservative. In developing low-cost microsatellites, we concentrate most of our effort on improving the return from the mission. This means putting resources into hardware or software and testing—not into bureaucracy, planning, and documentation. Of course, bureaucracy, planning, and documentation have their place, so we include the least we can for a safe mission. Reaching these minimum necessary levels is one of the keys to cost-effective mission operations in space.

## References

Allery, M.N. and H. Castelbert. "Microsatellite Operations in the First Year of the S-80/T Mission." Presented at the 2nd International Conference on Small Satellite Systems and Services. Biaritz, France. June 27–July 1, 1994.

Sweeting, M.N. "Sophisticated Microsatellites for Space Exploration and Applications." Presented at the International Workshop on Digital Signal Processing Techniques Applied to Space Communications. King's College, London. September 26–28, 1994.

# Chapter 20

# Human Space Flight Operations

Carolyn Blacknall, *NASA Johnson Space Center*
Felix Godwin, *Teledyne Brown Engineering*

Human space missions are a special topic because of their high public interest and other distinctive features: their mission phases, sub-phases, subsystems, trades, and program concerns are all new. Our emphasis is on what is unique to human missions, so we won't duplicate other chapters.

## 20.1   Rationale for Human Missions

Human missions are the subject of much debate. Some scientists perceive them as costly and risky efforts that displace smaller, science-oriented activities. They argue that robotic spacecraft are now so capable that expanding the human presence in space isn't necessary.

The three reasons for using crews involve politics and economics as much as engineering and science:

1. Using crews is valued as an end in itself
2. Humans have unique abilities
3. An available launch vehicle happens to use a crew

Having people in space in itself was the goal of the earliest human missions, and enormous prestige came with being first in Earth orbit and first on the Moon. A thousand years from now, history books and other media will still chronicle those national triumphs.

After the initial public excitement, the second reason for human flight emerged: learning what humans can do in space, how long they can work, and how far they can journey. Increased payload capabilities combined with miniature parts began to provide space crews with extensive instruments and laboratory equipment, allowing missions to take advantage of unique human adaptability, insight, and improvisation. As a result, planning and developing activities are much more flexible and can change quickly.

Humans are valuable in space for routine and unexpected situations. Astronauts can aid experiments by observing and promptly altering procedures based on these observations. They can position and activate cameras, change film, replace samples, and discuss experiments with Earthbound payload specialists. They can also exercise common sense to detect any erroneous ground commands that a robotic spacecraft might execute obediently with disastrous results. These abilities add tremendous flexibility to payloads flown aboard crewed vehicles.

Astronauts can also rendezvous and dock spacecraft payloads, which are difficult for robotic spacecraft. They can operate external devices such as the Space Shuttle's Remote Manipulator System, whose control from the ground would be hampered by time delays. They can retrieve spacecraft payloads, such as the Long Duration Exposure Facility and the Wake Shield Facility, and return them to Earth for laboratory examination.

Activities are designated *intravehicular activity* (IVA) when performed in the shirt-sleeve environment of a spacecraft's pressurized and habitable modules. *Extra-vehicular activity* (EVA) takes place outside—in space. This environment requires pressure suits and life support, as well as propulsive aids, such as the manned maneuvering unit (MMU), for free movement in zero gravity.

Humans in space can also respond to novel and unexpected opportunities for scientific observation and are especially valuable when the payload doesn't operate as expected. Using crews also allows in-flight maintenance. The crew can change fuses, reset circuit breakers, tighten connectors, and install stored spare parts. For some payloads, we can't foresee and therefore can't automate necessary repairs. Crew members have modified hoses and creatively used spare parts to repair payloads; an example is the "fly swatter" device developed during a Shuttle flight to flip a switch on a spacecraft. Such unplanned activities may also involve EVAs which have been used to capture, repair, and redeploy spacecraft.

Beyond Earth orbit—on planetary missions—the rationale for crewed missions also remains strong. Despite advances in robotics, machine vision remains severely limited, and a robotic rover (surface explorer) is a very clumsy substitute for an alert human with a camera and training in field geology.

The third reason for humans in space concerns politics and economics. It is noteworthy for its influence on the history of US space operations, as described by an eminent student of US space policy [Logsdon, 1986]. He explains why a space payload with no need for direct human presence may ride on a vehicle which happens to have a crew, even if this method costs the payload user more money.

This approach became the US's space policy with the 1972 decision to authorize a crewed vehicle, the Shuttle, as the future principal means of accessing space. The policy was confirmed in 1982 by National Security Decision Directive (NSDD) 42 and it lasted until 1986. Then, losing Challenger brought the painful realization that, in its risk level, space flight is still more akin to exploration than to travel by scheduled airline.

## 20.2　Implications of Humans in Space

Using crews doesn't change the 13 mission functions defined in Chap. 3, though it does change the details of how they're executed. More importantly, human flights also introduce new mission phases and subphases:

- Crew selection and training
- Abort and rescue options
- Return—Deorbit, coasting descent, reentry, landing, and recovery
- Activities after landing

The Europeans of past centuries launched repeated voyages of exploration even though these voyages sometimes cost the lives of 20 per cent or more of their crews. In the 20th century, the life expectancy of the first US Air Mail pilots was scarcely longer. But such reckless ventures are no longer acceptable. Our value on human life now requires extensive provisions for crew survival. An example is the Shuttle's abort option in the form of suborbital diversion toward an alternative airfield. Maintaining this option reduces the Shuttle's payload on missions to high-inclination orbits because of the distance to alternate airfields.

Thus, abort and rescue options create payload penalties and alternative branches for the mission timeline. Some of these options apply after the spacecraft is in orbit, but the mission plan must provide for unscheduled early return to Earth. This return may use a special reentry vehicle with minimum capability—functionally equal to a lifeboat.

If the mission proceeds as planned, reentry and landing occur as scheduled events that are peculiar to human missions. Like the rescue options, they create new mission subphases with their own demands on communication architecture, command and tasking structures, and the system's ground element. One effect may be to constrain the timing of earlier mission phases, due to the Earth's rotation relative to the orbit plane and the need to have a designated landing area within reach of the re-entry vehicle. Another effect is to require certain ground resources, such as rescue teams and communications relays, to be on standby, perhaps at remote locations.

Finally, crewed missions have unique phases after landing. Physiological rehabilitation of the crew may be brief or extended, depending on the length of time spent in zero gravity and the rigors of landing and recovery. A final phase is using crew members as consultants and instructors for later studies and missions.

Notice the short list of new mission phases doesn't include surface exploration of the Moon or another body. Surface exploration isn't peculiar to human missions because it also occurs in missions without crews. Sometimes, these missions include rovers that can journey away from the immediate landing site. In human missions the significance of surface exploration is not that it's a new mission phase but that it creates new trades and risks, as well as much more complex functions for mission control and payload planning.

Human spaceflight also introduces new subsystems. Besides rescue equipment, a major additional subsystem is environmental control and life support (ECLS). We must minimize the total weight of subsystem equipment plus consumable resources so that, in a very long (e.g. interplanetary) mission, the ECLS processes will be almost closed-loop. If fuel cells are used for electric power they may also supply water for the ECLS. In addition, the spacecraft's oxygen atmosphere requires a fire-suppression subsystem because fire is an extreme hazard in the confined space and many combustion products are toxic to humans.

The ECLS must do more than provide oxygen and water and maintain temperature and pressure within safe limits. It must guard against any biological contamination, which could quickly infect the entire crew in their small, closed environment. Certain types of illnesses, such as Legionnaires disease, flourish in air-conditioning systems, where temperature and humidity favor the growth of microorganisms.

Humans in space face two major external hazards: zero gravity and radiation. The long-term effects of zero gravity are still not fully understood, but they include loss of calcium from the bones and atrophy of cardiac and other muscles, so all long missions include physical exercise in their daily cycle of activity. In the 1950s popular media pictured space stations as rotating wheels in which centrifugal force provided artificial gravity, but this rotation is uncomfortable unless the rate is very low. A typical rate would be two rpm, which for one g requires a radius of more than 200 meters and hence introduces the complicated dynamics of tethered bodies. Continuous rotation of a spacecraft would also complicate the pointing of solar panels, antennas, thermal radiators, and optical instruments.

For near-Earth missions, we can reduce radiation hazards by orbiting below the Van Allen belts or by transiting the belts rarely and rapidly. For long missions away from Earth, the main radiation hazard is solar flares, which cause rare but abrupt increases in the solar wind (mostly protons). Shielded storm shelters for temporary occupation can counter these effects. The shelters must permit the crew to continue all activities essential to their health and operations. We can lower the storm shelter's weight by using existing masses, such as stored water, for shielding.

Besides changing the way operators carry out a mission and requiring new mission phases and protective subsystems, human flight constrains payloads. If a crewed vehicle delivers a payload to orbit, the payload would remain onboard through the landing phase if a malfunction keeps it from being released in orbit. This means we must design the payload to withstand structural loads during landing as well as ascent. Also, any payload energy sources, such as propellants and electroexplosive ordnance, must be safe through reentry and after landing.

The crew may also affect a payload by being sources of undesired effects: contaminating optical surfaces, disturbing precisely pointed instruments, and varying acceleration on experiments with their movements. This possibility leads to operating some in-space experiments as free flyers. These experiments stay apart from the crewed vehicle, carry out their operations, and return for human access only for maintenance or replenishment.

The longer humans stay in space, the more complex and risky the mission. If we are to keep cumulative risk down, longer missions require more redundancy and provisions for contingencies, both of which increase a spacecraft's weight and cost. Missions and programs must adapt to these requirements. For example, mission planning must use human hours as efficiently as possible to keep costs down, so managers tend to micromanage crews with minute-by-minute instruction from the ground. Skylab IV had notable friction regarding tasking and workloads—a tension that could easily intensify in interplanetary missions with long durations and communication delays (see Chap. 17).

For mission managers, human flight creates new time pressure during malfunctions. With no-one onboard, ground operators may be able to place the spacecraft in a safe mode and take whatever time is needed to resolve the problem; even a month of limited operations isn't a serious loss out of a typical five- or ten-year operational life. If a human mission malfunctions, time pressure is far more severe because of risk to life and because the time available for the crew to complete tasks may be limited, such as about ten days in Spacelab. The need for more prompt response increases mission cost by requiring more elaborate provisions for contingencies.

A program-level consequence of human space operations is that they are rare, politically and financially visible, and increasingly international for cost sharing. International programs must continually adjust to their partners' changing priorities. Also, the rarity of human space operations means the spacecraft are few in number, with little ongoing production to replace losses, so a single loss can affect a program for a long time.

As you'll see below, human space flight has specific implications for the Space Shuttle, the Space Station, and lunar and planetary missions. Please see Chap. 18 for coverage of Russia's operations for crewed missions.

## 20.3    Planning and Analyzing Shuttle Missions

To determine whether a payload or experiment should fly on the Space Shuttle, we need to examine how people plan a Shuttle mission. Recognizing the complexity and scope of preflight and on-orbit operations permits us to understand the costs and time requirements of mission planning for the Shuttle. The following activities detail the milestones and requirements for launch. Current plans for launches on the Space Shuttle to the Space Station follow similar guidelines.

A necessary part of human space flight is planning of crew activities, which means analyzing and developing activities needed for a specific mission. First, we create and refine a flight plan which outlines specific crew activities and essential flight-support functions. This analysis results in a minute-by-minute timeline of each crew member's activities. Simulations and training then make sure the flight crew can do each needed operation.

Through operations-support planning, we analyze in detail the flight requirements and ground operations essential to support and control a proposed mission. Planning to support operations includes a comprehensive review of flight-controller documentation and the necessary updates as flight requirements and practices change.

Payload experimenters must say what they need to support flight operations in certain documents: flight rules, command plans, communication and data plans, mission-control and tracking-network-support plans, procedures for operating systems, procedures for operations and maintenance, handbooks on flight-control operations, and flight-software documentation.

Payloads with specific pointing requirements must become involved with operations-support planning, which includes developing the mission's flight profile and producing, analyzing, and designing mission-planning products. These activities include assessing the flight profile and ascent performance and thoroughly analyzing the flight design to produce flight-design ground rules and constraints. Details on payload weights and center of gravity, as well as any special requirements for payload deployment or rendezvous, are necessary to develop these mission-planning products.

We then develop crew checklists for mission-specific operations and timelines. Payload operators provide details on how much time their experiment needs to operate as well as any specific requirements such as power, cooling, attitude, and times for crew involvement.

Software is created, modified, and tested for guidance, navigation, and control of the onboard payload and laptop computers. Products to reconfigure the Mission Control Center and Shuttle Mission Simulators for operations are developed to reflect the specific flight. The simulators' configuration must reflect the characteristics of major payloads, so we can effectively train flight crews and ground controllers.

Because launch is the most critical period for a space mission's safety, a major part of preflight planning involves launch-safety planning. We plan extensively to keep the crew safe during this hazardous phase. Preflight planning also requires performance analyses—trajectory, navigation, and guidance design—for ascent, orbit, collision avoidance, reentry, and landing operations. Then, we analyze performance in detail to support specific payloads, including payload deployment, rendezvous, proximity operations, and payload retrieval.

### 20.3.1    Major Milestones Leading to a Crewed Launch

Organizations sponsoring major Shuttle payloads should be aware of the major milestones leading to the approval of a Shuttle launch: the Flight Readiness Review and activation of the Mission Management Team.

The *Flight Readiness Review* is the last major Shuttle review before launch. It verifies that all integration operations are complete and certifies that the mission is ready to proceed. The Flight Readiness Review occurs two weeks before launch. Each manager must assess readiness for launch based on hardware status, problems encountered during launch processing, launch constraints, and open issues. Each NASA project manager and the contractor's representative for each of the Shuttle's main components must sign a Certificate of Flight Readiness. Organizations sponsoring payloads have to work closely with the Mission Operations Directorate and the flight-payloads officer to document the operation and ensure the safety of their payloads. All concerns and issues should be resolved before this review.

The *Mission Management Team* consists of program- and project-level managers. This group provides a forum for resolving problems and issues outside the guidelines and constraints established for the Launch and Flight Directors. Two days before launch, the Mission Management Team starts assessing any changes to flight readiness since the Flight Readiness Review and gives a "go/no go" decision to launch. Most payload activities have no changes significant enough to concern the Mission Management Team.

### 20.3.2    Training for Human Space Flight

Astronauts are key ingredients in any crewed space flight. Before astronauts can begin work on a specific payload, they must train on Shuttle systems and capabilities. A payload developer who needs to know what humans add to a mission must understand this training.

Pilot astronauts serve as commanders or pilots. Commanders are responsible for crew safety, vehicle safety, and the success of the mission. Pilots are second in command, and their main duties are to control and operate the Shuttle. Commanders and pilots usually help deploy and retrieve spacecraft, using the remote manipulator arm or other payload-unique equipment, and they may support payload operations.

Mission-specialist astronauts conduct experiments and monitor payload activities. They may deploy spacecraft and may move payloads with the Orbiter's remote manipulator arm. Mission specialists must know the Shuttle systems that support payload operations. They also coordinate onboard operations involving crew-activity planning and monitor the Orbiter's consumables. A payload developer will likely work with two or more payload-specialist astronauts as the mission plan progresses.

Astronaut candidates must pass one year of training before they're qualified as astronauts. After becoming astronauts, they continue to train in order to maintain proficiency and develop skills for specific flights. Following the one-year basic training for astronaut candidates, new astronauts continue with advanced courses in Shuttle systems and enter phase-related training for a specific Shuttle mission seven months to one year before the scheduled launch date. At this time, the potential payload sponsor may meet with the astronauts scheduled to operate their payload. The organization developing the payload works closely with the astronaut office to ensure that the flight crew and backup crew are well trained in operating the particular payload.

As advanced training for the specific mission begins, a special management team directs the flight crew's training. Training now involves scripts and scenarios developed uniquely for the mission, and instructors now test the individual astronauts with planned operations and staged malfunctions. The flight crews also train together on the Shuttle Mission Simulator, which simulates all mission phases. The Shuttle Mission Simulator is designed to allow installing update kits for different missions and payload configurations.   The flight simulators and computers can also work with other simulators such as the European Space Agency's Spacelab Simulator.

The two simulated orbiter cockpits contain controls and displays identical to those of an orbiter and are similar to simulators for commercial airlines. A closed-circuit television display provides proper spatial views of objects from the aft window and orbiter cameras. Computer-generated sound simulations duplicate noises experienced during an actual flight, including the onboard pumps, blowers, mechanical valves, aerodynamic vibrations, thruster firings, pyrotechnic explosions, gear deployment, and runway touchdown.

### 20.3.3    Ground Operations and Support

The Launch Control Center at the Kennedy Space Center controls all of the Shuttle's launch activities. Because launch processing is automated, the launch countdown for a Space Shuttle takes only about 40 hours, compared with the 80 hours needed for an Apollo countdown. In addition, only 90 people now work in the control center during launch compared with the 450 needed for earlier crewed missions.

On launch day, the flight crew enters the Orbiter on the launch pad by means of its access arm, a pathway 1.5 m wide and 20 m long. Towering 45 m above the pad surface, this arm swings out from the Fixed Service Structure to the Orbiter's crew hatch. The environmentally controlled *white room*, where the flight crew undergoes final flight preparations in an ultra-clean environment, is at the end of the arm. The access arm remains in its extended position until about seven minutes before launch, so the crew can get out if needed. In case of an emergency after it has been retracted, the arm can be mechanically or manually repositioned in fifteen seconds to allow for an emergency evacuation of the crew.

People from the white room help the flight crew through the Orbiter's crew hatch. They close the hatch and check the cabin for leaks. Then, they evacuate the white room, and the closeout crew leaves the launch-pad area. The pilot and commander check communications with the launch-control center and with the mission-control center at Johnson Space Center.

At T minus 9 minutes, NASA's Test Director gets a "go for launch" verification from the launch team. From this point onward, the ground-launch sequencer automatically controls the countdown. At T minus zero, the solid rocket boosters ignite and the hold-down explosive bolts and umbilical explosive bolts release. The mission's elapsed time sets to zero and its event timer starts. After the launcher clears the tower, operation of the flight transfers to the Mission Control Center at Johnson Space Center. Most payloads are inactive or require no crew involvement during launch and orbit insertion.

The Mission Control Center at the Johnson Space Center in Houston controls all of the Shuttle's flight operations. Since becoming operational in June 1965 for the Gemini 4 mission, it has been the hub of the US's program for human spaceflight. The payload operator works with the payload office of the Johnson Space Center.

The Payload Operations Control Center at Mission Control is available to payload customers for use in operating their experiments. Operators can receive data and telemetry, see live and playback video of their experiments, and possibly talk to the flight crew about experiment procedures. The payload officer can receive changes in experiment timelines and send them up to the crew. Remote payload-operations facilities may also do these tasks. The Commercial Payload Operations Control Center at Space Industries near the Johnson Space Center provides additional buildings and equipment that experimenters need.

If NASA's Mission Control Center can't operate, NASA's ground terminal at the White Sands Test Facility in New Mexico handles emergency mission control. Flight controllers from the Johnson Space Center are prepared to fly to White Sands and continue supporting the mission at a reduced level. As soon as Mission Control becomes operable again, support of the flight goes to another shift of flight controllers at Houston.

The Mission Control Center controls the Shuttle until it lands, when the Kennedy Space Center resumes control. After landing, the safety-assessment team determines when the Shuttle is safe. After the area is cleaned of any toxic gases, the recovery convoy helps the flight crew leave the Orbiter. Then, another crew of astronauts enters the Orbiter and installs switch guards, removes time-critical experiments, and prepares the vehicle for ground towing. Time-critical experiments can then be taken to nearby laboratories. Refrigeration, cooling, sample storage, and data handing are available to the experimenter. Astronauts remove payloads that are less time critical from the Orbiter after it is towed from the landing area.

In the post-flight debriefing, the crew's observations and comments feed into payload-data analysis and future changes to the payload. Comments from the flight crew, ground controllers, and others who participated in the payload operations during the mission can be invaluable. Lessons learned become part of future flights of the experimenters payload.

# 20.4   International Space Station

### 20.4.1    Current Concept for an International Space Station

The Space Station will remain in orbit with a permanent staff for scientific and technological research. It is an international undertaking by the US, Russia, members of the European Space Agency, the National Space Development Agency of Japan, and the Canadian Space Agency. Political debate about national economic priorities has caused repeated major revisions in the Station concept, and it remains subject to change; we describe it here based on our understanding of sources available in late 1994 and checked in mid-1995.

The station is a very large, open-truss structure supporting numerous solar panels, instruments, and modules for habitation, laboratory, and service work. The Space Station represents an evolutionary approach to space science, with the Station's size, function, and abilities designed to grow over time. The Station will have at least one inhabitable core module, with other modules for scientific research added as the station evolves. Unmanned platforms and free-flyers may also be a part of the total concept. The Station's nominal orbit is at 52° inclination and 380 km altitude.

The Station works like a spacecraft with unique features of on-orbit buildup, permanent habitation, and special subsystems for electrical power, thermal control, data management, guidance and navigation, attitude control, communication, and propulsion (in addition to environmental control and life support). The laboratory modules and their attached instruments yield all the advantages of direct human access as described in Sec. 20.1, so they'll support research far beyond that of any combination of LEO spacecraft without humans onboard.

Typical laboratory equipment includes rooms for experimental plants and animals, instruments for observing the Sun and the Earth, and furnaces for research on solidifying materials and crystals growing in zero gravity. Routine flights to resupply the Station also deliver and retrieve new payload elements, which astronauts either install in racks in the pressurized laboratory modules or mount externally on the truss structure.

## 20.4.2 Mission Phases

Figure 20.1 summarizes the planned orbital life of the Station; it weighs nearly 400 tons, so we must launch it piecemeal and assemble it in space. This assembly phase is expected to require about five years and 32 flights by the US's Shuttle and Russia's Soyuz, Proton, or Progress vehicles [International Space Station Alpha, 1994]. The Station will add modules mainly through mating (docking), but astronauts will need to work outside in order to connect interfaces including fluid and electrical systems. Figure 20.2 gives typical steps for adding one module.

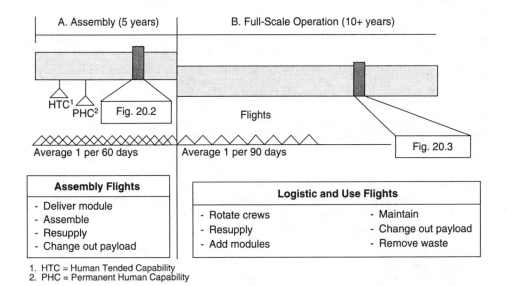

**Fig. 20.1.** **The Space Station's Phases.** In the assembly phase, the station grows by delivering and adding modules. During full-scale operation, flights may deliver new payload modules as well as crews and their supplies.

During Station assembly two notable milestones are the beginning of Human Tended Capability and of Permanent Human Capability. The tended phase begins when the first pressurized modules provide habitable space to help crews deliver and assemble the Station's elements. The permanent human phase begins with the delivery of the first Assured Crew Return Vehicle, which provides a means of returning to Earth without depending on a subsequent visit by a launch vehicle—crew members won't stay at the station until a return vehicle is available.

During assembly, progressively adding habitation, laboratory, and supporting services modules allows the Station to start its scientific-research mission well before completion. After completion, launches continue at a reduced tempo for

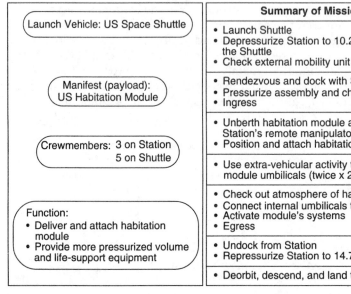

| | Summary of Mission Timeline |
|---|---|
| Launch Vehicle: US Space Shuttle | • Launch Shuttle<br>• Depressurize Station to 10.2 PSI to match the Shuttle<br>• Check external mobility unit |
| Manifest (payload): US Habitation Module | • Rendezvous and dock with Station assembly<br>• Pressurize assembly and check out atmosphere<br>• Ingress |
| Crewmembers: 3 on Station 5 on Shuttle | • Unberth habitation module and hand over to Station's remote manipulator system<br>• Position and attach habitation module |
| | • Use extra-vehicular activity to connect habitation module umbilicals (twice x 2 crewmembers x 6 hours) |
| Function:<br>• Deliver and attach habitation module<br>• Provide more pressurized volume and life-support equipment | • Check out atmosphere of habitation module<br>• Connect internal umbilicals to habitation module<br>• Activate module's systems<br>• Egress |
| | • Undock from Station<br>• Repressurize Station to 14.7 PSI |
| | • Deorbit, descend, and land the Shuttle |

**Fig. 20.2.   Excerpt from Assembly Timeline.** This figure describes one of the assembly flights (See Fig. 20.1) which delivers a habitation module. Modules are attached by docking but their electrical and other connections require crew work in space. The typical crew time on-orbit totals several days per assembly flight.

crew rotation and other sustainment purposes, with the Station permanently staffed with four to six crew members serving 90-day shifts.

Maintenance divides into on-orbit, on-ground, and depot (on-ground). On-orbit maintenance may be inside the Station or outside the Station using teleoper-ated robotics, humans, or a combination of the two. Routine system housekeeping is largely automated so the crew can devote their time to support research. Remote control from ground stations carries out some payload operations.

Atmospheric drag slowly reduces the Station's altitude, but propulsive reboost periodically restores it, so altitude varies cyclically as shown in Fig. 20.3.

### 20.4.3   Mission Management and Operational Tasking

Because the Station is an international undertaking, its mission management undergoes negotiation and change. Figure 20.4 shows a tentative concept for the basic management structure, headed by the mission-management team. This team consists of program-level managers from the international partners. The Mission Control Center in Houston handles all Space Shuttle launches, oversees assembly activities, and manages overall system operations and safety. The Mission Control

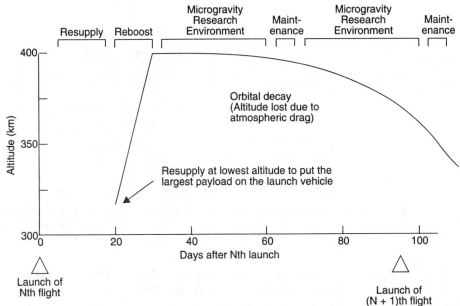

**Fig. 20.3.    Excerpt from Operational Timeline.** The Station maintains an altitude margin and a reboost propellant reserve to ensure that its altitude won't become dangerously low even if
resupply flights are late. Between reboosts, the freedom from propulsive disturbances allows microgravity research.

Center in Moscow handles Russian launches and related assembly activities. The
Payload Operations Integration Center in Huntsville, Alabama, integrates payload
operations; carries out and coordinates all payload operations in real time; and is
the voice, data, and command interface between the Station and its clients or payload users, e.g. researchers, at their operating locations and remote sites.

To develop tasking and execution procedures, the Station's program doesn't
use terms from the 13 functions defined in Chap. 3, but it does use these terms to
carry out the procedures. Station tasking begins with mission profiles, which planners split into scenarios (event sequences) for different facets of operation
[International Space Station Alpha, 1994 and International Space Station Data,
1994]. The scenarios then divide further. For example, the scenario for on-orbit
operations covers nominal, maintenance, and other operations. Once planners
have defined each low-level scenario by constraints and by starting and finishing
conditions, they develop functional flows to describe logical sequences. Processes
include inputs, functional allocations, and outputs. A matrix structure relates these
processes to the scenarios.

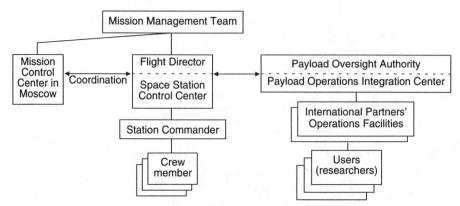

**Fig. 20.4.**    **Structure for Mission Management.** The management structure depends on international negotiation. We expect the final structure will allow Earthbound researchers to manage on-orbit payloads from their normal workplaces in various countries.

Table 20.1 suggests trades, issues, and concerns for cost-effective operation of the Station, based purely on our judgment. Soon, the Station program will replace this tentative list.

**Table 20.1.**    **Getting the Most from Space-Station Missions.** Continually replacing crews and payload modules will allow flexibility for learning and for improving efficiency during the years of station operations.

| **Crew mix, selection, and training** |
|---|
| • Train crews together to develop team cohesion<br>   – International program requires mix of nationalities and cultures—conflicts with usual criteria (shared background and culture) for small groups in prolonged isolation<br>• Decrease the training burden (allow lower fidelity training facilities)<br>   – Crew need not be experts on payloads<br>   – Familiarize them on the ground and continue on-the-job training in orbit |
| **Ground control team (payload and systems specialists)** |
| • Human missions require continuous staffing of safety-critical positions—for other positions, plan to "fail safe" and await temporary, on-call staffers who will resolve the problem |
| **Scheduling and tasking** |
| • Be prepared to change based on the priorities of international partners<br>• Exploit the flexibility humans provide<br>• Find the best balance between efficiency and cost<br>   – Don't try for 100% use—too costly, complex, and rigid<br>   – Allow margins for anomalies and benefit from the crew's judgment<br>• Use assembly phase lessons learned in the operational phase<br>• Allow for changes in crew-training criteria, scheduling, and payload recycling—be prepared to change the plan of operations (including resupply and microgravity) and maintenance cycles |

# 20.5　Human Lunar Missions

On July 20, 1989, President George Bush asked the National Space Council to determine what was needed to return to the Moon and to continue on to Mars. The *90-Day Study on Human Exploration on the Moon and Mars* [NASA, 1989], describes the main elements of future human missions. The planned approach has three stages: the Space Station, a return to the Moon, and then a trip to Mars. The study defined the general mission objectives and key program requirements. As a result, regardless of the exact approach selected—heavy launch vehicles, space-based transportation systems, surface vehicles, underground habitats, or other concepts of support systems—returning to the Moon is a crucial step in the evolution of space exploration.

A lunar outpost is an excellent platform from which to do research in astronomy, physics, and life sciences. The crew can use rovers to explore the Moon's geology and geophysics by taking soil and rock samples for analysis in a lunar laboratory. The Moon is a unique site for synoptic viewing of the Earth's magnetosphere, which can provide long-term information about the distribution of solar energy and the consequent warming and cooling of Earth's upper atmosphere.

At just three days from Earth, the Moon is an ideal location for learning to live away from Earth with increasing self-sufficiency. The lunar outpost will serve as a test-bed for validating critical mission systems, hardware, technologies, human abilities, and self-sufficiency—all of which we can apply to interplanetary exploration. The life-support systems tested on lunar missions essentially have undergone a full-scale operations test, after which they can evolve incrementally before developing a complete system for a Mars mission.

## 20.5.1　Significance of Past Missions

From 1969 to 1972, the US's Apollo spacecraft landed six times on the Moon. Pairs of astronauts spent up to three days on the Moon and explored its surface for many hours. During the last three landings, astronauts used rovers built like dune buggies to transport themselves about 30 km across the lunar surface.

These history-making successes show that human lunar missions are well within technological reach. But we should also remember that problems had to be overcome by system redundancy combined with prompt and astute action by the astronauts and the payload specialists at Mission Control. Among the six landing missions, these incidents included a computer glitch due to lightning strike during booster ascent, premature shutdown of a booster engine, difficulties in docking, computer false alarms, control-system malfunctions, a landing-radar malfunction, and a parachute damaged by contact with propellant. The explosion of an oxygen tank on Apollo 13 prevented lunar landing, and the crew survived only through prompt and ingenious measures. These events remind us of the versatility and the risk of human presence in space. Case [1993] and Wright [1993] describe typical early-1990s concepts for a return to the Moon.

## 20.5.2    Mission Phases

Figure 20.5 shows the phases of a human lunar mission. Translunar injection is the propulsive maneuver that gives the spacecraft enough velocity to get near the Moon. At the velocity needed for initial injection, the flight to the Moon takes about five days each way, but operators usually use slightly higher velocities, which reduce the time to three or four days. (See Chap. 10)

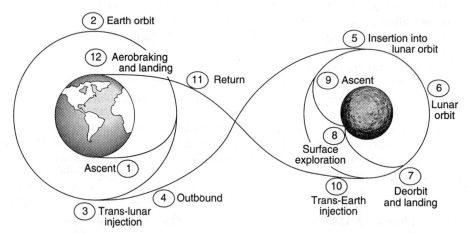

**Fig. 20.5.    Profile of a Lunar Mission.** The drawing is not to scale—the altitudes of the orbits around the Earth and Moon are greatly exaggerated for clarity. In each of these orbits, the spacecraft may do one or more revolutions to obtain flexibility in mission timing and maneuvers.

Injection could be direct (combined with ascent from Earth in a single propulsive maneuver), but a short time in a low parking orbit routinely separates the ascent and injection. This technique doesn't penalize propulsion and it gives the flexibility of a hold period to adjust mission timing and orbit phasing, as well as to verify proper system operation or correct problems.

These considerations also favor a parking orbit at the Moon, with the major extra advantage that propellant and structure for return (trans-Earth injection) and for landing on Earth don't have to decelerate, land on the Moon, and then be raised back to an altitude and speed for lunar orbit. Instead, only a capsule for the landed crew members and their lunar samples needs propulsion back to lunar orbit.

Without an atmosphere, the lunar orbit's lowest point or periapsis (sometimes called perilune or pericynthion) can be just high enough to safely clear terrain. The Apollo missions used altitudes as low as 17 km. Landing consists of decelerating from orbital speed and then transitioning to a descent which must reach near-zero vertical and horizontal velocity at the same time that it reaches zero altitude. This requires throttling (variable-thrust) engines and, preferably, direct human control to locate a clear and level landing area.

After landing, the crew prepares the ascent propulsive stage for return and then starts exploring the surface. Because the Moon is always about the same distance from the Earth, lunar round trips have flexible timing. The time spent on the lunar surface depends mainly on available supplies, reliability of equipment, physical endurance, and perhaps the onset of the two-week lunar night. The surface activities for Apollo included surveying the local area, documenting and collecting mineral samples, and emplacing long-lived automated instruments (Apollo Lunar Surface Experiment Packages). Thus, payload planning and data processing continued long after the astronauts returned. Apollo also showed how astronauts could work efficiently and safely in this novel environment. The purpose of future landings will be to carry out equivalent activities on a larger scale.

Like most other phases of the lunar mission, at least on the near side of the Moon, surface exploration benefits from the short distance to Earth. Voice and video links work continuously in near real time, with subsystem specialists at Mission Control available for immediate advice. However, as lunar operations mature, lunar explorers must refine ways of handling problems and not rely on a full shift of ground-control people working around the clock for their support.

After exploring the lunar surface, astronauts take the capsule up and rendezvous with the lunar-orbiting spacecraft, unless the mission uses direct trans-Earth injection. Near the Earth, the spacecraft could do an inverse of its earlier translunar injection and thereby enter an orbit around the Earth. But propellant for this maneuver would have to go to the Moon and back. Thus, an alternate approach is to take an atmospheric-reentry capsule to the Moon and back and use aerobraking for direct reentry. This reentry does require precise navigation into a reentry corridor, which must be neither too shallow to shed enough of the spacecraft's velocity nor so steep that excessive deceleration, heating rate, or impact speed destroy the spacecraft.

Future lunar missions may help human exploration of space to move into a more ambitious phase by identifying usable resources. Sources of oxygen or water would be most interesting, and processes for extracting oxygen from lunar surface materials have been described and compared, for example in Taylor [1992]. If we can get oxygen on the Moon for propulsion back to Earth, future missions could bring only fuel and not oxidizer. Thus, we could sharply increase their useful payloads to the lunar surface. In terms of propulsive effort, a low-Earth orbit is easier to reach from the Moon than from Earth—hence, spacecraft on interplanetary missions could eventually use lunar resources.

Table 20.2 is a basic checklist of issues for cost-effective lunar operations. The mission-management structure might be conceptually similar to that for the Space Station (Fig. 20.4) because the missions are similar in time scale and ease of real-time communication.

**Table 20.2.  Getting the Most from Missions to Explore the Lunar Surface.** Unlike Space Station operations, a lunar mission involves only one round trip. Thus, it allows much less chance for refining operational efficiency on the job. Careful preparation best uses the limited time available on the lunar surface.

| Get the Most From a Mission |
| --- |
| Ask smart questions<br>• Assign only tasks for which human presence is essential<br>• Logically rank and sequence these tasks |
| Provide the best possible resources<br>• Crew mix, selection, and training<br>• Ground support team (consultants/users)<br>• Exploitation of lunar orbiters (crewed or not)<br>• Efficient means to capture data (instruments, vehicles, communications)<br>• Automation of routine activities |
| Leave something there<br>• Fixed stations and tele-operated rovers<br>• Ground team to operate emplaced equipment long-term |
| Use the results<br>• Data conversion and distribution—accessibility to users<br>• Operational lessons defined and preserved to benefit later missions |
| **Choose most effective length of time on surface** |
| Balance more exploration (perhaps sustained by automated resupply flights) against<br>• Increased weight of consumables, electric power supply, tools, and spare parts<br>• Declining human efficiency and increasing risk (crew injury, equipment failure)<br>• Reaching limits of mission's plan and equipment (time to apply what has been learned) |
| **Get the most from a series of missions** |
| • Choose landing sites as part of coherent long-term plan<br>• Search for resources and begin experiments toward using them<br>• Build up equipment on the Moon—design for dormancy, reuse, adapting, or cannibalizing<br>• Maintain continuity of core operational and design teams (corporate memory)<br>• Maintain continuity of hardware lineage (retain and refine proven subsystems) |

# 20.6   Human Interplanetary Missions

The first interplanetary mission will be to Mars, the solar system's most Earth-like planet. Initial missions to Mars will prove the systems and equipment and further survey selected landing sites. Later missions will establish a Mars outpost with the objective of experimenting and exploring in an extraterrestrial environment.

## 20.6.1   Features of Interplanetary Flight

This section assumes the technology for a human Mars mission extends straightforwardly from what has already been developed for lunar missions. As early as the 1960s, NASA studies projected the first humans would land on Mars in 1984 [Manned Mars Surface Operations, 1965]. But the distances and times involved make interplanetary missions enormously more challenging than lunar missions. Going from the Earth's surface to an orbit of Mars doesn't require much more total velocity than going to the Moon, but we must also time the flight so Mars is at the correct point for arrival and return. The result is that, compared with lunar missions, interplanetary launch and return windows are sharply constrained. Also, because of the distances involved, multi-year missions are inevitable. Long Earth-Mars transit times have risks due to zero gravity and sporadic solar-flare radiation. A long stay on Mars means long exposure to an environment which may have unforeseen hazards. And, finally, long total mission time means increased probability of human or mechanical failure.

The conceptual designer of an interplanetary mission seeks to reduce the mission's total $\Delta V$ or propulsive requirement and therefore enlarge the useful payload that will arrive on Mars. Or the designer may try to shorten the trip to decrease risks to human life. Figure 20.6 shows three general types of trajectories for a Mars mission. Their times and total velocity budgets start from LEO and they use aerobraking instead of propellant for maneuver 4 (arrival at Earth). The graphs show the trajectories with the lowest velocity requirements entail lengthy stays at Mars. A further constraint on mission timing is that the orbit of Mars is notably eccentric. This feature combines with the phasing of the movements of Earth and Mars to produce a decades-long cycle of favorable and unfavorable launch windows. For example, the years 2001 and 2018 are favorable for most types of Mars missions. See Young [1988], TRW Space Data Book [1992], and the Space Flight Handbook [1963] for details on these and other trajectory trades.

Numerous studies have proposed ways to improve this time and payload trade. One alternative is simply to wait for better technology, such as nuclear thermal rockets [Emrich, 1991]. Technologically, nuclear rockets will work. They were built and demonstrated on the ground in the NERVA program in the 1950s. A revival of research on nuclear or other non-chemical rockets and their large-scale application would depend on a degree of commitment and interest that may occur at some time in the future but is certainly not evident now.

Another alternative is to "live off the land," exploiting Mars resources as described in the next section.

### 20.6.2    Mission Phases

A Mars mission begins like a lunar mission—with a launch, a parking orbit, and trans-Mars injection. The parking orbit is nearly always assumed because the size of the spacecraft requires multiple launches with final assembly in orbit.

Upon arrival at Mars, the spacecraft has expanded mission options because an atmosphere is present. Instead of needing a major propulsive maneuver, it can use aerobraking to shed most of the arrival velocity. The atmosphere's low density makes the entry corridor very narrow (Fig. 20.7), especially if preceding months of

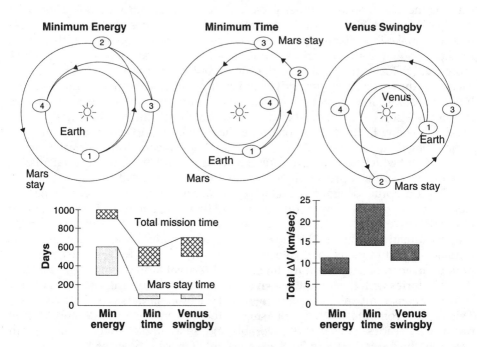

**Fig. 20.6.   Trajectories for a Mars Mission.** The scale is only approximate. The basic trend is that shortening the mission time has a big price in required ΔV. The Venus Swingby obtains some free ΔV by using Venus's gravitational field to change the direction of the space-craft's velocity vector. The phases of the mission correspond to: (1) departure from Earth; (2) arrival at Mars; (3) departure from Mars; and (4) arrival at Earth. The times and the velocities shown in the two graphs vary (shaded regions) depending on the launch date.

zero gravity demand that the crew receive only moderate g-loads. Lynne [1992] gives entry-corridor trades in some detail with a typical corridor width of 1°, some of which we'd have to allocate to uncertainty about the actual atmospheric-density profile at that time over that region of Mars.

As with lunar missions, total payload increases if we keep the Earth-return stage in orbit around Mars as part of an orbiter module rather than bringing it to the surface and back. For this purpose, higher orbits are better. The orbiter might be continuously staffed both to ensure its readiness for return and to do remote sensing, navigation, and communications relays to support the surface explora-tion. If staffed, the orbiter could operate at or near one of the moons, Phobos or Deimos, to investigate their potential resources for future missions [O'Leary, 1992]. The orbit of Deimos happens to be close to the Mars equivalent of a geosyn-chronous orbit, where a spacecraft would appear fixed in the Martian sky.

If the orbiter is designed to withstand aerobraking, the lander and orbiter can aerobrake and enter parking orbit as one unit. Alternatively, the lander or Mars

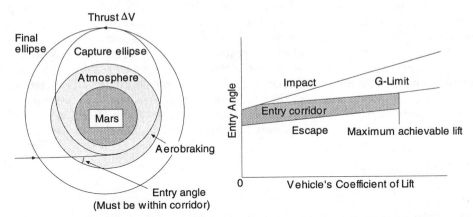

**Fig. 20.7.** **Aerocapture from Interplanetary Trajectory.** The depth of the atmosphere is exaggerated for clarity. Aerobraking sheds enough velocity to bring the spacecraft into an orbit around Mars. At apogee a small propulsive maneuver stops further aerobraking by raising perigee above the atmosphere.

Excursion Module may separate early and enter the atmosphere directly from its interplanetary trajectory. Meanwhile, the orbiter expends propellant for insertion in Mars orbit but need not withstand atmospheric heating. In aerobraking any large vehicle at Mars, one concern is the physical size of the aerobraking heat shield. The mass to be landed includes an ascent vehicle as well as everything for survival and exploration on the surface, so the heat-shield diameter may need to be tens of meters, which exceeds the limits of Earth-launch vehicles. Thus, we have to mechanically extend the shield or assemble it in space, with corresponding concerns about verifying its readiness for the mission.

Surface exploration of Mars works like exploration of the Moon. As on the Moon, a high-priority objective is locating resources that could make later missions much easier. The Martian atmosphere is about 95% carbon dioxide, with the rest mostly nitrogen, and Mars is believed to hold large quantities of water as permafrost. More advanced missions could combine carbon dioxide, water, and nitrogen to manufacture propellants for rocket and surface vehicles, as well as to sustain the crew.

Godwin [1960] was among the first to point out that, if carbon dioxide is available on a planet and if we bring one ton of hydrogen from Earth, theoretically we could manufacture nearly 20 tons of rocket propellants (fuel and oxidizer) for the return journey. Recently, others have more thoroughly examined this possibility. Zubrin [1992] describes several reactions that can convert the carbon dioxide and hydrogen to methane and water, which we could then electrolyze to get oxygen. The propellants methane and oxygen will store as liquids.

Zubrin describes a very interesting mission (Fig. 20.8) in which an uninhabited vehicle for propellant manufacturing and Earth return lands on Mars to produce propellants for a return flight before the crewed vehicle departs from Earth. He contends that his chosen manufacturing process is already developed and proven to a much greater extent than some of the other technology we need, so it's a reasonable part of even the first human mission. Given the enormous advantages of this propellant production, Zubrin envisages human missions to the Moon and Mars, using relatively modest masses launched from Earth. Bruckner [1993] describes a mission conceived recently by others applying Zubrin's ideas.

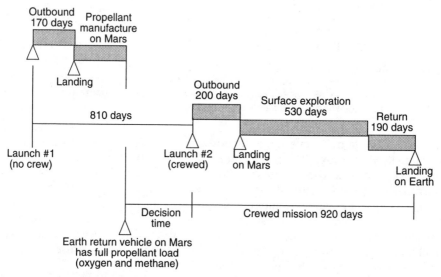

**Fig. 20.8.** **Zubrin Mission Using Martian Resources.** The first vehicle (no crew) uses nuclear power and Martian carbon dioxide to produce roughly 200 kg of propellants per day. The first vehicle accumulates enough propellant for the crewed return flight at Mars before the crewed vehicle departs from Earth.

When surface exploration is complete, the Mars lander ascends to either the parking orbit or directly to trans-Earth injection. If the orbiter has a crew, it can rendezvous in the parking orbit. Trans-Earth injection could use either a single module carrying orbit and surface crews, or two separate modules with extra capacity to provide redundancy (either one could return both crews).

Upon arrival at Earth, the vehicle may land by direct re-entry or by aerobraking to help maneuver itself into a parking orbit, which may be highly eccentric. In either case the payload for this maneuver need not include the habitation module that supports the crew during the long interplanetary flight; it needs only a capsule that can support them for the few hours until rendezvous with Earth-based vehicles or recovery teams. This approach again reduces the mass of propellant and heat shield that must go to Mars and back.

An alternative to aerobraking would be to have the approaching spacecraft met by an Earth-based recovery vehicle, which would maneuver to match interplanetary-approach velocity and allow docking with the capsule; a further maneuver would then return the vehicle to its original orbit. Or the spacecraft could land by direct entry and aerobraking.

All the items in Table 20.2 apply even more emphatically to interplanetary missions because of their long duration and other added challenges. Table 20.3 lists other issues for cost-effective operations for a Mars mission.

**Table 20.3. Getting the Most from Interplanetary Missions.** Add these items to the ones in Table 20.2 for lunar missions, emphasizing the challenge of enormous distances to even nearby planets.

| |
|---|
| Mission managers must allow astronauts unprecedented independence<br>  • Communication round trip time 7 to 44 minutes<br>  • Unknowns of the planetary environment |
| Task Mars orbiter (if used) and its crew (if onboard during Mars stay)<br>  • To best use the orbiter's advantages<br>  • To maintain the orbiter crew's morale during their stay at Mars |
| Continuous communication with Earth is very desirable, but difficult<br>  • Distance and power requirements<br>  • Other missions competing for the Deep Space Network<br>  • Planetary rotation limits viewing between surface sites on Earth and Mars<br>    – A Mars orbit best for communication relay is unsuitable for other functions<br>    – Add a Mars-synchronous relay satellite?<br>    – May lose communication when Sun is between Mars and Earth |
|   • Earth-Mars transit, typically 200 days each way, must include crew tasks<br>    – To get the most from the mission<br>    – To maintain crew morale |
|   • Contingency of early departure from Mars surface creates special requirements<br>    – Ascent vehicle's propulsion increased for flexibility in launch window<br>    – Redundant environmental control, life support, and consumables in orbiter |

Mission managers must adjust to the long communication times and the astronauts' resulting sense of isolation. Total mission time will be 500–1000 days, cost will be high, and public interest will be intense (at least for the first mission). Thus, conflict will increase between the goals of international representation (mix of crew nationalities) and team cohesion (shared background and culture).

One of the most important mission-design trades is selecting altitude and inclination for the orbiter module. Although an orbiter is valuable for relaying communications, Mars mapping, and Mars-ascent rendezvous, these functions have different orbit requirements.

Finally, Fig. 20.6 shows how orbital mechanics favors long stay times. Long stay times mean more risk and more research opportunities. After a 300-day stay, would we know Mars ten times as well as after 30 days? Where is our region of diminishing returns?

As our discussion has shown, human spaceflight introduces extra mission phases and concerns into space operations. But human missions still require the functions defined in Chap. 3 and the procedural steps and trades described in other chapters.

## References

Bruckner, A.P., M. Cinnamon, S. Hamling, K. Mahn, J. Phillips, and V. Westmark. *Low Cost Manned Mars Mission Based on Indigenous Propellant Production*. AIAA 93-1010, Aerospace Design Conference, Irvine, California, Feb. 16–19, 1993.

Case, Carl M. *The Lunar Campsite Mission Concept*. 30th Space Congress, Canaveral Council of Technical Societies, Cocoa Beach, Apr. 27–30 1993.

Emrich, W.J., A.C. Young, and J.A. Mulqueen. *Nuclear Thermal Rocket Propulsion Application to Mars Missions*. 28th Space Congress, Canaveral Council of Technical Societies, Cocoa Beach, Florida, April 23–26, 1991.

Godwin, Felix. 1960. *The Exploration of The Solar System*. Plenum Press. p. 81.

Logsdon, John. *The Space Shuttle Program: A Policy Failure?* Science, May 30 1986, pp. 1099–1105.

Lyne, J.E., A. Anagnost and M.E. Tauber. *Parametric Study of Manned Aerocapture*. Part II: Mars Entry, Journal of Spacecraft and Rockets, Vol. 29 No. 6 pp. 814–819, Nov.–Dec. 1992.

NASA. *Concept of Operation and Utilization*. International Space Station Alpha Program. NASA Johnson Space Center, 1994.

NASA. *International Space Station Data Book*. Space Station Program Office. NASA Johnson Space Center, 1994.

NASA. *Manned Mars Surface Operations*. Final Report RAD-TR-65-26, Contract NAS 8-11353, AVCO Corpn, Wilmington, Massachusetts for NASA Marshall Space Flight Center, 30 Sept. 1965.

NASA. *Report of 90-Day Study on Human Exploration on the Moon and Mars*. Nov. 1989.

NASA. *Space Flight Handbook. Vol. 3—Planetary Flight Handbook*, NASA SP-35, NASA Marshall Space Flight Center, 1963.

O'Leary, Brian. *International Manned Missions to Mars and the Resources of Phobos and Deimos*. Acta Astronautica, Vol. 26 No. 1 pp. 37–54, 1992.

Taylor, Lawrence. *Production of Oxygen on The Moon: Which Processes are Best and Why*. AIAA 92-1662, AIAA Space Programs and Technologies Conference, Huntsville, Alabama, Mar. 24–27, 1992.

TRW. *Space Data Book*. TRW Inc., Redondo Beach, California, 1992, pp. 4–2 and 8–21 through 8–31.

Wright, Michael J. *Conceptual Design of a Cargo Lander for the First Lunar Outpost*. 30th Space Congress, Canaveral Council of Technical Societies, Cocoa Beach, Apr. 27–30, 1993.

Young, Archie C. *Mars Mission Profile Options and Opportunities*. 25th Space Congress, Canaveral Council of Technical Societies, Cocoa Beach, Florida, April, 1988.

Zubrin, Robert. *Mars and Lunar Direct: Maximizing the Leverage of In-Situ Propellant Production*. AIAA 92-1669, AIAA Space Programs and Technologies Conference, Huntsville, Alabama, Mar 24–27, 1992.

# Appendix A

## Communications Frequency Bands

**Table A.1.** **Communication Frequency Band Designation and Frequency Range.** *The International Telecommunications Union (ITU)* and the *World Administrative Radio Conference (WARC)* regulate use of frequency band, transmission bandwidth, and power-flux density. [Larson and Wertz, 1992]

| Frequency Band | Frequency Range (GHz) |
|:---:|:---:|
| VHF | 0.03 – 0.3 |
| UHF | 0.3 – 1.0 |
| P | 0.225 – 0.39 |
| L | 0.39 – 1.05 |
| S | 1.05 – 3.9 |
| C | 3.9 – 6.2 |
| X | 6.2 – 10.9 |
| K | 10.9 – 36.0 |
| $K_u$ | 15.35 – 17.25 |
| $K_a$ | 33.0 – 36.0 |
| Q | 36.0 – 46.0 |
| V | 46.0 – 56.0 |

# Appendix B

## Mission Summaries

This appendix contains mission summaries for 14 space missions managed by the NASA Goddard Space Flight Center. The summaries include a mission statement, launch information, a description of the spacecraft bus, a description of the payload and science instruments, the orbit phases, a description of the ground system, an overview of mission operations, and a schedule. The following missions are included:

- ACE      Advance Composition Explorer
- COBE      Cosmic Background Explorer
- ERBS      Earth Radiation Budget Satellite
- FAST      Fast Auroral Snapshot Explorer
- FUSE      Far Ultraviolet Spectroscopic Explorer
- GRO      Gamma Ray Observatory
- ICE      International Cometary Explorer
- IMP      Interplanetary Monitoring Platform
- IUE      International Ultraviolet Explorer
- SAMPEX      Solar, Atmospheric, Magnetic Particle Explorer
- SOHO      Solar and Heliospheric Observatory
- SWAS      Submillimeter Wave Astronomy Satellite
- TRMM      Tropical Rainfall Measuring Mission
- XTE      X-ray Timing Explorer

# ACE - Advanced Composition Explorer

MOM: Frank Snow (GSFC)                    PROJ MGR: Donald L. Margolies (GSFC)

| Rev Date: 22 Feb 94 | **MISSION STATEMENT** | Current Mission Phase: C/D |
|---|---|---|

Observe particles of solar, interplanetary, interstellar, and galactic origins, spanning the energy range from that of solar wind to galactic cosmic ray energies. In particular, ACE will provide the first extensive tabulation of solar isotopic abundancies based on a direct sampling of solar material.

| Launch Date: August 1997 | Launch Vehicle: Delta II 7920 | Launch Site: ESMC |
|---|---|---|

## Spacecraft

**ACS**
- **Sensors**
  Sun Sensor Systems (2)
  Star Tracker (1)
  Accelerometers (2)
  Nutation Dampers (2)
- **Actuators**
  Spin Stabilized at 5 + −.1 RPM
- **Control**
  Attitude determined and
  corrected from ground

**Thermal**
- Spacecraft heaters
- Passive radiators

**C&DH**
- Autonomy cmd.
  capability
- 2 solid state recorders
- 3 selectable TLM formats
- Convolutional and Reed Solomon
  encoding

**Weight: 765 kg**

**Power**
- Total Power:
  548 Watts (BOL)
- Batteries:
  One 12 Ah Battery
- +Z side always
  within 4-20 deg of
  the S/C-Sun line

**RF**
- 4 Helical antennas
- 1 Parabolic dish
- 1 KBps Command
  rate
- Coherent 240/241
  ratio

**Prop**
- Twelve 1 lbf thrusters
- Hydrazine-Nitrogen
  blowdown system

## Instruments

| Inst. | Power | Mass | PI |
|---|---|---|---|
| CRIS | 17.8 W | 26.4 kg | E. C. Stone (CIT) |
| SIS | 16.8 W | 15.8 kg | E. C. Stone (CIT) |
| ULEIS | 15.1 W | 17.0 kg | E. C. Stone (CIT) |
| SEPICA | 8.8 W | 19.5 kg | E. C. Stone (CIT) |
| SWICS | 6.0 W | 5.0 kg | E. C. Stone (CIT) |
| SWIMS | 8.0 W | 9.0 kg | E. C. Stone (CIT) |
| EPAM | 5.1 W | 6.8 kg | E. C. Stone (CIT) |
| MAG | 2.4 W | 4.3 kg | E. C. Stone (CIT) |
| SWEPAM | 6.4 W | 6.7 kg | E. C. Stone (CIT) |
| RTSW | 57.4 W | 9.1 kg | (NONE) |

## Science

**Science Instruments**
- **Cosmic Ray Isotope Spectrometer (CRIS)** - Designed to provide statistical significant measurements of all stable and long-lived isotopes of galactic cosmic ray nuclei from He to Zn over the general energy range from about 100 to 600 MeV/nucleon.
- **Solar Isotope Spectrometer (SIS)** - Designed to provide isotopically resolved measurements of the elements from Li to Zn over the energy range 10 to 100 MeV/nucleon.
- **Ultra Low Energy Isotope Spectrometer (ULEIS)** - Designed to measure ion fluxes over the charge range from He through Ni from about 20 keV/nucleon to 10 MeV/nucleon.
- **Solar Energetic Particle Ionic Charge Analyzer (SEPICA)** - Designed to measure the ionic charge state, Q, the kinetic energy, E, and the nuclear charge, Z, of energetic ions above 0.2 MeV/nucleon.
- **Solar Wind Ion Composition Spectrometer (SWICS)** - Designed to determine the elemental and ionic-charge composition and the temperature and mean speeds of all major solar wind ions from H through Fe at solar wind speeds ranging from a minimum of 145 km/s (protons) to a maximum of 1532 km/s (FeIX).
- **Solar Wind Ion Mass Spectrometer (SWIMS)** - Designed to provide an accurate determination every few minutes of the abundances of most of the elements and a wide range of isotopes in the solar wind.

**Monitoring Instruments**
- **Electron, Proton, and Alpha-Particle Monitor (EPAM)** - Designed to measure solar an interplanetary particle fluxes with a wide dynamic range and covering nearly all directions.
- **Magnetic Field Monitor (MAG)** - Designed to measure both the strength and direction of the interplanetary magnetic field at L1.
- **Solar Wind Electron, Proton, and Alpha Monitor (SWEPAM)** - Designed to measure low energy solar wind electron fluxes from 1 to 900 eV and ion fluxes between 0.26 and 35 KeV.
- **NOAA Real Time Solar Wind (RTSW)** - Processes data received the EPAM, SPEPAM, and MAG instruments and transmits a composite reduced set of data to NOAA ground stations for the purpose of continuously monitoring and reacting to solar events.

## Orbit

### Orbit Parameters:

**Parking Orbit:**
ACE will spend only about 1/2 revolution in it's parking orbit

**Halo Orbit Vector (GCI):**
x = 410879.0385177396 km
y = −1249787.3655573796 km
z = −275439.04331810248 km
$V_x$ = 0.3827291085863497 km/s
$V_y$ = 0.1047969639593863 km/s
$V_z$ = 0.01996648734520219 km/s

**Fig. B.1.    Advanced Composition Explorer (ACE).**

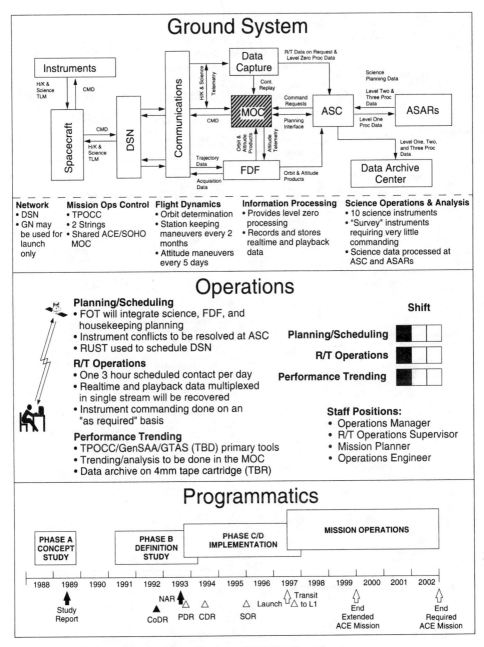

**Fig. B.2.** **Advanced Composition Explorer (ACE) Continued.**

# COBE - Cosmic Background Explorer
POD: Jim Williamson (GSFC)

| Rev Date: 7 Jan 94 | MISSION STATEMENT | Current Mission Phase: N/A |
|---|---|---|

Test the Big Bang theory of a primeval explosion 15 billion years ago. In 1992, the results from analysis of the first two years of data revealed that COBE's observations supported the BIG BANG theory.

Launch Date: November 18, 1989     Launch Vehicle: Delta 5920 ELV     Launch Site: VAFB

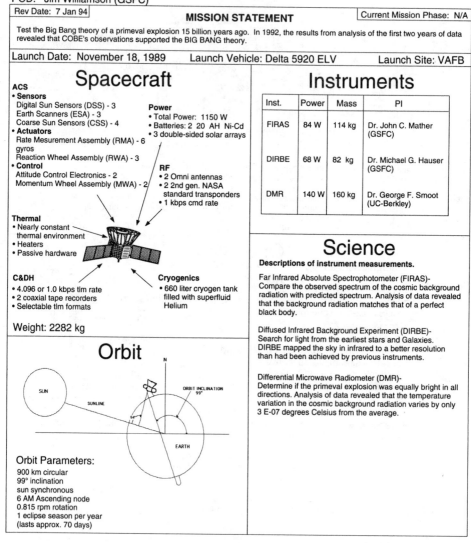

## Spacecraft

**ACS**
- **Sensors**
  Digital Sun Sensors (DSS) - 3
  Earth Scanners (ESA) - 3
  Coarse Sun Sensors (CSS) - 4
- **Actuators**
  Rate Mesurement Assembly (RMA) - 6 gyros
  Reaction Wheel Assembly (RWA) - 3
- **Control**
  Attitude Control Electronics - 2
  Momentum Wheel Assembly (MWA) - 2

**Power**
- Total Power: 1150 W
- Batteries: 2  20  AH  Ni-Cd
- 3 double-sided solar arrays

**RF**
- 2 Omni antennas
- 2 2nd gen. NASA standard transponders
- 1 kbps cmd rate

**Thermal**
- Nearly constant thermal environment
- Heaters
- Passive hardware

**C&DH**
- 4.096 or 1.0 kbps tlm rate
- 2 coaxial tape recorders
- Selectable tlm formats

**Cryogenics**
- 660 liter cryogen tank filled with superfluid Helium

Weight: 2282 kg

## Instruments

| Inst. | Power | Mass | PI |
|---|---|---|---|
| FIRAS | 84 W | 114 kg | Dr. John C. Mather (GSFC) |
| DIRBE | 68 W | 82  kg | Dr. Michael G. Hauser (GSFC) |
| DMR | 140 W | 160 kg | Dr. George F. Smoot (UC-Berkley) |

## Science
Descriptions of instrument measurements.

Far Infrared Absolute Spectrophotometer (FIRAS)- Compare the observed spectrum of the cosmic background radiation with predicted spectrum. Analysis of data revealed that the background radiation matches that of a perfect black body.

Diffused Infrared Background Experiment (DIRBE)- Search for light from the earliest stars and Galaxies. DIRBE mapped the sky in infrared to a better resolution than had been achieved by previous instruments.

Differential Microwave Radiometer (DMR)- Determine if the primeval explosion was equally bright in all directions. Analysis of data revealed that the temperature variation in the cosmic background radiation varies by only 3 E-07 degrees Celsius from the average.

## Orbit

**Orbit Parameters:**
900 km circular
99° inclination
sun synchronous
6 AM Ascending node
0.815 rpm rotation
1 eclipse season per year
(lasts approx. 70 days)

**Fig. B.3.     Cosmic Background Explorer (COBE).**

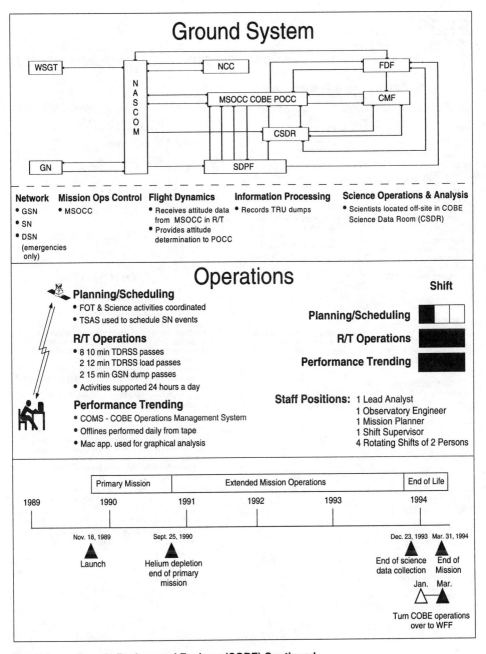

**Fig. B.4.    Cosmic Background Explorer (COBE) Continued.**

# ERBS - Earth Radiation Budget Satellite
POD: Jim Williamson (GSFC)

| Rev Date: 21 Apr 94 | **MISSION STATEMENT** | Current Mission Phase: Extended Ops |
|---|---|---|

To study the Earth's atmosphere and its radiation environment.

| Launch Date: October 05, 1984 | Launch Vehicle: STS-41G | Launch Site: KSC |
|---|---|---|

## Spacecraft

**ACS**
- **Sensors**
  Sun sensor - 2
  Earth Horizon Scanner - 2
  Inertial Reference Unit (IRU) - 2
- **Actuators**
  Attitude Control Electronics (ACS)
  Magnetic Control Electronics (MCS)
- **Control**
  Electromagnets - 4
  Momentum wheel - 1

**Thermal**
- Thermostatically controlled film heaters
- Active louvers
- Passive devices

**Power**
- Solar arrays: two fixed panels, 14.6 sq. meters
- Total power: 2400 W
- Batteries: two 22-Cell 50-Ah Ni-Cd

**RF**
- 1 high gain ESSA antenna
- 2 low gain omni antennas
- 2 NASA standard TDRS/GSTDN transponders
- 1.0 kbps command rate

**C&DH**
- 1.0, 1.6, and 12.8 kbps tlm rates
- 4 Tape Recorder Units (TRU)
- 4 selectable tlm formats

**Prop**
- Blowdown hydrazine with nitrogen pressurant
- 2 140 kg propellant tanks
- 8 0.5 lbf thrusters - 4 for ΔV, 4 for yaw

Weight: 2250 kg

## Instruments

| Instruments | Power | Mass | PI |
|---|---|---|---|
| ERBE Scanner | 28 W | 29.0 kg | Dr. R. Wy (LaRC) |
| ERBE Non-Scanner | 22 W | 32.0 kg | Dr. R. Wy (LaRC) |
| SAGE-II | 14 W | 33.6 kg | Dr. W. Chu (LaRC) |

## Science

**Descriptions of instrument measurements.**

**Earth Radiation Budget Experiment (ERBE) Scanner/Non-Scanner:**
Two instrument package, also being flown on NOAA-9 & 10, to provide readings of the Earth's solar-absorbed radiation and emitted thermal radiation. Analysis of this data will provide:
- the solar output constant.
- the equator-to-pole energy transport gradient.
- the average monthly radiation budget on regional, zonal and global scales.
- the average diurnal variations, regionally and monthly, in the radiation budget.

**Stratospheric Aerosol and Gas Experiment II (SAGE-II):**
Second in a series, this instrument measures atmospheric attenuation of solar radiation during spacecraft sunrise and sunset. Analysis of this data will be used:
- to map vertical profiles of stratospheric aerosols, $NO_2$, water vapor, and $O_3$.
- to determine seasonal variations and the radiative characteristics of the stratospheric aerosols, $NO_2$, water vapor, and $O_3$.
- to identify sources and sinks of aerosols, $NO_2$, water vapor, and $O_3$ and to observe natural transient phenomena such as volcanic eruptions, tropical upwellings, and dust storms.

## Orbit

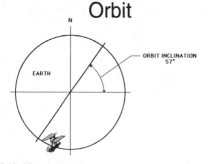

ORBIT INCLINATION 57°

EARTH

**Orbital Parameters:**
- 580 km circular
- 57 inclination
- 96 minute orbit
- 4 full-sun periods/year
- 10 180 yaw maneuvers/year to provide solar array illumination

**Fig. B.5.    Earth Radiation Budget Satellite (ERBS).**

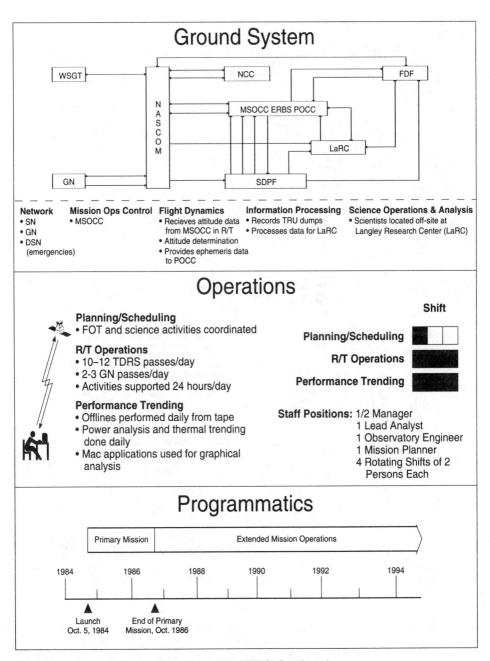

**Fig. B.6. Earth Radiation Budget Satellite (ERBS) Continued.**

# FAST - Fast Auroral Snapshot Explorer

MOM: John Catena (GSFC)                    PROJ MGR:  Orlando Figueroa(GSFC)

| Rev Date: 27 Jan 94 | MISSION STATEMENT | Current Mission Phase: C/D |

Investigate the plasma physics of auroral phenomena at extremely high time and spatial resolutions utilizing fast data sampling, a large burst memory, and triggering on various event types to collect the specific data needed to characterize different events. FAST is the 2nd mission of the Small Explorer program.

Launch Date: August 1994        Launch Vehicle: Pegasus XL        Launch Site: Vandenberg AFB

## Spacecraft

**ACS**
- Orbit norm spinner typically
- Closed loop spin ; open loop precession
- HCI(1), Spinning SS, 3axis mag
- Torquer Coils (2)
- Sun avoidance req for power & thermal reasons

**Power**
- Total Power: 50W orb avg
- Battery: 9 AH super NiCad
- Body mounted GaAs Solar Arrays

**Thermal**
- Primarily passive
- 2 sets of heater strings
- Thermastatically controlled heaters

**RF**
- NASA near Earth Xpndr
- 2kbps uplink
- Directed omni coverage (360°x ±45°about spin axis)

**C&DH**
- 4k & .9, 1.5, 2.25 Mbps downlink
- CCSDS packet standard full compliance
- 1 Gbit SSR (inst provided)

**Prop**
- N/A

Weight:  187 kg

## Orbit

Direct insertion

Perigee = 350 km
Apogee = 4200 km
i = 83°
Arg of perigee = 164°
RAAN = ~137

## Instruments

| Inst. | Power | Mass | Developer |
|-------|-------|------|-----------|
| EFE | 9.3W | 5.0Kg | UC- Berkeley |
| MFE | 2.2W | 3.0Kg | UCLA |
| TEAMS | 4.5W | 7.5Kg | UNH, Lockheed- Palo Alto |
| ESA | 11.9W | 13.2Kg | UC- Berkeley |
| IDPU | 2.8W | 12.7Kg | UC- Berkeley |

INVESTIGATORS:
U-C Berkeley - C. Carlson (PI), C. Catell, F. Mozer, J. McFadden, M. Temerin, R. Ergun
Lockheed - D. Klumpar, W. Peterson, E. Shelley
UCLA - R. Elphic
UNH - E. Mobius

## Science

**Descriptions of instrument measurements.**

**Fields Instruments:**
- Electric Field Experiment (EFE). Electric Field Languir probe instrument comprised of 3 orthogonal boom pairs with associated electronics and sensors which will proivde information on the plasma density and electron temperature.

- Magnetic Field Experiment (MFE). Fluxgate magnetometer will provide 3 axis mag field info for DC to 100 Hz. Search coil magnetometer will provide mag field data over 10 Hz to 250 KHz range.

**Plasma Particles Instruments:**
- Time-of-Flight Energy Angle Mass Spectrometer (TEAMS). This high sensitivity, mass resolving spectrometer will cover the core of all plasma distributions of importance in the auroral region (.01 to 10 keV/charge). TEAMS combines the selection of incoming ions according to their energy per charge with a time-of-flight analysis for incoming electrons above 25keV/e.

- Electrostatic Analyzers (ESA). Quadrispherical ESAs are used for electron and ion measurements to provide 360° field coverage while performing energy sweeps covering 3eV to 30keV (20keV for ions) up to 16 times per second.

**Fig. B.7.     Fast Auroral Snapshot Explorer (FAST).**

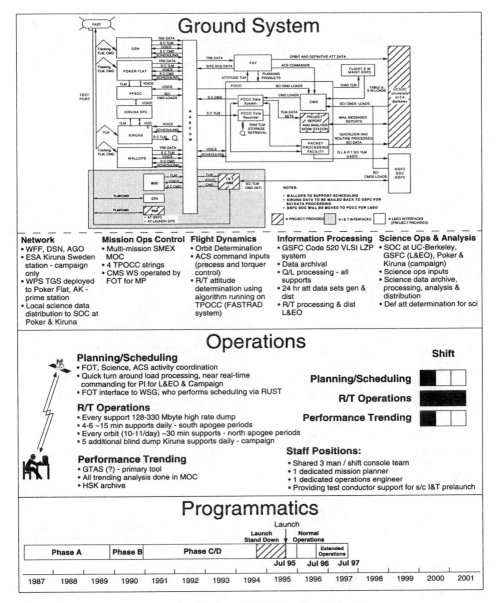

# Ground System

**Network**
- WFF, DSN, AGO
- ESA Kiruna Sweden station - campaign only
- WPS TGS deployed to Poker Flat, AK - prime station
- Local science data distribution to SOC at Poker & Kiruna

**Mission Ops Control**
- Multi-mission SMEX MOC
- 4 TPOCC strings
- CMS WS operated by FOT for MP

**Flight Dynamics**
- Orbit Determination
- ACS command inputs (precess and torquer control)
- R/T attitude determination using algorithm running on TPOCC (FASTRAD system)

**Information Processing**
- GSFC Code 520 VLSI LZP system
- Data archival
- Q/L processing - all supports
- 24 hr att data sets gen & dist
- R/T processing & dist L&EO

**Science Ops & Analysis**
- SOC at UC-Berkeley, GSFC (L&EO), Poker & Kiruna (campaign)
- Science ops inputs
- Science data archive, processing, analysis & distribution
- Def att determination for sci

# Operations

**Planning/Scheduling**
- FOT, Science, ACS activity coordination
- Quick turn around load processing, near real-time commanding for PI for L&EO & Campaign
- FOT interface to WSG, who performs scheduling via RUST

**R/T Operations**
- Every support 128-330 Mbyte high rate dump
- 4-6 ~15 min supports daily - south apogee periods
- Every orbit (10-11/day) ~30 min supports - north apogee periods
- 5 additional blind dump Kiruna supports daily - campaign

**Performance Trending**
- GTAS (?) - primary tool
- All trending analysis done in MOC
- HSK archive

**Shift**

Planning/Scheduling

R/T Operations

Performance Trending

**Staff Positions:**
- Shared 3 man / shift console team
- 1 dedicated mission planner
- 1 dedicated operations engineer
- Providing test conductor support for s/c I&T prelaunch

# Programmatics

| Phase A | Phase B | Phase C/D | Launch Stand Down | Launch / Normal Operations | Extended Operations | |
|---|---|---|---|---|---|---|
| 1987 1988 1989 | 1990 | 1991 1992 1993 | 1994 | Jul 95 | Jul 96 Jul 97 | 1998 1999 2000 2001 |

**Fig. B.8.    Fast Auroral Snapshot Explorer (FAST) Continued.**

# FUSE - Far Ultraviolet Spectroscopic Explorer

MOM: Bob Nelson (GSFC)                                      PROJ MGR:Frank Volpe   (GSFC)

| Rev Date: 22 Feb 94 | MISSION STATEMENT | Current Mission Phase: B |
|---|---|---|

- Study distribution and abundance of trace species in interstellar and intergalactic gas, using absorbtion spectroscopy of faint distant sources such as active galactic nuclei and quasars
- Study physical processes in interstellar material, stellar explosions and mass loss, active galactic nuclei, and planetary magnetospheres
- Study the mechanisms by which stars and planetary systems form and evolve

| Launch Date:  October 2000 | Launch Vehicle: Delta II 7925 | Launch Site:CCAFS |
|---|---|---|

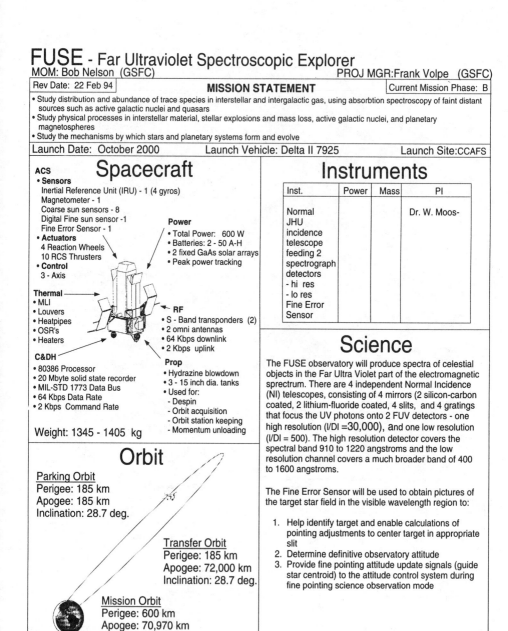

## Spacecraft

**ACS**
- **Sensors**
  Inertial Reference Unit (IRU) - 1 (4 gyros)
  Magnetometer - 1
  Coarse sun sensors - 8
  Digital Fine sun sensor -1
  Fine Error Sensor - 1
- **Actuators**
  4 Reaction Wheels
  10 RCS Thrusters
- **Control**
  3 - Axis

**Power**
- Total Power:  600 W
- Batteries: 2 - 50 A-H
- 2 fixed GaAs solar arrays
- Peak power tracking

**Thermal**
- MLI
- Louvers
- Heatpipes
- OSR's
- Heaters

**RF**
- S - Band transponders  (2)
- 2 omni antennas
- 64 Kbps downlink
- 2 Kbps  uplink

**C&DH**
- 80386 Processor
- 20 Mbyte solid state recorder
- MIL-STD 1773 Data Bus
- 64 Kbps Data Rate
- 2 Kbps  Command Rate

**Prop**
- Hydrazine blowdown
- 3 - 15 inch dia. tanks
- Used for:
  - Despin
  - Orbit acquisition
  - Orbit station keeping
  - Momentum unloading

Weight: 1345 - 1405  kg

## Orbit

Parking Orbit
Perigee: 185 km
Apogee: 185 km
Inclination: 28.7 deg.

Transfer Orbit
Perigee: 185 km
Apogee: 72,000 km
Inclination: 28.7 deg.

Mission Orbit
Perigee: 600 km
Apogee: 70,970 km
Inclination: 28.7 deg.
Period: 23 hr, 56 min.
Perigee Longitude: 145 deg W

## Instruments

| Inst. | Power | Mass | PI |
|---|---|---|---|
| Normal JHU incidence telescope feeding 2 spectrograph detectors - hi  res - lo res Fine Error Sensor |  |  | Dr. W. Moos- |

## Science

The FUSE observatory will produce spectra of celestial objects in the Far Ultra Violet part of the electromagnetic sprectrum. There are 4 independent Normal Incidence (NI) telescopes, consisting of 4 mirrors (2 silicon-carbon coated, 2 lithium-fluoride coated, 4 slits,  and 4 gratings that focus the UV photons onto 2 FUV detectors - one high resolution (l/DI $=30{,}000$) and one low resolution (l/DI = 500). The high resolution detector covers the spectral band 910 to 1220 angstroms and the low resolution channel covers a much broader band of 400 to 1600 angstroms.

The Fine Error Sensor will be used to obtain pictures of the target star field in the visible wavelength region to:

1. Help identify target and enable calculations of pointing adjustments to center target in appropriate slit
2. Determine definitive observatory attitude
3. Provide fine pointing attitude update signals (guide star centroid) to the attitude control system during fine pointing science observation mode

**Fig. B.9.     Far Ultraviolet Spectroscopic Explorer (FUSE).**

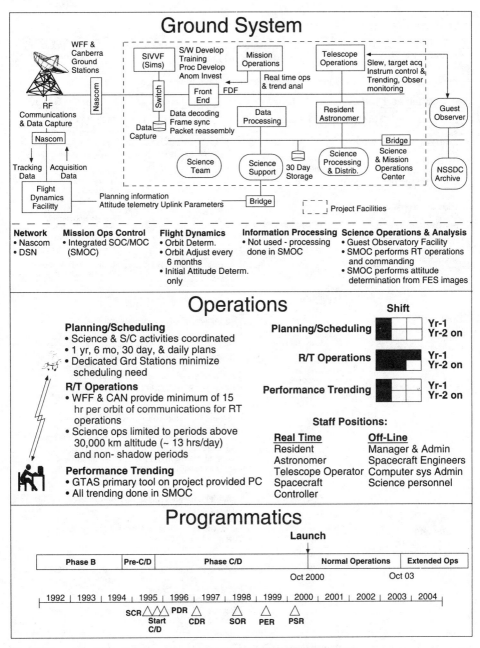

# Ground System

WFF & Canberra Ground Stations

SIVVF (Sims)

S/W Develop Training Proc Develop Anom Invest

Mission Operations

Telescope Operations — Slew, target acq Instrum control & Trending, Obser monitoring

Real time ops & trend anal

Nascom

Switch

Front End — FDF

RF Communications & Data Capture

Nascom

Data Capture

Data decoding Frame sync Packet reassembly

Data Processing

Resident Astronomer

Guest Observer

Tracking Data | Acquisition Data

Flight Dynamics Facilitty

Planning information Attitude telemetry Uplink Parameters

Science Team

Science Support

30 Day Storage

Science Processing & Distrib.

Science & Mission Operations Center

Bridge

NSSDC Archive

Bridge

Project Facilities

| Network | Mission Ops Control | Flight Dynamics | Information Processing | Science Operations & Analysis |
|---|---|---|---|---|
| • Nascom<br>• DSN | • Integrated SOC/MOC (SMOC) | • Orbit Determ.<br>• Orbit Adjust every 6 months<br>• Initial Attitude Determ. only | • Not used - processing done in SMOC | • Guest Observatory Facility<br>• SMOC performs RT operations and commanding<br>• SMOC performs attitude determination from FES images |

# Operations

## Shift

**Planning/Scheduling**
• Science & S/C activities coordinated
• 1 yr, 6 mo, 30 day, & daily plans
• Dedicated Grd Stations minimize scheduling need

**R/T Operations**
• WFF & CAN provide minimum of 15 hr per orbit of communications for RT operations
• Science ops limited to periods above 30,000 km altitude (~ 13 hrs/day) and non- shadow periods

**Performance Trending**
• GTAS primary tool on project provided PC
• All trending done in SMOC

**Planning/Scheduling** — Yr-1 / Yr-2 on

**R/T Operations** — Yr-1 / Yr-2 on

**Performance Trending** — Yr-1 / Yr-2 on

### Staff Positions:

**Real Time**
Resident Astronomer
Telescope Operator
Spacecraft Controller

**Off-Line**
Manager & Admin
Spacecraft Engineers
Computer sys Admin
Science personnel

# Programmatics

Launch

| Phase B | Pre-C/D | Phase C/D | Normal Operations | Extended Ops |
|---|---|---|---|---|

Oct 2000    Oct 03

| 1992 | 1993 | 1994 | 1995 | 1996 | 1997 | 1998 | 1999 | 2000 | 2001 | 2002 | 2003 | 2004 |

SCR △△△ PDR △    △    △    △
Start CDR SOR PER PSR
C/D

**Fig. B.10.** Far Ultraviolet Spectroscopic Explorer (FUSE) Continued.

# GRO - Gamma Ray Observatory

POD: Robert E. Wilson (GSFC)      PROJ MGR: Paul Pashby (GSFC)

| Rev Date: 7 Jan 94 | **MISSION STATEMENT** | Current Mission Phase: |
|---|---|---|

Perform extensive study of Gamma Ray sources throughout the universe.

Launch Date: April 5, 1991     Launch Vehicle: STS-37 Atlantis     Launch Site: KSC

## Spacecraft

**ACS**
- **Sensors**
  Fixed Head Star Trackers - 2
  Sun Sensors - 6
  Inertial Reference Unit - 1
  (contains 3 gyros)
  3 Axis Magnetometer - 3
- **Actuators**
  RWA - 4
  Magnetic Torquers - 3
  SA Drive Assembly - 1
  High Gain Antenna Drive - 1

**Power**
- Total Power: 4304 W
- Batteries: 6 50 AH Ni-Cd
- Positionable Solar Arrays

**Thermal**
- Instruments thermally isolated from each other and the spacecraft
- Heaters
- Passive hardware

**RF**
- High gain antenna - 1
  1 Kbps cmd rate
- Low gain antenna - 2
  0.125 kbps cmd rate

**C&DH**
- 1, 32 and 512 kbps tlm rates
- Selectable tlm rates

**Prop**
- Propellant tanks -4
- Orbit Adjust Thruster - 4
  (100 lbs thrust each)
- Attitude Control Thrusters - 8
  (5 lbs thrust each)
- Propellant distribution modules - 2

Weight: 15,909 kg

## Instruments

| Inst. | Power | Mass | PI |
|---|---|---|---|
| BATSE | 122 W | 766 kg | Dr. G. J. Fishman (MSFC) |
| COMPTEL | 216 W | 1451 kg | Dr. V. Schoenfelder (Max-Planck Institue, Germany) |
| EGRET | 180 W | 1779 kg | Dr. C. E. Fichtel (GSFC) Prof. R. Hofstadter (Stanford U. , CA) Dr. K. Pinkau (Max-Planck Institute, Germany) |
| OSSE | 161 W | 1802 kg | Dr. J. D. Kurfess (Naval Research Laboratory, DC) |

## Science

**Descriptions of instrument measurements.**

**Burst and Transient Source Experiment (BATSE) -**
Continuously monitor a large segment of the sky for the detection and measurement of bursts and other transient sources of Gamma Rays. It is able to measure time variations in such events to a fraction of a millisecond, allowing for a detailed analysis of emission mechanisms.

**Imaging Compton Telescope (COMPTEL) -**
It has two detector arrays, one above the other, to aid in determining the direction of arrival as well as the energy of Gamma Ray photons from 1 to 30 million electron volts (MeV) of energy. It has a wide (1 steradian) field of view and the capability of rejecting background events.

**Energetic Gamma Ray Experiment Telescope (EGRET)**
It measures the highest energy Gamma Rays, up to 30 billion eV. It can measure the direction of a point source to a fraction of a degree and determine the spectrum of Gamma Ray emissions.

**Oriented Scintillation Spectrometer Experiment (OSSE)**
It consists of four identical instruments, each mounted on a gimbal allowing it to rotate through 180°. One of a pair can measure the background while the other is observing a Gamma Ray source. Detectors are optimized for detecting Gamma Rays of up to 10 MeV.

## Orbit

Orbit
Parameters:
450 km circular
28.5° inclination
93 minute orbit

**Fig. B.11.    Gamma Ray Observatory (GRO).**

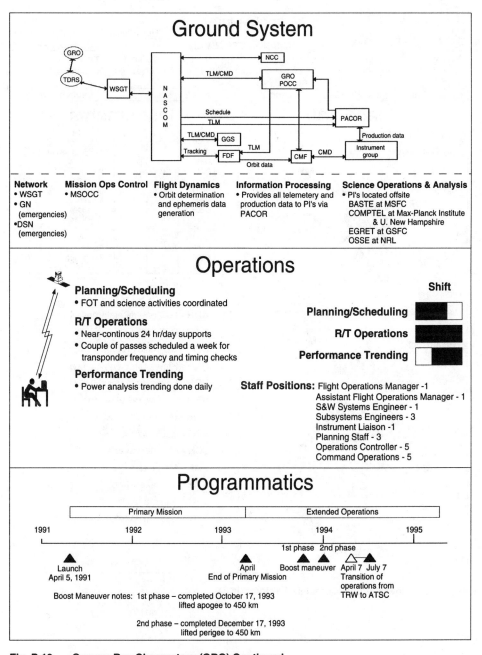

# Ground System

**Network**
- WSGT
- GN
  (emergencies)
- DSN
  (emergencies)

**Mission Ops Control**
- MSOCC

**Flight Dynamics**
- Orbit determination and ephemeris data generation

**Information Processing**
- Provides all telemetry and production data to PI's via PACOR

**Science Operations & Analysis**
- PI's located offsite
  BASTE at MSFC
  COMPTEL at Max-Planck Institute
  & U. New Hampshire
  EGRET at GSFC
  OSSE at NRL

# Operations

**Planning/Scheduling**
- FOT and science activities coordinated

**R/T Operations**
- Near-continous 24 hr/day supports
- Couple of passes scheduled a week for transponder frequency and timing checks

**Performance Trending**
- Power analysis trending done daily

**Shift**

Planning/Scheduling

R/T Operations

Performance Trending

**Staff Positions:** Flight Operations Manager -1
Assistant Flight Operations Manager - 1
S&W Systems Engineer - 1
Subsystems Engineers - 3
Instrument Liaison -1
Planning Staff - 3
Operations Controller - 5
Command Operations - 5

# Programmatics

| Primary Mission | Extended Operations |

1991    1992    1993    1994    1995

1st phase   2nd phase

Launch
April 5, 1991

April
End of Primary Mission

Boost maneuver

April 7   July 7
Transition of
operations from
TRW to ATSC

Boost Maneuver notes: 1st phase – completed October 17, 1993
lifted apogee to 450 km

2nd phase – completed December 17, 1993
lifted perigee to 450 km

**Fig. B.12. Gamma Ray Observatory (GRO) Continued.**

# ICE - International Cometary Explorer

POD:  Gilbert Bullock (GSFC)          PROJ MGR:  Paul Pashby (GSFC)

| Rev Date: 22 Feb 94 | MISSION STATEMENT | Current Mission Phase: |

Original - Study Earth's magnetosphere and changing conditions in the solar wind.
Extended - Study geomagnetic tail, comet Giacobini-Zinner, coronal mass ejection coverage
          - Observations of the solar wind through 2104.

Launch Date:  August 12, 1978      Launch Vehicle:  Delta          Launch Site:  KSC

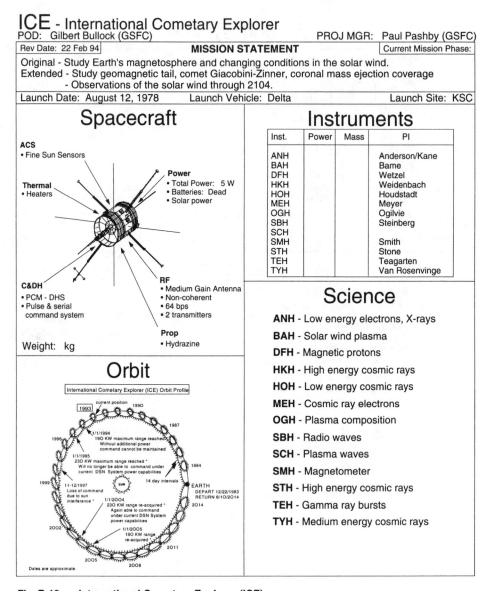

## Spacecraft

**ACS**
• Fine Sun Sensors

**Thermal**
• Heaters

**Power**
• Total Power:  5 W
• Batteries:  Dead
• Solar power

**C&DH**
• PCM - DHS
• Pulse & serial
  command system

**RF**
• Medium Gain Antenna
• Non-coherent
• 64 bps
• 2 transmitters

**Prop**
• Hydrazine

Weight:   kg

## Instruments

| Inst. | Power | Mass | PI |
|-------|-------|------|-----|
| ANH | | | Anderson/Kane |
| BAH | | | Bame |
| DFH | | | Wetzel |
| HKH | | | Weidenbach |
| HOH | | | Houdstadt |
| MEH | | | Meyer |
| OGH | | | Ogilvie |
| SBH | | | Steinberg |
| SCH | | | |
| SMH | | | Smith |
| STH | | | Stone |
| TEH | | | Teagarten |
| TYH | | | Van Rosenvinge |

## Science

**ANH** - Low energy electrons, X-rays

**BAH** - Solar wind plasma

**DFH** - Magnetic protons

**HKH** - High energy cosmic rays

**HOH** - Low energy cosmic rays

**MEH** - Cosmic ray electrons

**OGH** - Plasma composition

**SBH** - Radio waves

**SCH** - Plasma waves

**SMH** - Magnetometer

**STH** - High energy cosmic rays

**TEH** - Gamma ray bursts

**TYH** - Medium energy cosmic rays

## Orbit

International Cometary Explorer (ICE) Orbit Profile

current position
1993          1990
                              1987
1/1/1994
190 KW maximum range reached
Without additional power
command cannot be maintained
1/1/1995
230 KW maximum range reached
Will no longer be able to command under
current DSN System power capabilities
                              1984
1996
                        14 day intervals
1999                                    EARTH
11-12/1997                              DEPART 12/22/1983
Loss of command      sun               RETURN 8/10/2014
due to sun
interference          1/1/2004
                      230 KW range re-acquired
                      Again able to command       2014
                      under current DSN System
                      power capabilities
2002
              1/1/2005
              190 KW range
              re-acquired
                                    2011
        2005          2008
Dates are approximate.

**Fig. B.13.    International Cometary Explorer (ICE).**

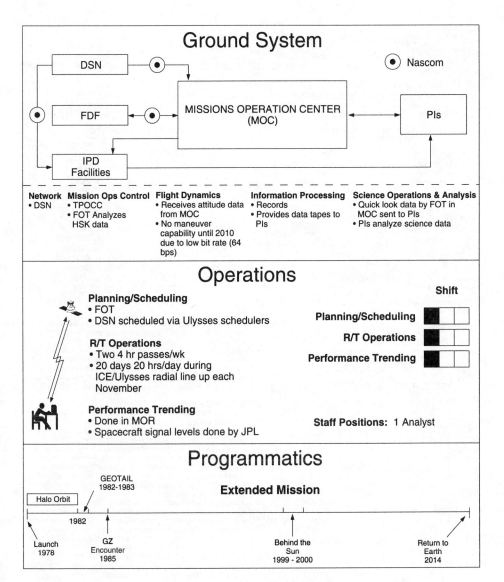

Fig. B.14.    International Cometary Explorer (ICE) Continued.

# IMP - Interplanetary Monitoring Platform

POD:  (GSFC)                                                      PROJ MGR:  (GSFC)

| Rev Date:  22 Feb 94 | **MISSION STATEMENT** | Current Mission Phase: |

Study solar and galactic cosmic radiation, solar plasma, solar wind, energetic particles, electromagnetic field variations, and interplanetary magnetic fields.

Launch Date: October 26, 1973     Launch Vehicle: Delta                    Launch Site: KSC

## Spacecraft

**ACS**
- Sensors
  Sun/Earth
- Actuators
  None
- Control
  None

**Thermal**

**C&DH**
- Low rate tlm (3.2 kbps)
  + 800 bits

**Weight:  258.7 kg**

**Power**
- Total Power:  W
- Batteries: Dead
- Solar Array

**RF**
- 2 Diplexers
- Hybrid circulator
- 8 Element monopole turnstile array (4 active, 4 passive)

**Prop**
- Cold gas
- Monopropellant (Freon-14)
- Empty

## Instruments

| Inst. | Power | Mass | PI |
|-------|-------|------|-----|
| GNF |  |  | Dr. N. F. Ness |
| GAF |  |  | Dr. T. L. Aggson |
| IOF |  |  | Dr. D. A. Gurnett |
| GME |  |  | Dr. F. B. McDonald |
| CHE |  |  | Dr. J. A. Simpson |
| GWP |  |  | Dr. D. J. Williams |
| APP |  |  | Dr. S. M. Krimigis |
| CAI |  |  | Dr. E. C. Stone |
| MAE |  |  | Dr. G. Gloeckler |
| IOE |  |  | Dr. L. A. Frank |
| LAP |  |  | Dr. S. J. Bame |
| MAP |  |  | Dr. A. Lazarus |

## Orbit

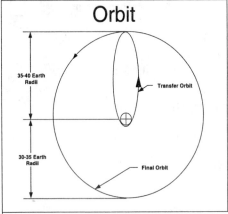

35-40 Earth Radii          Transfer Orbit

30-35 Earth Radii          Final Orbit

## Science

**Fields**
- **Magnetic Fields Experiment (GNF)** - Measure the vector magnetic field in three dynamic ranges by using three mutually orthogonal fluxgate sensors.
- **DC Electric Fields (GAF)** - Make vector measurements of the DC electric field with a sensitivity of about 0.1 mV/meter.
- **AC Electric and Magnetic Fields (IOF)** - Measure spatial and temporal characteristics of both electric and magnetic AC vector fields and their polar relationships along the IMP-J orbit.

**Energetic Particles**
- **Cosmic Ray Experiment (GME)** - Study solar and galactic electrons and nuclei throughout the solar cycle.
- **Cosmic Ray Experiment (CHE)** - Study the solar flare particle acceleration and particle containment in the vicinity of the sun.
- **Energetic Particles Experiment (GWP)** - Study: propagation characteristics of solar cosmic rays through the interplanetary medium over selected energy ranges, electron and proton patches throughout the geomagnetic tail and near and through the flanks of the magnetopause, and entry of solar cosmic rays into the geomagnetic field.
- **Charged Particles Experiment (APP)** - Measure protons, alpha particles, $Z \geq 3$ nuclei, and X-rays in a wide energy level.
- **Electron Isotopes Experiment (CAI)** - Study local acceleration of particles, solar particle acceleration processes and storage in the interplanetary medium, and interstellar propagation and solar modulation of particles in the interplanetary medium.
- **Ion and Electron Experiment (MAE)** - Determine the composition and energy spectra of low energy particles observed during solar flares and 27-day solar events.

**Plasma**
- **Low Energy Particles Experiment (IOE)** - Study the differential energy spectra of low energy electrons and protons measured over the geocentric radial distance of 40-earth radii.
- **Los Alamos Plasma Experiment (LAP)** - Make a comprehensive study of electrons and positive ions in the regions of space traversed by the spacecraft and to coordinate them with magnetometer and other scientific data.
- **Plasma Experiment (MAP)** - Measure the properties of the plasma in the interplanetary region, transition region, and in the tail of the magnetosphere.

**Fig. B.15.     Interplanetary Monitoring Platform (IMP).**

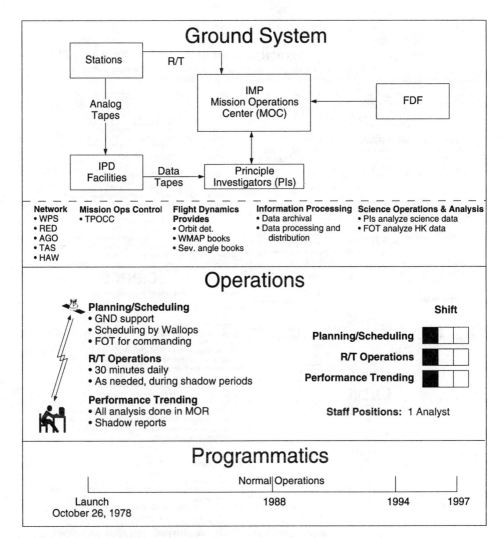

# Ground System

| Stations | | |

Stations —R/T→ IMP Mission Operations Center (MOC) ←— FDF

Analog Tapes

IPD Facilities —Data Tapes→ Principle Investigators (PIs)

| **Network** | **Mission Ops Control** | **Flight Dynamics Provides** | **Information Processing** | **Science Operations & Analysis** |
|---|---|---|---|---|
| • WPS | • TPOCC | • Orbit det. | • Data archival | • PIs analyze science data |
| • RED | | • WMAP books | • Data processing and | • FOT analyze HK data |
| • AGO | | • Sev. angle books | distribution | |
| • TAS | | | | |
| • HAW | | | | |

# Operations

**Planning/Scheduling**
• GND support
• Scheduling by Wallops
• FOT for commanding

**R/T Operations**
• 30 minutes daily
• As needed, during shadow periods

**Performance Trending**
• All analysis done in MOR
• Shadow reports

**Shift**

Planning/Scheduling
R/T Operations
Performance Trending

**Staff Positions:** 1 Analyst

# Programmatics

Normal Operations

Launch
October 26, 1978          1988          1994          1997

**Fig. B.16.    Interplanetary Monitoring Platform (IMP) Continued.**

# IUE - International Ultraviolet Explorer

POD: Gil Bullock (GSFC)       PROJ MGR: (GSFC)

| Rev Date: 22 Feb 94 | MISSION STATEMENT | Current Mission Phase: OPS |
|---|---|---|

To serve as a general facility for observing the ultraviolet spectra of astronomical sources, leading to a better understanding of the elements, the temperatures, and the pressures within stars, and a better understanding of interstellar matter in general.

| Launch Date: January 26, 1978 | Launch Vehicle: DELTA 2914 | Launch Site: KSC |
|---|---|---|

## Spacecraft

**ACS**
- **Sensors**
  - Digital Fine Sun Sensor (FSS)-2
  - Analog Coarse Sun Sensor (CSS)
  - Fine Error Sensor (FES)-2
  - Inertial Reference Assembly (IRA)
  - Panoramic Attitude Sensors (PAS)-2
- **Actuators**
  - Reaction Wheels - 4
  - Low Thrust Engine (LTE) - 8
  - High Thrust Engine (HTE) - 4
- **Control**
  - On -Board Computer (OBC)
  - Control Electronics
    Assembly (CEA)

**Thermal**
- HAPS Heaters
- Scintific Instrument Heaters
- Passive Hardware

**C&DH**
- Data Multiplexer Unit (DMU)
- 1.25, 5, 10, 20 Kbps Telemetry Rate
- Selectable Telemetry Formats
  - Fixed ROM Formats
  - Variable Programmable Formats

**Weight: 700kg at Launch**

**Power**
- Total Power: 424W
  @ß=67.5 (at Launch)
- Batteries: 2 - 6Ah
- Sun must be in the
  (X, +Z) 1/2 Plane

**RF**
- VHF Transpnder -2
- Turnstile VHF
  Antenna System
- Sband Transmitter -2
- Sband Power Amplifier
  and Antenna -4

**Prop:**
**Hydrazine**
**Auxiliary**
**Propulsion System**
**(HAPS)**
- Monopropellant Catalytic
  Hydrazine Blowdown
  System
- Nitrogen pressurant
- 0.2 lb LTE -8
- 5 lb HTE -4

## Instruments

| Inst. | Power | Mass | PI |
|---|---|---|---|
| LWP | 20.6w | 15.525kg | IUE is operated with a guest observer (GO) program. Each GO is the principle investigator for their project. |
| LWR | 20.6w | 15.525kg | |
| SWP | 20.6w | 15.525kg | |
| SWR | 20.6w | 15.525kg | |
| FES1 | 4.3w | 4.3kg | |
| FES2 | 4.3w | 4.3kg | |

LWP - Long Wavelength Prime camera
LWR - Long Wavelength Redundant camera
SWP - Short Wavelength Prime camera
SWR - Short Wavelength Redundant camera
FES - Fine Error Sensor, #1 or #2

## Science

**Descriptions of instrument measurements.**

**Ultraviolet Spectroscopy**

**Mission Objectives:**
- To obtain high resolution (~0.1Å) spectra in the ultraviolet region of the spectrum from 1150Å to 3200Å of stars and planets brighter than 7th visual magnitude, for detailed analysis of stellar and planetary atmospheres in order to determine more precisely their physical characteristics.

- To obtain lower resolution (~6Å) spectra over the same wavelength range for both stellar and extended objects as faint as 12th magnitude or fainter (15th magnitude desired) as a function of observing time for investigation of peculiar objects such as quasars, Seyfert galaxies, pulsars, X-ray sources, and variability phenomena to shed light on questions of cosmological significances.

**Spectrographs**

(1) Type: echelle
(2) Detector: Proximity focused converter and SEC vidicon camera
(3) Entrance Aperatures: 3 arcsec circular
                                     10x20 arcsec elliptical

| (4) High Dispersion | Short l | Long l |
|---|---|---|
| (a) Wavelength Range | 1191-1925Å | 1893-3032Å |
| (b) Resolving Power | 0.1Å | 0.1Å |
| (c) Limiting Magnitude* | 7 | 7 |
| (5) Low Dispersion | | |
| (a) Wavelength Range | 1141-2084Å | 1750-3032Å |
| (b) Resolving Power | 6Å | 6Å |
| (c) Limiting Magnitude* | 12 | 12 |

* Estimated for 30 minute exposure on BO V star.

## Orbit
( Current )

Greenwich
Meridian
At Launch

Elliptical
Geosynchronous
Orbit

Transfer
Orbit

Apogee
Motor
Burn

Parking
Orbit

Third
Stage
Ejection

Perigee = 30268.55 km
Apogee = 41315.53 km
Velocity at:
  Perigee = 3.5074 km/s
  Apogee = 2.6950 km/s
  Period = 1436.3748 min

KSC 80.6°
West Meridian
At Launch

Sun
January, 1978
At Launch

| a = 42170.17 km | e = 0.130981 | i = 34.4062° |
|---|---|---|
| Ω = 97.3677° | w = 31.3987° | M = 296.1916° |

**Fig. B.17.    International Ultraviolet Explorer (IUE).**

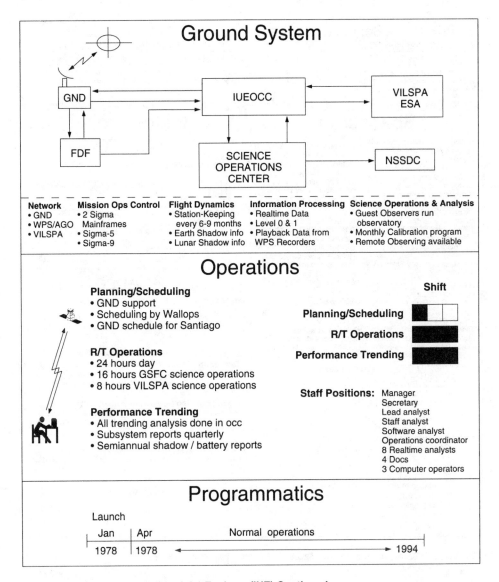

Fig. B.18.    International Ultraviolet Explorer (IUE) Continued.

# SAMPEX - Solar, Atmospheric, Magnetic Particle Explorer

POD: Jim Williamson / Code 513 (GSFC)        PROJ MGR: Paul Pashby / Code 602 (GSFC)

| Rev Date: 25 May 94 | **MISSION STATEMENT** | Current Mission Phase: On-orbit |
|---|---|---|

The primary scientific objectives of SAMPEX are to measure the elemental and isotopic composition of solar energetic particles, anomalous cosmic rays, and galactic cosmic rays over the energy range from -1 to several hundred MeV per nucleon. SAMPEX is the 1st mission of the Small Explorer program.

| Launch Date: July 3, 1992 | Launch Vehicle: Scout ELV | Launch Site: Vandenberg AFB |
|---|---|---|

## Spacecraft

**ACS**
- Stabilized with bias momentum using single momentum wheel aligned with sun-line.
- S/A maintained within 5 deg. of sun line during normal science modes.
- Minimize orbital debris exposure to instruments.
- Instrument zenith pointing over poles.
- 1 Fine Digital SS, 5 Coarse SS, 1 Magnetometer.
- 1 Momentum wheel, 3 Mag. Torque Rods.
- Independent Analog Safehold Control

**Thermal**
- Passive in nature
- White and black paint coatings
- Silver Teflon and Kapton MLI
- Thermistors
- Thermostatically controlled heaters, survival and operational
  - Each set powered by its own bus, switchable off the essential bus.

Weight: 350 lbs.

**Power**
- Total power req.: 65 Watts
- Total power gen.:100 W
- Battery : Single 9 Amp-Hour Super NiCd battery.
- Four deployed non-articulating S/A panels - 4 panels of 8 strings (32 strings total).

**RF**
- NASA S-band Xpndr
- Coherent 2039/2215 MHz (Uplink/Downlink)
- Diplexer allowing the 2 identical, quadrifilar helical antennaes to both transmit and receive.
- Antenna gain : -5.0 dBi

**C&DH- SEDS**
- CTT/RPP/1773 Software Bus
- 26.5Mb Solid State Recorder
- 2 kbps uplink
- 900, 16 or 4 kbps downlink
- CCSDS packet standard

**Propulsion**
- None

## Instruments

| Inst. | Power | Mass | PI |
|---|---|---|---|
| DPU | 3.4W | | DAN MABRY |
| MAST/PET | 3.4W | 8.8kg | JAY CUMMINGS |
| LEICA | 4.6W | 7.4kg | GLENN MASON |
| HILT | 5.2W | 22.8kg | BERNDT KLECKER |
| Total | 16.6W | 39.0kg | |

## Science

**Descriptions of instrument measurements.**

- **Data Processing Unit (DPU)** - Provides an interface between the instruments and Small Explorer Data System (SEDS).

- **Mass Sprectrometer Telescope/Proton Electron Telescope (MAST/PET)** - Measures isotopic composition of elements from Li (Z=3) to Ni (Z=28) ranging from 10 Mev to several hundred Mev/nucleon.

- **Low Energy Ion Composition Analyzer (LEICA)** - Measures ion fluxes over the charge range from He through Ni from about 0.35 to 10 MeV/nucleon and 0.8 MeV above the mass of Ni.

- **Heavy Ion Large Telescope (HILT)** - Measures heavy ions from He to Fe in the energy range from 8 to 310 MeV/nucleon.

## Orbit

- Semi-major axis : 7008 km
- Eccentricity : 0.0099
- Inclination : 82 deg.
- Precession : 1 deg./day (with respect to vernal equinox)
  2 deg./day (with respect to sun line)
- Rotation of Argument of Perigee : -3.25 deg./day
- Orbit Period : 97 min. 19 sec.
- Perigee : 6938 km
- Apogee : 7078 km

**Fig. B.19.    Solar, Atmospheric, Magnetic Particle Explorer (SAMPEX).**

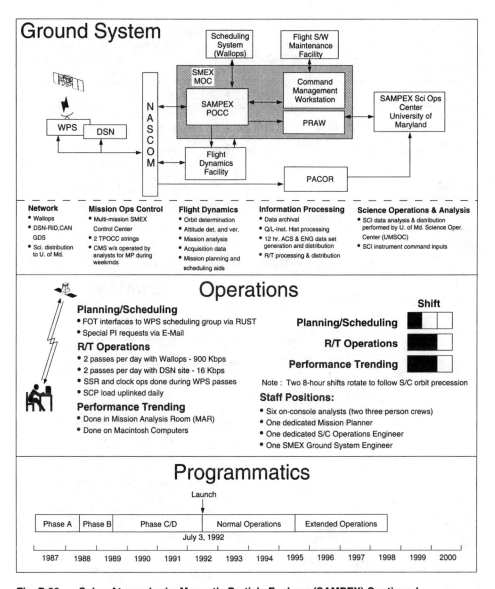

**Fig. B.20.    Solar, Atmospheric, Magnetic Particle Explorer (SAMPEX) Continued.**

# SOHO - Solar and Heliospheric Observatory

MOM: Dan Muhonen (GSFC)                    PROJ MGR: Ken Sizemore (GSFC)

| Rev Date: 22 Feb 93 | MISSION STATEMENT | Current Mission Phase: C/D |
|---|---|---|

Study solar wind, solar seismology and coronal dynamics, with emphasis on probing the interior structure of the Sun, characterizing strong and weak magnetic field regions in the chromosphere and corona, and investigating the outflow of plasma and the solar wind origin.

| Launch Date: July 1995 | Launch Vehicle: Atlas II AS | Launch Site: ESMC |
|---|---|---|

## Spacecraft

**AOCS**
- **Sensors**
  Fine Pointing Sun Sensor (FPSS) - 2
  Sun Acquisition Sensor (SAS) - 3
  Star Sensor Units (SSU) - 2
  Inertial Reference Unit (IRU) - 2
  Attitude Anomaly Detector - 1
- **Actuators**
  Reaction Wheel Unit (RWU) -
  electronics and 4 reaction wheels
  Control Actuation Electronics - 1

**Power**
- Total Power: 2240 W
- Batteries: 2 - 20 Ah
- +X-side always in full sun

**Thermal**
- Mode control by S/W
- Heaters
- Passive hardware

**RF**
- High Gain Antenna
- 2 Low Gain Antennas
- 2 kbps cmd rate
- Coherent

**DHSS**
- Low, medium, high rate tlm (1, 54, 246 kbps)
- Two magnetic tape recorders
- Selectable tlm formats

**Prop**
- Blowdown hydrazine
- Helium pressurant
- 8 (redundant) 4N thrusters

Weight: 1875 kg

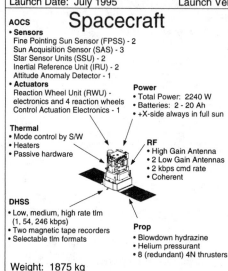

## Instruments

| Inst. | Power | Mass | PI |
|---|---|---|---|
| MDI | 80.0W | 60.6 kg | Dr. P. Scherrer |
| GOLF | 53.5W | 27.35 kg | Dr. A. Gabriel |
| VIRGO | 29.6W | 14.6 kg | Dr. C. Frohlich |
| CDS | 70.0W | 95 kg | Dr. R. Harrison |
| SUMER | 80.2W | 95 kg | Dr. K. Wilhelm |
| UVCS | 60.0W | 117.3 kg | Dr. J. Kohl |
| EIT | 27.8W | 14 kg | Dr. J.P. Delaboudiniere |
| LASCO | 99.2W | 65.4 kg | Dr. G. Brueckner |
| SWAN | 16.3W | 13.25 kg | Dr. J.L. Bertaux |
| CEPAC | 27.3W | 20.5 kg | Dr. D. Hovestadt |
| CELIAS | 25.0W | 29.2 kg | Dr. J. Torsti |

## Science

**Helioseismology**
- **Michelson Doppler Imager (MDI)** - Study radial stratification and latitudinal variation of pressure, density, compostion, sound speed, and the internal rotation of the sun
- **Global Oscillations at Low Frequencies (GOLF)** - Study the solar eigenmodes of low degrees and the measurement of the global magnetic field
- **Variability of Solar Irradiance and Gravity Oscillations (VIRGO)** - Study the variability of solar irradiance and gravity oscillations

**Solar Atmospheric Remote Sensing**
- **Coronal Diagnostics Spectrometer (CDS)** - Study the heating of the corona and the acceleration of the solar wind
- **Solar Ultraviolet Measurements of Emitted Radiation (SUMER)** - Study plasma density, temperature and flow in the upper chromosphere, the transition zone, and lower corona
- **Ultraviolet Coronagraph Spectrometer (UVCS)** - Study mechanisms for accelerating the solar wind, for heating the coronal plasma and establish the plasma properties of the solar wind
- **Extreme-Ultraviolet Imaging Telescope d'Astrophysique (EIT)** - Study space-time evolution of coronal structures and coronal heating and solar wind acceleration
- **Large Angle Spectrometric Coronagraph (LASCO)** - Study coronal heating, solar wind acceleration and coronal transients
- **Solar Wind Anisotropies d'Aeronomie (SWAN)** - Study anisotropies of solar wind and temporal variations of latitude distribution

**"In Situ" Solar Wind**
- **Costep-Erne Particle Analysis Collaboration (CEPAC)** - Study physical conditions of the sun, processes responsible for emission and affecting emitted particles, and the recording of interplanetary and galactic cosmic rays
- **Charge, Element and Isotope Analysis System (CELIAS)** - Study the elemental and isotopic abundances and the ionic charge state and velocity distribution of ions originating in the solar atmosphere

## Orbit

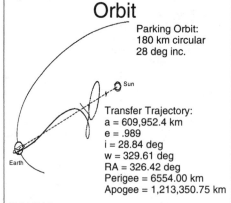

Parking Orbit:
180 km circular
28 deg inc.

Transfer Trajectory:
a = 609,952.4 km
e = .989
i = 28.84 deg
w = 329.61 deg
RA = 326.42 deg
Perigee = 6554.00 km
Apogee = 1,213,350.75 km

Halo Orbit:
$A_x$ (in ecliptic sun direction): 206,448 km
$A_y$ (in ecliptic perpendicular to $A_x$): 666,672 km
$A_z$ (normal to the ecliptic): 120,000 km

**Fig. B.21.    Solar and Heliospheric Observatory (SOHO).**

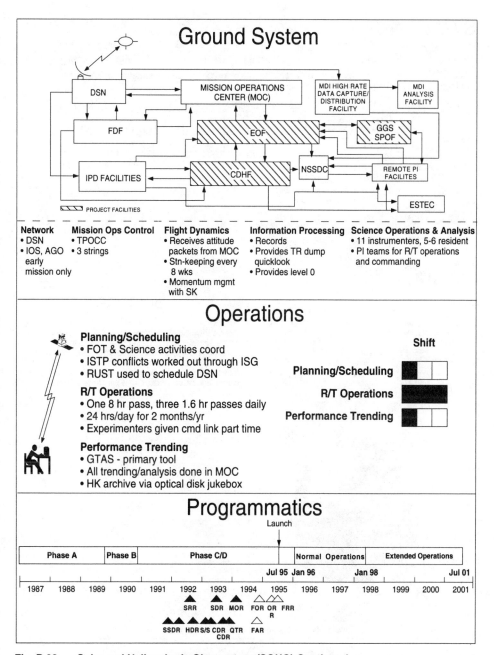

**Fig. B.22.  Solar and Heliospheric Observatory (SOHO) Continued.**

# SWAS - Submillimeter Wave Astronomy Satellite

MOM: John Catena (GSFC)                    PROJ MGR: Orlando Figueroa (GSFC)

| Rev Date: 5 Aug 94 | **MISSION STATEMENT** | Current Mission Phase: C/D |

The SWAS program is a pathfinding mission to study the chemical composition of interstellar galactic clouds to help determine the process of star formation.

Launch Date: June 1995        Launch Vehicle: Pegasus - XL        Launch Site: WWF

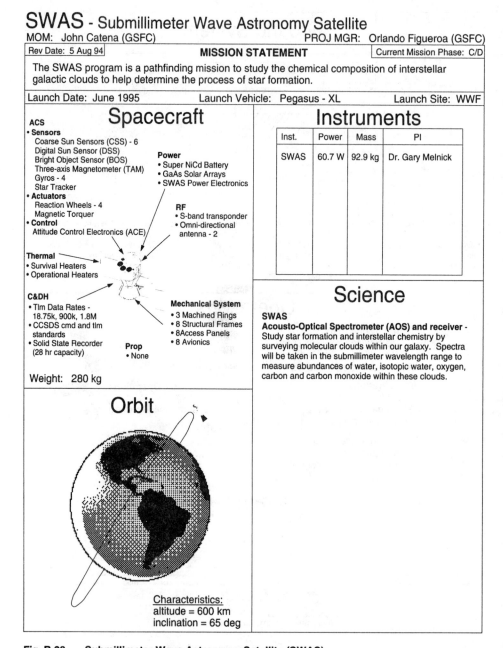

## Spacecraft

**ACS**
- **Sensors**
  Coarse Sun Sensors (CSS) - 6
  Digital Sun Sensor (DSS)
  Bright Object Sensor (BOS)
  Three-axis Magnetometer (TAM)
  Gyros - 4
  Star Tracker
- **Actuators**
  Reaction Wheels - 4
  Magnetic Torquer
- **Control**
  Attitude Control Electronics (ACE)

**Power**
- Super NiCd Battery
- GaAs Solar Arrays
- SWAS Power Electronics

**RF**
- S-band transponder
- Omni-directional antenna - 2

**Thermal**
- Survival Heaters
- Operational Heaters

**C&DH**
- Tlm Data Rates -
  18.75k, 900k, 1.8M
- CCSDS cmd and tlm standards
- Solid State Recorder (28 hr capacity)

**Prop**
- None

**Mechanical System**
- 3 Machined Rings
- 8 Structural Frames
- 8Access Panels
- 8 Avionics

Weight: 280 kg

## Instruments

| Inst. | Power | Mass | PI |
|-------|-------|------|-----|
| SWAS | 60.7 W | 92.9 kg | Dr. Gary Melnick |

## Science

**SWAS**
**Acousto-Optical Spectrometer (AOS) and receiver -**
Study star formation and interstellar chemistry by surveying molecular clouds within our galaxy. Spectra will be taken in the submillimeter wavelength range to measure abundances of water, isotopic water, oxygen, carbon and carbon monoxide within these clouds.

## Orbit

Characteristics:
altitude = 600 km
inclination = 65 deg

**Fig. B.23.    Submillimeter Wave Astronomy Satellite (SWAS).**

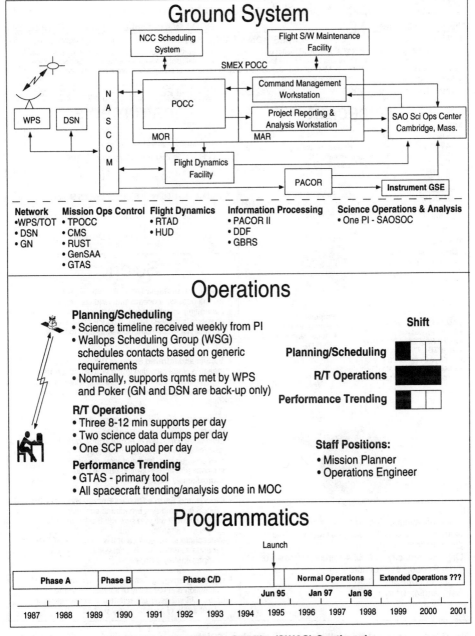

**Fig. B.24. Submillimeter Wave Astronomy Satellite (SWAS) Continued.**

# TRMM - Tropical Rainfall Measuring Mission

MOM: Karl Schauer (GSFC)                        PROJ MGR: Tom LaVigna (GSFC, Code 490)

| Rev Date: 13 Aug 94 | **MISSION STATEMENT** | Current Mission Phase: C/D |
| --- | --- | --- |

TRMM is an integral part of the NASA Mission to Planet Earth Program. As an Earth system science mission, TRMM is designed to advance our understanding of rainfall and to determine the rate of rainfall and total rainfall over the tropics and subtropics between the north and south latitudesof 35°. TRMM will study the distribution and variability of precipitation, latent heat release, Earth's radiant energy budget, and lightning over a miltiyear data set.

| Launch Date: August 1997 | Launch Vehicle: NASDA H-II ELV | Launch Site: YLC |
| --- | --- | --- |

## Spacecraft

**ACS**
- Sensors
  ESA (1)
  DSS (2)
  CSS (8)
  IRU (1)
  TAM (2)
- Actuators
  RWA (4)
  MTB (3)
- Control
  ACE (2)

**RF**
- TDRSS 2nd Generation Transponders (2)
- High Gain Antenna (1)
- Omni-Directional Low Gain Antenna (2)

**Power**
- Total Power: 1100 W
- Batteries: 50 A-hr Super NiCd (2)
- Solar Array (2)

**Thermal**
- Operational Heaters
- Survival Heaters
- Heat Pipes
- Thermisters

**C&DH**
- 80386 Processor
- 2 Gbit RAM Solid State Recorder
- MIL-STD 1773 Data Bus
- 32 Kbps Data Rate (I-channel)
- 2.048 Mbps Data Rate (Q-chanel)
- 1 Kbps Command Rate

**Propulsion System**
- Hydrazine fuel
- Delta-V and Roll Thrusters
- EVD (1)

Weight: 3620 kg

## Orbit

Insertion Altitude: 380 km ± 10 km
Nominal Mission Altitude: 350 km ±1.25 km
Orbit Inclination: 35°
Orbit Eccentricity: 0.00054 (circular, frozen orbit)
Orbit Period: 91.5 minutes
Argument of Perigee: 90°
Beta Angle: 0° to ±58.5°

## Instruments

| Inst. | Power | Mass | PI |
| --- | --- | --- | --- |
| PR | 250 W | 475 kg | Dr. T. Kozu |
| VIRS | 40 W | 50 kg | Dr. Bill Barnes |
| TMI | 50 W | 65 kg | Dr. Jim Schiue |
| CERES | 45 W | 45 kg | Dr. Bruce Barkstrom |
| LIS | 45 W | 21 kg | Dr. High Christian |

## Science

- **Precipitation Radar (PR)** - PR will measure the rainfall rates over oceans and land surfaces, generating scientific and housekeeping data at a rate of approximately 93 Kbps. The PR is the first quantitative radar instrument of its kind to be flown in space.

- **Visible and Infrared Scanner (VIRS)** - VIRS will measure cloud radiation in the visible and infrared regions of the spectrum, genereating scientific and housekeeping data at a rate of approximately 50 Kbps.

- **TRMM Microwave Imager (TMI)** - TMI is a multichannel, dual-polarized, passive microwave radiometer which will measure rainfall rates over the oceans and, with less accuracy, over nonhomogeneous land surfaces, generating scientific and housekeeping data at a rate of approximately 8.8 Kbps.

- **Clouds and Earth's Radiant Energy System (CERES)** CERES is a precision broadband scanning radiometer that will measure the Earth's radiation budget and atmospheric radiation from the top of the atmosphere to the surface of the Earth. CERES will generate scientific and housekeeping data at a rate of approximately 9 Kbps.

- **Lightning Imaging Sensor (LIS)** - LIS measures the distribution and variability of lightning over the Earth. LIS will generate scientific and housekeeping data at a rate of approximately 6 Kbps.

**Fig. B.25.    Tropical Rainfall Measuring Mission (TRMM).**

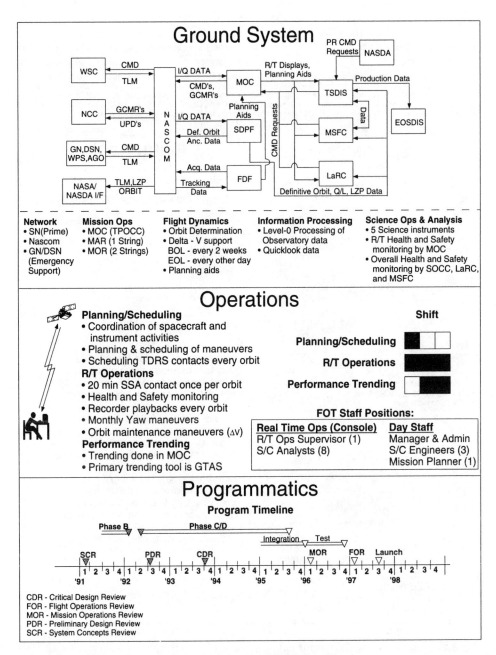

**Fig. B.26. Tropical Rainfall Measuring Mission (TRMM) Continued.**

# XTE - X-ray Timing Explorer

MOM: J.B. Joyce (GSFC)                    PROJ MGR: Orlando Figueroa (GSFC, Code 410)

| Rev Date: 06 Sept 95 | **MISSION STATEMENT** | Current Mission Phase: C/D |

The objective of the XTE mission is to design, develop, launch, and successfully operate an observatory capable of meeting the scientific objectives of measuring astrophysical X-ray source characteristics with high temporal resolution over a broad energy range for a period of two years. XTE will study a variety of X-ray sources including white dwarfs, accreting neutron stars, black holes, and active galactic nuclei. Measurements will be made over a wide range of photon energies from 2 to 200 KeV. The XTE is designed to study the intensity variations and spectra of these objects over time scales as short as microseconds and as long as years.

Launch Date: October 1995    Launch Vehicle: Delta II 7920    Launch Site:KSC ER

## Spacecraft

**ACS**
- **Sensors**
  DSS (2)
  CSS (8)
  IRU (1)
  TAM (2)
- **Actuators**
  RWA (4)
  MTB (5)
- **Control**
  ACE (2)

**Thermal**
- Operational Heaters
- Survival Heaters
- Heat Pipes
- Thermisters

**C&DH**
- 80386 Processors
- 1 Gbit RAM Solid State Recorder
- MIL-STD 1773 Data Bus
- 32 Kbps Data Rate (I-channel)
- 1.024 Mbps Data Rate (Q-channel)
- 1 Kbps Command Rate

**Power**
- Total Power:  800 W
- Batteries:  50 A-hr Super NiCd (2)
- Solar Array (2 panels)

**RF**
- TDRSS 2nd Generation Transponder (2)
- High Gain Antenna (2)
- Omni-Directional Low Gain Antenna (2)

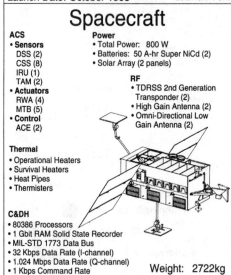

Weight:  2722kg

## Instruments

| Inst. | Power | Mass | PI |
|-------|-------|------|-----|
| PCA | 57.6 W | 1274 lbs. | Dr. Jean Swank |
| ASM | 18 | 78 | Dr. Hale Bradt |
| HEXTE | 66 | 890 | Dr. Rick Rothschild |

## Science

**Proportional Counter Array (PCA)** - Conducts extended observations, obtaining broad-band spectra with energyresolution of 18% at 6 KeV and high time resolution.

**All Sky Monitor (ASM)** - Obtains a measurement of every source in the sky above its sensitivity threshold and in the accessible parts of the sky every 1.5 hours.

**High Energy X-Ray Timing Experiment (HEXTE)** - Observes the PCA targets simultaneously to extend the sensitive coverage to higher energies with energy resolution of 18% at 60 KeV.

## Orbit

Nominal Mission Altitude:  580 km
Orbit Inclination:  23°
Orbit Eccentricity: 0.00195
Orbit Period:  96 minutes
Argument of Perigee: 210.96°

**Fig. B.27.    X-ray Timing Explorer (XTE).**

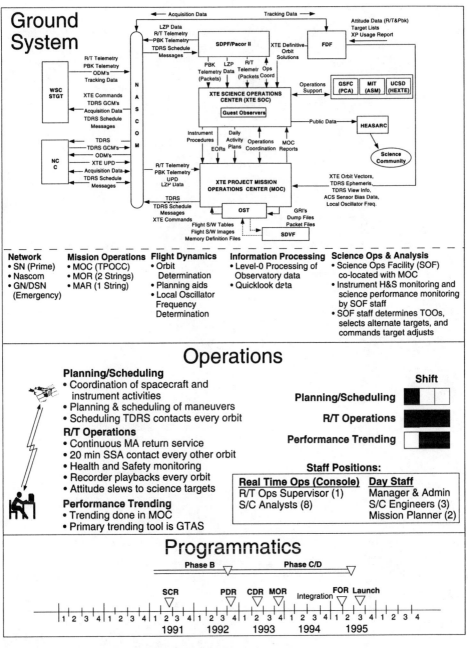

**Fig. B.28. X-ray Timing Explorer (XTE) Continued.**

# Index

## H

# P